This book: Mount Shasta Sightings is © by Brian Wallenstein December 2012. All Rights Reserved.
No part is used without permission of the author. Reviewers may quote passages from the book and reproduce some pictures with consent depending upon content or photographer.

DISCLAIMER: this book is based upon events that have been conveyed to us as true from seeming credible and reliable sources. In circumstances such as existence of UFOs and extraterrestrial life, we do not take liability if these premises prove to be false. Witnesses agreed to use their real names, or remained anonymous for privacy reasons.

Credits: Author Brian David Wallenstein
Assistant and Assistant Researcher: Pamela S. Padula
Proofreading, Editing: Brian David Wallenstein, Pamela S. Padula
Cover Art: Brian David Wallenstein
Marketing: Brian David Wallenstein
Marketing Assistant: Pamela S. Padula

ISBN: # 978-0-615-68295-2
Made in the United States of America

MOUNT SHASTA SIGHTINGS

A Local and Regional Chronology of UFO Sightings and More

By Brian David Wallenstein © 2012

"Man is an artifact designed for space travel. He is not designed to remain in his present biologic state any more than a tadpole is designed to remain a tadpole."
William S. Burroughs

Dedications and Acknowledgements

If not for the encouragement of my support group, which consists of my Mom and Dad, Al and Irene, and my two children, Ali and Shae, with their inspiration and resolve to my success, I would not have completed this book. They never criticized, demeaned or judged the idea, but stood at my side in confidence, keeping my cup filled, as they ripped with frank honesty. I humbly thank you guys for that.

In my career as a writer, spanning 36 years from my first paid work; I've always worked alone. This time, the book required a team. At first, I did not know that. Many people wished to be involved and assist, and I appreciated that immensely, so "Thank You" to those who offered help. I was a little leery of the idea until it fell into place and just worked. I was given the gift of a person who works as hard as I do and that has offered more than she was asked. Like me, she wore the hats of at least 20 people. She is Pamela Padula, my assistant and assistant researcher. She helped this book take on some of the complexity that would not have been possible to accomplish without her. She has been a prayer answered! Her shining belief in my work and how she fit right into the flow like clockwork has been miraculous. I offer gratitude and thanks for her contributions, help and presence and going the extra mile in getting some of the interviews we received. With much appreciation, thanks girl.

I also would like to thank the community of Mt. Shasta, California, for their bravery, help and support in this project, along with "coming forward" to my posters, ads, and personal requests. This book is for you and of you and would not exist if it were not for you. I appreciate the openness and cooperation of the St. Germain Foundation; Peter Mt. Shasta; the employees of the U.S. Forest Service; Chief Parrish Cross and the Mt. Shasta Police Department; Ashlyn, of Shasta Visions, for her open heart and helpfulness; and all others for walking their talk and sharing. I also want to thank all those who lent constant support.

A personal thanks to the following: Those individuals from the past contributing to PROJECT BLUEBOOK; All participants of the National Investigations Committee on Aerial Phenomena (NICAP); Reuben Uriarte, President of the Mutual UFO Network, West Coast (MUFON), special acknowledgement to his generous wealth of information and encouragement of this book, along with his selfless acts of service and dedication to the proliferation of non bias information about UFOs; Peter Davenport, who mans the role for National UFO Reporting Center (NUFORCwww.nwlink.com/~ufocntr/) and the use of their sightings, the people working at the Computer UFO Network (CUFON); and finally, thanks to the 'Off-World' residents who have kept us curious, questioning and self-examining. I'm hoping we all come to a peaceful resolution with your existence here and our own enigma. Bless you all.

PLEASE NOTE:

The author must stress this fact implicitly: While all the sightings before the closure of Project Blue Book were flawless due to a team of workers, funding and proofreaders, current circumstances require reports submitted to NICAP, MUFON, NUFORC, or CUFON need careful scrutiny by you or at least the use of SPELL CHECK. DO NOT use abbreviations or the small (i) for your identity. Please DO use proper grammar; otherwise a good sighting will be at risk of being tossed. Individuals on the receiving end must translate what is being said and are restricted for time and assistance. While working a timeline, it took 1000+ hours to CORRECT simple punctuation, grammar, and spelling errors in the sightings considered for use in this book. Please take this to heart. (The Author)

UFO Reporting Agencies: NUFORC Hotline-206-722-3000
http://www.nwlink.com/~ufocntr/

MUFON: http://www.mufon.com/reportufo.html

There are more agencies to report to, but you may have a local chapter of MUFON in your Neighborhood, or report to either of the above agencies over the Internet.

TABLE of CONTENTS

Dedication -- 4
Table of Contents --- 6
Acronyms -- 7
Preface -- 9
Introduction --- 13
   Disclosure is happening -- 15

Chapter One-Why Mount Shasta? ---------------------------------- 17
   The Mountain -- 19
   Local Indian Tribes -- 26
   Local Indian Lore -- 29
   Other Myth, I AM -- 33
   Mount Shasta Today -- 40

Chapter Two-Our Approach to Myth & Religion --- 41
   Spirituality -- 47
   Changing Paradigms -- 50
   A Little About NASA --- 55
   Mars Makes a Statement --- 61
   President Obama, Siskiyou County and Mars Visitations ------ 64

Chapter Three - Sightings Descriptions, Kinds of Flying Craft ------- 69
   Kinds of Aliens --- 71
   The UFO Sightings -- 72
   Bigfoot/Yeti Sightings --- 73
   Inner Sightings --- 74
   Anomalous Sightings --- 78

Chapter Four-Local Sightings ------------------------------------- 80
   A Little History about UFO Interviewing ---------------------- 80
   Part A-Mount Shasta UFO Sightings --------------------------- 85
   Part B-Bigfoot Sightings -- 148
   Part C-Inner Sightings -- 155
   Part D-Anomalous Sightings ------------------------------------- 164

MAPS Section* --- 169--201

Chapter Five-Historical Chronology, Intro. ------------- 202
   Section-1: Ancient UFO Sightings 300 B.C. to 1900 --------- 209
   Section-2: 20th Century Sightings 1900-1935 ----------------- 226
   Section-3: Second World War Years 1936-1945 -------------- 238

Section-4: Post War Technology Years 1947 ---------------------------------249
Section-5: 1950-1959 ---282
Section-6: 1960-1969 ---324
Section-7: 1970-1979 ---379
Section-8: 1980-1989---416
Section-9: 1990-1999---434
Section-10: 2000-2012 ---493

Conclusion --618
Bibliography --625

ACRONYMS

Some of you may know what the acronyms mean we use throughout the book and for those of you who do not, here is a list of terms that will be helpful in your understanding of the information.

DEFINITIONS:
UFO: Unidentified Flying Object
Off-Worlder: our description of extra-terrestrial beings
MUFON: MUTUAL UFO NETWORK
NICAP: National Investigations Committee On Aerial Phenomena
APRO: Aerial Phenomena Research Organization
PROJECT BLUE BOOK or (BB): Organization Established to Study UFOs (Originally classified)
CIA: Central Intelligence Agency
FBI: Federal Bureau of Investigation
AF: Air Force
AFB: Air Force Base
BBU: Blue Book Unknown
CUFON: Computer UFO Network
UFOCNTR: (NUFORC) National UFO Reporting Center
CUFOS: Established by J. Allen Hynek Center for UFO Studies
DARPA: Established by the DOD (Department of Defense) The Defence Advanced Research Projects Agency. This agency was created for development and deployment of new technology for use by the military. DARPA was responsible for finding the funds to develop many technologies which have had significant effects on the world, including computer networking (starting with the ARPANET; which eventually grew into the Internet), as well as the NLS, which was both the first hypertext system, and the predecessor to the graphical user interface.

Its original name was simply Advanced Research Projects Agency (ARPA), but it was renamed DARPA (for Defense) on March 23, 1972, and back to ARPA on February 22, 1993, and back to DARPA again on March 11, 1996.

The Soviet launching of Sputnik inspired the U.S. military to keep arsenal and technology ahead of its enemies, like the space race in 1958. The DARPA group is independent (like black ops) rather than above board, and report directly to senior Department of Defense management. Their R&D department obviously is one of immense secrecy. The Air Force used the CIA before 1947 like this and then created within the air force, its own investigative branch for UFO research and development. It evolved into this after collecting artifacts, crash site parts, and information from Hitler's collection, to start building hi-tech weapons. DARPA employs over 200 personnel (about 140 technical) managing their own $2 billion yearly budget. Since DARPA focuses on short-term (two to four-year) projects run by small, (need to know) focused groups, its disclosed yearly budget is 2 billion dollars, but that is only on the surface. The secret section behind DARPA and NASA are one in the same and are not known to the public or what it has stored. It is also funnelling tax dollars from our overinflated defense budget.

NASA: National Aeronautical and Space Association

www.blackvault.com site since 1996 devoted to freedom of information about UFO documents from project blue book and other government documents, great resource

MUFON - Mutual UFO Network - Seguin, TX mufon.com mufon.org
CUFOS - J. Allen Hynek Center for UFO Studies - Chicago, IL cufos.org
FUFOR - Fund for UFO Research - Maryland fufor.com
Computer UFO Network (CUFON) cufon.org
Richard Hall's Home Page hallrichard.com
Project SIGN Research Center - Wendy Connors (Pending relocation)
Project Blue Book Research Center - David Michael Hall nicap.org/bluebook/bluebook.htm
International Center for Physical Trace Research - Ted Phillips angelfire.com/mo/cptr
NARCAP - National Aviation Reporting Center On Anomalous Phenomena narcap.org
NUFORC - National UFO Reporting Center - Peter Davenport www.nwlink.com/~ufocntr/
Project1947 - Internet UFO Group - Jan Aldrich project1947.com
Temporal Doorway - Mark Cashman homepage.ntlworld.com temporaldoorway.com
Project FT - Project Flying Triangle studies in the UK- Victor Kean larryhatch.net
U UFO Database Website - Larry Hatch westol.com
UFO Anomalies Zone - Stan Gordon
www.nicap.org- the most comprehensive information and is a great resource we used.
www.ufoevidence.org has some sightings used

PREFACE

THERE ARE UFOs IN MOUNT SHASTA. THIS BOOK WILL PROVE IT. The following information and brief chronology depicts the presence of UFOs in Northern California, Southern Oregon, throughout the country, world and our human history. The material aims to allow the reader to grow comfortable with this notion. Our awareness of UFOs may be part of a natural course of events in life to further acquaint us and broaden our view of our galactic environment; a universe that is teeming with life.

Perhaps like the salmon, we grow in the safety of the river only to naturally migrate to the ocean when we are old enough to navigate it competently. To think that we humans are the only conscious life form in an infinite universe is not only trite but also egotistical and inaccurate. It is for us to assert and see the truth through the plentiful discoveries made over the millennium in sightings recorded on rock drawings, archeological digs, anthropological data, writings, pictures from telescopes and the advanced technology found, as our roadmap home.

We present many possibilities in this book. Included in the chronology are significant events in human history that could have a connection to UFOs. Feasible explanations may become clearer as to how many ancient sights were built if the academic community accepted 'off-world' intervention as a sound variable.

We use **'Off-Worlder'** as an alternative to extra terrestrials (ETs). Since our neighborhood goes beyond the confines of just this planet; we identify our native territory on and off-world.

The frequency and duration of sightings can be answered by the simple connections between the development and distribution of technology, like nuclear weapons sites, military bases, secret research and development facilities, which become a focus for UFO activity and the jumps in human thinking that foster these outpicturings. For example, when the minuteman missile silos were built, there were sightings connected with all of them. Maybe the occupants of these UFOs take interest in the significant leaps our species make and risk revealing themselves, because they are concerned about the impact our weaponry has on the larger picture. Maybe they hope we muddle through these anxious times in our evolutionary leaps with deeper critical thinking skills. It seems they have been here all along the way of our development early on as you will see.

There are people however that somewhere decided that they can make decisions on our behalf without our consent and oversight, shoveling misinformation, ridicule and conspiracy theories into the mix and keeping this information classified.

With the lack of disclosure by the scientific community and the military industrial complex, it is no wonder that the general population remains in the dark about this subject and have been unable to respond and contribute needed viewpoints. One could speculate the reasons for lack of disclosure that include the fear of mass panic, or "National Security," a lack of centralized control, or that the ETs are possibly hostile, implying that we are incapable of adapting to change. Could it be all of us are being kept in the dark because the undisclosed knowledge threatens the benefits a few are reaping at the expense of us all? All these are good subjects that the book sheds some light on. At this point of time, over 85% of the global population know and believe that ETs exist. Millions of people have seen UFOs and can deal with it. The bias and doubt deliberately placed in our collective cultures about Off-Worlders, is only meant to stop disclosure, leaving those in power undisturbed control over a carefully structured hierarchy/caste system. The exposure of existing technology could spring us from our dependence on oil/energy and the large companies that own the capitalistic systems of the world that regulate it with food and medicine, are the ones with the most to loose and the most to gain from keeping ET technology and our own advanced discoveries suppressed. Have we really evolved, or are we still slaves to a few with the bigger guns?

As Franklin D. Roosevelt aptly observed,

"The liberty of a democracy is not safe if the people tolerate the growth of private power to a point where it becomes stronger than the democratic state it itself. That, in essence, is fascism; ownership of government by an individual, by a group, or any controlling private power."

There is a long and suspicious list of missing inventions and inventors who succeeded with working models in about every subject. A guy in Oregon demonstrated on his website his 250 mile a gallon steam and fuel powered car. He mysteriously vanished into thin air two weeks before making the applied technology public. Does this imply any beneficial invention or innovation, take a back seat to obsolete ideas, only because they generate massive revenue for a few, who refuse to adapt at the expense of everyone else? Space exploration is only for the privileged and wealthy private industries now, as well?

The saving grace is we can't stop progress. We as a species will not go any further unless ALL the variables are evident and explored. We have come too far and are far too inquisitive, to exclude the distinct possibility of UFOs having a place in our history. It is our birthright to know the truth of our relationship with the stars and our origins. The leaks are too many to stop the dam of lies from breaking. If we don't help break the monkey-grip of power we have inadvertently allowed the self-serving arbitrary ruling class to have over us, we all may not have much of a future.

Even without disclosure, the collaboration of the multi-disciplines of science and religion is fast closing the gaps in our understanding. When the biological sciences sat down in a round table discussion with physicists and computer sciences, they realized that principle stays the same from a single atom, single cell, to multifunctional systems, ecosystems, solar systems, computer systems and spiritual systems. It never occurred to these individuals to communicate with and among each other, before now, or did someone know, 'A house divided falls?' However, things are moving forward.

INTRODUCTION

If one studies history and those driven to explore like Christopher Columbus, Thomas Jefferson, Nikola Tesla and many others, they set out to discover new horizons. Searching for Atlantis, the fountain of youth to the Garden of Eden, these explorers were inspired by myth, stories, religious ideas, and through these inroads, many other discoveries were gained. It has been a spiritual push, the desire and urge to find…. Something.

This push of discovery is in our approach to the subject of UFO sightings, too. It is not our intent to demean or negate any possible connection between the Myth of Ancients living inside the mountain, or of a base, but to remove speculation or sweeping generalizations associating the presence of UFOs with any specific occult belief. If we can show that the sightings are valid in themselves, then it will be an easier task to confirm something is, or is not in the mountain, based on combined evidence, and proceed from that point and allow the reader to draw their own conclusion. It has been hard enough to gain any reputable credibility as it is for the existence of UFOs beyond the all out denial, deliberate humiliation of witnesses and debunking of evidence by the arbitrary powers that be.

Associating UFOs and sightings with cults, which the science community uses as convenient excuses to dismiss them, makes it easier to deny them entirely. It all confounds any credibility and adds an element of doubt from assumptions that seem as fantasy to the scientific and rational mind. The military and other organizations for undisclosed reasons, want UFOs to not exist. If admissions were made, the looks of doubt and fear many receive upon comment on this subject would disappear. The confirmed existence of advanced technologies and reverse engineering, could then avail benefits to all. No longer would there be leverage over our survival.

We have been carefully primed and made to doubt our intuitions and our own perceptions by the subtle and subliminal promptings of those who want to keep secrets, used through the mass media and other mechanisms. This is very evident by the way those who have had sightings are treated and treat themselves. The very doubt and ridicule that amasses around the entire subject of UFO-logy demonstrates the cultural manipulations we have had installed on us. Unfortunately for those who support this secrecy, the data, evidence and testimony is indisputable. When highly trained fighter pilots and police officers, versed in the observational arts, experienced so many confirmed and legitimate sightings, it is hard to dismiss.

How can the public be expected to swallow these professionals in question are incompetent lest they (the military) are poor judges of character? If the military really felt their pilots and or commercial airline pilots were as incompetent or crazy as they imply, then why are they allowed to fly these billion dollar pieces of precise equipment, and be fully entrusted to protect us? The doubt is all smoke and mirrors.

Einstein said, "The only difference between a scientist and a home maker is that the scientist writes down their findings!" We are all credible witnesses.

We are taking about hard core data and sightings collected over the last 60 years from civilians, police, professional military and federal officers, that illustrate the numerous and consistent sightings over the globe and over our area. The craft were photographed and in many situations confirmed on radar and witnessed by dozens and analyzed.

Be it that there are threads of truth within the myths to the character of our alien visitors is entirely another subject at this point. Verifying the observations of UFO activity on and around Mount Shasta first will make the task of unraveling the mythologies much easier.

Regardless of how little we know or don't know of our galactic neighbors intent, or how long ETs have been here and how instrumental they have been in matters of our evolution, they are here, and disclosure has been suppressed. It is the hope of this book to offer objective evidence of the existence of life anywhere in our universe, and arrive at this conclusion without denial, because it ruffles some feathers or exposes deceit and non-disclosure. So we have several issues to address: Non-disclosure by humans and why, and our link to the aliens.

DISCLOSURE IS HAPPENING

A group of high-ranking military officers, including CEOs from technology companies, formed a group to do just that, disclose classified information.

Over 460 officials including current and former CIA, NASA, DARPA, Naval Air Force Brass, Generals, aeronautic engineers, aerospace, weapons and technology companies, swore under oath, recorded on film at the NATIONAL PRESS CLUB, of our current involvement with Off-Worlders, their long standing presence on our planet, their bases on the Moon and Mars and their diffusing of nuclear weapons, our military launched into space. The Off-Worlders informed the General of Canada, that we must cease and desist the use of nuclear weapons or testing in space, or for that matter use at all, on or off-world, as it is not good for anyone. The group, "The Disclosure Project" and the 460 individuals were to speak under oath to congress scheduled for December 2011. The meeting obviously remains suppressed.

Fortunately there is a youtube web link were you can see the Press Club releases.
http://www.youtube.com/watch?v=9fueo2TZT4&feature=player_embedded
http://www.youtube.com/watch?feature=endscreen&NR=1&v=IeLtVp2FOqQ

The 460 individuals exercised courage in the face of imminent danger. They represent those officials whose allegiances are to the common good, peace and welfare of sentient beings that outweigh oppression, greed and terrorism. They are all nationalities speaking in one resounding voice, and they represent the highest good of our collective races ideals. Their efforts need to be acknowledged and the torch carried. These high profile individuals would not put their private and professional lives at risk unless they felt what they are supporting is of the highest importance.

CHAPTER ONE
WHY MOUNT SHASTA?

What mystery and attributes link UFOs, Off-Worlders and extra terrestrial activity to Mount Shasta? Is it the alpine allure, its distance from civilization? Does it possess a rare element, or do the natural caverns inside make it an ideal place to be concealed? Is it sentiment, functionality, an unsuspecting place? What could it be? Maybe it is a combination of things? What quality coined it, one of the 'Seven Holy Mountains of the World' and compared it to Mount Fuji in Japan?

Many people claim to have seen UFOs here including the author. Some have watched the ships go literally into the mountain, and out of its sides. People in many walks of life from civilians to Forest Service employees and police have viewed star gates and portals with craft arriving and departing.

Mount Shasta is an amazing geological feat of nature. It is gorgeous to hike, climb, and ski, abundant in the bounty of nature and sacred to many, but there must be a logic to the sightings and there must be something these craft fly into unless they just fly into the side of the mountain for fun like the Subzero 'Kamikazes'.

Why are they here and why is its considered a hot spot ranking number 13 out of 300 hot spots on the globe? (NICAP.ORG)

Perhaps the myth of ancient Telos (Meaning mound in Sumerian) has some credibility or relation to it, but that link only became relevant in the early part of the 20th century with the publication of Fredrick Oliver's book. Maybe we have several different alien races holed up in the mountain including our own military, since there are places (we aren't supposed to know but they put up signs) to stay away. Perhaps we are already collaborating with these Off-Worlders? Are they good or ill willed, teaming up with ill willed or good-hearted humans? What about abductions?

There have been a few sightings with missing time. A group of loggers missing for days in Oregon had claimed to be abducted. The event was later turned into a movie. We have left this subject for the abduction specialists.

What secrets does our mountain tenaciously hold, that makes her significant like the pyramids, Findhorn or Stonehenge? Is it a geological or electromagnetic kind of vortex that amplifies cosmic energy, creating ley line grids globally? Is it a very ancient space portal, put here by our space progeny or just a really high point on the west coast as a convenient landmark from space?

People have claimed to hear machinery underground resonating through the lava tubes, not just in the immediate vicinity of Mount Shasta but in a few different locations around the county. Some noises were so loud the residents have been kept awake for hours. Maybe there is machinery down there holding the planet together or perhaps with the abundance of gold in this county, it is ideal for an underground foundry, needed for the advanced alien electronics. It is all a mystery.

Again, perhaps it is just because it's a really cool place to set up an outpost. If you were an explorer, what kind of place would you choose to set a base camp and would it be like this?

These are all questions asked by many people. We hope to unravel some of these mysteries.

THE MOUNTAIN

Mount Shasta as seen from the southwest, looking north. (BDW)

Unless you know of Mount Shasta, it is a shock to see a volcanic formation with its twin peaks that lies 70 miles south of the Oregon border in Northern California. Standing proudly at 14,179 feet, Mount Shasta is part of the Cascade Mountain Range. She is one in several volcanic mountains spotting the Pacific Northwest and part of the Pacific Ring of Fire. Mt. Lassen lies to the southeast where Mt. McKinley, Mt. St. Helens and others, lie to the north.

A little northwest of Mount Shasta is Black Butte, peaking at 6,334 feet; its classic cinder cone shape illuminates this profound and ominous mound.

Black Butte as seen looking north at its south face. (BDW)

About four decades ago the long-standing fire lookout was removed from Black Butte Summit. It was one of the first in the area.

Being so treacherous to get to, the lookout became impractical to keep in good repair. The forest service had to fly it out in pieces, with a logging helicopter.

Castle Crags viewed from the east side looking west. (BDW)

Due south of Mount Shasta is another eruption called Castle Crags. Mystic and foreboding, a rocky castle Mother Nature created. Jutting up and within the Crags, the beautiful pristine Castle Lake sits atop and within her crater. Heart Lake is a very small and intimate lake, hides on a higher plateau within the crags, is just magnificent, with dozens of other lakes spotting the landscape in all directions around the Crags.

Castle Lake and inside of crags looking south at the north side of it's southern inner face. (BDW)

View from Old Stage Road by the original Berryvale Post Office of Mt. Shasta (BDW)

Mount Shasta was originally discovered and viewed by Spanish and Russian explorers, the Laperouse Expedition in 1786 and then by Gabriel Moraga in 1812. Peter Skeene Ogden, Alex McLeod, and a slew of others as part of the Hudson Bay Company, also set foot in the county. The company fur trappers and other trappers just about wiped out all the beaver, otter and martins in the area, to send back to England for the haberdasher.

There was also an early report about a Spanish ship seeing Shasta erupt. This has been proven to be an error. What the Spanish seaman saw, was the glow from forest fires in the Salmon-Trinity area and the Humboldt Hills. The mountain did not erupt in the 1768 as once thought. Predicted to erupt about every 350-400 year, it seems we are overdue for one. It is considered inactive, however current research claims Mount Shasta is a super volcano spanning a large area and is quite active.

Mount Shasta also carries with it the headwaters of the Sacramento River. With many confluences and feeder streams and rivers, the San Joaquin Valley, the Bay Area and Southern California are provided the invaluable commodity of water. It is a valuable and political bombshell that occasionally explodes.

Both John Muir, naturalist, and Joaquin Miller, adventurer, have extensively written about the majesty and mystery of the mountain and the area, its natives and its weather.

The surprising and deadly storms that are generated out of nowhere near the peak, the incredible world-class fishing that is experienced here, with all the native flora and fauna and unique wildflowers, helped influence President Roosevelt to designate the area a National Forest decades ago to try and protect it. Logging has been a large industry here for both construction and for paper production, as the late William Randolph Hearst would attest. His family and he invested millions, into thousands of acres of land for wood products, close to a century ago.

Some clear cutting is still going on, and you can see the scarring of its presence from the past along with the damage from wildfires. The Indians used to employ "slash and burn" and the U.S.F.S. (United States Forest Service) used this procedure as well, although too long ago. These practices essential to forest burn management have been stopped probably from cutbacks and policy changes, but it has turned the area into a tinderbox. Simple clean up, trimming and recycling of the lower branches could be used for fuel, mulch, and could create massive infrastructure jobs and renew forest health, but the obvious falls on deaf ears and eyes.

The neglect of forest management is promoting these deadly wildfires of expanding magnitude, and the evidence speaks for itself in yearly losses of life and limb.

When a permanent town was established here along the Old Stagecoach Road, it was named "Strawberry Valley", but too many Strawberry Valleys existed in the Postal System, so it was changed to Berryvale into Sisson, to Mt. Shasta, back to Sisson but Mt. Shasta from its frequent use, finally won.

Adored by many naturalists and those who could afford to travel here, the area was appreciated for its natural beauty, pristine waters and purity of spirit and healing energies. Ney Springs and other healing spas emerged, drawing affluence here from the cities. Mossbrae Falls has natural soda water percolating from the volcanic springs that is sweet and quenching. Stewart Springs has heavy volcanic water that is famous for its healing properties by the Indians, settlers and current residents.

Once gold fever hit, the area and its people were forever changed. Treaties with the local tribes were dishonored, their sacred fisheries exploited and the land destroyed by the hoards of gold hounds. With this betrayal of the Sacred Lands the Native Tribes went ballistic. The forest was their church that also sustained them. These deliberate acts of the white settlers were a desecration to the Great Spirit only to be resolved with blood. With gold and greed, the local Indians were just in the way, and a strategy was being formed in Sacramento, to fumigate the Natives and move them onto reservations permanently.

Fremont, a young 'ne'er do well', nephew of a prominent Washington politician bent on making a name for himself, set out to make history by staging an alleged 'peaceful resolution' banquet. Over 3,000 Indians attended, and were served strychnine-laced food that killed them all. Since the tribes were considered savages and as low as animals, and did not believe in the Christian god, as Fremont saw it, there was no crime committed in the murder of these natives, or breaking of treaties and laws. With the presence of GOLD at hand for the white settlers, any reasons were acceptable. "The ends justified the means", as the saying goes. Fremont got his job in the Capitol and a city named in his honor.

Several huge multinational paper and logging companies still hold deed to many acres of property, and several have been somewhat environmentally responsible, and some not. Many logging companies moved to the Amazon rain forests to play out the free-range strip mine practices they used to employ here with no oversight, until the tree hugging hippie environmentalists helped bring into law responsible logging practices, that exist here and other places now.

The disdain of the area are the countless loss of lives generated by the white settlers, the gold hounds and military who wanted the locals gone so the area would be owned and possessed by the white politicians, the railroad and other entrepreneurs. The Local Indian Tribes were either forced off their land onto reservations, wiped out by the military, absorbed or driven away, and with them, their rich past and wealth of history, scattered like the wind. There are areas that are cursed with so much violence and conflict, you can feel the tears of sorrow. Be it the injustice of a Native American, a railroad worker, or an oppressed group caught in the greed of human lusts, this area's purity seems to contrast these extremes noticeably. This is not mystical but sad.

Many books are available from local authors on the history of the many groups that lived here.

One thought to consider in the treatment of differing cultures is: The North American Indians, The Incas, The Mayan Culture, The African nations and more, welcomed the arrival of the White man with open arms. Although an "alien" with differing beliefs, these people without forethought helped the newcomers at every turn letting down their formidable means of protection. They opened their hearts, culture and homes to them only to be labeled "savages", demeaned, robbed, poisoned with whiskey, their women raped by the newcomer aliens, who justified it with a holier than thou religion and stole their resources, culture and gold. Contrary to being savages, the natives practiced the Christian ethic (without it being called that or knowing it was the Christian way) simply and were hoodwinked by those who could preach and not do.

Perhaps this is why we are so paranoid about alien invasions. Perhaps we fear the bad behavior we previously demonstrated in approaching open and trusting individuals, being used on people by a more advanced race. We still steal from third world nations, exploit and carry racist attitudes against neighbors who live across any arbitrary boundary line, alienate and discriminate as an excuse to take advantage of less fortunate people, no matter where they live and without valid reasons. What is gained in the larger picture? Maybe it is to learn good manners and live the credo "Do unto others as you want done to yourself", and this is what it feels like in return. Hopefully correction, remediation, forgiveness and collaboration are gained.

THE LOCAL INDIAN TRIBES

Some of the tribes that still reside here such as the Shasta and the Karuk Tribes have websites available to visit. Bill Miesse, who assembled the "The Annotated Bibliography" from the College of the Siskiyous, delves into the general history quite succinctly and points you to many books written about the area, and the research done on the Indian Tribes.

The five/six Tribes that were here up to the introduction of the white man were, Shasta, Modoc, Achumawi (Pit River Indians), Wintu, Klamath and the Okwanachu. The Okwanachu no longer exist at all, as a tribe, but are scattered and denuded.

It is interesting to note the original tribe coined the Shasta Indians, (The debates still exist how and where the name Shasta came from and how it evolved) got pushed out by the Native American Indians from East and Plains, obviously from the crunch and push of the white settlers, close to extinction.

No tribe occupies the mountain itself nor has it ever. It had 5 territories divided up among the tribes but Mount Shasta itself was uninhabited and considered sacred. Above tree line is where the dead go and the Wise men reside for counsel. The Klamath Tribe occupy territory a little further north and into Oregon and were not considered as part of the local tribe, or the Karuk who also occupy nearby territory, but further to the west.

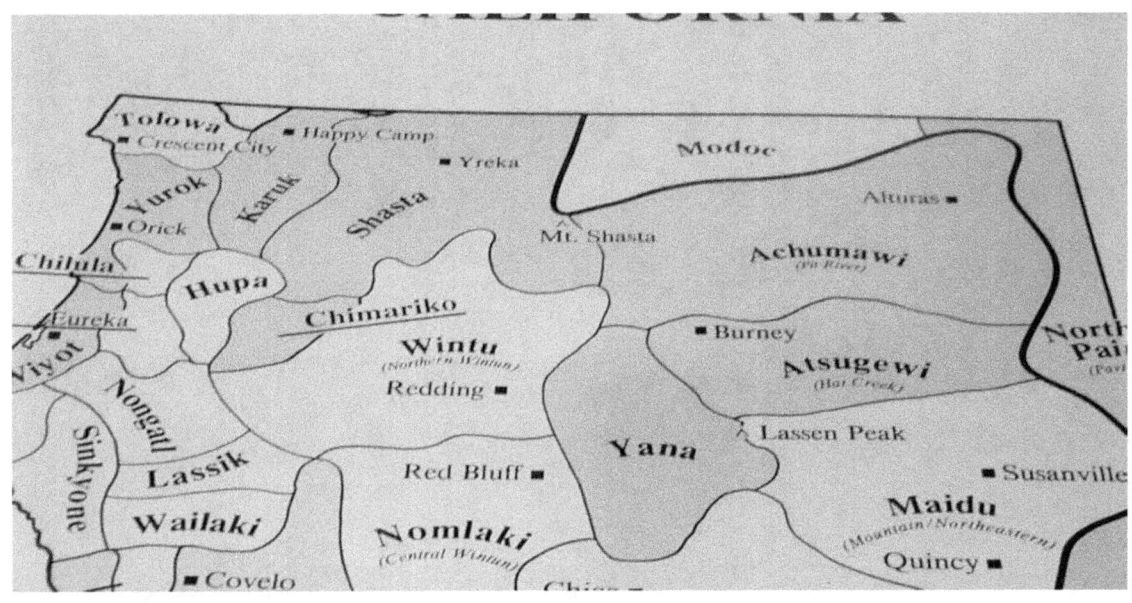

Map from Tribal Areas of California (Pacific Western Traders, Folsom, Ca. 95630)

Many researchers wrote in-depth books about each tribe, Dixon, Olmstead, Steward, Muir, and Miller. Each had recorded the specific tales of each tribe, like Coyote and Turtle, or Bluejay, Wildcat, Lizard. These animal relationships and the stories were used to impart social skills among the children of the various tribes including the Achumwami. Other tales carried through the oral tradition from the Shasta, about Yellowjacket, Eagle and Winds Daughter, Grizzly Bear and spoke of universal creation myths and origins. I highly encourage those of you who want to delve deeper into the local Native Indian lore and culture, to visit this extensive website as it offers most of the authors and books available on the area. Much information was gleaned from these comprehensive sites.

www.mtshastaspirit.com is and awesome resource put together by several local authors/scholars, such as Bill Miesse, who painstakingly organized the corresponding books and chronology located at COS. The Mount Shasta Collection onsite offer the most complete materials relevant to the mountain and you can access the bibliography by going to:

www.siskiyous.edu/shasta/bib.htm

Petroglyphs from Modoc Tribe (Laurie Moore)

LOCAL INDIAN LORE

Before the discovery of gold in Siskiyou County by white settlers, the local Indians had an intimate and harmonious relationship with the area. The fur trappers and the Indians actually had a mutually respectful relationship and worked together in trading. The gold rush changed everything, and greedy gold prospectors, which precipitated the excuse to cause war on the tribes, compromised the sacred fishing grounds. It was, in fact, the disrespect of the white man breaking agreements on the fishing sites that started the problems.

The original 5-6 local tribes that claimed the region dealt with some scrapes but were mostly a cooperative community.

Shasta Indian Legends boast that The Great Spirits from the skies created this planet from the heavens with flowing lava and fire, then put out the great fires with the great snows pushed out a hole from the heavens with a spinning rock. (Sounds like pushing through the atmosphere) As the floodwaters of the deluge subsiding, great marshlands developed.

The Gods and the Head Great Spirit, (more than one lived in the cold parts of space) in their loneliness he (they) descended, adding the rivers, trees and rocks, the Grizzly People, and populated it with all types of critters. The Great Spirit created by pointing his finger to the planet and making the animals and flora.

The Head Great Spirit and his youngest daughter lived in the great mound (Telos, meaning mound), which protected them from the floodwaters that put out the great fires that forged the planet, which finally froze and receded. In his joy of this reprieve, he asked his daughter to open the door of the mound, but to be careful to not let her long flowing red hair get caught in the wind, as it would pull her out of the hut. He instructed her to ask the wind gods to not blow so hard, so the trees, their mountain and hut not be destroyed. His daughter however careful, was carried out of the hut by an errant wind cloud that caught her hair and took her away. She then landed hard in the dirt, rolling down the hill into a bush, unconscious. She was rescued by the head grizzly, which took her home for his wife to nurse back to health and then decide her fate.

The kind and compassionate Grizzly People raised her, into adulthood. She then willingly, married the eldest son, mated and made the Red Man Indian tribes.

With much guilt, the Mother Grizzly went up to the Hut and told the God that she had found his daughter, but kept her and still alive, now the mother of her grandchildren, and the Sky God's, she was sorry for her selfishness. Mother Grizzly hoped and begged forgiveness, but the god in his rage, killed and burnt up the woman grizzly, anguished for the unauthorized CROSS BREEDING with his daughter's seed. He then with a cold heart, forbade the entire grizzly species from walking on two legs anymore and had them use their claws like an animal instead of the club and fingers of a human, cursing them to a life of a four-legged creature.

The Grizzly, still with longer arms and no tail like a man, shows the remnants of their earlier nobility. (This is similar, in a way, to Adam and Eve's fall from Eden).

This was the creation theory passed down from one of the tribes. Although it has been said there was no cohesive religion among the Indians, they all did honor the Great Spirit and their families. There are many mythological stories the local tribes used about birds, beaver and the like, passed down to the children instructing on how to conduct their lives.

Like many Native American Creation Theories, the Great Spirit and their families came from the sky, an orb or a cloud, and intermingled with the native peoples. Unlike many myths on the planet, in Mount Shasta, the sky female is the one who mated with the natives and not the sky male like in many far eastern cultures, including Christianity. However there are matriarchal goddesses that carry the seed of the earth male to fruit a mix. Women also constituted, 'The Oracle of Delphi.' Unfortunately the patriarch and its hard ass tactics, still manages to call the shots and does not forgive in this story.

The area tribes all acknowledged the great Flood, however, the Modoc Tribe in particular have been attributed to a longer standing, perhaps genetic collective memory that predates the deluge and as the other tribes profess, the existence of their Great Spirit takes residence within the mountain.

All the local Native tribes had Shaman. Either gender was acceptable, but the female had to be postmenopausal. Payment was withheld if the patient died or if they failed in their charms to heal. If too many patients died, they would be put to death. Good malpractice insurance, eh? The Shamans, especially the Achumawi who considered Mount Shasta to be a power spot, sought out the guardian spirits or "Tinihowi" who resided on the mountain for their power and guidance. The Tinihowi had been described as tall, longhaired, etheric beings serving as mentors for the Shaman. These beings lived somewhere on the mountain above tree line. (Could they have been from the inner city?)

The local tribes seemed to all practice sucking out the cause of pain, using their breath to heal but also could use it to inflict pain on others. In fact, one tribe believed the little people or fairies of the woods "Axeki" would inflict pain on unsavory people who disrespected nature and other beings, so they would learn respect. These little guys supposedly live in the rocks and cliffs throughout the region. Sounds like the Axeki dispensed instant karma.

The Modoc and Pits believed that consciousness and awareness is intrinsic in all matter. They talked to the rocks, trees, and animals as equals and thus had a collaborative relationship with their environment, very similar to Pagan beliefs as well, all unified in consciousness and intent. Not too far off from current beliefs and science documenting our relationship with our world being one of an energy connection, the observer and observed being one reciprocal exchange, which evolved into the science of Quantum Physics.

Perhaps the local tribes with their lore of the Gods dwelling within the mountain, the Shamans' guides occupying the upper part of the mountain above tree line and the minions who dispense karma upon those who need a wake up call, have roots in long standing observations of inner mountain dwellings, Off-World activity, and is part of their innate connection and brotherhood with the Universal Self after all.

OTHER MYTHS & THE I AM PHILOSOPHY

Fredrick Oliver's book, "A Dweller on Two Planets", written in the late 1800s and published in 1905 by his mother, Mary Elizabeth Manley-Oliver, six years after Oliver's death in 1899, deals directly with the beings that allegedly occupy the inside of Mount Shasta and their philosophy. Oliver wrote the book at the age of 17 in Yreka, California, 37 miles north of Mount Shasta.

Oliver claims that an entity named Phylos the Thibetan who lived countless lifetimes including in Atlantis and Lemuria (Mu), tens of thousands of years ago, would literally seize his writing arm and do automatic writing through him. It had scared and stunned the young Oliver, but after a while he had no choice but continue writing, becoming proactive when he understood what he was doing.

The book is a great read and available online for free or on Amazon in hardcover. Phylos the Thibetan explained in the book that his last incarnation was of a gold miner in the Siskiyou County Gold fields. He was guided to spend time on Mount Shasta at the promptings of his close friend, a Chinese shaman, prophet and laborer.

He and his companion were led into the vicinity of where Mud Creek and Clear Creek are located to an opening on the side of Mount Shasta and brought into a group of Higher-Beings not from this planet. These beings were highly advanced in all matters of science, spiritual practices and teleportation. These Masters educated him to the secrets and origins of the human race and those who cared for the planet. They acted as our guides for the further evolution of our species, leading us closer to the Christ Consciousness. This ancient race, many years ahead of us, claimed ownership to a city in the Mountain called Telos (according to the entity).

The book is rich with imagery, scientific invention, prediction and details of other planets, cultures along with how consciousness works and how we fit in. Inventions including radar and wireless were mentioned before they even came into existence years later. It is uncanny how many innocent individuals with no formidable sophistication act as conduits for universal knowledge such as Fredrick Oliver possessed.

His book added to the upwelling of mystical interest emerging in the midst of the exploding industrial revolution and technological childhood of both Europe and the U.S. Oliver's book sets a precedence for many copycats and other mystics to capture the spotlight and add a new allure to Mount Shasta as a spiritual center of the west coast. It set the stage for Mount Shasta's fame in the circle of metaphysics and high mysticism. The mountain now became the next seventh spiritual mountain of the world, compared to Mt. Fuji in Japan, among other attributes.

Next on the scene was Guy Ballard who started the St. Germain Foundation with his wife, Edna, who acted as his manager, banker and promoter. Ballard claims that he was channeling the spirit of St. Germain who contacted him while he was hiking on the mountain. Germain, a count in France who lived in the early 1800s, was also accompanied by Jesus Christ who lives in the same higher realms, along with a heavenly host of guides for humanity. They actually desire people to find the "GOD" within rather than play god to their followers. They act as familiar, non-threatening images to people until they can accept their own intrinsic divinity according to the Theosophical Beliefs.

According to Ballard, his consciousness was dilated by these experiences on the mountain and began to write channeled books under the name of Godfre Ray King.

Ballard's channeled books also took on a coincidental similarity to the late Fredrick Oliver's work. With the in-roads made by Fredrick Oliver the St. Germain Foundation beginning in the 1930s, other I Am cults came forward in the early 20th century, (coined Theosophical thinking) which buzzed up to Shasta. Ballard's group drew a great following worldwide of over 400,000 people and still enjoys a strong following. Madam Blavatsky, a Theosophist from Europe, was also instrumental in influencing followers of the occult to migrate to Mount Shasta's newfound spiritual fame. Many very wealthy people were involved. Both Oliver and Ballard's teachings also correlated to some principles of metaphysics contained in the holy books of the east and the west.

Guy Ballard and wife: Edna below (Photo credit unknown)

St. Germain lodging at Shasta Springs in Dunsmuir, Ca. Note the Violet aura that appeared in photo. (BDW)

The Theosophical Society also had branches in The Rosicrucian Fellowship, Elizabeth Claire Prophet and others. Of course it has been found the Bible is a synthesis of many stories many much older and reminiscent of Genesis, and the tale of Gilgamesh, offering the hope of eternal life, life in the literal heavens and performing miracles.

The depictions of these religious and esoteric practices actually contain practical, applied scientific methods and methods of meditation, allowing one to control their ability to channel life energy over a broad spectrum. This results in many seeming miraculous acts that are latent in all people, all connected to love, according to the participants of the Theosophical practices.

Although to many, these participants seem a little unusual but they do no harm and even help instill a sense of joy around others. It is no wonder that the insecurity of life during the depression and the two world wars may have pushed the human psyche to consecrate and bless some higher purpose to life. Frankly, it is no different or more or less valid than any existing religion on the planet. If it helps fill people with hope rather than harm, then fine. Any religion should find its devotees making choices to learn self-responsibility from an inner connection rather than from dictates of self-acclaimed Masters.

If we in fact, by default, exist, we must by this intrinsic nature, have a direct connection to the essence of life, ergo, have a dialog with this omniscience. It is a reciprocal experience. Many of these stories and teachings in practice, are archetypal and correspond with Gnostic teachings. Gnostic beliefs claim the essence and power of the Universal Consciousness, (The I AM Presence) exists within everyone and everything and is the glue that holds reality together. It is the invisible essence of all that is and its origins are present within our mind and our experience. With practice, our being-ness set free through understanding and inner trust, can manifest fulfilling acts of love, creativity, healing, purpose and peace. The ultimate is to see God in everyone and everything and to do no harm.

The Hermetic Doctrine, the Nag Hamadi writings and the ancient Kabala also depict the same principles of consciousness. Humans' destiny is to evolve into being, and apply this altruistic principle into practice. This allows a more humane and perfected world to emerge where love, abundance and harmony exist as the norm. I am not condoning or condemning any religion, but illustrating the beliefs.

The presence of UFOs on Mount Shasta then can be viewed without any connotations so every individual can then arrive at their own conclusions. It would seem our Off-World relations would also embrace some spiritual notion as well. It would be curious to find out what that could be.

Modern archeology has discovered and revealed the "in your face" evidence of extra terrestrials visiting our planet and being chronicled in the Bible. Some of these stories came from much older stories contained in the Sumerian Text, while others were added.

Indian holy books of Krishna include blueprints for making flying machines explaining that Krishna, in fact, was an ET, and chronicled aerial wars between opposing factions of these Off-Worlders. Many of these stories have been confirmed through recent discoveries of ancient cities under water, around India. Until we can get all the facts about our past, misinterpretations abound and problems arise, confusing the facts.

It is also most unfortunate some of the "leaders" of these cults go off the deep end, half cocked and start declaring war on the gentiles. Elizabeth Clare Prophet planned for the apocalypse underground in Montana with a complete arsenal and years of food. So much for living the dream of peace applied. Arming ones self to the teeth and excluding those outside the sphere of the chosen, doesn't seem like the graceful and refined idea of heaven on earth.

The Branch Davidians, Al Qaeda, Jim Jones and Heavensgate all took the lives of their people, by twisting the premise of love and making it exclusive. It is still occurring in the Near East, Middle East, and in the U.S. on the premise of predatory capitalism, using different ideologies, such as "Greed is Good", "Give your life to Allah or Jesus", or" Kill a commie for Christ", by wiping out your enemy, instead of seeing the enemy is our own biased projection on others. Let us not forget to demonize another culture and its peoples through propaganda, is a great way to fleece them of their resources, under the guise of democracy and religion. The U.S. has done this for oil in the recent past, and implemented this in our very own country against the Native Americans and other third world countries, as well as the middle class.

Today, many zealots are imitating this pernicious behavior and plan to bury themselves in bunkers with machine gun sentries protecting their installations. It is tragically amusing those who claim to follow the Prince of Peace are harming others, justifiable in emotional times, instead or rising above one's fear aggression circuits of fight or flight and trusting in the 'Truth." Thankfully, benign and peaceful cult groups have moved up to Mount Shasta over the years. They quietly appreciate the mountain and the people, operating with dignity and respect and don't impose themselves on anyone. They do worship the dwellers within the mountain, but choose to keep their beliefs to themselves.

Petro glyphs in Siskiyou County (Laurie Moore)

The petro glyphs around the Modoc area and many sightings close by, raise questions about this area being a UFO hot spot. Most certainly, I would have a plethora of questions for a more advanced race of beings, and what their spiritual beliefs are and what spiritual values are held by those they have met throughout the galaxy(ies).

Turning focus to North and South Americas, the indisputable reference to alien intervention is undeniable, from the petroglyphs in the American Southwest and all over South America, to the Nasca lines in Peru, (lines that form beautiful images of birds, snakes only visible from flying in the air like they were made for flying craft to see, built over 1500 years ago). The Ancient cities and pyramids were built with ingenuity beyond the capability of 21st century technology. They continue to baffle modern scholars. The civilizations, structures and ruins found in South America, Siberia and off the coast of India are proving to be 50,000 years old or older. Current research has revealed references made in the Mayan tablets to a time 900,000 years ago when one of the Space gods (Viracocha) came to help create mankind. Puma Punku by Bolivia is a stunning example of extra-terrestrial architecture built 13,000 feet up, which cannot be duplicated with modern tools and has no evidence of human built intervention.

MOUNT SHASTA TODAY

Our area has dramatically changed in the last 40 years. The logging industry now gone caused a loss in our local identity. It has been a difficult time trying to reinvent our community. It has gone through several reincarnations most currently a Bay Area retirement community and in doing so, lost some of its hometown flavor.

Several logging companies still exist, although some 'on the slide' private companies still clear-cut, as it seems, in the dead of night, destroying the alpine beauty. The artificially induced weather changes and the experiments our so-called scientists are using on the atmosphere, along with the drought, have killed many trees, species of wildlife and have driven out all but the heartiest souls.

The real estate boom that flopped has left many empty homes in decay and disrepair with staggering human cost. The lack of money, both federally and statewide have left many facilities closed.

On the bright side, camping, hunting and fishing is still excellent, as well as culinary mushroom hunting. The historical sites are still a great a draw as is skiing, but nothing matches stargazing and UFO watching with the unfettered crystal clear skies and abundant astronomical events.

CHAPTER TWO
OUR APPROACH TO MYTH AND RELIGION

Even though it is not the subject of our book, Mysticism and Theology has shrouded the UFO subject with controversy. Here are some of the historical premises both Religion and Myth have played out in the arena of our civilization. What we were unable to fill with fact or have forgotten, religion and mythology have amply offered postulates as valid as science, but in some cases, fictionalized, and have failed miserably.

Since the beginning of recorded time, Off-Worlders, Gods and Mythological Characters have had monuments, events, stories and legends devoted to them that contradict well-known theories. The presence of seers, shaman, oracles and mystics, performed duties of channelings and healings for their leaders, the tribes and kingdoms and did so to insure bounteous harvests, dispense first aid and healings, both psychic and physical, to all that were in need. They also gave counsel in matters of diplomacy in war and peace. It is still practiced among our nations leaders. Genghis Kahn had a mystic mentor who informed him that he would be a world leader and warlord. The Druids performed what is called magic. They named the earth and nature spirits, made peace with them resulting in a working relationship. They originated what was called Paganism and even though each spirit is looked upon as a deity, they are part of unified network, all harmoniously working together for a peaceful world. Having balance, harmony and peace has been the aim and ideal of many civilizations, and still is. Many fell short, being compromised by greed and tyranny and those civilizations died out. The Incan, Zuni and others mysteriously vanished. Some scholars speculate some of these groups actually ascended to a higher realm, finishing their learning here, while others were destroyed, and fated for another try.

Science has proven that a symbiotic relationship in nature keeps it balanced and abundant, the same way our body maintains its homeostasis. Perhaps this weeding out process in relation to human civilizations is on the same order with nature.

Most ancient cultures had gods that came from the heavens in fiery chariots to help guide the destiny of man. Continuing into the Latin and Greek periods were evidences of higher learning, cultivation, refinement, as well as debauchery. The word of an advanced alien culture spoken of by Plato and known as Atlantis further influenced the path of man. Many myths of these gods like Apollo, Neptune, Thor, Andromeda and Athena have spawned, inspired and lifted humanity out of a mental fog, to one that put our progeny and origins in the stars.

Higher mathematics, music, astronomy and writing manifested into the art of physics, engineering and medicine which inspired people like Leonardo Da Vinci, William Blake, Jules Verne, H.G. Wells and others, who further applied, created, cultivated and refined, spawned from these ancient myths and stories. Our history has benefitted from their inventiveness and continues to spark our imagination. Because of these mythologies, our identity has dilated as we trudge over the path of time and curiosity to find our place in the universe.

There have been baffling and monumental discoveries made in the last 100 years about the pyramids in Egypt and Peru, and in the near and Far East. While the European aristocracy was fighting over the Flat Earth concept, these Ancient Civilizations had already discovered heavenly bodies were spherical and applied mathematical formulas to astronomy, engineering and physics that modern day scientists cannot match. The Church demonized European scientists such as Galileo, for proving the earth and all other heavenly bodies were spherical.

Batteries made from vinegar and copper in jars were discovered in Egypt, jet engines developed by Hero of Alexandria and steam locomotives used in southern Italy dot the tides of times.

Many other artifacts have been recently discovered. We would be hard-pressed with the most current technology to even begin to build the underground cities found in deep caverns and caves in Afghanistan, connected with Ahora Mazda and the Zoroastrians, (alleged to be an extra-terrestrial race) a city built to endure the ice age and not be detected from aerial surveillance. Models of working aircraft carved in both gold and wood, predating Christianity, were found in both the near East and South America.

The cities spoken of in the Indian Vedic writing in the Bhagavad-Gita, where Krishna actually lived, contains blueprints for flying craft, lay only 70 feet off the shores of India, and date back 56,000 years. There is also concurrent evidence of nuclear blasts from these flying Indian craft, engaged in small conflicts in certain places spoken about in the Vedic texts. Physicists in the Sinai have also verified geological evidence of a large nuclear battle from radioactive remains, including piles of human remains in many abandoned cities tens of thousands of years old.

Found in buried libraries, the Sumerian Text verified these battles and their locations that the geological evidence confirms. The text was found to also contain older versions of biblical stories that all correlate to historical events already confirmed. The writings describe in detail, the rulers of Sumeria, their home planet, their spacecraft and the high civilization.

It reveals they instructed the human race in math, language, agriculture, engineering, medicine, astronomy, cosmology, education, politics, metallurgy, manufacturing and technologies still far more advanced than ours. It also describes mans accelerated origin on this planet. The Sumerian Civilization, which claimed, from translated records, kick started the human civilization 56,000 years ago, far longer than formerly thought. From agriculture to Ziggurats, the original texts of our ancient history has been found, collaborated and proven. If anyone is genuinely interested in these historic facts, Zecharia Sitchin has written extensively on the translations of this authentic and genuine text and script in his book "The 12th Planet." Sitchin has brought this information to the mainstream.

The churches understated the significance of the Sumerian Culture and have deliberately manipulated full knowledge of our origins over time, most obviously by leaving this information out. Many of the Christian stories go back way further than Biblical times. Thankfully, the essence of the stories lie somewhat intact even though the stories were fashioned to compliment whatever King was in power at that time. There is much deletion in the bible of 'dangerous knowledge' that revealed more than one god, in fact, there were groups of immortals, and events that contradicted the current Judeo-Christian concept of life on this planet. Christ himself revealed that every person was as capable as he was, in performing what seemed to be miracles, and also have the same connection to the source God as he does.

The Epic of Gilgamesh, originally found in Sumerian writings, has all the creation myth, the story of the Great Deluge and more. It exposes the history of our origins as being part of extra-terrestrial events, which in itself unravels the current earth-based concepts of human reality. Not only would the monotheism of Christianity be threatened but the institutions, rituals and control the church dominates over the people be shaken. The Sumerian text not only describes in detail the layout of the solar system graphically, but it tells of the formation of the earth millions of years ago and how it was primed for life.

It further explains how they (Anunaki) purposely accelerated the evolution of the Homo erectus to Homo sapiens (which they are classified as, too) through simple genetic manipulation and cloning. It makes one wonder why many religious organizations want to keep the secret of cloning illegal. If we ever found out that our predecessors and part of our parent race were only an advanced versions of ourselves, it would reveal that the authority religion enforces is but a sham and that the genuine spiritual connection we have, is the same for us humans as it is for any species alike, no matter where they live in the universe. The same essence in other words is universal. The Off-Worlders may have done us a favor and saved us a lot of time and misery with the jumps they allegedly helped us make.

Even though it may seem that the Elohiem (a group of gods) that were spoken of in the Bible were a more advanced culture, does not make them god(s), nor impair our connection to the one Universal Self. It only accelerated our timeline, and did not corrupt it.

Several Pyramid sites under the ocean off the coast of Cuba date farther back than 25,000 years. Stone carvings predating anything ever found of human faces, animals and lizards were found in South America and span 100s of thousands of years.

Mayan writings spoke of an Alien Leader (Viracocha) claiming he and his group had visited earth as far back as 900,000 years ago and recent discoveries off the coast of South America confirm these dates through carbon dating. Landing sites have been found in Croatia, Siberia, under the ocean in Japan as more sites are coming into focus every day. Thirty-eight newly discovered pyramids, several larger than Giza in Egypt, occupy Mainland China. Their Government is planting trees on them to hide and cloak their presence. No one is allowed to explore them. While some are ancient tombs of dead leaders, others are of unknown origins.

Of course the NASCA lines of Peru could only be aircraft runways, and markers from space, as they would serve no other purpose. Rather than conjecture, these have been proven to be science fact by the finest Academic communities on the globe including Oxford, Dartmouth, Massachusetts Institute of Technology (MIT), NASA, Stanford, Princeton, Cambridge and many others. All along the road of our history, UFOs and otherworldly beings have occupied our domain. One day we may confirm why, or accept why, but in the mean time, it would be a good thing to keep an open mind. The only real glitch in the premise is that if the ETs were from off our world, they must be gods. Not so.

There had seemed to be no middle ground and that in itself has been used for fuel for many human wars. There are men and Gods and no in between. Some of these Gods may have been just as bloodthirsty and hungry for power as the most tyrannical humans. Not such good attributes to model civilizations after, or our own behaviors after. Perhaps our Off-World relations are also evolving as we humans are, so we all must rise above these unbecoming instincts and habituations, to gain wisdom and compassion.

On the lighter side, the Catholic Church now embraces the possibility of alien life, and they are first in line to convert any willing ET over to Catholicism. The Catholic Church has a very expensive sophisticated telescope array in which they observe the heavens daily.

The "Ancient Alien" Series on the History Channel is a stunning resource for hard fact, solid science and exposure to many points of view ranging from Theologians to Astro-Physicists, with reputations ranging from Rhoades Scholars, Nobel Prize winners, High Ranking Military officials and has been a great help in the creation of this book. The series is available on DVD and available through the History Channel.

SPIRITUALITY

One of the fulltime occupants of Herd Peak Lookout on Miller Mountain (PSP)

What is spirituality? Humanity has assumed that there is matter and there is spirit, our soul, and the material of things. Science and atheism postulate that life begins and ends from a random set of accidents with no order or intent that virtually leaves life meaningless unless it can be proven. However, all that we call matter is based upon a self-evident fact that is overlooked: an idea. For example, every living creature has a set of instructions, a template that instructs matter how to behave in a certain order and that instruction set, which is also a self-correcting open system, is called DNA.

As environmental and mental conditions change, we adapt, which if you think about it, is intelligence and evolution in action.

Adaptation is ever changing in an ever-changing universe. Even the wind blows everywhere and fluid flows. So, all living and non-living tissues have to adapt. The scientific community may then argue "Well, what about non-living matter?" Well, all non living material has an instruction set which is arranged in its chemical compounds that is regulated by atomic, chemical and electromagnetic attraction or repulsion, stability and instability which manifests its character as an element or compound. They may seem fixed but even atoms can decide to migrate.

In electricity, it is expressed as waveform signatures, which translates to a musical scale, a spectrographic scale of colors going on infinitely in octaves. This arbitrary arrangement of groupings cannot be answered by science alone, as it cannot answer why DNA is how it is. The omission of consciousness, since it is intangible, is the only driving force of life (creative force) that constitutes intent or movement. Adaptation would not exist unless the system was flexible and open ended. Alchemy would have never emerged if there were not the possibility of flux and transformation.

It is like trying to answer why love is. This is not a learned fact but experiential knowledge.

If we look further, every element, every atom and atomic particle that exists is pressed out of a seeming nothingness. Much, if not all has been forged in the fires of stars, and from the intensity of black holes. Quantum physics embraces this notion; from nothingness to motion.

The general rejection of GOD, is not dismissive of a universal intelligence, but of a man on a cloud. Maybe the cloud was a flying saucer, and the 'God' a more advanced individual. Those silly stories then could be viewed as a chronology of aliens in spacecraft, which leaves room for both an intrinsic spiritual identity and advanced cultures, existing in concert without these ideas being in conflict. The entire idea could be laughed at as well, with a large guffaw of outlandishness.

We are fueled by the same energy within the nucleus of every atom in our body as everything else is, in the universe. It has been described as our essence; of what our spirit is. It is found to be omnipresent and infinite. Some call it the electronic body, our spiritual self and being the glue that holds the atomic bonds of matter together, makes matter an electronic substance. There are times that every person gets out of their own way enough to allow, to receive without any forethought, this experience and is flooded with a quickening and a deeper understanding and connection to life.

In fact, Gallop did a poll in the 1970s on how many people at that time had what they coined a "Religious or Spiritual Experience." Out of 5,500 people, 77% described very similar experiences of relatedness, unity, a sense of peace and connection with nature and the universe. This is very significant.

Even though we are approaching the subject of UFOlogy in a fairly scientific manner, we cannot negate the fact that we are conscious beings aspiring to a clearer picture of our identity and what our part in this universe is. No matter how one cuts it, all questions do lead back to identity.

Consider this idea: In order to observe light, we must be going as fast or faster than light to see it. To take that further, in order to identify anything, we already must be part of it in order to recognize it. Whether it is, food, music, order, harmony (most people cringe when someone is singing out of key) it is so blazingly obvious, we never take it into consideration. Why would we cringe unless we already know what harmony, or being in tune, atoned is? This self-evident fact is without value judgments or projections, on our part. We use it everyday, taking it entirely for granted.

In a booklet the author wrote called, "The Universal Template" the point focuses on the possibility the 95% of unidentified material in our double helix is in actuality the template that give instructions to the environment we are always responding to, like: gravity, definition, our co-identity with food, (what it is) conscious identification of all the waveform signatures, such as sight sound, touch and other senses we are not aware we have yet. It is our veritable blueprints for the amusement park we live in called the universe. It also must contain the further progression of our own latent traits. We may find we have the ability to change our form consciously, or heal our bodies instantaneously, or be able to travel at the speed of thought. Imagine wanting to get pastrami from New York, and just think yourself to that location. That would be really beneficial.

We only use 10-20% of our intelligence and brain according to scientists. Imagine developing the rest.

Jon Anderson, from the band "YES" sings the lyrics in one of his songs that aptly describes this; "All Complete Inside the Seed of Life of You."

Maybe our intuition is what is driving us to seek and find the knowledge that will fill our longing, to consciously understand our unity with the rest of the universe. Our instincts now require our conscious participation.

CHANGING PARADIGMS

The quagmire for the theologian is; if an alternative historical theory doesn't fit in the seven days of creation, it doesn't exist. For the scientist, if it doesn't fit into a Darwinian theory of evolution, and that our species demonstrates higher intelligence evidenced as, high civilization agriculture, education, organized structures of government and commerce, earlier than the 12,000-year mark, it can't be valid.

Maybe there are other alternatives. Perhaps the seven days of creation in Genesis were from another solar system calendar where a day is 4,000 years or 25,000 years in length. Perhaps if the calendar were corrected, by knowing where the calendar originated, it would make both time spans large enough to take on other theories and perspectives. Perhaps the thought of what man has called God(s) was the mistaken identity of a more advanced race of people and not the spiritual essence, of all that exists. Perhaps God in general, does not fit the definition of a man on a cloud but the energy inside us, and within all that is.

Gold mines dating back 450,000 years have been discovered in both Africa and South America. How can that be explained unless there was intelligent life here that long ago with the technology to mine? We try and make history fit our concepts instead of letting historical and archeological findings inform us. Much information has been glossed over or covered up as an outlier in scientific method if it doesn't fit into the context the theorist is proposing.

Some sound variables that would bridge the missing link in our origins are dismissed too quickly for their simplicity, or they are preposterousness. They are no more or less incredible than what some consider valid. Why is it incomprehensible to consider, we have been genetically accelerated, when there is possible evidence to that effect? Do we not also employ the same genetic procedures in improving food crops, medicine for a higher quality of life? We can't keep trying to fit square pegs in round holes if we expect to get an accurate and genuine picture of our presence on this planet, even if it makes some people uncomfortable.

To add to this problem, we-the-public may be kept in the dark in regards to knowledge and discoveries about our origins by those who think they can make informed choices for the masses. If we cannot gain access to genuine information, how can we arrive at a logical and accurate conclusion, if some of the variables are deliberately concealed? It is a form of cruelty, to keep us all in the dark.

In 2005, the human genome mapping process was finished by the scientific community. If you recall or not, it was big news as they discovered we all go back to one set of parents, genetically. It was ascertained through the mitochondrial DNA record that is passed down through the mother, that a single genetic mother is the mother of us all. It was discovered this type of DNA, records severe geophysical events that cause adaptations from extreme and stressful environmental conditions. The presence of a global flood, the deluge, depicted in many holy books as historical reference is in fact evident in our DNA. It shows up as a population bottleneck at that time, supposedly to less than 3,000 people, 36,000 years ago. The scientific communities were able to ascertain the location of the one set of parents from the DNA as well. It pointed to the "cradle of civilization", as it is referred to, the alleged Garden of Eden, in Iraq. How this is extrapolated is through environmental and chemical markers that directly relate to latitude and longitudinal coordinates of exactly what was occurring biologically at the time in that region, recorded in adaptive DNA that altered instincts to accommodate a flood, chemically.

These discoveries correlate to historical data found in the Ancient Sumerian Libraries, as well and again, documented on brass cylinders and stone tablets dating back to 56,000 years ago and earlier.

Francis Crick, the researcher who discovered the double helix, speaks of these adaptive discoveries in several articles, published in Nature, (the science magazine), added a theory of how some of our DNA arrived on the planet. He believes some of the components could have been transported from meteors, space dust, mold or other cosmic flotsam and jetsam, not just here but everywhere. The author remembers as a child, thinking how flower seeds are transported into unlikely places, like Dandelion seeds, flying through the atmosphere, to land where they are blown. Nature has intelligently designed these mechanical means of procreation, so maybe the same mechanics is prevalent for intergalactic transmission, and the success of life as well.

Interestingly enough, Circadian Rhythms (a biological time clock that adapts life forms location in space to a particular planets rotation around the sun, seasons, etc.) for optimal survival, is also part of the Mitochondrial DNA package. It adapts and regulates sleep patterns to the particular planet, the metabolism to the seasonal changes in temperature, to insure warmth, food availability, reproduction patterns to the best environmental circumstance and other instinctual survival functions.

Sea Mollusks natural and natal rhythm is regulated to MARS! Its rhythms are indigenous to the solar day and night cycle of the Martian clock. They also are underwater creatures, which reveals that Mars once had oceans on it. This information is not new, but most people don't connect the dots. It would be curious to start checking if this rhythm is different in other species as well. How did the sea Mollusks get here? This is another question that needs an honest answer.

How did then some other species that lived on other planets arrive here? There is no reason we should surmise that UFO and UFO activity doesn't exist. First off, we live in an infinite universe teeming with life, and the likelihood of life being as abundant as it is on this planet elsewhere, would follow suit.

Things seem to follow a pattern and that pattern is evident in the pattern of intelligence and of the UFO sightings. They seem to correlate with some kind of evolutionary driver that could have helped accelerate our development beyond the hypothesis of science, and beyond mental prejudice.

It seems part of how nature works. The bee helps pollinate the flowers and vegetables, sometimes dogs and wolves raise a lost human child, and dolphins help African fishermen on shore, fish, and share the catch while other species adopt other species. Why does it seem out of the question we are being aided? It is totally natural for a big sister or brother to aid their younger siblings, or to adopt lost pets. What would be such a problem if a parent race that has for millennium, gathered information and observed the progression of our species unaided, found that accelerating it could help us leap over blocks of time that could be dangerous? Perhaps by splicing the best of successful trials, this improvement has saved us from extinction and still allowed us the genetic memories and adaptations, that abbreviating it would not diminish and make us more autonomous and self reflective, thus useful and quicker. How about giving us a bowl of milk when we need it? It is not such a big leap is it?

In our own civilization, such procedures are used and have been used to improve food crops, create virus and bacterial resistant strains of inoculations and improve the robust qualities of life in general. Why would aliens help us and do this? What do they want from us? What do we offer them? Perhaps the answer is as simple as purpose, friendship, love and relatedness.

Science must take into consideration the bond of love, community and family as a driving force in the purpose of any species, as it is self-evident. Science can be very weak when it comes to the subject of intangible intent if the subject is non-mechanical. A car without a driver is no better than a doorstop, as a cell is without a network, an organ or organism, to express its intent into, does not add up.

A LITTLE ABOUT NASA

Sightings abound all over the globe, but the emergence of the 'Space Age' born within the duration of both World War One and World War Two, all eyes were on the Heavens.

In 1915, only a year after Robert Goddard applied for two patents on liquid and a three-stage solid rocket fuel propellant, the National Advisory Committee for Aeronautics, (NACA) was formed. NACAs inception helped accommodate the rising rocket and aerospace technologies emerging and being used in battle during the First World War and beyond. In order to control and stay informed, the necessity for surveillance became necessary in this emerging Space Age.

Orson Wells, who recited "The War of the Worlds" in 1938, caused mass panic on the globe for his convincing radio broadcast. Those listening believed the story was real.

The United States Space Program did not exist, as we know it, until the 1950s. What happened leading up to the emergence of what we think we know of NASA, happened on several fronts.

We all know about the Roswell Incident where an alien spacecraft crashed in the desert. It was disclosed, then immediately covered up in total denial, with the wreckage being shipped to Wright Patterson Air Force Base in Dayton, Ohio. However, there were other similar classified incidents in the United States and around the globe happening in the same time period. A small but significant sample of relevant occurrences is brought to the forefront in our chronology. Strategies in denial had to be implemented immediately, and sustained in this sensitive nuclear age.

During the Second World War, an alien spacecraft was supposedly retrieved from the Black Forest in Nazi Germany (circa 1936-1938). The Nazis and their scientific team began working on reverse engineering the technology they found, and succeeded. The Pinamunda (sic) Research Center, buried deep in the German foothills, (Dorsten Mines) was center for much of the advanced weapon technology gleaned in this period. Werner Von Braun, credited for the creation and development of the V2 Rocket, which evolved into the still used Saturn V, candidly admitted in an interview, "I did in fact get help from extra terrestrials." Tesla type generator coils were also retrieved from remote locations that they powered in Nazi Germany. U.S. G.I.s found these and were told to keep it a secret.

After World War Two, the Americans in a classified event called, 'Operation Paperclip,' shipped over all the Nazi Scientific team to continue working on what is The United States Space Program. This Branch broke into the three plus branches: research; White Sands, Jet Propulsion Labs (JPL) and NASA. Armed with a storehouse of artifacts from all over the world, including ancient relics Hitler and the Nazis stole from the Near and Middle East, cutting edge secret scientific communities, and research from the war, including time travel findings, the U.S. gained ultimate supremacy. Von Braun, James Webb, head of JPL, and other high ranking officials some Nazi and some Americans who ran and still run White Sands, Aberdeen Proving Grounds in Maryland for electronics, Fort Bliss in Texas for ballistics, Huntsville Alabama, and other 'non existent' agencies set out the private secret agenda of the space program with unilateral support from the White House. The agenda was to retrieve alien artifacts from the Moon and Mars, to further validate our alien origins and bring them back to use.

We now have 500-750 primarily ex-Nazi run scientific programs. The state department did consider Von Braun a serious risk and criminal but so were the employees that worked at Auschwitz on humans for experiments. Operation Paperclip is still considered "classified."

So (NASA) was created to formulate a plan to make a space agency designed on the surface to get U.S. Satellites in orbit to compete with Russia's Sputnik, while their covert research entities operated from within.

The agency housed the secret team's research in broad daylight with no one knowing except the select few. There was great competition in the Senate and Congress to have the agency also develop and deploy weapons, a manned space program, all-military lead spearheaded by Senator Lyndon B. Johnson. Eisenhower signed into law on July, 8 1958, the NASA (National Aeronautics and Space Act) that on the surface handled non-military issues, while the military aspect remained totally classified, including its plans for alien technology retrieval and UFO study, all signed into law and in the by laws and articles of the space act. The Brooking's Report depicts the authorization of total denial, secrecy and non-disclosure in great detail, approved, in closed sessions with Congress.

Both agencies, the former NACA and the NASA that handled public matters, occupied the same building, but not all the employees knew each other's business. NASA the classified entity, with its cover behaving as an information service, keeps the public up to speed with its space discoveries and satellite launches, worked for a long time pulling the wool over the publics' eyes.

As stated, in 1958 the hidden agenda NACA held was to retrieve alien technology on the moon and on Mars, under the guise of keeping up with the Russians leaps in technology with Sputnik, when it was devoted to finding the origins of humanity and extra-terrestrial influences.

What the 1958 Brooking's Report established in article 271 is, any and all information gleaned by NASA in discovering of any alien artifacts, advanced technology and alien life is automatically classified, blacked out and denied to the general public.

Eisenhower felt that any public ET awareness would destroy the fabric of civilization when in reality it could help spring humanity from the death grip of tyrannical overlords keeping our civilization prisoners in ignorance and archaic thinking. Pretty slippery work, and it is all funded through the military budget, undercover.

This is a very significant point as it answers the question why there is non-disclosure from our governments. It reveals shadow organizations that are funded by our tax dollars under the heading of "military and defense" when in reality it is not. The actual space missions came to pass in 1961, with Kennedy promising our country getting to the Moon within 10 years, keeping the U.S. supremacy in space. Within that interim time, advanced technology already existed on the planet discovered by both the U.S. and Russia, while a secret collaboration actually existed, under the space race for the publics' benefit.

Russia quietly, since the turn of the century, was also developing their space program that was in ways, further along than the U.S. agencies. They too kidnapped German scientists, during and after the war, for their programs. They also had discovered artifacts and ancient ruins in Siberia, Bosnia, Croatia and the Gobi Desert area, of advanced races that precluded us. With the mutual collaboration of both superpowers, our collective race to space was guaranteed.

It should now be clearly obvious the reasons for non-disclosure. A secret space program being fueled by individuals who feel they are more entitled than the rest of us to know their origins, have use of high tech advanced technology and reap the benefits we all are inheritors of.

Much more evidence exposes the deceit the Secret Space program employs. There has been a question brought up regarding why Project Blue Book was disbanded in 1970. Perhaps these organizations found what they were looking for and have competent use of it. The boys found their toys and no longer need the ruse, to confuse the public eye? There is also a question about why George Bush Senior, decided to cut funding to NASA. Maybe the secret space program became volatile, independent and self sustaining through the Military Industrial Cartel, like Hughes Aircraft who also receives subsidies and direct contracts in exchange for its own technological benefits for its engineering successes? Perhaps speculative, but it still is a legitimate question.

It is also suspicious why the Saturn V rocket has not been innovated, upgraded and evolved into something better. It still is in use globally and has not been changed since 1961. The same spare parts are still servicing it; it still is inefficiently using fuel and is obsolete. In all probability, it is for the publics' benefit and the benefit of those agencies outside the realm of secrecy.

A featured article in Wired Magazine dated, November 2012, is about rocket scientist Elon Musk and his breakdown of the Saturn Rocket. He points out the obsolescence of the Saturn V and his solution. Musk actually has upgraded the entire propulsion system in hopes of using it to propel us to Mars.

The fact remains, any smart business upgrades its product for efficiency, from automakers to refrigerators and computers. To think a military contractor deliberately keeps its products inefficient, such as a power supply for a space hungry world, is purely naïve and deceptive.

All of this points to the distinct possibility that what we will call the Secret Space Program (SSP) has taken over all the significant advances of reverse engineering and invention we are not privy to. They are already using the hyperspace drives and electromagnetic neutral gravity motors the ET saucers employ making navigation effortless and fluid. Until we can touch them however, these devices remain conjecture.

Author's Note: Even though this book is devoted to UFOlogy, the mythos of the individuals involved in the Space Program in both Hitler's Elite scientific community and our own, cannot be ignored. It has been discovered that Werner Von Braun, James Webb and other staff directors are all 33-degree Masons. It is no accident the runway at White Sands and the launch pad at Cape Canaveral are both designated '33' for this esteemed Masonic degree.

Isis, Osiris and Horace of whom the insignia for the early space Mission of Mercury was modeled, is part of the Egyptian and Aryan Race connection to Off-Worlders they claim to aspire in meeting. Von Braun himself is a devotee of Horace and has allegiances to the belief systems of the Egyptian connection to the cosmos. He feels he is destined to meet his predecessors in the glory of space and that it has privilege to orchestrate this endeavor.

The Nazi connection was linked to Isis, Osiris and Horace as well with the Aryan race, is no secret. If anyone reads the fine print of this philosophy, it actually includes all of humanity, not just some of us, but all of us. However like Hitler, these individuals believe they are the "chosen" exclusively, as many self-involved, self-serving individuals feel. The danger in this is, in order to be exclusive; others must be excluded. What about the rest of us, and our connection to the universe? Creating another holier than thou "brotherhood or gang," is counter to our evolving global and galactic mentality. Why is it that some of these geniuses have at their disposal, the privilege of pursuing their own agenda at our expense? This just reinforces the superior, inferior, master, slave relationship we are attempting to tear down and put to rest. It would be one thing if they were not public servants, but they are, and they are serving themselves. Von Braun himself felt part of this ancient race and has planned on meeting his predecessors in space sometime, with the use of the restored technology that has been found. I am all for freedom of belief, but if it is at the expense of others, there is a fundamental error in interpretation.

It is a subject worth pursuing, but is not in the scope of this book. To go into a deeper investigation of this at this point, will have to wait. It does though have great significance in the direction our country has gone and answers why things are the way they are.

MARS MAKES A STATEMENT

In 1990, The "Phobos Event", which was hushed up, appeared publicly for the first time in Zecharia Sitchin's Book "Genesis Revisited". It concerned the loss of two Soviet spacecraft sent to Mars to explore the Martian moon, "Phobos." Named after the Martian moon, Phobos #1 and #2, both satellites were sent off in 1988 to reach Mars in 1989. Both NASA and the European Space Agencies collaborated on these projects with the Soviets. Phobos #1 just vanished without explanation and no photos were ever taken by it.

The second probe; Phobos #2 did send pictures back to earth by the two cameras aboard it while in the Martian skies. The Soviet Mission analysts described that there was an object "something some may call a flying saucer" and another picture of its shadow of a cigar shaped object flying in the planets skies between the Soviet craft and the surface of Mars. Immediately the spacecraft was directed to shift from Mars orbit to approach the moonlet and from a distance of 50 yards, it was bombarded with laser beams. The image shows a missile approaching Phobos #2 from the moonlet side where the UFO was situated, taken with the infrared camera just before it stopped transmitting, obliterated by the alien missile.

Sitchin surmises that the Anunaki still have active bases on Mars awaiting their return, protected by these robotic systems. Pictures below.

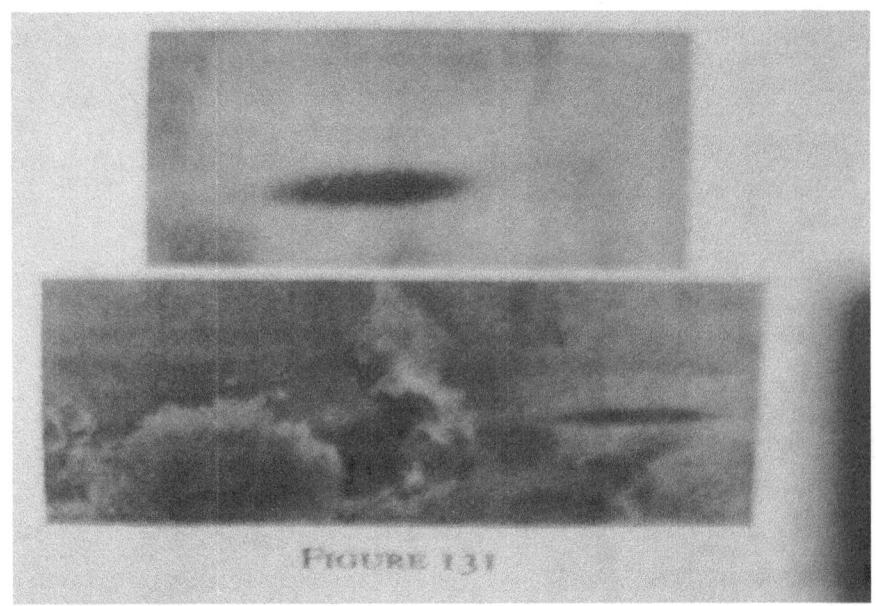

Figure 1 Images of UFO that later destroyed Soviet Mars Orbiter taken by Orbiter

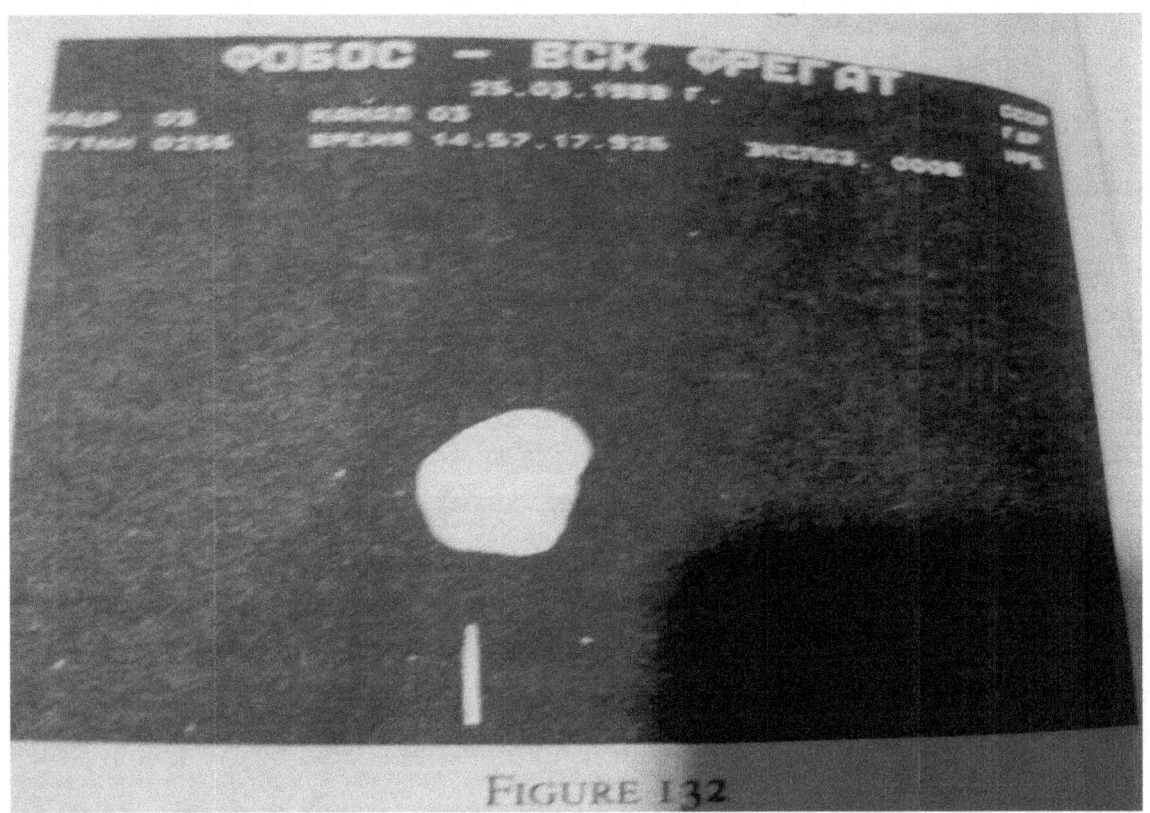

Figure 2 of Alien missile shooting at orbiter on infrared camera (Soviet and NASA pictures used in Sitchin's book "The End of Days"

The powers that be officially call this an "unexplained accident."

From an earlier discovery in 1983 of a 10th planet, (Nibiriu) the geopolitical powers formed a secret organization from Infrared Astronomical SATELLITE (IRAS) (NASA'S property) that discovers heat emitting celestial bodies, i.e. planets. The IRAS has scanned this planet every six months and it is confirmed to be moving in our direction! In fact, the group's main intention was to observe the 10th planet they confirmed finding, and allowed the news media in on it. The planet currently continues its elliptical orbit getting closer every year.

This discovery of the 10th planet appeared in every major paper; The Chicago-Sun Times, December 20, 1983 Detroit News, The Philadelphia Enquirer, on and on. The news of this was so shocking (and exciting to most, in a good way) the space fairing nations agreed to force a retraction the very next day; they blatantly lied and played it off as a misunderstanding.

This event created a significant change in U.S.- Soviet relations. A meeting ensued where The U.S. and Soviets made an unconditional agreement and cooperation in Space relations that Gorbachov and Reagan shook on publicly, to seal the deal. With Gorbachov at his side, Reagan made his famous remark pointing to the skies, "Just think how easy his task and mine would be in these meetings that we held if suddenly there was another planet outside in the universe… I occasionally think how quickly our differences would vanish if we were facing an alien threat from outside the world." Sitchin Z., (2007) End of Days, p303-306 Harper, New York

Mars could support life; in fact Richard Hogeland a former NASA research scientist who worked on photographing visual evidence, went public before NASA could suppress it all. Hogeland published some of these documented pictures of structures on Mars and the Moon. The images show broken robot parts, crashed spacecraft littering the surface of the moon and buildings on both the Moon and Mars, now suppressed and denied.

Hogeland went forward and published several books about these subjects with pictures that skirted the censors, with documented data from these internal NASA projects. NASA officials, under threat of deadly force, are restricted from disclosure while most public photographs have UFOs airbrushed out. There are allegedly hundreds of craft about our planet. Armed staffs observe every conversation; many employees do not have clearance and is on a "Need to Know" basis. There are employees who resent information is kept from the public, and leak out what they can, potentially putting their own freedom at risk from these unauthorized disclosures. Many employees have no idea of the classified entity within NASA.

In 1989, a set of guidelines known as "the Declaration of Principles Concerning Activities Following the Detection of Extraterrestrial Intelligence," was formed, establishing the procedures to follow in the event we receive a signal or other evidence of ET intelligence, which is to delay disclosure for at least 24 hours. It did not give consideration for any immediate encounters, or for any remedial action. It is another way to give time for those in charge to fabricate a plausible lie to feed the public and covet their findings privately.

PRESIDENT OBAMA and MARS VISITS

There is a story closer to home (Siskiyou County) about one of our presidents; Barack Obama. Keep in mind some stories, like many UFO stories, get to the public because they seem so amazing, ridiculous and silly that no one would believe them. The powers that be, know that the people who disclose likely belittle and ridicule themselves by judging their own observations as "ridiculous" thus use that self depreciation to discredit the stories, diffusing any possible credibility.

Unfortunately, our current politicians lie and ridicule to hide their faux pa, like the Wall Street fiasco and the lack of weapons of mass destruction in Iraq, to name a few. With that in mind, the author was personally contacted and received an email and link from www.exopolitics.com sent from a person highly involved in Defense Advanced Research Project Agency (DARPA), and the prominent officials who have participated. They asked me to help them disclose the following information. They claim President Obama, participated in a Space Camp in 1980, at the College of the Siskiyous (one of Author's Alma Mater, also) preparing him (Obama) for a journey he allegedly took to Mars from "The Jump Room" located at a Hughes Aircraft facility in El Segundo, Ca. while still a juvenile.

The Jump Room is a Tesla warp coil portal attenuated to the corresponding receiver on Mars. Simplified, it's like a two way radio/fax machine. Obama attended the class under the name of Barry Soetoro. He was in class with eight other teenagers one of who, Regina Dugan, is the current head of DARPA.

The class taught by Major Ed Dames who acted as the trainer, also served as a scientific and technical intelligence officer and advisor for the U.S. Army. Both Andrew Basigo, 50, a Washington Lawyer served on DARPA'S Time Travel program, and William Stillings, 44, who also participated in the Mars program for his technical genius, mutually confirmed that Obama did in fact enroll and attend the training class in Weed in 1980 and went to Mars to visit the U.S. facilities there in 1980 and 1983. These programs of course, were and are all part of the CIA, "Mars Visitations Programs". Most of the participants have parents in the CIA.

In 2011, while on George Nooray's radio show, 'Coast to Coast', Major Dames called in while Basaigo and Stillings were being interviewed about the Mars project. Major Dames was obviously furious at their disclosures and denied any authenticity to the story, adding another splinter of doubt to its credibility. It is this standard operational procedure of denial used in every sighting that has made it to the public view, effective in discrediting the event and the people involved.

However, there is a great deal of information and corresponding evidence to this story and the three whistleblowers that include President Dwight Eisenhower's great granddaughter; Laura Magdalene Eisenhower. She refused a covert attempt to recruit her into what was a secret colony on Mars. Former Department of Defense (DoD) scientist Arthur Neumann, testified publicly that he was transported to a U.S. facility on Mars for a DoD project meeting.

There seem to be some incidents in this disclosure that could deliberately discredit some of the events. It is the author's opinion that it's been doctored and embellished (wild creatures) to keep it in a state of ridicule, glossed over and dismissed to stop further examination or to protect the actual truth behind the tales, rather than have it removed entirely.

The author did go up to the College of the Siskiyous, and in fact found that in the summer of 1979 there were aerospace programs offered, but the 1980 summer catalog was not found. We found three different articles from three different local newspapers on the summer program for 1979 that ended up being finished at Beale Air Force Base! There is possible truth to this story.

A ruse can act as a cloak. We won't know unless we all get to go there and honestly, the idea of wild Martian beasts that eat people is a great ruse.

Perhaps, they are possibly mechanical or biological sentries (boogie men) made to ward off trespassers. Humans make and use them in their gardens designed as scarecrows and or mole stakes that blast moles in the potato patch and scare off any other wildlife. Humans also use Land Mines, drone aircraft and other deterrents as well; showing the likelihood there could be some truth to this. The following website contains the entire story in great detail about Mars, 'The Jump Room", our presence on the red planet and more. It is highly encouraged for you to visit the site and leave your opinion on their blog.

The web address is:
http://www.exopolitics.blogs.com/exopolitics/2011/11/mars-visitors-basiago-and-stillings-confirm-barack-obama-traveled-to-mars-1.html

While on the subject of non-disclosure and distortion of facts to the general public, in December of 2010, in the late afternoon, a Saturn V rocket was launched from Vandenberg Air Force Base. Its trajectory was over the Pacific Ocean and it malfunctioned, smoked, separated into several pieces and crashed into the ocean. Over 1 million people witnessed the event. The authorities from Vandenberg immediately fabricated a lie trying to pawn this event off on a well-educated public, as a Boeing 747 jet flying to Hawaii, and not a rocket. Its "vapor trail was being mistaken for a rocket, since it was flying into the sunset," the military claimed. Most people have grown up watching jets take off and land from Los Angeles, have witnessed countless rocket take offs both live and on TV, and the public is NOT STUPID. The author grew up like a lot of us, very familiar and comfortable with what a rocket is, what a vapor trail is, and the difference between that and a jet, including the sound.

They further insulted the public, by manufacturing a youtube video with very sharp, young and savvy guys that were obviously military, attempting to misinform the public with shabby, (you got to be really dumb) unconvincing videos of how this was really a jet.

An unconvinced public just blew it all off, (the other effective measure, apathy) from the disgusting stupidity of it all. Most newspapers and news stations later admitted it was a malfunctioning rocket.

In the April 2011 issue of Popular Mechanics (Thank God for them) was an in-depth detailed article about the Air Force's screw up that crashed their Saturn V Rocket into the Pacific Ocean. It was carrying a satellite (one of two) they sent up, that deployed itself prematurely. This premature release caused the rocket to crash, and yes this was the event they tried to cover up. It was further disclosed in the article that the satellite pair, one of which was already in orbit were sent up to occupy the Lagrange Points (points invisible to earth but stable to not move or be affected by gravity) to actually spy on the dark side of the moon. The article did not reveal why they wanted to observe the non-visible side of the moon, but there are allegedly manned bases already there with the ancient bases as well.

The U.S. has been trying to remove these valuable relics and evidence by bombing them and was observed by our other global countries that stopped them. They all fervently disapproved, since it was non-consensual to start with and the U.S. has no right to make those decisions unilaterally. China also wants to explore the moon, which they do have a right to do, and preserve any history that may be beneficial to the inhabitants of our planet.

It is good other countries do not carry the same selfish intent in this matter, as we all have a right to this information.

CHAPTER THREE
SIGHTINGS DESCRIPTIONS: KINDS OF FLYING CRAFT

Over the span of time and sightings, we are left with a bag full of different flavor UFO configurations, sizes and colors that imply some specific purpose. By and large most sightings up to the advent of our own terrestrial contraption of flight, the cigar shaped, silver saucers, silver balls, glowing balls, globes with an aura, etc., lack much of the aeronautic wings, rudders and friction flaps terrestrial craft use. It was not until after we were flying that aircraft with wings and fins appeared and have been reported.

My dad was in the Air Force when I was a kid and we lived only five miles away from Wright Patterson AFB. We used to go there and I was cut loose to explore and board older aircraft, the B-52s and go into hangars and look around. There was not the security there is today in the 1950s and 1960s and there was a lot going on there.

Supposedly, there had been over 58 different extraterrestrial craft identified, although the number is probably higher. Earth hybrid technologies that employ the reverse engineering of downed UFOs could have contributed to some of the secret experimental craft our own military industries have been using, confusing the sightings over the last 30 years and maybe longer.

I have been privy to some secret information that happened, like the Philadelphia Experiment that involved teleportation of live beings over fax lines in the late 1940s into the 1960s, along with experimentation with time travel.

Keep in mind, when the chronology of reports goes into and after the 1950s, the governmental agencies would and did use deception, including the occasional admission of UFOs to cover their own secret operations.

With experimental aircraft ranging from the U-2 spy plane that looked like a hanging cross, to the Stealth bomber and the SR-71, which beginning in the late 1970s was misidentified as a UFO consistently, it is no wonder that there were conflicting reports.

The addition of Iridium flares and weather balloons of aluminum all added to the mystique of UFOlogy when in fact they were not UFOs but used as decoys.

Those unfamiliar with stargazing also had a hard time discerning the path of Venus and Jupiter in their visible seasons as UFO sightings. The powers that be also used that as an advantage to debunk UFOs and cover their own secret operations.

At the beginning of each decade subsection, we try to give some relevant information as well as historical background in what occurred in respect to our aerial technology.

We also include movie releases like "Men in Black", as there is an adage that goes like this, "The best way to keep some things concealed, is put it out in the open."

With the closure of several well-run and efficient programs like project Blue Book that ended in the early 1970s and NICAP, you will also notice the degradation in the quality of the reports. We have painstakingly removed the irrelevant ones, however anyone can visit the NICAP, Project Bluebook, MUFON, or the CUFON archives for complete reports across the globe if the necessity arrives.

KINDS of ALIENS

Current evidence taken from the Kepler telescope shows the abundance of habitable (Goldilock zone) planets. Just within its depth of field, which is a small fraction of the universe, lie tens of thousands of solar systems and millions of planets that could support life. It makes logical sense that we may have been visited by more than one alien culture.

From ages of testimony written and drawn, many ancient cultures depicted varied sky or star beings from different locals about the galaxy. The Dogan tribe in Africa, possess a star compass given to them in antiquity, (circa-1800s), by Off-Worlders that came from the Sirius constellation. The compass accurately reproduces the location of their solar system. The star cannot be seen unless one possess an electron telescope, as it is invisible to the naked eye, and this star and corresponding solar system were discovered 100s of years later, so the compass is genuine. The Anunaki that arrived and established ancient Sumer, to Viracocha who arrived in South America, all these differing Off-World races have been noted.

The "Grays" are the alleged race of small, gray, off-white, large-eyed space-fairing tribe who took responsibility for the cattle mutilations and abductions involving genetic splicing and experimentation. They claimed to be peaceful and have only been forced to attempt these experiments to avert extinction caused by extended deep space travel resulting in physical degradation. They also claim, as all the aliens seem to claim, to have "originated" life on the planet.

Perhaps they all contributed to the collective gene pool, and in result we have benefitted and also paid with genetic anomalies that cause illness.

With the blanket of secrecy our own shadow government has over disclosure, it would take all of us becoming genuinely plugged into our Universal Knowing to make heads or tails from all the information, or some brave Off-Worlders to take a stand on our behalf, in the name of integrity, to speak-up and disclose the truth, without puffing themselves up. The easiest method would be having a government we all thought we had; honest, and willing disclosure takes place.

The author is of the opinion that we are a synthesis of several Off-World races, homogenized together, no differently than what we as an earth born global community already partake in. We all mix and match already. In essence, what does it matter? I would rather be a universal mutt with the best of all possible worlds than limited to just one☺

THE UFO SIGHTINGS

This book covers several different kinds of sightings that have occurred in Siskiyou County and other surrounding counties. The UFO sightings that we have received from local residents are from within the last 30 years or so. Most were not reported to any federal agency, so you are hearing about them first.

We also researched the NICAP, MUFON and other UFO agencies records to find relevant and interesting sightings that have been reported in this region since the early 1900s. We included some going back to biblical times to the dawn of humanity and developed this into a chronology so you can see how far back and how ubiquitous the presence of UFO has been in human history.

It is said only one out of ten sightings have actually made it to reporting agencies such as the Air Force, Mufon, the Police and other branches of government. Some never leave the minds of the witnesses and were never recorded. Some we found have been blacked out all together but we were able to find a brief reference to some of these sightings.

Folklore, early writings, ancient libraries containing cuneiform tablets, petroglyphs and pictographs have sufficed in primitive and early sightings. We have chosen to list the local sightings that came to us before listing the chronology. There are also many reported sightings that will amaze you.

BIGFOOT/YETI SIGHTINGS

We have included sightings of Yeti or Bigfoot in their own section. There is an inordinate amount of documentation, interest and publicity of Bigfoot. Every year, the town of Happy Camp celebrates their mascot: Bigfoot. In fact, they have a giant wooden sculpture of a Bigfoot inviting people into town.

Countywide sightings have been occurring since the local Native American Tribes have been here, along with the existence of small wood type fairies the Indians called the Axeki. They were mentioned briefly (wood fairies, Axeki) in some limited sightings, but too sparsely. The Tinihowi spirits were also seen, and were sacred to the tribes, being guides for the Shaman. Some equate them with the divine guide beings described by Fredrick Oliver, Guy Ballard and others.

The Karuk Indian Tribe (Klamath River Basin) report common Yeti sightings on their reservation as the Yeti know it is a safe haven. During hunting season, deer know the safe humans to gather by, the Yeti know the Karuk Tribe respect them and allow them safety on their reservation.

With all the research available and the existing books and websites about Bigfoot, we decided to include a few outstanding sightings that occurred on the west side of Mount Shasta proper, reported to us by locals.

Many local residents choose to keep their sightings private to protect these creatures from the potential exploitation of Bigfoot hunters.

The Ancient Alien Theorists surmise that the Yeti are a possible left over of genetic experimentations conducted by some of the early Off-Worlders, to create a sustainable, intelligent adaptation to this world. Some theorize that they are possible left overs of Cro-Magnons and Homo sapiens hybrids. There are some that postulate the Yeti could have been sentries left here to protect some of the Off-Worlders bases they left behind. However, if you have seen a Yeti up close (as the author has, three of them with four other witnesses) those theories take on an entirely new meaning when you stare into their bright amber eyes that hold a depth of intelligence beyond "theory." Wherever or however their origins exist, they possess a substance of mind similar to our own.

INNER SIGHTINGS

Some individuals claim the mountain has acted as a catalyst, drawing them here, through psychic and telepathic communication. They claim the occupants within the mountain operate on higher dimensions and call them here for spiritual significance. We refer to these as inner sightings.

Some remarks by the late Dr. Carl Jung regarding UFO sightings are included. The archetypal significance and skepticism finally precipitate into a story, from one of his clients that had visitations similar to Guy Ballard and Fredrick Oliver's book characters. It seems that many of those who occupy inner space (or space that exists at higher frequencies than our eyes and ears can sense) use spacecraft to travel to other planets and other dimensions among their myriad of aptitudes, including bio-teleportation, according to the participants.

The spectrum shows matter is not solid; it goes in all directions subsonic to ultrasonic, visible light into gamma then out of sight and beyond, which we will eventually learn to master. We only use at most, 20% of our brain capacity as it is. The teachings reveal that we actually occupy a realm of omnipresence, that we too can become aware of and master.

The entities that supposedly occupy this realm have been reported to appear as a ball of light throughout history, with no mechanical assistance. These beings apparently have evolved beyond the need for spacecraft, however, use them for the benefit of species that would be scared off by such an entrance.

Those who reported alien abductions claim that the medium of communications has been telepathy. It is also the same for those who had inner sightings regarding how the conversations took place. Visual images, audible voices and tactile impressions occurred. Our nerve endings do not end at our skin. It is not a far stretch to examine our own modes of communications and start to understand that we humans use telepathy on a daily basis and are not conscious of it and take it for granted. Unconscious body language and chemical and hormonal communication happens on subtle levels, beyond our skin.

It has also been refined into a weapon called mind control. It is another capacity that we possess that when we begin to become aware of it, we can refine it, for beneficial use. We already call it synchronicity, when we think of some one and they call us, or visa versa. We must realize our neurons don't end at our skin or else our senses would not receive any long distance communication like sight, sound, etc. Imagine a field you drive by everyday contains poppies or marijuana and since you never knew what was in this field, you only viewed it as a field of weeds.

Someone showed you the plants and that field never looked the same again. The point being is that once we recognize something, we always will be able to recall and target it in our world. In essence, this is how we learn. We view the universe as a huge field of weeds and are not conscious of all the things in it, or our talents or aptitudes until we are introduced and tuned into them. It is not that the knowledge or contents did not exist, we just had not been aware of it. Like a flashlight beam we expand to enlarge the field of focus, and as we dilate it, we become aware of what has always been there, that was out our field of view; including UFOs. When we grow accustomed or 'attenuated' to new skills, they become commonplace.

This capacity was being used and refined by both the Russian and American spies in the Cold War. It was called PSI. Many experiments including hypnosis, auto conditioning, and power of suggestion and waveform manipulation were refined. (The use of radio waves that run on the same frequency the brain uses, to remotely control people.) These subliminal conditionings were all used and obviously have validity. Watch TV commercials and learn how you are being controlled and your kids, and expand it to the shows you watch to see if it controls your mood, or your appetite.

The problem is in the misuse of these talents, such as mind control and are dangerous, manipulative and in short, abusive. They can and will burn those who exploit and trifle with them. Mind control is not inline with choice, free will, or educated conscious decision-making, asserted from a heartfelt desire of mutuality. It is important for everyone to be informed in how these talents are used for slavery, and how to counteract the influence. Saying "NO" is a powerful start. It is high time all of us gain our trust back in our own senses and give it a voice.

We already dismiss too many things we ought listen to within and act upon, rather than gloss over from fear or of ridicule or apathy. It may be that our consciousness dances over our nervous system like the fingers over the strings of a lute. The more we realize that we have more capability to communicate and interact on many bandwidths of energy, we become aware of the simultaneous effects it has on our environment. It is like when we throw a rock in a pond it creates a ripple effect. This is the same as our thought waves operate. They ripple from our mind and emotions into everything in our environment. It is a larger, reciprocal action, reacting on many octaves of energy.

It may seem abstract but if you understand how a radio works, how it is set on a channel, and how that channel corresponds to higher wavelengths, it creates a harmonic resonance, like playing a chord. It is scary to take responsibility for our thoughts, but the more we learn, the easier it gets.

The reason this is brought up is that these capacities are science and proven facts, where they once may have been occult or esoteric, they are out in the open, and in that we need to know when and how others may be using these principles to harm.

Anger, for instance, is a note or notes on a musical scale, and depending on how much volume is given it along with bass and treble, it is broadcast through our environment. Even repressed anger carries, and those around us pick it up and are uncomfortable. Your children may pick it up and behave stressfully. Love also is broadcast. In fact, everything is broadcast. It is the quality that translates into a bandwidth that we are either conscious of or not conscious, that we inject in our world. It is most uncomfortable when we become aware of the emotions we broadcast. We all of a sudden realize we have feelings we feel out of control about. Fortunately, when we get to the point of recognition, the process of correction is evident with self-awareness. Love is part of that self-control that fosters self-examination.

All of us use our telepathic powers and don't even realize how often and how much we rely on it to interact, or react in our world. There is machinery that emulates our emotional energy, and is used as weapons, including depression. There also may be other intelligences that operate on a different bandwidth, or channels just like cell phones operate on G3-G4. If we are not tuned in or 'attenuated' or conscious that are operating on these frequencies, we miss out on a great deal of communication.

The Local Indian tribes were tuned into the guides that operated on different frequencies and were able to see these beings, the little impish ones and the Shamanic beings. Be it one believes in these beings or not, does not negate the sciences behind the science and frequencies of the spectrum that all energy expresses through, from dense matter to gamma and beyond. This embraces some of the beliefs behind the inner sightings, and the premises that these individuals base their experiences on.

ANOMOLOUS SIGHTINGS

Mount Shasta offers anomalies, like black waters, electromagnetic disturbances, warp time that visibly manifest, cloud banks that appear and disappear within minutes (over and around the mountain). There are not many of these stories, but the ones reported were interesting enough to include.

There are many left over hauntings that left traces in the places they occurred. More than an anomaly, there seem to be incidents in history that need to be revealed and rectified. Unjust deaths and cruelty to groups of supposedly free people used as slaves, occurred by greed driven industrialists hoarding gold and absconding property for the railroads, have blemished the local history.

There are stories all around the area that hold this sad and violent energy, that in itself would be a tragic chronology if written. The solution seems for this issue is awareness so it does not repeat, and a continuous prayer for and of forgiveness and wisdom for all those involved. That is another entire book.

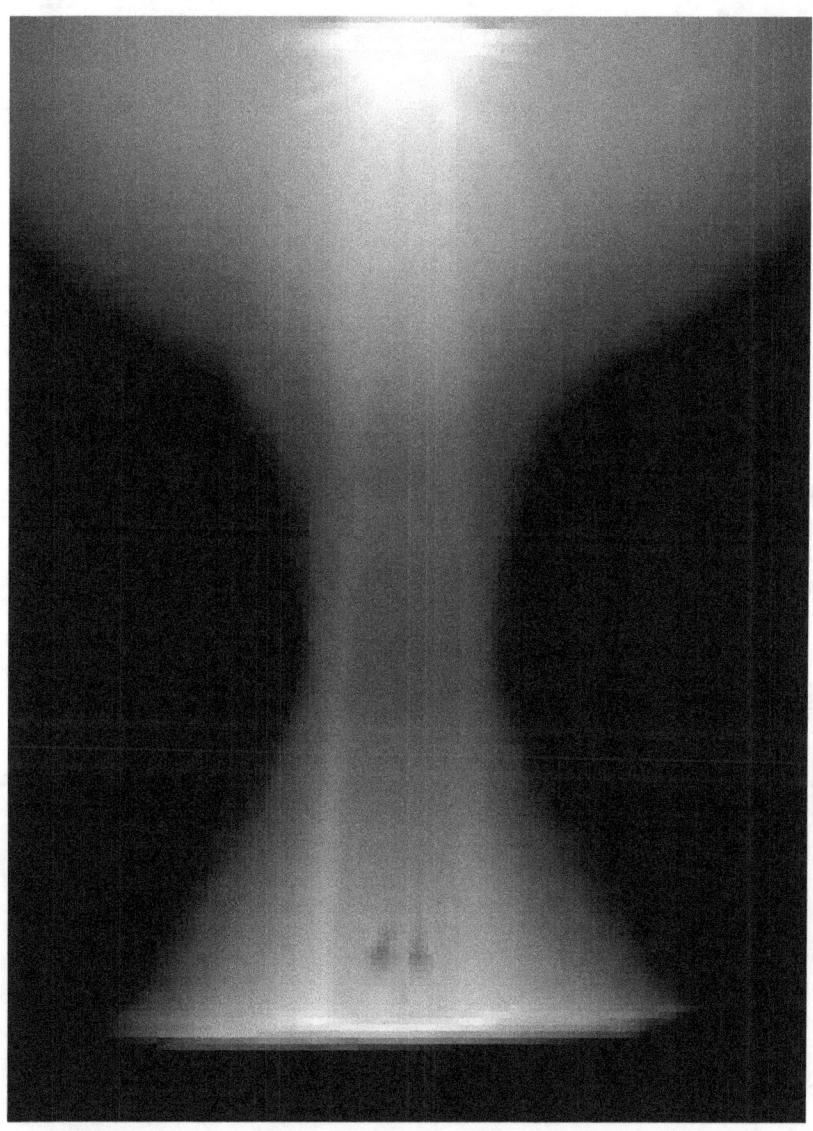

CHAPTER FOUR
LOCAL SIGHTINGS

A Little History about UFO Interviewing

Over the last 30 years, I attentively listened to the stories of individuals who either just happened to bring up UFOs or some conversation evolved into the question of life in the cosmos beyond earth. In such a place as Mount Shasta, the subject of UFOs would seem to be an easy proposal to get a response to, but it is not so.

All in all, many locals were tight lipped in fear of ridicule and the ones brave enough to come forward, did so requesting anonymity. Very few participants did not care and allowed me to use their name, but I chose to protect some of these individuals in most cases for good measure, but know their identities if there ever was a need to disclose it. Several insisted I use their names, so I did.

It was surprising to see the vast array of individuals who witnessed these otherworldly events. Some were attorneys, doctors, police, loggers, housewives, mechanics and the like. Few of them would be questionable as far as mental competence or integrity. Taking note, I personally asked them to interviews over the last two years. Some had died, but I added their stories from memory or relayed by relatives, if they had merit. Posters were put up over the span of six months last year, and ran four months of newspaper space, requesting stories about the paranormal and UFO sightings and Bigfoot. I chose to limit the Bigfoot section by the sheer size of the UFO sightings and the already impacted subject regarding Sasquatch/Yeti in our area. It has come to the point that snack food companies are using the Bigfoot for a sales pitch and once again demeaning the creature. What you will see are genuinely amazing verbal sightings of Yeti on the mountain although only a handful.

Some of the sightings were received via email. I also conversed with the witnesses. Some sent me pictures, and some personal interviews were conducted at my location with my assistant, questioning each witness over and over again for consistency of events.

We requested simple details such as the date, time, location, how many witnesses were involved, how many UFOs, how long the incident occurred, craft details, any audible disturbances or environmental changes, magnetic and such. I attempted to get pictures of most the locations as they are today and included maps in the center of the book to make it easy to refer to the locations as you leaf through the sightings.

Through critical examination, any stories that were not of merit were deleted. Most the witnesses interviewed were familiar to us and in a small community most individuals choose to be honest or risk being knocked out of the talent pool for having no integrity, or being a liar. Small communities do have their advantages.

We attempted to get each witness comfortable enough in their own skin to speak freely and feel that they were being respected, which allowed them to let their guard down and give the best details of their sighting as possible. We have had cultural expectations placed on all of us about paranormal events and UFO sightings subliminally. These tactics were deliberately placed on us to stop the flow of information, but the damaging effects to ones own self-credibility, mixed with the fear of being viewed a liar, a nut or a space case, has put a paranoiac reaction in most witnesses psyche.

We turned this around with some of the more sensitive individuals, and in essence, de-programmed them from that stigma. It is truly insulting to see a mature, competent, well educated, versed, articulate and mentally present individual who work or have worked for the Federal Government for their skills in non subjective observation, doubt and demean themselves when it comes to UFOs from the nasty work of psychological warfare our own government installed on us. It is very angering and wrong.

When competent fighter pilots, who gave all their heart, soul and mind to protecting their world and did it excellently are mocked and ridiculed by their commanding officers for the very thing they were hired to do, (observe and report, investigate and perhaps mitigate) is an abomination and an insult to the very credo they are living.

If anything, these brave, and exquisitely trained individuals/witnesses need to be vindicated, expunged of any false testimony and NEVER be exploited, used and abused by their superiors, and or used as patsy, scapegoats or the like. The more we all stand up and state what we see, hear, witness, and stand up for the truth, the more likely others will not lie about it.

The acute faculties of observation we use daily, without any effort, allow us to operate high-speed automobiles effectively, use sophisticated machinery, program computers, perform surgeries, succeed skillfully in all the arts, adapt to temperature changes, stop at stop lights, avoid fatal mistakes, breathe, keep our heart beating and run our bodies in multiple changing environments, effortlessly. That in itself qualifies us to observe events that is out of the ordinary. Einstein once said, "The difference between a homemaker and a scientist is that the scientist writes their observations down." However writing recipes, or following them would fit in the same context.

The government agencies employed the services of The Rand Corporation, Military Intelligence, The University of Colorado, and various psychological entities to define uniform interviewing techniques and create practices to conduct these investigations. Dr. James McDonald, who worked on project Blue Book, was instrumental in recording these interviews with all witnesses and assembled an audio archive. The practices used are illustrated in a lengthy document available online on the NICAP.org site. The Rand Corporation, which assembled the documents of interviewing strategies and standards, denied ever creating it. Investigative agencies for the government within the Air Force and Project Bluebook, originally deemed classified entities, did not formally exist in the 1940s, per se. There was however, an investigative branch within the Air Force that evolved into the CIA in 1947. There were also other sub groups in deeper cover that were investigating and removing artifacts even before some of the other groups had a chance to convey the site of the sightings. They carried out investigations if the sightings had significant characteristics of appearing on radar, accompanying electromagnetic anomalies i.e. affecting communications, telephones, killing automobile engines, affected humans, or left salvage or alien bodies. This early branch, NACA evolved into NASA that took quiet note of these sightings, apart from even project BlueBook.

The dual role of NASA, the classified entity, and a public relations and information service, kept the public up to speed with its surface projects of manned space flight and satellite placement and general space discoveries. Project Blue Book commenced in 1952, one of the busiest UFO sighting years on record.

If authorities called in the UFO reports with officers trained in observational techniques such as police officers, air force pilots or military agencies the credibility and interest escalated. To eliminate unwanted attention from the civilian world and the media, they would discredit and ridicule the witnesses no matter their stature. Many of these organizations were set up to trip themselves and other entities to maintain the blanket of secrecy over the entire system.

We have left a little relevant information regarding the formations and branches the U.S. Government created to investigate UFOs including the FBI, CIA, MUFON, NICAP, CUFON, NUFORC, DARPA and included some (very little) in-house correspondence revealing the backbiting, denials, black ops, general manipulation, withholding and proliferation of disinformation of UFO related matters and the reasons why. (Just to give you a flavor of the stupidity of it all.) This is in the main body of the historical chronology, not our local sightings. These letters are also available online at the NICAP site.

From 1940 until 1970, in the main historical sighting chronology after this section, you will notice the serious and focused demeanor witnesses were treated with. The attention to detail is quite obvious. After 1970 and the dissolution of Project Blue Book, the sighting format becomes less detailed that spread inaccuracies although an unintentional by product of a volunteer staff overworked. Without the oversight that Project Blue Book offered, which kept the powers in check, there was little care about the concern of a well-educated public. I tip my hat to NICAP, MUFON, and NUFORC, et al, for their efforts. At this point MUFON and NUFORC is doing their best to keep the accurate protocols of investigation intact.

Mount Shasta's Local Sightings
Part A: UFOs

The following sightings are the ones received to my posters and newspaper ads, word of mouth and local research. Few were reported to NICAP, MUFON, NUFORC, and CUFON, etc. We've probably only heard 10% of the unreported local sightings, and some we found were just about erased from history. We did discover some amazing incidents that are included. The more lengthy chronology sections are the reported sightings to the authorities that were classified up until 1975. There are some unavailable sightings still considered classified and top secret. We included a sample of the cattle mutilations and crop circles that have been occurring in the Red Bluff area for decades and a landmark case in 2009, in the other section. The regional chronology also includes some landmark sightings, globally.

From the research done we found a very early sighting in 1956 by an individual named David Williamson in Mount Shasta, but it lead to unconfirmed details. It is included but we cannot verify the story. Coincidentally, in November 29, 1966, a lone pilot with the name of Jack Brown observed a saucer from his single engine plane over Mount Shasta, but that is all the report indicated. He did sight a disc while flying, going from east to west at an estimated 300 m.p.h. (Not the same person as Capitan Jack Brown from the Mount Shasta Police Department).

Occurred: October 12, 1956

Location: Mount Shasta, CA

Witness: 1
Object: 14+ craft
Duration: minutes

A sighting occurred of 14 craft on Mount Shasta. All available information, possible classified case. In October 1956, Mt. Shasta resident David Williamson observed 14 lights over Mount Shasta making unusual sky maneuvers. Then one appeared in the sky right on top of Mt. Shasta.

{An interesting documented case of a UFO sighting over Shasta occurred on October 12, 1956. David S. Williamson, a resident of Shasta, left his sister's cottage to walk home to his apartment. As he rounded the corner of the cottage, an unusually brilliant star just above the summit of the mountain peak attracted his attention.

He wondered what star it could be as it shone brighter than any star or planet he had ever seen. Then it began to move. Williamson yelled to his sister to come outside and see the strange light. After a few moments, he retreated to the warmth of his apartment and continued his observation from there using a pair of binoculars. With the magnified view, he was able to observe that the light was actually being caused by a set of four lights arranged in a diamond shape and connected to one another by lines of light "something like a neon tube".

Two nights later, he observed the same group of lights just before 8:00 p.m. and in the presence of four other people.

Seven nights later, at almost the same time, he again observed lights in the sky, this time 14 lights in grouped in two rows and having the appearance of "looking at a lighted room through windows". Over the next several minutes, he noticed that the formation of the lights in these rows changed four times. A few minutes after this observation, Williamson saw a UFO descend from the sky, come below the top of the Shasta peak and hover there.}

Author's Note: The above "documented" details; we cannot find documentation for, from either NICAP or MUFON. In the above bracket bars {} my research assistant stumbled on from a book by David Hatcher Childress <u>Lost Cities of North & Central America</u> ©1992 Adventure Unlimited Press

Mount Shasta's Police Department UFO Run-Ins Almost Lost to History

Police Officer/Captain Jack Brown of The Mount Shasta Police did have, as we can reconstruct, two sightings, both of substantial significance. A retired Police Chief (name undisclosed) who worked with Jack Brown for several years, confirmed in a telephone call with me (the author) that Jack Brown did have two sightings, one in 1960 and one in 1966.

The first one occurred in August of 1960, as his name is mentioned with 13 other police witnesses with Officers, Pete Chinca, and George Kerr. This rash of sightings started in the famous Corning and Red Bluff sighting (still a very HOT UFO spot), replete with a great deal of detail, newspaper articles and testimony of UFOs performing aerial feats, discharging electromagnetic interference that affected radios and autos. It was tracked on radar as it worked its way up the Interstate through Corning, Red Bluff, Redding, Dunsmuir and Mount Shasta and up into Oregon, starting August 13th through the 18.th The sightings actually consisted of at least two craft, but only one was seen in Dunsmuir and Mt. Shasta. This sighting is included in our local sightings.

According to a retired Police Chief (undisclosed), the military presence in Mount Shasta was long standing and intense for several years. Perhaps after the late 1980s by a consensus, they decided to conveniently loose the documents in order to frustrate any further research, attention or hassle. Both Captain Brown's sightings were allegedly recorded in the logs and reports kept at the PD, and now, misplaced, lost or missing. It makes it look like nothing ever happened and suspicious to those who recall the events. They were blacked out to non-existence except for the one 1966 official report we accidently found disclosed in its entirety to the Mutual UFO Network (MUFON) periodical, in our research.

The facts show Captain Brown had significant visual information regarding the crafts behavior, performance and EMI (electrical magnetic interference) that affected his auto in the 1960 case, and in some instances affects people's bodies. His first encounter possibly fell into significant value of the inner circle. We discovered Brown's interviews were recorded onto tape but we were unable to get any copies. We know where the copies reside and perhaps, at a later date, they will be available for public viewing.

The facts remain that Captain Brown had several sightings, not just one and a lot of frustration from the constant involvement of the military on-site. The fact he is mentioned in the 1960 reports in APRO, as part of the Red Bluff sightings, which had a great deal of similarity in behaviors, (i.e. lumbering craft doing aerial feats, static discharge on close proximity vehicles) is no coincidence.

Another important fact is the Air Force tried denying the entire five-day happening, rationalizing it was star or vapor light refraction. That sighting is listed after both of Captain Brown's sightings. The scientific community emphatically proved the contrary and the event did show up on radar. Astronomers also proved the stars the Air Force tried to use as excuses were not even visible in the sky at all during the event and that no vapor light could interact with empty space to create what they tried to use as an excuse. 14 police officers witnessed the event in vivid detail independent of each other with exacting testimony.

Jack Brown's testimony just fueled the reality of the event, so trying to wipe his testimony from history would help these denials and disclaimers by the USAF that UFOs don't exist.

The smoking gun is that the later Police Chief, who was also a co-worker with Jack Brown, shared some information. The Author interviewed him and even in his brevity he admitted, "Jack Brown had more than just one sighting, and one in 1960 as you brought up. It was also in the logs." He confirmed the call from Pete Chinca from the steel bridge as well, but it was obvious that he still felt uncomfortable disclosing any information, and perhaps he had been threatened at one time.

He did mention that the community at large became a little tired of all the Military attention. The retired Chief (name unavailable) also admitted that his quoted statement about the city businesses liked the Lemurians since "they were generous and paid in gold", was made up and a joke that he and an undisclosed reporter for the L.A. Times used on a live TV News report, just for fun in the late 1960's. That report somehow ended up as a historical statement many people believed. It was said in jest and never meant to make it to the publics' ears as any kind of fact, but his reporter friend recorded the call, and went ahead and gave it to the public. Below are the cases and details. Enjoy.

Occurred: August 17-18 1960; Approx. 1930hrs.
RECONSTRUCTION of a Missing and Possibly Classified Sighting
Location: Mount Shasta, CA
Sighting 1960 (Jack Brown) (Officer-Captain)
Object:1 shiny disc 15 meters
Witness: 2 Mt. Shasta PD employees and numerous citizens
Duration: 45 minutes
Author's Note: (Jack Brown was a Mt. Shasta Police Department employee. He worked his way through the ranks over the years from Officer to Captain. The sightings he was part of, spanned in his career from Officer up to Captain, so for simplicity's sake we will refer to him as Captain).

Captain Brown's first sighting in 1960 (when he was an officer) from the evidence we have reconstructed, from family, and other reliable sources, indicate Officer Pete Chinca had spotted a UFO on the steel bridge (Lake Siskiyou Dam) and radioed into George Kerr who informed Jack Brown and asked him to look out the window in town and tell him what he saw. Captain Brown in fact observed a UFO to the east that several other citizens called in to report. According to those who had read the now missing files and eyewitnesses, there was a small crowd watching the saucer flying and hovering above the National Guard Armory, close to 45 minutes with Captain Brown who arrived after the radio call with Pete Chinca, now on scene. It then began to move northerly so Jack Brown followed it out north up the mountain in his car. According to Brown's kids, and an unidentified witness, Brown had this craft in his sights about 50 feet above his car and his car died near the railroad tracks on Everitt Memorial Highway, where he got out and watched it fly away towards the mountain northeastward. All of this was entered in the August 1960 police reports with statements from other witnesses. One of the dispatchers read this sighting in its entirety to her daughter spanning in time, somewhere from the late 1970s to the late 1980s when the daughter was both an older teenager and young adult. Now the logs cannot be located or are missing entirely, but the evidence coincides with the stories Captain Brown's children conveyed, the testimony the retired police Chief disclosed to the author, in May of 2012 and the incident the daughter of the dispatcher recalls.

The logs the city of Mt. Shasta possesses go back to 1914 and up to 1975 but somehow the logs that had been intact in the 1980s are now mysteriously missing for the time period from 1956 to 1975, the years many of the local sightings were reported. Both the author and his assistant researcher went on location to the P.D. and viewed the logbooks from 1914-1955 but no others were available even upon request and multiple queries. They were there and are now gone.

Between the retired Chief's admission to the several sightings Jack Brown had, his children's testimony, and his name and identity in the 1960 NICAP and APRO logs with the Corning/Red Bluff incident, the Dispatchers reading of this report to her daughter, all collaborate and confirm the information as we have researched, to be accurate and true. The retired police Chief also disclosed that he personally had gotten tired of the military using his office as an outpost during these long and drawn out investigations of Brown's experience. He also confided candidly in the attention Mt. Shasta received as a by-product including the production of a movie using locals and their cars as extras for some earthquake catastrophe. The town and its traffic were filmed as testimony for its fleeting moments of fame.

One of Jack Brown's children said that his dad was pretty fed up with the prodding and poking he constantly endured from the Military with their invasive questioning, day in and out, that lasted for years. He also mentioned that he remembered seeing drawings of UFOs in newspapers that probably came from the Corning/Red Bluff papers, but any mention of Jack's sightings were not found on microfiche or hard copy. They must have been deliberately removed.

"The "Steel Bridge" location (now underwater) where either Kerr or Chinca called in to Jack Brown about sighting. This is now Lake Siskiyou Dam. (BDW)

South side of Armory with Mt. Shasta in background (BDW)

MT. SHASTA ARMORY where Officer Brown saw craft over it for about 45 minutes in 1960. (BDW)

After painstakingly searching microfiche and querying local papers, it was obvious that any local reporting was not executed in print, or removed by someone's orders, or later removed.

It is the author's opinion, that this incident had some significant and strategic value and may still be considered classified. Therefore, any references to it could have been removed being too much of a security breach to disclose to the public. Perhaps it brought too much attention and controversy to this town already mired in mystical chatter. The fact that it was a group sighting, confirms its validity, the fact that it had electromagnetic interference that stalled Officer Brown's car, adds scientific and possible strategic credibility and verifies the significant nature of the sighting. Some parts of the federal government have this notion that this kind of information is dangerous left unchecked in the hands of the public, posing a threat to national security, or provoke a possible invasion. It may had validity at that time, in the cold war, but the excuses of deniability are incredible, not the sightings.

Jack Brown had another sighting in 1966 (below) in which his treatment by the investigating bodies involved, was offensive and indicates that he already had long standing dealings with the government in this earlier sighting. We found the original 1966 MUFON report and the late Dr. James McDonald's audio files of Jack Browns 1966 testimony, are housed in the Library archives at the University of Arizona.

Brown's testimony recorded by James McDonald of which we tried to get a copy from this archive, are protected under the auspices of the University of Arizona. These audio files supposedly were to be available to the public and those interested in UFOlogy, but the copyrights that his wife holds have not yet been decided, and are still unavailable. Unfortunately, Dr. Mc Donald died from a reported suicide a year after Project Bluebook was disbanded, (which he was a controlling member) and the CIA took over operations. Imagine that.

After assiduously searching for weeks, we stumbled upon a written interview of Jack Brown's 1966 event by MUFON. It suddenly appeared for us, that you now have privy to read replete with all its notations.

The author feels so fortunate to be able to give back to the Mount Shasta Community and Jack's family these invaluable experiences, as they are far reaching. Bless you, Jack.

We included the second sighting next in line, and included the Corning/Red Bluff incident of 1960 after that, as it is necessary for you to view the magnitude of the original sequence of events the lead up to the 1960 sighting and the later 1966 sighting.

There is a mystery to all of this in the confidentiality, classified and top secret labeling. It seems that in a world where the illusion of disclosure and public participation, that had allegedly driven the space program, are contradicted with this cast iron fist of restriction/contradiction. Mix this with the occult and metaphysical theories attached to the mountain that were brewing here since the publishing of Fredrick Oliver's book "<u>A Dweller on Two Planets</u>" and you may get a sense of the attention Mount Shasta has received during and after the Gold Rush. It is really sad that notions such as "trust the government, and don't question authority," were really used to free unworthy agents to corrupt the underpinnings of the great potential of a thriving democracy into a tyranny, secretly. Where we should have been critically thinking and observing, there was a fox in the hen house.

Many missile silos popped up in our local hills, along the freeways, filled with Minutemen, Nikes, and personnel beginning in 50s and early 60s. Radar relay stations stood invisibly in plain sight in Yreka along with classified structures. You will notice a pattern as we did of the UFO activity with Air Force Bases, and other sites like in Red Bluff, and Corning that occur year in and year out, along this corridor, up to and through Redland Oregon. The fact the UFO hovered over the Mt. Shasta Armory is a curious idea. Did you know an emergency airstrip was built in the 1940s along North Old Stage Road?

Armory, location of Jack Brown's **first** sighting (BDW)

JACK BROWN'S SECOND SIGHTING

Mufon Report: Historical Mt. Shasta Case
Taken from MUFON NEWSLETTER OCTOBER 1981
Occurred: July 3, 1966; 0300hrs.

Location: Mount Shasta, CA

Witness: 2 plus others
Shape: disc
Duration: 10-15 minutes

In an interview on May 27, 1967, Paul Cerny, then chairman of the Bay Area subcommittee of NICAP, learned that Captain Jack Brown of the Mount Shasta, California Police Department, had encountered a peculiar craft-like object in the early morning hours of a July morning in 1966. The following article is based on Mr. Cerny's report.

At 3:00 a.m., while on duty and patrolling the northwest city limits of Mount Shasta, Captain Brown first observed the object through his car's windshield. The UFO closed with him rapidly, apparently on a collision course with the patrol car. To avoid a crash, Brown swerved violently, barely missing a ditch, and came to an abrupt halt. Fear and plans to escape by foot in case the UFO exhibited any further hostilities prompted the police officer to get out of the car and lean on the open door, one foot planted on the door frame.

The alleged craft, a smooth disc, approximately 30 to 40 feet in diameter and 10 to 12 feet thick, hovered silently 20 to 30 feet above the car and emitted a blue-white glow, which seemed particularly bright at the disc's rim. The intensity of the light hurt Brown's eyes as he gazed upward at the disc. The object had no visible windows or openings; however, on its underside were two thick tubular protrusions, curved and tapered, about eight inches around at the ends, skid-like in appearance, but with no perceived function. Also, Captain Brown commented that he detected no exhaust trails coming from these structures.

Brown stated that after the apparent craft lingered for 3 or 4 minutes, it drifted off, moving erratically across the road and adjacent fields "as if nonchalantly surveying the immediate area." Suddenly, the object shot off toward the slopes of Mt. Shasta where it stopped and hovered again before accelerating a second time. It climbed eastward over the mountain and disappeared.

The patrol car's headlights and two-way radio functioned normally throughout the encounter. In fact, Brown contacted his dispatcher, David Vacari, who after dashing to a window sighted the glowing object in Brown's vicinity. Furthermore, Officer Mazzeri, 8 miles north of Mount Shasta, in the town of Weed, California, spotted the UFO heading for the mountain but did not observe the object hovering over the car because the terrain blocked his view. (Mr. Cerny saw fit to point out the absence of electro-magnetic effects on Brown's body and automobile; the Captain experienced no paralysis or tingling sensations, and as previously stated, the car performed as usual during the incident.)

According to Mr. Cerny, Captain Brown, 42 and a family man, seemed well above average intelligence, physically fit, in good mental health, and, in short, a competent observer. *Captain Brown claimed that he would be reluctant to report any future sightings because of the way the Air Force had handled his first report, indicating that this already occurred once before (the first sighting in 1960 perhaps?).* Evidently, the Air Force insisted on his completing many tedious and irrelevant forms and report sheets before informing him that what he and his fellow officers had seen was something other than what they had described.

Author's Note: THIS IS THE SECOND TIME CAPTIAN BROWN HAD A SIGHTING. LOOK AT AUGUST 1960 FOR THE RECONSTRUCTION OF HIS CASE. (Listed first before this one) At one time the sighting was reported to the Mt. Shasta Police Department, with paper copy in their logs. It is acknowledged by NICAP in their testimony of the Corning/Red Bluff Flap of 1960 but now Brown's testimony is missing, blacked out and perhaps classified. The Mt. Shasta Police Department logs are MISSING now. A witness who was read the report in 1980 up to the 1990s, in its entirety by a the dispatcher onsite at the PD, (a relative), and collaborating stories from kids and admission of a retired Mt. Shasta Police Chief who worked with Jack said that "Brown had several sightings one in 1960 in 1966." All confirm these facts, but it is obvious the paper logs once onsite are unaccounted for, and no one apparently can verify their whereabouts.

Author's Additional Note:
Below is the 1960 Corning/Red Bluff Incident on up the I-5 corridor. All the data was taken from NICAP and Project Blue Book and excerpts they used from the local Corning/Red Bluff papers, the police reports of actual testimony, with a chronology. All other sources are included. This event was one of the first genuine sightings with recorded EMI, radar, visual testimony with multiple witnesses correlating stories that could not be denied. It was the beginning of what landed in Mount Shasta's lap, of which this book discloses to you.

(Copied exactly as it appeared in NICAP)
Corning, California, August 13, 1960 (Historical Report)
11:50 p.m. (UFOs on their way up to Mount Shasta)

California Highway Patrol Officers Charles A. Carson and Stanley Scott were patrolling on Hoag Road, east of Corning, California when they saw what looked like a huge airliner descending from the sky in front of them. Thinking that a plane was about to crash, they stopped and got out of the car to get a better look. They watched as the object descended in complete silence to about 100 to 200 feet from the ground, then suddenly reversed and climbed back to about 500 feet from the ground and stopped. Officer Carson described it in a police teletype report: At this time it was clearly visible to both of us. It was surrounded by a glow making the round or oblong object visible. At each end, or each side of the object, there were definite red lights. At times about five white lights were visible between the red lights. As we watched the object moved again and performed aerial feats that were actually unbelievable.

The officers radioed the Tehama County Sheriff's Office and asked Deputy Clarence Fry to contact the local Air Force radar station at Red Bluff. Deputy Fry reported back that the radar station verified that an unidentified object was visible on radar. As they continued to watch the object:

On two occasions the object came directly towards the patrol vehicle; each time it approached, the object turned, swept the area with a huge red light. Officer Scott turned the red light on the patrol vehicle towards the object, and it immediately went away from us. We observed the object use the red beam approximately 6 or 7 times, sweeping the sky and ground areas.

The object then began to move slowly to the east, and the officers followed. When they had reached the Vina Plains Fire Station, a second object that came from the south approached the object. The second object moved near to the first and both stopped and hovered for some time, occasionally emitting red beams. After a time, both objects vanished below the eastern horizon. They had observed the first object for a total of about two hours and fifteen minutes.

When they returned to the Tehama County Sheriff's Office, they found that the object had also been seen by Deputies Fry and Montgomery, as well as by the night jailer. All described the same thing.

The next day, Officers Carson and Scott drove to the Red Bluff Air Station to discuss the sighting and to speak to the operator that had seen it on radar. The Air Force, however, now denied that the object had been seen on radar, contradicting what the radar operator had told Deputy Fry the night before. The visit was completely unproductive.

Walter N. Webb, NICAP advisor however obtained more productive information. He contacted Carson and received a copy of the report, drawings of the object, and a letter from Carson.

In the letter, dated November 14, 1960, Carson said: We made several attempts to follow it, or I should say get closer to it, but the object seemed aware of us and we were more successful remaining motionless and allow it to approach us, which it did on several occasions.

In the original report, Carson also mentioned that: "Each time the object neared us, we experienced radio interference. "

Again, from the letter to Webb: The object was shaped like a football, the edges, or I should say outside of the object were clear to us...[the] glow was emitted by the object, was not a reflection of other lights.

The Official Explanation

What was the "official" explanation for the sighting? In a letter to a NICAP member, the Air Force said: The findings [are] that the individuals concerned witnessed a refraction of the planet Mars and the bright stars Aldebaran and Betelgeux. . . [Temperature inversions] contributed to the phenomena as the planet Mars was quite low in the skies and the inversion caused it to be projected upward.

They also said: A contributing factor to the sightings could have been the layer of smoke, which hung over the area in a thin stratiform layer. This smoke came from the forest fires in the area hung in layer due to the stable conditions associated with the inversions.

When NICAP pointed out that Mars, Aldebaran, and Betelgeuse were all below the horizon at the time of the sighting, the Air Force changed the star involved to Capella, which was slightly above the eastern horizon at the time. They neglected to explain the fact that, as the sighting progressed; Capella would have risen in the sky, whereas the objects disappeared below the eastern horizon at the end of the sighting.

Officer Carson had this to say about the Air Force explanations:

I have been told we saw Northern lights, a weather balloon, and now refractions. I served 4 years with the Air Force; I believe I am familiar with the Northern lights, also weather balloons. Officer Scott served as a paratrooper during the Korean Conflict. Both of us are aware of the tricks light can play on the eyes during darkness. We were aware of this at the time. Our observations and estimations of speed, size, etc. came from aligning the object with fixed objects on the horizon. I agree we find it difficult to believe what we were watching, but no one will ever convince us that we were witnessing a refraction of light.

A check of the meteorological records of the area for that night by atmospheric physicist James E. MacDonald failed to find any evidence that would indicate the presence of a temperature inversion.

Over the next week, similar sightings were reported, including another sighting the very next evening by Deputies Fry and Montgomery that were also seen by a Corning police officer. Dr. James E. McDonald

Source: Greatest Scientific Problem of Our Time? (April 22, 1967) NICAP.org BOOK

SECTION VII: Officials & Citizens

The reports of technically trained observers, military and civilian pilots, in themselves are sufficient to make a strong case for UFOs. However, when we also realize that a broad cross-section of reputable citizens have described identical phenomena, it seems incredible that UFOs are not an acknowledged fact. The disc-shaped, elliptical and other main types of UFOs observed by pilots and such responsible persons as judges, civil defence officials, professors, lawyers and clergymen have reported to scientists with great frequency. Some of these individual observer categories could fill another complete section of this report. From the hundreds of cases on file, the following have been selected to provide a survey of what has been seen by officials and private citizens of various backgrounds.

LAW ENFORCEMENT OFFICERS

Police switchboards are the first to be swamped with calls during concentrations of sightings, since there is no established procedure for citizens to follow when they see a UFO. Examples abound of cases in which police responded to citizens' reports of UFOs, and saw the objects for themselves. Police Officers on patrol duty, too, have observed unexplainable objects maneuvering overhead.

During a six-day concentration of UFO sightings in northern California, August 13-18, 1960, at least 14 police officers were among the numerous witnesses. At 11:50 p.m. (PDT) August 13, State Policeman Charles A. Carson and Stanley Scott were patrolling near Red Bluff when they noticed an object low in the sky directly ahead of them. (Their report of the sighting was put on the police Teletype, a copy of which was submitted to NICAP confidentially by a police source. Later, NICAP Adviser Walter N. Webb contacted Officer Carson and was sent another copy of the teletype report, a sketch of the UFO, and a letter giving additional information.)

Verbatim text of the police Teletype report to the Area Commander:

"STATEMENT MADE BY OFFICER CHARLES A. CARSON CONCERNING OBJECT OBSERVED ON THE NIGHT OF AUGUST 13,1960.

"Officer Scott and I were E/B on Hoag Road, east of Corning, looking for a speeding motorcycle when we saw what at first appeared to be a huge airliner dropping from the sky. The object was very low and directly in front of us. We stopped and leaped from the patrol vehicle in order to get a position on what we were sure was going to be an airplane crash. From our position outside the car, the first thing we noticed was an absolute silence. Still assuming it to be an aircraft with power off, we continued to watch until the object was probably within 100 feet to 200 feet off the ground, when it suddenly reversed completely, at high speed, and gained approximately 500 feet altitude. There the object stopped. At this time it was clearly visible to both of us. It was surrounded by a glow making the round or oblong object visible. At each end, or each side of the object, there were definite red lights. At times about five white lights were visible between the red lights. As we watched the object moved again and performed aerial feats that were actually unbelievable.

At this time we radioed Tehama County Sheriff's Office requesting they contact local radar base. The radar base confirmed the UFO - completely unidentified.

Officer Scott and myself, after our verification, continued to watch the object. On two occasions the object came directly towards the patrol vehicle; each time it approached, the object turned, swept the area with a huge red light. Officer Scott turned the red light on the patrol vehicle towards the object, and it immediately went away from us. We observed the object use the red beam approximately 6 or 7 times, sweeping the sky and ground areas. The object began moving slowly in an easterly direction and we followed. We proceeded to the Vina Plains Fire Station where it was approached by a similar object from the south. It moved near the first object and both stopped, remaining in that position for some time, occasionally emitting the red beam. Finally, both objects disappeared below the eastern horizon.

We returned to the Tehama County Sheriff's Office and met Deputy Fry and Deputy Montgomery, who had gone to Los Molinos after contacting the radar base. Both had seen the UFO clearly, and described to us what we saw. The night jailer also was able to see the object for a short time; each described the object and its manoeuvres exactly as we saw them. We first saw the object at 2350 hours and observed it for approximately two hours and 15 minutes. Each time the object neared us we experienced radio interference. We submit this report in confidence for your information. We were calm after our initial shock, and decided to observe and record all we could of the object. Stanley Scott 1851 Charles A. Carson 2358."

Extracts from Officer Carson's letter of November 14, 1960, in answer to Adviser Webb's questions:

"We made several attempts to follow it, or I should say get closer to it, but the object seemed aware of us and we were more successful remaining motionless and allow it to approach us, which it did on several occasions.

"There were no clouds or aircraft visible. The object was shaped somewhat like a football, the edges (here I am confused as to what you mean by edges, referring to the outside visible edges of the object as opposed to a thin, sharp edge, no thin sharp edges were visible) or I should say outside of the object were clear to us . . . [the] glow was emitted by the object, was not a reflection of other lights. The object was solid, definitely not transparent. At no time did we hear any type of sound except radio interference. "The object was capable of moving in any direction. Up and down, back and forth. At times the movement was very slow. At times it was completely motionless. It moved at high (extremely) speeds and several times we watched it change directions or reverse itself while moving at unbelievable speeds.

"When first observed the object was moving from north to south [patrol car moving almost due east]. Our pursuit led in an easterly direction and object disappeared on eastern horizon. It was approximately 500 feet above the horizon when first observed, seemingly falling at approximate 45-degree angle to the south.

"As to the official explanation [See Section IX.], I have been told we saw Northern lights, a weather balloon, and now refractions.

"I served 4 years with the Air Force, I believe I am familiar with the Northern lights, also weather balloons. Officer Scott served as a paratrooper during the Korean Conflict. Both of us are aware of the tricks light can play on the eyes during darkness. We were aware of this at the time. Our observations and estimations of speed, size, etc. came from aligning the object with fixed objects on the horizon. I agree we find it difficult to believe what we were watching, but no one will ever convince us that we were witnessing a refraction of light.

/s/ Charles A. Carson
Calif. Highway Patrol." [1]

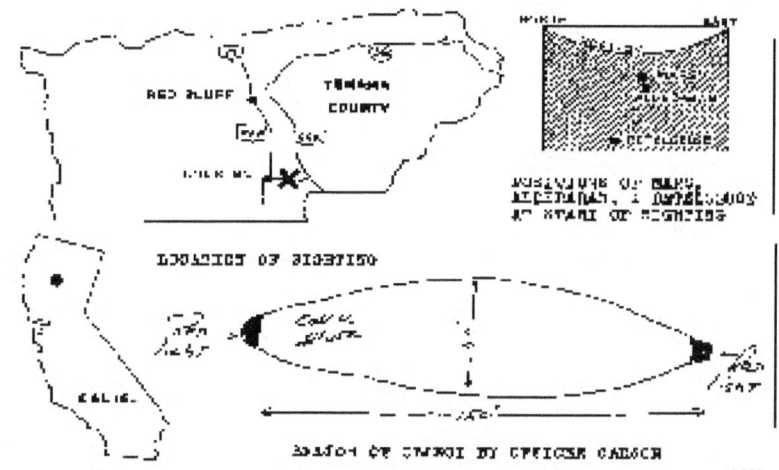

On 13th August 1960, near midnight, while travelling east of Corning, California, state police officers Charles Carson and Stanley Scott witnessed a lighted object drop from the sky. In fear that the object might fall upon their vehicle, they exited the vehicle and ran for cover. The object continued falling, to an altitude of approximately 100 feet. At that point, it reversed direction suddenly and climbed to an altitude of approximately 400 feet. It stopped at that point and began to hover. In his official report, Carson states, "At this time, it was clearly visible to both of us. It was surrounded by a glow making the round or oblong object visible. At each end, or each side of the object, there were definite red lights. At times about five white lights were visible between the red lights. As we watched, the object moved again and performed aerial feats that were actually unbelievable."

The two officers contacted the Tehama County Sheriff's Office by radio, asking the dispatcher to contact the Air Force Base in nearby Red Bluff. Radar operators there confirmed the presence of the object on their radar screens. The UFO remained in view for more than two hours, during which time two Tehama County deputy sheriffs and a county jailer also saw it, according to Carson's report.

Carson's report continues, "On two occasions the object came directly towards the patrol vehicle; each time it approached, the object turned, swept the area with a huge red light. Officer Scott turned the red light on the patrol vehicle towards the object, and it immediately went away from us. We observed the object use the red beam approximately six or seven times, sweeping the sky and ground areas. The object began moving slowly in an easterly direction and we followed. We proceeded to the Vina Plains Fire Station where it was approached by a similar object from the south. It moved near the first object and both stopped, remaining in that position for some time, occasionally emitting the red beam. Finally, both objects disappeared below the eastern horizon. Each time the object neared us we experienced radio interference.

August 1960: Northern California An intensive concentration of UFO sightings occurred over a six-day period in northern California. Dozens of witnesses, including at least 14 police officers, reported typical disc, elliptical and cigar-shaped UFOs. The state police sighting of a highly manoeuvrable ellipse, which shone red beams of light toward the ground the night of August 13, was reported on the front page of state newspapers and on the newswires.

Chronology of Main Cases:

August 13-14 Hollywood: 10:30 p.m. Red elliptical UFO passed overhead, hovering once. Willow Creek. After 11:00 p.m. Circular red UFO approached, circled, dove, climbed away. Red Bluff: 11:50 p.m.- 2:05 a.m. State policemen reported reddish elliptical UFO that made "unbelievable" maneuvers. [See Section I.] Second UFO reported during latter part of sighting.

August 16-17 Corning: 8:30 p.m. Two cigar-shaped objects flashing red and white lights passed from E to NE.
Eureka: 9:30 p.m. Group of 6-8 white and red lights maneuvering in formation. Air Force explanation: aircraft refuelling mission.
Corning: About 9:50 p.m. Boomerang-shaped UFO passed from SW to NW, twice emitting bursts of white light.
Mineral: About 11:00 p.m. Dozens of witnesses, including Tehama County police officers, watched six brightly lighted objects "dipping and diving and moving at simply unbelievable speed" in the southern sky. Objects alternately hovered, speedily changed position.
Concord and Pleasant Hill: 11:40 p.m. to 12:15 a.m. Circular UFO flashing red and blue lights manoeuvred over area, hovering, moving up and down, side to side.
Near Healdsburg and Santa Rosa: Early A.M. Deputy Sheriff observed "flattened ball, dull red and crimson on the edges," hovering and moving slowly about 5 degrees above horizon.

August 15, 1960: Policy letter to Commanders, from office of Secretary of Air Force: The USAF maintains a "continuous surveillance of the atmosphere near Earth for unidentified flying objects -- UFOs."

August 17-18 Roseville Night: Two oblong-lighted objects bobbed around in sky for an hour; witnesses included police captain and sergeant.
Folsom. UFO with two bright white lights on front, red lights at rear, maneuvered over area off and on for two hours at night; whining noise "like spinning top" heard.

August, **Dunsmuir: 12:10 a.m.** Oblong reddish UFO with associated smaller yellow light descended, then rose and sped away. High-pitched sound "like rushing wind" heard.
Redlands: 1:45 a.m. Oval-shaped UFO with dome and row of red lights on edge, maneuvering slowly in sky. August 18: Honeydew: (Humboldt Co.) 9:54 p.m. The postmaster watched a delta-shaped object, clearly visible for more than 2 minutes. UFO approached, made sharp turn and moved away. Red glow visible on front lights on inside of Vehicle.

LIST OF CALIFORNIA POLICE WITNESSES: Aug. 13-18, 1960
State Highway Patrolman: Stanley Scott Charles A. Carson
Tehama County Sheriff's Office: Deputy Clarence Fry Deputy Montgomery
Chief Criminal Investigator: A.D. Perry Deputy Bill Gonzalez
Sonoma County Sheriff's Office: Deputy William Baker Deputy Lou Doolittle Plumas County Sheriff's Office: Deputy Robert Smith
Roseville, Placer County: Captain Hugh McGuigan Sergeant James Hill
Mt. Shasta Police: Officers, Jack Brown Pete Chinca, and George Kerr

1. From Air Force Intelligence Report
2. London Daily Sketch; July 27, 1955
3. Associated Press; May 15, 1959
4. Michel, Aime, The Truth About Flying Saucers. (Criterion, 1956), ppg. 206-207
5. From Air Force Intelligence Report
6. Ruppelt, Edward J., Report on Unidentified Flying Objects. (Doubleday, 1956), p.217
7. Ibid., p.201
8. Ibid., p.204
9. International News Service; Atlanta, July 25, 1952
10. From Air Force Intelligence Report
11. From Air Force Intelligence Report
12. United Press; July 27 & 28, 1952
13. From Air Force Intelligence Report
14. From Air Force Intelligence Report
15. United Press; Washington, July 28, 1952

16. United Press; July 29, 1952
17. From Air Force Intelligence Report
18. From Air Force Intelligence Report
19. From Air Force Intelligence Report
20. United Press, Associated Press; September 15, 1952
21. Reuters; September 20, 1952
22. Ruppelt, op. cit., p.257
23. Michel, op. cit., p.133
24. Ruppelt, op. cit., p.258
25. United Press; Stockholm, September 29, 1952
26. United Press; October 8, 1952
27. Writer's Digest; December 1957

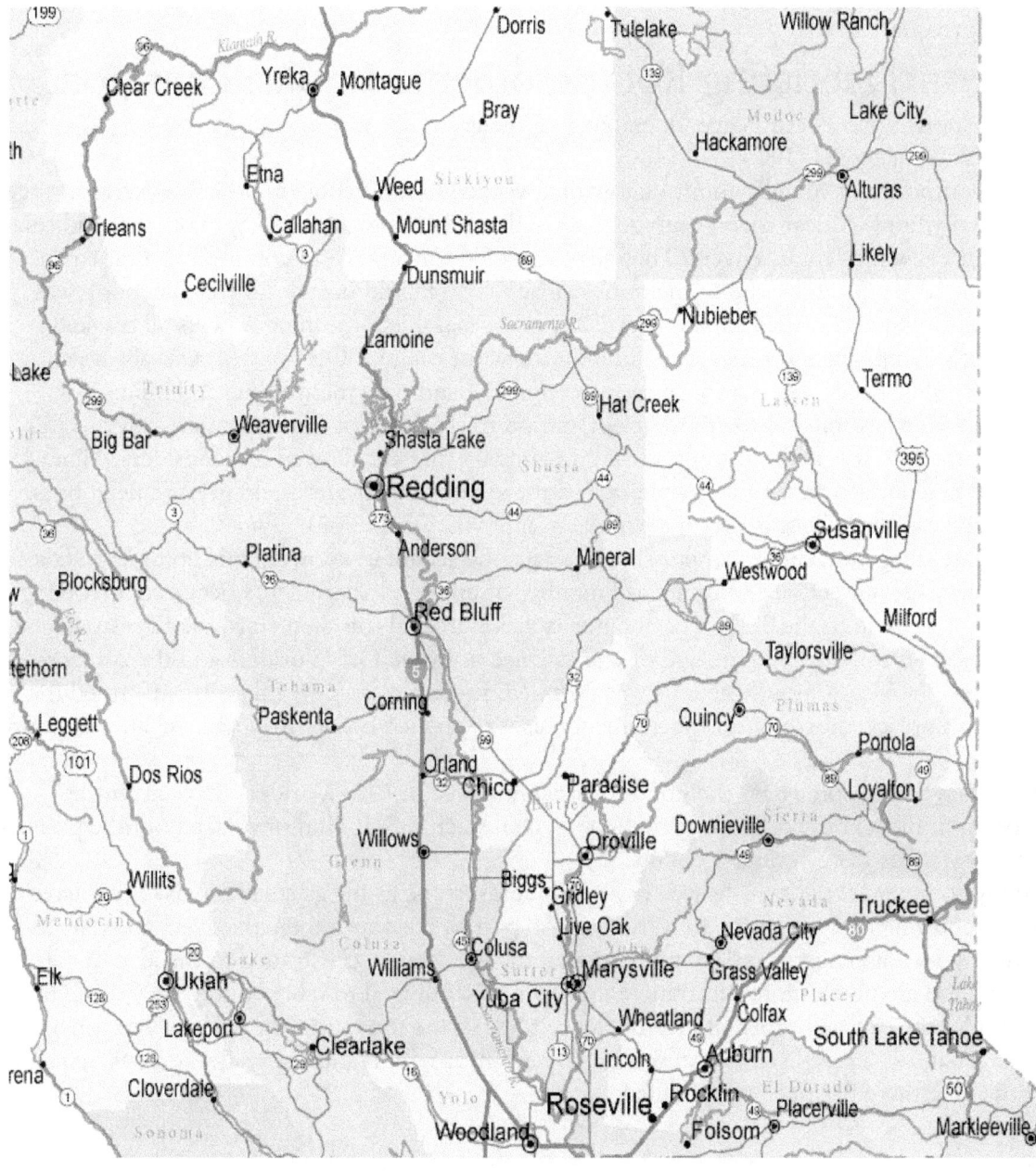

(Map courtesy of unknown,)

CORNING, CA: On June 30th, 1966 right before officer Brown's second sighting another sighting occurred in the Corning area. Look in the main chronology listings further in the book at this date for details.

CORNING, CA: On July 4th, 1967 another sighting occurred in Corning with 2 CHP's 3 other witnesses and a low flying and hovering craft also included in the main chronology.

Occurred: 1968-1974
Location: Debriefing Rooms, other locations: Pentagon
Witness: Anonymous ex-Air Force Operative in Pentagon
Object: Varied occupied UFOs
This is a composite of friends whom I know that worked in the Pentagon. For safety reason they remain anonymous. A dear friend of mine used to debrief Canadian and U.S. pilots that had run-ins with UFOs while on maneuvers. They also debriefed Canadian and commercial airline pilots who witnessed UFOs and some of their unusual activity. He said before Project Bluebook was made public, (which classified situations will always be classified, meaning now its all blacked out) a pilot's competence was never in question for witnessing UFO activity. It was always an excuse for public deception. Sightings happen regularly, and now many pilots and military personnel are accustomed to the presence of extra-terrestrials. Not all of them are privy to full disclosure though. It is a select group within the military that are allowed to be insiders. "They have been here longer than most people even want to know. Consider them galactic neighbors, and in some cases relatives. If they were in plain sight you would not know it."

There are things the author cannot share, but just be secure in knowing it is up to us to take responsibility for our actions. Only a small number of unintended accidents occurred, involving jets getting too close to the Tesla type antigravity fields around the alien craft, the terrestrial jets would loose control, discorporate and or crash. Once both the Off-Worlders and the Air Force realized this, the idea of hostile actions was dropped. In fact, the off-world group informed the military with great remorse of this phenomenon as it happened before they had a chance to warn the pilot.

There have been some days the non-terrestrial humans (or Off-Worlders) have given the thumbs up to fliers in the days we started going over Mach 2-3. It kind of freaked out the pilots at first. Heard this from several accounts.

Many commercial pilots have seen dozens of ships, some as big as small towns. Many have stopped reporting because of the heat they get, on part to not alarm the passengers and part to keep their jobs. Passengers now are more interested, less afraid, accustomed to the idea of ETs being real and are much more accepting. Pilots were discouraged to talk about it; they were hung to dry under the excuse of mental incompetence to hush the topic up for excuses of national security. That was a ruse used to control the flow of information and the reality of how frequent and how long there has been alien contact.

Perhaps we do have a genetic relationship or a spiritual one, with our off-world counterparts, but until we all get the facts, our own dishonest military, government, DARPA and black-ops disclose what we pay them to do, we wont know anything collectively until the Off-Worlders introduce themselves publicly. That will be the day if we ever mature to the point someone won't try to shoot them, or believe that they want us as food or slaves, or our planet. That is totally doubtful. We have polluted our bodies and our planet so much we're probably inedible as it is. With the vast quantity of alcoholic, drug infested, bug infested, incompetent neurotics, would you want to eat you, or how reliable are you to show up to work??? We have made ourselves so undisciplined, undesirable, dumbed-down and disinterested; it may be that no invading alien race, if they were to invade, would find us desirable or useable and suitable. "Beam me up Scotty, there is no intelligent life on earth," doesn't seem so comedic anymore. They would not have to invade; they just have to show up. Ha, what a joke, on us! Those depressed, or strung out on money, greed, power and materials would not be reliable or trustworthy; the aliens would be left with slim pickings, horses, oxen, and sheep. Perhaps it was better in the 50's? Just kidding.

As far as the author is concerned, it is a little like the ETs don't want to interfere with our own development except if we threaten life, liberty and the pursuit of intergalactic happiness. We as a species can't even drill for oil without greed influencing the decisions to invest in proper safety gear or stay sober at the ship's wheel of a tanker and starve each other for our own financial benefit. Sure seems a far goal to have the behavior, etiquette and responsibility to be a galactic citizen, as it is. When we can clothe, free, and house everyone as to create a sense of security that people can produce to their maximum because they live in a world that regards them as valuable, productivity will happen naturally. Contrary to antique beliefs, every person wants to contribute, give, produce especially if they are not slaves and have energy to dream, create and live without fear, hunger and illness. Our very own bodies operate on this premise as you read this now. It does not worry or have to bully, intimidate, beg or lord over the cells and organs like humans believe they must do to squeeze work out of others. That is created out of slave labor for tyrants, and is not based on any universal maxim of good. Imagine traveling across the vastness of space with insecure power demanding egos, it would not work.

Also contrary to popular belief, to prepare to go to space is constructive to our species. The refinement to ponder and apply it to the challenges of cooperation is applicable to living on earth among each other. It was a deceitful ruse our leaders used to break the spirit and deactivate our collective vision to explore space as a global goal of mankind in a peacetime venue, by shutting down the Space Program. Instead, having no incentive to expand into our universe, we are overpopulating a small cage with too many rats and no way out. We need our space program and it needs to be for we the people, not run by private industry. Let us make it a public collective.

Anyway, off subject, you think? No, it goes hand in hand that when a crisis occurs, upgrades, adaptations and inventions appear out of the stress and need for a fresh solution outside the revolving door. Just a thought to ponder while you read about freedom, knowledge and responsibility…

Occurred: 1960; before I-5 built
Location: Hwy.99 Mt. Shasta, CA
Witness: 1+
Object: 5 craft
Duration: 2-3 minutes
Road was still Hwy. 99, the subject was driving in late afternoon or early evening south of Black Butte when he noticed people pulled off the road on both sides, pointing into the sky at something. He stopped his car and with a crowd of other people, observed 5 saucer craft heading north to Black Butte, high enough to clear Mount Shasta. No other information is available.

Occurred: Summer 1964; Approx. daytime
Location: Dunsmuir, CA
Witness: 4
Object: disc
Duration: 2-3 minute
I've been a journalist and worked in TV. I am independent now. Never published my encounters. But it was Mt. Shasta where it all started. I lived in Dunsmuir 1960-1969. During this period I was very observant. One sighting was a craft observed hovering above the power lines south of the freeway heading south from Mt. Shasta city. Observed several minutes as the car with 4 of us just drove by. Diameter of craft was 30-50 feet.
You had to be blind to not see it. Others must have seen it that day in the mid 1960's. I have one or two others but this was the most striking, and viewed by a group. My mother who is now 96 also remembers it. I spoke with her about 3 months ago. "What was that thing?" she asked. I went over my thoughts at the time and it was hovering right above a pole with a transformer on it. I think there were wooden poles at that time. One other night in Dunsmuir a friend and I saw a light go over us and disappear into Mt. Shasta about 4 am one summer when we were out when we shouldn't have been. (Or were supposed to be depending on your take.) We waited about 10 minutes, as I can recall, and a different color light emerged from the top of the mountain shooting straight up and disappearing in the infinite but not traveling around the globe like a satellite. It was up and gone. Rather slowly too. Ba

Occurred: September 15, 1975; 2200hrs. (Author's sighting)
Location: Bonny Doon, CA
Witness: 2
Object: Saucer with windows on top dome and dome on bottom 75+ feet across
Duration: 10 minutes

Excerpts from 'If You Wanna Get to Heaven This event occurred on the way to a lecture at Foothill Junior College near Mount Diablo, California. An argument ensued between my partner, Kala and myself, Athens. We were almost to Bonny Doon north of Santa Cruz. It was an amicable argument among friends, but we stopped to resolve it fully in the fresh evening air. It had just rained and the sides of Hwy. 1 down to the ocean was terribly muddy. We pulled off the road to park by a big bluff that overlooked the Pacific Ocean. The bluff itself towered up, so it had to be climbed to view the ocean on the other side, or drive further down on a muddy trail. Kala insisted we drive down the wet, muddy 500-foot road to a lookout where the bluff leveled out for a great ocean view. I begrudgingly complied against common sense, because she was cold. We made it half way down this dirt road and then sank up to the axels in a mud puddle to the point the bumpers were touching the ground. Arguing and pissed off, I blurted out, "We are so stuck in the mud, and spinning our wheels in our thinking. NOW we manifested it. We're not going anywhere until we resolve this. I hope you're satisfied." I realized I should have just stated the danger it put us in, but I was tired of arguing and just gave in. We got out of the car, took a breath and came up with a plan that would resolve the other situation first. We were to part our ways from this lengthy trip we'd been on together. The trip was a spiritual vision quest that actually became a genuine learning and awakening experience for us both, but I was done. I was ready to go home having a renewed outlook on life. Kala was still in deep search mode, and I was exhausted of the life on the road. Originally, I was to keep the car but then Kala had no way to hook up with other members of this journey, unless we went to the appointed rendezvous point, Mount Diablo Campground. Kala could then secure another vehicle and be with the existing group. I felt the attending seminar organizers (The Two) were going to be a no show. We only had contact with 'The Two' less than a half dozen times in 8 months and in that time, they had abused the public. They were arrested for improper use of a credit card, and incarcerated. They were not genuine in their behavior or their vision. They (these erratic cult leaders) were showing signs of arrogance and dangerousness. I was convinced The Two would abandon their group unless the individual members tracked them down, which came to pass. On the other hand, I lived on trust and faith of what was emerging, and our needs were met honestly and with integrity. We were being shown the quantum physics of our spiritual identity from the inside out. It happens when you drive across the desert over 800 miles on an empty tank of gas, among other unusual events. It seems sincerity has a lot to do with enlightenment.

The Two sent a press release that they were going to commit suicide and resurrect. This was not part of what we understood as their credo. It was a shock that put many followers, even ones that were devoted to them, the public and concerned family members in a frenzy of worry. It was totally selfish, insane and not at all in line with unconditional love, evolution, transcendence or metamorphosis. It was violence for sensationalism. A Jesus the martyrs repeat. It had already been done.

Even if by some miracle they were released from jail, which was unlikely, Kala still believed they were going to show up at the college. Buying that possibility, especially after my epiphany of not following any gurus or self-acclaimed masters of the ethers or of chaos, was not an option. The Two were on a ruddy path and Kala's mind was made up, and no convincing changed that.

Unless awakening originates from within, in intuition and insight, there is no awakening. I really did not want to go up there, but go back to Santa Cruz, and let her get a bus to Mount Diablo. I gave in though, and would take her up to Mount Diablo, let her take the car and also have all the remaining money we shared. I would then leave with the clothes on my back and go home. I knew it was going to be ok. Kala agreed with this decision since she was not ready to give up the road or her belief in these two con artists. I was, on the other hand, in resolve with my leap of faith, and decided to let it all go and follow my heart.

 I would go ahead and make sure she reached the meeting point safely. She could hook up with the other members at the campground and then I would hitch hike to Santa Cruz. The correct agreement had been reached. Everyone's needs were met, and that is a good feeling.

At that time, we were outside the car enjoying the clean, ionized air, and the new decisions we made on this profound journey. A moment later, a comforting but unusual buzzing began to resonate in the ground and the air around the area. A good-sized spacecraft, 75 feet across with bright white with blue ones on the edges, rose quietly above the bluff that made this humm. The craft had a dome on the bottom, and a dome on top with portholes, but the windows illuminated so brightly masked any occupants.

The contrast against a misty black sky with the gleaming hull of this magnificent ship and bright lights would have hidden any forms to the naked eye, as it was.

The craft peacefully floated over to the automobile, and stopped above it. I turned to Kala and yelled, "RUN" since I did not know what the ship was going to do. We could feel the mass of the ship even though it floated so weightlessly that added to its power. You could feel the electricity like a magnetic bubble around it. It was awesome.

We ran almost as far as the pavement of HWY. 1. We turned around breathing hard, exhilarated and scared, but it was not a scary experience. It was just significant. We watched the graceful craft float away from the direction of the car and out towards the ocean. It toddled gracefully about 700 feet off shore where it stopped.

It then ascended vertically, and was gone in a flash! The instantaneous absence of the ship left a void that was filled with peaceful but pregnant silence. The fresh ocean air and the profound thing that just happened, filled the moment with total wonder.

We approached the car and noticed it had been turned entirely around and moved out of the mud puddle it sank in, ready to drive. It was picked up or levitated and softly put down 50 feet up away on solid ground. There were no tracks from it being rolled, or dents and the tires had not moved at all. A muddy handprint on the front tire confirmed that. It was my print. The UFO had done this while we were running to the road. We were very thankful for the favor since we didn't have the extra money for a tow truck.

We made it to the appointed campsite that night, the day before the alleged meeting that did not happen. I gave Kala the car, the money and good wishes.

The next sighting is at the campground.

AUTHOR'S NOTE: *This is one of the author's personal experiences. It is in a book called, <u>If You Wanna Get to Heaven</u>. The third edition will be available within the next year.*

This probably is one of the most unusual and wonderful sightings ever, and up until now except in the book about "The Two" and has not been disclosed to the public until now, and gets better. (BDW)

As unbelievable and magnificent as it was, we both recounted the experience several times to each other to confirm that it actually happened. I actually contacted my x-partner in this journey after 37 years. That was two years ago. She had written a book (not about this) that she was being interviewed for on National Public Radio (NPR) and hearing this interview, I realized she was still alive and was able to contact her. We chatted about many miraculous events and her recollection of this event is exactly as we both experienced it.

Occurred: September 16, 1975; 0830hrs. (Author's Sighting) second in sequence

Location: Mount Diablo, CA (East Bay Area)

Witnesses: 31
Object: 1 Silver saucer with dome and windows on top and bottom dome 75+ across
Duration: 7-10 minutes
Excerpts from 'If You Wanna Get to Heaven'

We drove to Mount Diablo State Park the next day and pitched tent. This time I had my own tent, was on my own with no expectation of going further. It was a beautiful night, it was safe, quiet and there were laughing people close by in other camps that added a level of comfort that I was able fall fast asleep, too. That night I dreamt about literally leaving the face of the earth in a spacecraft. I felt the vertigo of leaving the atmosphere; then I was alone in it, sitting on the floor by a window watching the face of the planet disappear, it made me dizzy and sob, as I did not want to leave. I woke up thankful I was still earthbound to the smell of fresh coffee.

It was about 8:30 and there were 15 people from the group anticipating seeing their cult leaders and 15 nature-loving campers not from any group, and me; just visiting. We were all outside enjoying the gorgeous, clear warm bay area weather with a stunning panoramic view so clear over the bay. It was like a picture taken by Ansel Adams.

Everyone was smiling, inhaling the fresh brewed coffee, admiring nature's artwork when quietly and lazily a gorgeous UFO 75+ feet across, shining its beautiful metal disc with a bulge on top and the bottom, a classic stately looking spacecraft with square portholes on the top section, seen by us all absolutely as clear as it was day light with the sun shining on it, came lumbering weightlessly towards our campsite. It was very deliberate as it was targeting us to visit. In fact, its altitude was about 4,500 feet, with an attitude on the same plane as we were and came about 800 feet away hovering and just stayed. We all looked at it, and it just placidly reciprocated a curious kind of nonthreatening welcoming demeanor. I think they smelled our coffee! Every one of us had our mouths gaping, holding our coffee cups in one hand looking at the ship and then to each other pointing, until we all laughed. It was absolutely gorgeous, benign, letting us observe it for several minutes. It then shot up about 1,000 feet stopped, then shot east and disappeared. No one said a word but the entire energy, young and old alike just accepted it and began to just be happy. It was simply beyond words for us all.

I felt they brought me a message that I made the right decision and no doubt it was probably the same one that helped us the night before un-stick our car from the mud puddle in Bonny Doon. I could feel them, their warmth and concern. It is like when you know something is correct. I knew they were communing with me. I could feel it. It was a very sacred experience. I chose to be of use here on earth to help bring people into their own levity. We don't need to go to space to do that. We need to find it first inside us.

This ship was also identical to pictures from a book written by a retired Air Force Commander who took genuine pictures in the 1970s of UFOs while on duty. Allegedly, they were from the Pleiades. Two weeks later I found myself in the city of Mount Shasta, a place I knew I would return to that I knew absolutely nothing about except it was really cool, commanding and there was something here for me. Had no idea about the myths, the lore or anything, if I had, I may had avoided it.

(Girard Ridge, Castle Crags, and Mt. Eddy behind, sighting below)

Occurred: November 6, 1970 or 1972; Approx. 2000-2100hrs.
Location: Girard Fire Lookout (McCloud, CA)
Witness: 5
Object: 3 Disc/Spheres lit
Duration: 7-10 minutes

Mike Thompson, Dennis Derr and a few others (unable contact to use their names or corresponding testimony) were all on the Fire Lookout on Girard Ridge. The day was over and they were off the clock, for a while. Mike estimated the time was about 8:30 or 9:00 pm. It was a bit chilly so the woodstove was loaded in the tower and got so hot it drove the occupants out on the catwalk for fresh air. Something caught Mike's eye south east of Castle Crags, south of the lookout. There were three lights in the sky between the ridge and the lookout about two miles away. These visible lights one in front of the others, or perhaps in a stacked formation, looking like a triangle, was brighter. He did not remember the colors. Mike noticed a telepathic connection with the craft, as it responded to their attention, by moving in a northwest direction. Mike was also amazed at the surprise that both Dennis and another friend demonstrated as these two were skeptics, hard-wired to the sciences, and they were amazed that they were seeing these things. When I asked Mike the approximate size of the craft, he said from his distance, perhaps a quarter size of his thumbnail looking down his arm, and eye level with the tower, which would have made the craft about 6-7,000 feet in elevation. Every time they noted the craft verbally, the craft would move again, to the west and would hover for a while. They tracked them up the canyon until they disappeared west. He said they watched them for about 10 minutes.

(Second witness testimony)
Location: Girard Ridge Fire Lookout
Witness: 5
Object: 3
Duration: 7-10 minutes

Dennis Derr, part of the off-duty group of fire service workers, working on the Pacific Crest Trail at Tom Dow Creek and on Girard Ridge Fire Lookout in the Fall of either 1970 or 1972, reported this sighting to us from his memories independent of Mike Thompson, during the same time we interviewed both. The crew was enjoying their off time playing cards, drinking a beer and basking in the warmth of the fire ablaze in the wood stove. The wood stove was cooking them out of the building, so they opened the door and stepped out onto the catwalk for some fresh air. He and Mike and others stood on the west side of the catwalk facing Castle Crags and Mount Eddy. The lights came over Mount Eddy, in a stacked formation. One of the craft perhaps the lead one zipped a mile north and then back. "Remembering that we went out on the catwalk on the northwest corner to take a look around late evening and saw three light above the horizon of Mt. Eddy in the form of a triangle."
Mike and Dennis's story appear to commence differently, however a simple explanation is the difference of time they were both on the catwalk and took notice of the craft is proportional to where the craft were in the sky.

Dennis said, "The single craft on the left went way ahead and stopped still over the horizon of Mt. Eddy. I believe it went down behind the Mountain and came backup. The other two then went toward the single on stopped and then they went together with the single and separated back into the triangle formation. The two went down behind the mountain as the first one had and came backup. All three stayed there for a few moments, still in the triangle formation, and then it appeared that they all sped away to the North and disappears in a matter of a few moments. Dennis said the craft were lit with lights of every color in the spectrum, but each ship lit in different color. Green and Yellow, White, Red and Purple, lights. The craft stopped straight out from them, eye level. Together, apart, together, they dashed."

"Are you seeing this?" they asked each other confirming their view. Do you see that? Questioning, Then the craft left, in an instant, gone warp speed, like star trek. A small pinpoint of light, far, far away." Meanwhile a loud crash was heard from inside the lookout that made them all jump." It's just a log ember that rolled out of the stove."

Looking from Girard ridge, North at Castle Crags and Mt.Eddy behind

Occurred: Mid July 1974; Approx. 2000hrs.
Mount Shasta/Whaleback Mountains north of Weed, CA
Witness: 1
Object: 1 cloaked but audible craft
Duration: 2 minutes

Around 1974, a forester working in Mount Shasta's Whaleback region planting trees claims to have seen many UFOs in evening and had run close to one which was cloaked but it was very noisy taking off and making a whoosh noise, moving bushes in close proximity. It blew his hair when it rose with a warm rush of air as he heard it take off. He also viewed one scanning north to south on the west face of Mount Shasta. This was in the twilight hours of the summer.

Occurred: Late summer 1976; Approx. dusk
Location: Windy Point, Mount Shasta Mountain
Witness: 1
Object: 1 glowing disc
Duration: 5-7 minutes

DF is a seasoned railroad engineer. Relaxing from a tense week, he and a friend went up to Windy Peak to view Mount Shasta, Sergeant's Ridge and Grey Butte. The two saw a large craft ablaze in yellow with a white rim, traipsing south towards Grey Butte. He said it looked like a bee bouncing along the peaks slowly for a few minutes until at the south end of the butte, it shot into space at a high rate of speed at approx 45-50% angle and was gone out of the atmosphere in seconds.

Occurred: Late October 1976; Approx. midnight
Location: Mount Shasta Ski Bowl
Witness: 4
Object: Lumbering mysterious light, UFO
Duration: 10 minutes

A group of four went up to the old Ski Bowl at the end of Everett Memorial Highway, to look at stars and satellites. They all observed what they thought was perhaps a climber coming down Red banks to the east with a light in a strait line coming toward them. This is not only impossible as it is a ridge, the light would brighten then go out. It also seemed to be off the ground. With the mountain as the backdrop, the light would increase in size then go out and it was closing in on them at a very rapid pace. It would take a climber in good health to get from where the light started to its location at least two hours what it did in a matter of several minutes. The group felt alarm, and panicked and left as fast as they could since it was not normal.

UFO sighting reported

By Jim Schultz

Weed police officer Pat Hogue dubbed it as "a close encounter of the worst kind."

That's how he felt about a UFO sighting early Thursday morning over the southern shoulder of Mt. Shasta.

Loren Gelwick, a service station operator at the Woodside 76 gas station at the south end of Weed reported a bright light at about 5 a.m..

He was not the only one.

The Weed police reported receiving several calls from residents who also said they saw the object.

Weed merchant Dennis Tinsman told police he saw the object and studied it for a while through the scope on his rifle.

Officers in Mount Shasta, Dunsmuir and Yreka were alerted by radio, but none claimed to see the bright object.

Officer Hogue did see the object, however, but said all it looked like to him was a bright star.

The above sighting occurred in the mid 1980's with the event possibly published in Redding Record Searchlight. We could not find the source paper. Ex-officer we interviewed agrees with article but did not add anymore to it. He was a little reluctant to disclose. The dispatcher also was informed but no exacting dates or details were available.

Occurred: Mid-September 1978; Approx. 2000hrs.

Location: Mt. Shasta, CA-South Old Stage Road

Witness: 6
Object: 3 Triangles
Duration: 15 minutes+ (Slow retreat)

In the early fall of 1978, Pam McGaugh and friend Raymond Zanni were returning to Pam's house after a drive. It was dark, about 8:00 p.m. but still early in the evening to provide a little light. They pulled the pick-up into the west-facing driveway, at South Old Stage Road in Mt. Shasta, parking the truck facing south. They noticed some lights in the sky a couple miles to the southwest, directly above the Cantara Loop area. They hopped out of the truck, never diverting their attention from the odd lights. They instinctively met at the front of the vehicle to examine the objects more attentively. The lights were moving very slowly their way. Pam ran to the front door of her house calling to her sister, Laurie. Laurie followed Pam back outside, glancing up to see the display herself. The clusters of lights were now closer, allowing them to determine there were three separate craft, each shaped like a triangle and flying in a triangular formation. The lead one maintained its forward position.

Laurie decided to gather more witnesses and ran across the driveway and field to the home of their neighbors, Ernie (a railroad employee) and Jane Moore. Ernie and Jane stepped outside walking over to Pam and Raymond. Laurie continued to gather neighbors running to the door of Bill Moore (a railroad employee, also). Bill and Laurie hurried to join Ernie and Jane with all four individuals joining Pam and Raymond.

By this time, the craft were directly above and holding their position, completely stopped. There were three lights emitting from the bottom of each craft, but the lights did not reach to the ground. There was no sound. A strong vibration could be felt, creating a rumbling feeling in their chests. Both Ernie, who served in Europe in World War II and Bill, who served in the Air Force in Vietnam, stated that neither had seen craft resembling these in their years of service.

As the craft remained stationary for what seemed to be five minutes or so, the witnesses, standing with their heads back, gazed straight up, gaped-mouthed. The witness estimates the distance above them to be a tad less than a football field or about 90 yards. The size of the objects seemed small, perhaps only 15 or 20 feet across. The witness recalls thinking the craft could perhaps be large enough to hold a couple of people at most. Witness states she had the impression of a dark metal hull, but is not sure what she actually saw, or could estimate the color due to the darkness.

Kicking the dirt and rubbing their kinked necks, they started to wonder what they should do next. The witness states it felt like the craft inhabitants sensed their observers' drifting attentions. The craft then reacted quickly with sudden movement and then slowly accelerated away. As all the neighbors set to return home, Pam and Raymond walked to the top of the rise on the hill, still south of the Pioneer Railroad Crossing, and observed the craft which were heading straight towards Black Butte, disappearing around the east flank, still in the triangle flight formation. Upon return to the home, the sisters called their mother, Patricia, dispatcher at the police department, to report what they all experienced. They do not think it was ever 'officially' reported outside the agency, if at all.

(Second witness to above sighting, Laurie Moore)
Pam ran in the house very excited seeing this triangle formation of lights still coming from the south. I went outside to look for myself, and saw three white colored lights to the south, a couple miles away south. They were easy to see, like planet sized and bright in the sky. Each craft had 3 lights on them. The craft were lower in the sky, like lower than a plane. I ran and got our neighbors to also witness this event, Jane, Bill and Ernie Moore. The craft were like 300 feet above us now, and stopped and hovered. They were very quiet. The group of 3 lights just stayed there for 15-20 minutes. We all got bored and it seemed the UFOs sensed that and they (the lead one and the two smaller ones) dispersed and headed towards Black Butte and disappeared behind it.

Occurred: Fall 1981; Approx.1900hrs.
Location: Gazelle, CA
Witness: 1
Object: Bright light
Duration: Unknown, missing time

On a fall night in 1981, I decided to drive to the home of some friends in Mt. Shasta City. I was residing in the South part of Scott Valley so I drove the Callahan-Gazelle Road to the '99 Junction then proceeded south on Highway '99 to a long straight stretch in the road, a few miles south of Gazelle. I pulled to the side of the road to take a brief rest at a wide spot where there was a dirt road heading west. As I got out of the rig and walked around the front to the passenger side, I saw a big, brilliant white light to my Southwest, estimating about four or five miles out. I kept watching as I completed the task at hand, noting the light had now traveled further North and East, and positioned itself directly to my West about half way between myself and the ridges to the South of the China Mountains. Then, while still looking, I traversed around the front of the car. As I reached for the handle of the door, diverting my eyes for a fraction of a second, and then looking back up, the light vanished. I drove off to Mt. Shasta and upon arrival noted the time, feeling disoriented and uneasy when the realization hit that the drive had taken 45-50 minutes longer than the norm. Had I stood there, by the roadside much longer than memory accounted for? I remember feeling like I just wanted to go home and that's what I immediately did, completing the trip in approximately one hour's time as was customary. The missing time the witness experienced still cannot be filled in, causing an uneasy feeling to this day.

Occurred: Spring, 1980 or 1981; Approx. 0700-0800hrs.
Location: Mt. Shasta, CA
Witness: 1
Object: 1 Saucer
Duration: several minutes

While hunting over by the Sewer ponds where the Mt. Shasta Resort and houses currently reside, on the northern part of the tract, Jeff Hough was hunting for ducks. He was driving his vehicle slowly flicking his spent cigarette ash out the window and caught sight directly above him about 130 feet above to the eastern edge of the road, a disc shaped saucer. He stopped the vehicle and observed the craft. It was a large silver disc with no discernible windows or entrance ports. Silently it hovered for about a minute, shot off sideways towards Black Butte, then changed course at a very high rate of speed, shooting directly up at "warp speed" and vanished.
Witness said the felt 'weirded-out' and shocked. It seemed to have affected his psyche, as he was shaken up, confused and disoriented. He again started driving and within a minute he saw a coyote standing in the field. The coyote also seemed greatly confused, staring in the air not moving away or even reacting to the witness being there, as the animal viewed him. Witness shot at the coyote, missed further confusing and rattling them both. **Author's Note:** Jeff Hough, a personal friend, died February 8th 2013 and we wish for you all to be as thankful as we are that he was able to disclose this event just weeks before his departure. We will miss you Jeff...

Occurred: September 1980; Approx. 0530hrs.
Location: Mount Shasta, CA
Object: -3
Witness-3
Duration: 12-20 minutes

My friend C. who's still alive saw the UFOs as well as my Dad and myself. I will recount the day. I was 15 years old on my way to the Mormon Church (seminary) at 5:30am fall of 1980. We turned onto Adams drive from Rockfellow Drive. I was sitting in the middle next to my dad who was driving. We saw three circular disc shaped UFOs. We had to pull over while they flew here and there in the sky for at least 12-20 minutes. It was quite shocking, amazing and weird all at the same time. I felt almost scared for minute as if some unknowns were invading us. At the time I really didn't understand. What really stands out was how different they are (the craft) compared to other flying aircraft. The movements they made were super fast, then they would hover in air but different than a helicopter. These airships were super steady and fluid in the way they flew about. They also had brilliant lights that were completely around the aircraft and the lights moved. It was like nothing I had ever seen. I remember we stared as if perplexed, almost stunned by the experience. We talked about it and we all came to the same conclusions. UFOs!

Occurred: July 1981; 1145hrs.
Location: Mount Shasta, CA
Witness: 2
Object: 1 cigar/disc
Duration: 20-25 minutes

Our witness is a seasoned individual who holds a master in psychotherapy. This is a sighting she witnessed along with a companion in 1981 in the middle of summer. Sometimes to escape the heat and star gaze many drive up the highway to the old ski bowl to take in a view of the surrounding valley having a gorgeous panoramic view of the southern and southwestern sky exists. The vista view where our observer goes is about 7,500 feet above sea level and offers a magnificent view.

In 1981 during a summer evening Both JD and a friend went up Everitt Memorial Hwy. to star gaze and enjoy the cool evening parked above the city so they could see the southern sky. The valley to the Castle Crags area was visible. A UFO the size of a large luxury ocean liner hung on the altitude just above the top of the crags as if looking for something, and lazily hung and scanned over the area. The night sky was littered with stars and a half moon but the craft was pervasively obvious apparently not concerned if anyone saw it. It was bright white like a maglite. It did not have any seeming intent on leaving and just hovered over the area. It was not a helicopter made no noise nor was choppy and hovered weightlessly unalarmed as to let the observers sit for over 20 minutes and take in the event. It slowly drifted side to side as if scanning for something but with no urgency. Her friend was very nervous and having seen a landing before was extremely excited and wanted to leave as quickly as they could to avoid and abduction.

Her son also who spend a lot of time in the Shasta Forests had observed a red and green light low to the ground and upon further inspection saw it was an alien type craft that had a spinning gyroscope like center that seemed to generate its motive of power. According to this witness, the craft was the size of 2 football fields, and the craft later appeared over Castle Crags where a coworker who happened to be camping there, viewed the craft. This is all she could remember, and witness is not available for further comment.

Mount Shasta seen from Herd Peak Lookout, northwestern side of mountain. (PSP)

Herd Peak Fire Lookout (PSP)

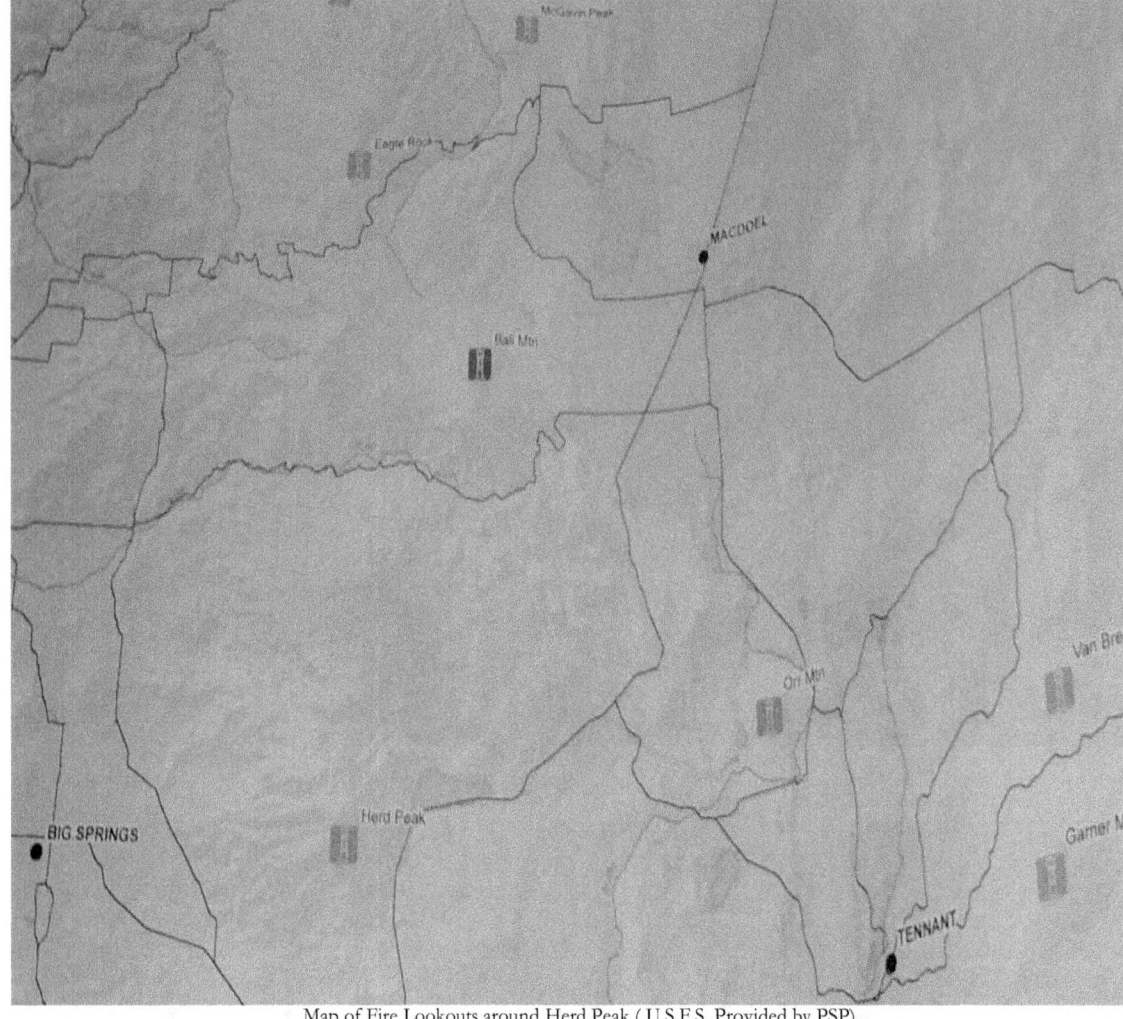
Map of Fire Lookouts around Herd Peak (U.S.F.S. Provided by PSP)

Occurred: 1964-1985, Approx. 30 years of various sightings
Locations: Siskiyou County USFS Lookouts
Reported 4/2012
Witness: 1 USFS Fire Officer Trained in Long Distance visual observation
Multiple sightings, multiple occasions from early 1960s thru 1980's
Durations: 2 to 15 minutes
Witness testimony:
Witness is a 53-year veteran of the forest service. She has been on fire watch in Siskiyou County all this time and has observed great fires and storms, helicopters and airplanes flying as well as satellites and the space station. She has keen awareness and specialized depth perception with extensive map reading capacities necessary to perform her duties; directing fire units to areas of smoke or flames and tracking lightning storms. She has had several sightings over the last few decades. Some are very significant and are included.

In 1964, stationed as vacation-relief for two weeks of the summer season at Herd Peak on Miller Mountain, our observer was looking towards Mount Shasta's, Northwest side, about twilight or dusk hours. She observed a ball of light above timberline by the Hotlum area of Mount Shasta. Suddenly, she viewed a door opening and closing on the side of the mountain. It was a rectangular door. The ball of light either entered or left and the door closed. This never repeated again during her stay.

Another incident in the summer in either the late 70s or early 80s while on the lookout by Lake Mountain, a craft appeared about 10,000 feet up. It had visible square openings on the side, and she said it looked like a train passenger car going around a curve, moving up and down and side-to-side, resembling the distortions of air occurring from heat, making the windows looked wavy. It appeared to be spinning clockwise while heading towards Mount Shasta. The portholes looked multi-colored; red, blue and white. The craft obviously had some type of aura or energy wave around it distorting vision like heat from an engine in a car, putting out heat waves on a cold day, making the air waver. She claimed the craft was enveloped in what seemed a kind of force field. The craft disappeared when it was near the topside of the Shastina cone of Mount Shasta.

She witnessed other various sightings throughout the years; some looked like an orange light, bobbing up and down and side-to-side, and then disappearing into the mountain.

She also claimed that while observing Dry Lake, and again while looking along the gullies below the Hotlum area on Mount Shasta, she saw other orange lights, one a craft that would just hang and observe. When the planes came by, it would cloak, and then when the planes would be far enough away, it would de-cloak.

One time, by Scott Bar Mountain in the Kelsey Creek area in broad daylight, she observed another craft with portholes but never saw the occupants as it lumbered by. During an evening while on Collins Baldy Lookout, N. received a call from Scott Bar Mountain Lookout, stating she should quickly look between the two towers. As she did, she observed a bright-red ball, large in size, shoot straight up in front of Scott Bar Mountain with the other lookout informing her it seemed to have come out of the ground. As the two watched from their different perspectives, the ball of light suddenly shot straight out, stopping half-way between both towers, then reversing course, heading straight back towards Scott Bar Mountain and disappearing into the ground. One night when she was home, her neighbor came running over claiming to have seen a light streak by, stop, scan the area and dart her way. When they both went back, sure enough the craft reappeared to observe them, then took off.

Occurred: Spring 1980; Approx. 1430hrs.

Location: Mount Shasta, CA

Witness: 2

Object: 3 spheres making a triangle

Duration: 2-3 minutes

In 1980, a homeowner off W.A. Barr Road, (S.H.) and a guest viewed three orange balls with white outlines, going over Mount Shasta at an estimated 300 mph, going from north to south. They called police to report it. They also said it was in a triangle pattern with a white outer rim. Said the viewing lasting several minutes.

Occurred: August 1982; Approx. 0300hrs.

Location: Mount Shasta, CA

Witness: 1

Object: disc-orb

Duration: 2-3 minutes

Subject could not sleep so they went outside to get some fresh air in late summer eve. While looking up to the east at mountain they saw a low-flying disc that was a bright as the full moon. It was just above the trees and hovered a while and then took off at a high rate of speed. It looked to be about 40 feet in diameter as it shot west out of sight. It made no sound at all. TW

Looking east at west side of Mount Shasta

In 1982, the economy was bad for several years, so we rented out our house in Coffee Creek and left to find work. For seven months we lived at **Shasta View Dairy [photograph]** and worked remodeling Stewart Mineral Springs. Once in a while I would stay at the house for the day by myself. One day I was standing by the two trees in the yard looking at the spring flowing by. I felt something there, looked up, and there was a being next to me **[picture below]**. We looked at each other for a few moments, and then it was gone. I had no idea at that time what it was, and certainly did not want to tell anyone

An ET or an "unquiet dead" or a wandering spirit drawn by witness.

Occurred: Late Spring 1986; Approx. 1245hrs.
Location: Burney to Redding, CA
Witnesses: 2
Object: 1 disc
Duration: 12 to 15 minutes approx.

Subjects were on their way to Redding from Montgomery Creek using Hwy.89 to Hwy.299 driving in their auto. Their car all of a sudden dies and will not restart. Occupants are stranded and concerned from being hit from behind since the car came to a halt in the middle of the road. The starter would not turn over even with repeated attempts. They had no prior problems with their car, and then they noticed an unusual sight over to their right. They rolled the car to the side of the road and noticed on their left, across the road the Pit River Power Station with a round glowing silver object hovering above it. The UFO was 150 feet away from them hovering about 60 feet above the plant. It looked like it was about 12 feet by 12-foot object, and the witnesses said it looked the size of a VW Van. Meanwhile the car still won't start. They did not recall how long they were stranded but it seemed that the UFO was leeching power off the plant, draining the area and creating electromagnetic interference with their car. It was at least 12 to 15 minutes. The car finally started and they got out of there in a hurry, without noting if there was any noises or auras accompanying the EMI (electrical magnetic interference) drain. The local paper, The Redding Searchlight reported that there was in fact a UFO sighting there that day and time, which many witnesses reported. The classic EMI effects in sightings like this, was also mentioned in the newspaper. The witnesses were relieved that they were not the only ones to have seen it.

From the Eastman's Originals Collection, Department of Special Collections, General Library, University of California, Davis. The collection is property of the Regents of the University of California; no part may be reproduced or used without permission of the Department of Special Collections.

Occurred: Late December 1986; Approx. evening
Location: Mt. Shasta City, CA (Ivy and Alder St.)
Witness: 1
Object: 1 large cross, 2 smaller ones
Duration: 2 minutes

A neighbor called the witness saying she was viewing unique lights in the sky right over their area and she should go look. Witness went outside her front door and looked to the mountain that was behind her to the east. In the sky moving at a slow rate of speed from that direction which is east, about 200 feet above her, and 60 feet in front of her, was what looked like a golden cross. It hung above Alder St. She said that the center of it was a golden translucent warm force, like it was alive, and two smaller cross-shaped craft emerged from it. She said that it felt very benign as the cross emanated a glow to the ground. It then the smaller ones rejoined the larger ship, hovered back towards the mountain and disappeared. She was amazed and unafraid.

Occurred: Summer 1987; Approx. dusk
Location: Grizzly Peak, McCloud to Round Mountain
Witness: 2
Objects: 4 white light cubes
Duration: 10-15 minutes
Subject 35-year-old Siskiyou County resident, working homeowner and her friend. In Summer of 1987 the witness was in Round Mountain in a pasture, intent on retrieving a horse in potential danger of abuse began calling horse when she observed four, 3 dimensional foot and half cubes floating, spinning and interacting, while putting out a white light, and flew around witnesses. This incident lasted about 10-15 minutes then the cubes just took off and flew away.

(Grizzly Peak and Round Mountain) (BDW)

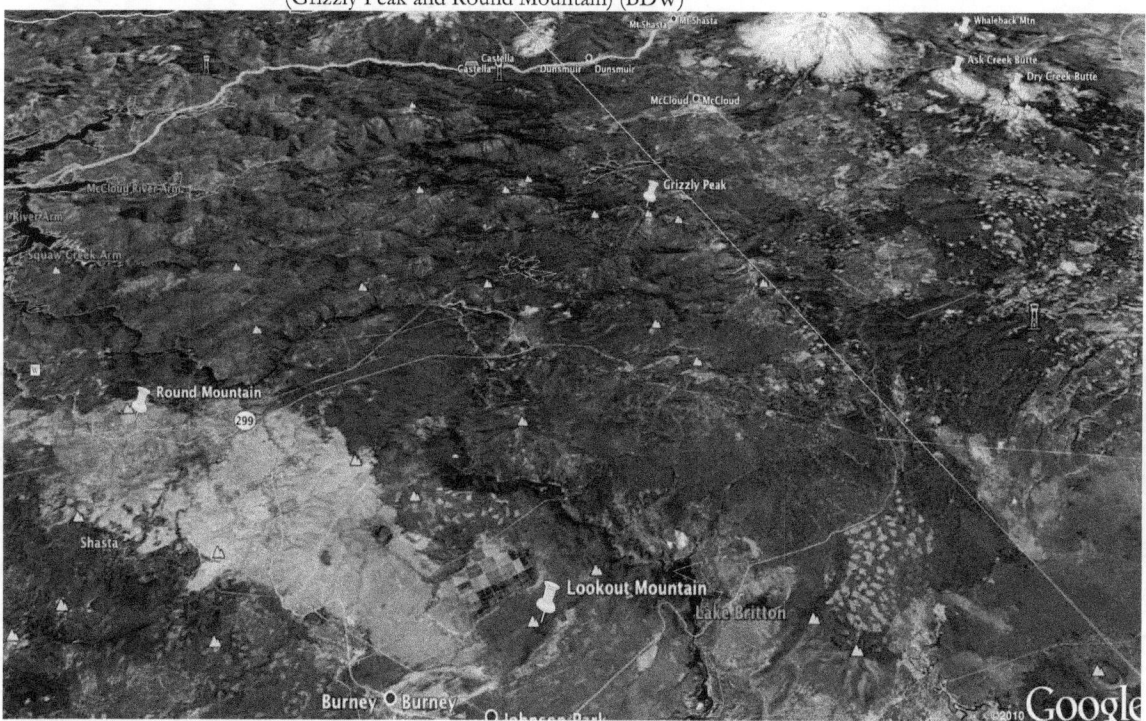

(Second sighting)
Occurred: Fall; mid 1990s; Approx. 2130hrs.
Witness: 3
Object: 5-6 craft
Duration: 25 minutes

Location: Dunsmuir, CA

The same subject living in Dunsmuir, viewed from west side of North Dunsmuir. While looking north and a little east, towards Mott Ave., up the hill or north on Interstate 5, saw a bright object flying, making an outline of Girard Ridge. Then five smaller objects joined it as intense bright lights. They (she, family and friend) watched these craft for about 25 minutes traipsing the mountain, scanning the ridge.

Witness and friend decided to jump in the car to get a closer view of this UFO activity, and drove up north to Mott Rd., which is about a mile and half further up hill with better view of Girard Ridge. After they reached the Mott Road interchange, they flipped around back south, noting the large craft and three lights were still positioned about the ridge. Two of lights apparently jumped the freeway as they were now to their right in the brush. Driving back south from the direction they came on Dunsmuir Ave., the two smaller flying lights that jumped over the freeway, spinning, undulating in an out of each other on their west (on their right) was dodging trees (picture below) and keeping up with their car as on a pursuit of them. In fact one was riding just above their trunk, they could not see until they noticed it in the rear view mirror. These craft aggressively matched their speed as they passed the gates of Shasta Springs. (St. Germain lodging) Frightened and panicked they tried to evade these small craft. Eventually the craft took off into the thicket by the river and disappeared.

Going south on Mott Road, the roadway lined with trees and western brush the craft flew through before disappearing.

Occurred: Jun to Aug. 1991; Approx. 1530hrs.
Location: Pit River Bridge, Redding, CA
Witness: 2
Disc: 230 feet
Duration: 1 minute

In the summer of 1991 our subject accompanied with J. O'Connor were driving over the Pit River Bridge in the afternoon. An 80-meter (225 feet diameter) UFO shaped like a disc came out of the water below the bridge and went above I-5 and took off straight up.

(Pit River Bridge over the arm)

Occurred: Early 1990; daylight
Location: Pit River Bridge, Redding, CA
Witness: 1+
Object: disc, several
Duration: 2 minutes

Several weeks ago, the author mentioned several sightings to an elderly man (sighting above) in a candid conversation at the market. The man has lived in Mt. Shasta for 70 years. Out of the blue he mentioned witnessing UFO water launches out of the Pit River from 60-80 meters wide in broad daylight at noon, late spring in the early 1990s. Witness claims the craft came out of the river, shot straight up, hovered, went both north and south as to investigate, and shot to the west and up out of sight. It shocked me as the person did not seem the type to disclose that sort of stuff, but he saw them and was matter of fact about it. He had run one of the mills over in McCloud during his life. He confided that there is no reason to doubt that life exists all over the universe and any curious explorer may find our planet to say the least: "interesting."

Occurred: Summer 1600, 1700 and 2100hrs. Respectively in 1986, 1990, 1992

Location: Lake Shastina, CA

3 sightings
2 by husband one by wife on 3 separate occasions
Object: 4 respectively
Duration: Variable
Both employed full time in technical music and real estate businesses.

1. Female witness was on deck peering to the west looking at lake about 4 p.m. on summer's day. She observed in sky an obviously damaged, swaying, wobbling UFO loosing altitude, measuring about 40 feet across, dripping and oozing molten metal and sparking. It looked like it was going to crash. She yelled at her husband to come and look and as she did, another larger craft appeared out of a cloaking device and hooked some type of tractor beam to it, stabilized it, and with its load secured took off to the north following Big Springs Road. The two tried to chase it but it was so far away it would have been futile to try and chase it. The incident took about 30 seconds over the lake then continued its path. (Picture Below)

2. The husband had had another sighting but the UFO was cloaked and it had landed close to their home in the empty lots. The ground was shaking, he could hear a distinct whoosh and the bushes were moving like a wind was spinning them. He could feel it take off and almost could make out a protoplasmic image of it taking off. The Doppler effect changed the pitch of the objects whine as it ascended. The witness being a studio sound tech adds a level of competency to this assertion. (Picture below)

3. In the 1990s witness was sitting on a deck at one of the local restaurants in Mt. Shasta with a mountain view and noticed a bright light between the peaks that got bigger. It looked like a bright planet but was in the middle of the day and moving about the mountain. It then began to drift away to the east and just disappeared before he had a chance to get others attention on it.

Lake Shastina taken from Herd Peak on Miller Mountain (P.S P.)

Occurred: September 1991; 1130hrs.
Location: Black Butte Summit on I-5, Mt. Shasta, CA
Witness: 1+?
Object: 1 metallic disc
Duration 1 minute

A Cal-Trans employee running a giant cement mixer just north, and a half mile above the Abrams Lake off-ramp on I-5, on what was the truck pull off, gazed down the interstate looking south and saw a football shaped shiny disc hovering 60 feet above the Abrams Lake overpass. He said it was approximately 30 feet in diameter and began to hover slowly towards him. Working the dangerous mixer, assembling a test strip, he peered down to look at his work and when he gazed back to the object, it had already vanished, either cloaking itself or exiting at a tremendous rate of speed. Having had thoroughly viewed the object inspecting it with enough time, he was thoroughly satisfied with its presence.

Abrams Lake overpass where saucer was seen hovering above. (BDW)

Occurred: Late summer 1991; 2200hrs. (Author's sighting)
Location: Mt. Shasta, CA
Witness-3
Object: 1 disc
Duration: 1 minute

My partner and I were sitting outside on the porch stairs looking up at sky on Mt. Shasta Boulevard observing satellites. We were facing west and my three-year-old daughter was lying on my legs, facing up. A UFO, located pretty far away at around 20,000 feet, zoomed in from the west horizon, stopping a third of the way closer and taking only a second to get there from the horizon's line. It hovered there for about 20 seconds, hanging and swaying a little. All three of us saw it, and all eyes fixed on the object. It shone very bright in the sky, then zigzags in both directions, north to south, stops again and takes off eastward. Within several seconds it clears the skyline.

My daughter sits up, looked into my eyes, points up and declares, "Daddy, I DID NOT see that, it was NOT there." An obvious acknowledgement and denial of what she did see as no one spoke the entire time until she made her declaration, which made us bust up and laugh.

Picture of house on Mt. Shasta Blvd, where Mom, Dad and daughter saw saucer (story above) (BDW)

Occurred: Sighting in October 1992, 2400-0300hrs. (Author's sighting)

Location: Military Pass, Mount Shasta, CA

Witness-2
Object: 9-10 large discs, 2 portals
Duration: 3 hours

My buddy and I decided to get some view of the sky from the northeastern side of Mount Shasta, traveling on Military Pass accessed from Highway 97, (a road reaching around the waist of the mountain) by Whaleback Mountain, over the hill through the snowmobile park and around to the east to the McCloud side. Driving a few miles on dirt after descending from the snowmobile park, lays a beautiful meadow that boasts a clear line of sight eastward and to the northeastern sky and the sheer wall of the Mount Shasta. Upon arrival, we jumped in the back of the pick-up, gazing up at stars we normally don't get to see from Mt. Shasta City.

Sitting and facing north, about 10 o'clock in front of us if 12 o'clock is above our heads, is a perfect equiangular triangle of equally bright stars, almost as bright as any planet that could be seen. After several minutes in succession with the lower left star, the star would blink then the top one, then the right one. We could see something slow down in the middle until it materialized fully into a disc, looking like it came out of a wormhole, or hyperspace. It then made a perfect right angle turn in its solidity and sped into the side of the mountain. I said, "Hey, did you see that, what just happened?" His reply, "Yeah, I saw it. It happened." As we discussed it, the experience repeated itself and another ship came out of hyperspace and did a right angle turn and went into the mountain. They appeared almost the color of a cup of coffee with cream as they reassembled out of hyperspace, changing to a silvery color. Very interesting phenomena: It looked as if you could see the atoms reassembling into a solid form, changing from a much faster rate of movement, and it was consistent in every re-entry. The craft would then enter the mountain through what looked like a cloaked entry. We watched it a total of nine or ten times, in the span of three hours. I was sure there would be more but we left, excited and fatigued. It was an exquisite experience. Seeing the activity so many times left no doubt of its authenticity.

Where it gets more interesting is about two to three years later, I was helping a friend in their cleaning business, at a client's house and on and their mantle piece was a picture of the same exact star configuration in front of which these occurrences happened. My jaw dropped when the owner of the house came in and said, "That's a star gate on the northeast side of the mountain." I said, "I know, I watched them." She said she has been watching ships come out of there since she was a little girl, living most of her life in McCloud. There are those who know and try and to protect the area from too much inquiry. Picture below of mountainside.

Looking west at the East side of Mount Shasta/Shastina with flattop is side where craft went into. (Military Pass) (BDW)

A BLACKED OUT CITY WIDE OCCURANCE

Occurred: Summer 1993; Approx. 1530hrs. (Author's Recount)

Location: I-5 Corridor; Dunsmuir to Mount Shasta, CA

Witness: hundreds
Object: 1 disc, 2 missiles
Duration: Lasted 5 minutes

In the summer of 1993, I was working in my auto repair facility, which faces west and was located on the backside of a building on the west side of Mt. Shasta Boulevard. It was around 3:30 in the afternoon. A friend of mine who worked heavy equipment on the freeway came in all excited about a missile and UFO sighting that just occurred. He said that two missiles were chasing a UFO up Interstate-5 corridor traveling north and the UFO rose above the missiles, got behind one of them, blew it out of the sky and then disappeared into the mountain. The other missile seemed to be disabled and crashed onto the southwest side of the mountain, undetonated. A minivan then pulled in with a mother and 9-year-old child, stating they pulled over along the freeway with hundreds of others to watch the spectacle. She also stated she saw two missiles and the one that did not get blown up by the craft, sailed off- course into the southwest side of the mountain and crashed without exploding.

The head of one of the search and rescue groups was fervently assembling his group to head up the mountain to investigate, and later told me, he was greeted by black ops vehicles in big trucks armed with heavy weapons that 'under deadly force' commanded the search and rescue team to leave or else be 'forcibly removed'. The head of search and rescue came back angry and pale from the ordeal, being treated so harshly he suffered from PTSD. He was unable to talk about it, as if he was threatened, if he did speak of it. How did the military get there so quick? Many eyewitnesses' stories were consistent, but the following week, the story in the paper claimed Vandenberg AFB said they sent two dummy missiles, as a test, to the Corridor Islands that went awry. They did not obviously launch the UFO to be chased, and neglected to include the UFO, anyhow. The story was a cover. Two weeks later, the paper read a plane crashed on the opposite side of the mountain, making no sense and convoluting it with other events. The incident was buried and the locals who saw it forgot. Upon remembering and sharing, my friend still recalls and so do several others. The travelers on the freeway are scattered, probably forgetful and are no longer even aware of the circumstances. This is one sighting that has been blacked out by both the government and by all media. The only ones who know it happened were the ones to witness it. It is unfortunate the manner in which this data is handled, causing collateral damage to those not privy to the clearances of the privileged groups of government officials, and stricken from the record.

Occurred: Autumn 1996; 2200hrs.

Location: Mount Shasta, CA

Witness: 2
Object: cloaked force
Duration: 2 minutes

Subjects had the feeling they were being followed and watched from above and warned to not go up the mountain. What started out, as a leisurely motorcycle ride up the mountain got strange very quickly. It began to get very cloudy as a Lenticular cloud descended down around the mountain, like a dense fog almost instantly. They were only going 30 mph and around Penny Pines the motorcycle they rode, began to literally get pushed backward and down into the ground. It was a new motorcycle with no problems, but they could not continue the climb. Idling and stopped they agreed they were being persuaded to leave, so they complied. As soon as they turned around and began to go back down the hill, all the pressure they felt pushing down on them disappeared.

Occurred: Summer 1997; Approx. 1930hrs.

Location: Mount Shasta, CA

Witness: 4
Object: flying craft 2
Duration: 2 minutes

Hi, I saw the article on you book in the paper. We have some interesting photos shot from our deck-facing Mt. Shasta. They were taken about 15 years ago. The person taking them did so thru his sunglasses to soften the light. They are amateur photos but show a lenticular cloud over the mountain with some type of airplane looking object arising from it. At first there is one object which shows up blue and then in another picture a second airplane looking object shows up in a pink color. I say airplane looking object for lack of a better word. They are not airplanes but have an aerodynamic look. I don't think I can transmit them over the computer, as they are just 4x6 prints on glossy paper. Let me know if you are interested. **S.** (S. pictures below)

Note the two craft, pink and blue in each picture, on the move.

Second picture for above sighting (pictures taken by witness)

Occurred; Mid-July 1999; Approx. 2200hrs.
Location: Mt. Shasta, CA (North Mt. Shasta Boulevard, Abrams Lake)
Witness: 2
Object: 1 really large craft
Duration: 45 minutes

The witness and friend are both from Mt. Shasta. When they were younger, they often went out on North Mt. Shasta Boulevard, climbing out onto the railroad trestle above I-5 or above the digital sign at Abram's Lake overcrossing bridge, to sit with their legs dangling above the freeway, watching the traffic go by while stargazing.

One night, either in Spring or Fall of 1999, they were doing just that. As they looked up at the stars, they realized there was an area with no stars showing or light coming through at all. Then they realized the darkened spot was actually moving very slowly. It was about ½ mile across and shaped like a triangle or delta. There were lights emitting below the craft from all three tips of the UFO. It seemed the craft was silent as neither witness heard a thing.

The craft was moving very slowly, about 1-2 miles per hour and appeared to be performing some sort of land survey or work. The two young men were able to view it for about 45 minutes as it meandered, slow and low, across the sky from east to west, covering the corridor south of Black Butte and north of Spring Hill before finally moving out of view as it dropped behind Mt. Eddy.

(Second sighting for above individual)
Occurred: Summer 2003; Approx. 2200-2400hrs.

Location: Yreka, Ca.-Fort Jones, CA

Object: Triangular lights, fast moving craft
Duration: 3 minutes

In 2003, a Mt. Shasta resident and his girlfriend were driving in the north county towards Fort Jones from Yreka along Hwy. 3 on a beautiful summer night. They pulled over on a dirt side road to go 'parking', exited their vehicle, sat on the hood, talking and enjoying the warm night stargazing. They noticed movement in the sky fairly high up but low enough to see a craft with three lights in a triangular pattern. They watched it fly across the sky taking about three minutes to reach the horizon. The witness stated, "While the shapes were similar to the triangle viewed by Mount Shasta, this one was much smaller in size."

Occurred: September 9, 1999; Approx. 0035hrs.

Location: Mount Shasta/Sand Flat, CA

Shape: Fireball
Duration: 7 seconds

After seeing a disc 35 minutes earlier that night, witness fell asleep to be reawakened. A large slow green fireball-moving west to east, 200 feet above, angling as it moved, struck trees, loudly cracking them and then vanished. I was awakened again to a very bright fireball slowly crossing Sand Flat west to east. It lit the entire Sand Flat area with a green glow. The fireball was about 200 feet high, 300 feet north away from me, about 30 feet in diameter It appeared to be a perfect sphere and translucent except for the core which was a solid yellowish orb about 3 feet in diameter. I could see a fine pattern like bicycle spokes radiating from the center 3 dimensionally. The fireball then vanished in the forest with a loud " crack" sound. The next morning I found a tree limb snapped off, and a large snag at the top of the adjacent tree, which was also snapped off. The batteries in my camera as well as my cell phone from this event were dead. I did manage to find a fresh battery and took photos (not available) of the injuries sustained to the trees that were obviously fresh. ((NUFORC Note: Date changed to 09/09/99, the presumed date of the sighting.))

Occurred: September 9, 1999; Approx. 2200hrs.
Location: Shasta/Sand Flat, CA
Witness: 1
Shape: Disk
Duration: 1 minute

A large disc floated over Sand Flat about 200 feet high, north to south direction. I viewed 3 large rectangular lights (like windows). The star field above it was blocked out. I was face up in my sleeping bag on the ground when I saw these windows with an aspect ratio of 7 to 1 horizontally curving around what appeared to be a large disc. It seemed at first that the angle of it was about 45 degree elevation due north and it had lights that blinked on and off in unison repeating three times as it floated over me. I attempted unsuccessfully to awaken my son whom was with me but he would not budge. There was no noise at all to this event and it flew directly over my head and vanished. If the lights were on the outer rim of the craft, I would estimate the size of the craft to be 2-300 feet in diameter. I was stunned and the battery in my camera was dead after the passage of this craft. I had hoped for a good UFO sighting but gave up after a while and the CSETI group had moved to another location since all of us were awake the previous night stargazing.
  ((NUFORC Note: Date corrected to 09/09/99, the presumed date of the sighting.))

Occurred: September 9, 1999; Approx. 2200hrs.
Location: Mount Shasta, CA
Witness: 1
Shape: globe fireballs
Duration: 5 hours

There were multiple anomalous lights, with white flashes, orange fireballs and white and red lights, on the side of Mt. Shasta, moving and signaling. My son and I ran across Steven Greer, CSETI Group, "We witnessed from sand flat, looking at Mount Shasta from 10 p.m. to 2:00 a.m., about five very bright white flashes, like a flash bulb, going off at high elevation. The multiple meteors left normal trails with white to red blinking lights on both sides of the mountain, below Shasta's Peak occurring for several hours. About 12:00 a.m., witnessed strange anomaly of interference occurring to star field over about 5 degrees by 10 degrees for 30 seconds. It was like looking through a heat wave. (Cloaking distortion) About 12:30 a.m., a slow moving orange fireball about 500 feet above moving northwest. After it passed, a sounded like a piston engine airplane was heard. There were no navigation lights. A camper reported a helicopter the previous day, hovering by this location lasting for about 30 minutes on the face of Mount Shasta in the daylight. There were also lights on the steep face of Mount Shasta shifting from time to time going to very bright on occasion. We were once followed in the daytime by a White and Blue Bell Jet Ranger helicopter for an hour.

Occurred: September 2005; Approx. 1330hrs.

Location: Mt. Shasta, CA (Mt. Shasta Boulevard and Alma Street)

Witness: 1
Object: Metallic, heavy looking
Duration: Quite some time. She walked away with object hovering in same area.

While walking from school to work, witness came to the above-mentioned intersection and was startled when she suddenly noted an odd object hovering in the sky above the area of the old railroad depot building (Shasta Burgers) and Mt. Shasta Boulevard and Alma St. The base of the object was approximately 15-20 feet above the nearby telephone pole. It seemed as if the object was made of metal and was heavy in appearance. The craft was bright yellow like a tractor or a school bus. It was chipped and dented, with a reddish-rusty tone apparent under the yellow paint in the damaged areas. There was neither sound nor any visible means of propulsion. She also noted no lines, tethers, cords, etc. though as the object was moving in an extremely slow manner.

The object was bell-shaped with some sort of apparatus on the top that looked almost like a sail or panel of some sort. The main body of the object was about 10 or so feet tall and considering the bell shape, perhaps 6 or so feet across on the top portion, wider at the base which was a cone-like bottom. The craft was moving very slowly, about the speed of a pedestrian. Could object been held up by a tractor beam from a cloaked craft?

The Witness noted that no one else seemed to notice it at all which she thought very odd. She wasn't sure if she should point it out to passersby at this very busy intersection or not, but couldn't believe everyone around was oblivious to the presence directly above them. She noted the object seemed like it could have been acquired from a 'Used UFO' lot or even an UFO wrecking yard. Sketch, below shows object's appearance to witness.

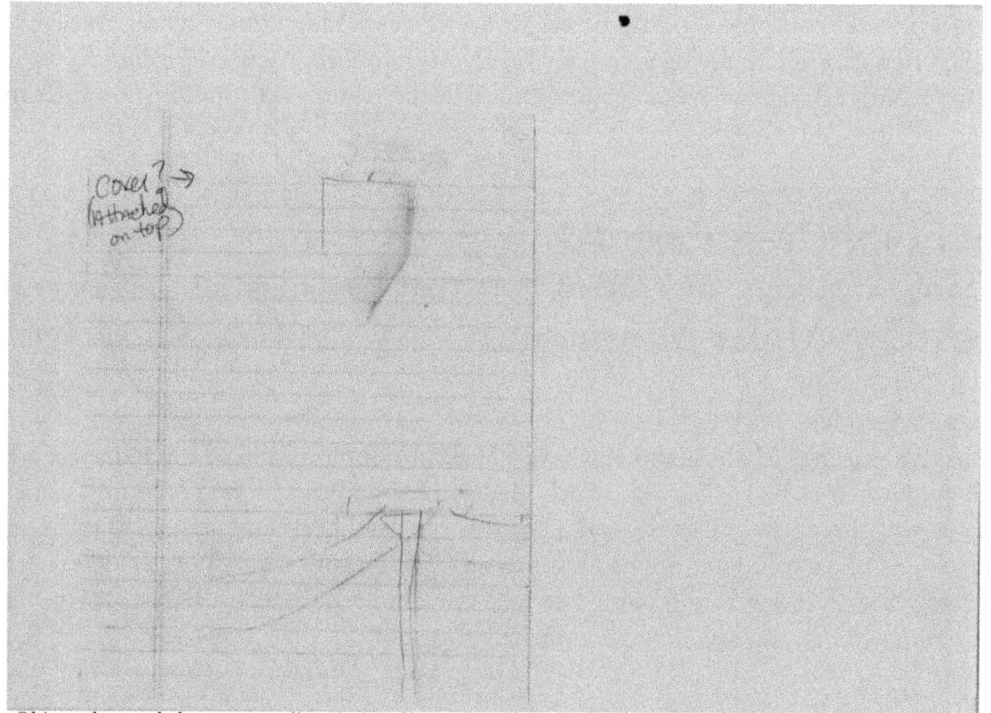

Object observed above power lines in Mt. Shasta City. Appeared to be on a tractor beam being towed.

Occurred: November 13, 2005; Approx. 0500-600hrs.
Location: Lakehead to Redding, CA
Upper Sacramento River Canyon
Witness: 4
Shape: Unknown
Duration: 30-35 minutes
Intense, White Light; sweeping movement similar to airport beacon in "Search" Mode.

   Very early in the morning, while still completely dark outside, four people were heading south on I-5 to first a class in Sacramento and then, that evening, a Rolling Stones concert in San Francisco. While rounding a corner below Delta, the witnesses viewed a light more intensely bright light than any light the witnesses had viewed previously in their lives. The light seemed to act in a searchlight sort of way, fanning out back and forth, sweeping the air above. The intense beam seemed to reach miles into the surrounding area, arcing over the top of the surrounding ridges.

   When first noted around the Lakehead area, the light was to the southwest of the witnesses as they motored down the freeway. After a bit, the beam was parallel to the witnesses, directly to the west. Finally, it was behind the witnesses, to the northwest. The light's source was always just out of view with the steep mountains, cut banks and ridges next to the highway obscuring the source. As the witnesses continued south, heading down the hill just south of Salt Creek, the light suddenly came downward, piercing directly into the car's back window, surprising the occupants and disturbing the semi-slumbering young ladies in the back seat to a sudden full awakening, evoking comments similar to, "Hey!" "What in the heck!" The source of the light could not be seen at all due to the intensity of the lumens directed into the vehicle, blinding any ability to view the source. The witnesses were smiling and puzzled, but not in any way alarmed. The light continued in this fashion for perhaps 25-30 seconds, then left the window and lifted off of the car, moving again to above, behind (north) and to the west of the motorists.
As the witnesses continued south, the bright beam was visible until reaching the area just to the south of Redding where the dawning day finally obscured the view of the light.
Author's Note: A similar event occurred listed in the regional chronology, on the same date.

MUFON CMS Case #14395
Occurred: September 29, 2007; 2345hrs. PDT   (Left in its original MUFON format)
Location: Miller Mountain, Herd Peak Lookout; 20 miles NE of Weed, CA
Shape: Disc
Duration: Seconds
Witness testimony: "My husband and I were deer hunting on Miller Mountain. It had snowed a bit the night before up on the top, which made us happy because we could now spot fresh deer tracks in the new snow. All of the top dirt roads are dead-ends and about 1-2 miles long. The snow clouds had cleared and we had a sunny sky with puffy clouds. We decided to take one of the "back side" western facing roads that take you along the steepest part of the mountain, pretty much straight up and down and heavily forested.

We were happy that morning to notice that there hadn't been any other vehicles up on the top back roads yet that day, because there were no tire tracks in the fresh snow. We proceeded to take our time along this road with my husband driving. We came over a rise about half way to the end and my husband said, "What's that?" I was looking out of my side window for deer tracks and turned my head forward to look out the windshield ahead. I noticed a "Very shiny and bright metal object maybe 8'-10' wide, in the road ahead, approximately 50 to 75 yards away. Looking at it, the first words out of my mouth were, "Oh, there's a pickup ahead of us." It looked like a new, extremely shiny chrome pickup roll bar with two lights on top…but I could see no "pickup", only what I thought was a roll bar. Then my husband said, "No, there are no pickup tire tracks in the snow ahead of us and we are on a dead-end road."

As I leaned down to get the binoculars and come back up to look it was gone. I said, "Well, let's go to the end just to see." We drove to the end and indeed, there were no fresh tire tracks in the snow at either end. We talked about what we had seen, and came to the conclusion that no pickup could have gone over the edge of the cliff or gone up the side of the mountain…it had to have been a UFO. I asked him what he saw as I was getting the binoculars off the pickup floor, he said, "It just kind of floated up a bit off the road and disappeared." We did not see it again, nor did we stop and get out of the pickup where it had cloaked itself."

Investigator Notes:

December 15, 16, 17, 2008.

Playing phone tag, December 15-17, 2008. Witness apparently owns & operates a business in Weed, California.

Emails exchanged. Mainly regarding sending motivational emails to witness for landing location information. We ran into delays in obtaining information due to Christmas Holidays.

January 5, 2009

Witness on January 5, 2009 finally emailed position of UFO landing (or near landing) location (position exact) (see jpg attachment of satellite image-'Google maps')-Latitude 41degrees 37'55.36"N, Longitude 122 degrees 14'10.82"W, Elevation 2097 Meters (6880 Feet).

Miller Mountain is roughly 12 miles due north of summit of Mt. Shasta, and is 15 miles NE of Weed, California.

Notes: Witness-used Google Maps. I checked and verified witnesses position-but Investigator used 'Google Earth'. My position of same exact spot as seen visually on map display reads: Latitude 41degrees 38'3.44"N, Longitude 122 degrees 14'11.86"W. Suggests a possible slight or negligible variance between 'Google Maps' and 'Google Earth'.

No GPS coordinates available by my Google Earth programming.

Never have spoke (yet) with other witness-Mr. ----------- ------------, (------ husband). - _____ still a bit reluctant to discuss sighting at present time.

This is a very interesting landing or 'near-landing' in this CE-1 Case. Witnesses did not 'see' from their truck close by minutes after sighting any apparent physical traces on ground (snow). No tracks or landing gear imprints seen. Witnesses keen experienced hunters familiar with this hunting area. So would they have been keen in seeing animal (or perhaps entity?) tracks or footprints (in snow)? Both witnesses remained in their car, for their safety. Perhaps out of curiosity, but also some fear and trepidation of what else might be nearby remaining unseen. Examining closely the Google Earth images of landing site, it does show a natural clearing along this isolated one way (dead end) mountain road, which a craft might naturally use to land by. Elevation of landing site 6880 feet altitude, and is a very remote location, though it would be accessible by 4-wheel drive vehicles without much difficulty (or it appears from satellite image).

Witness considered 'very credible'. Witness was reluctant to report event until over a year later. It affected them about what they might have seen. No missing time or other unusual experiences. Listening to witnesses, I would tend to agree with this assessment. I might like to visit witnesses and see if they would take investigator to site, retracing their movements some time.

General Area Research:

------------ reported that many persons in their area (general Shasta area) have often seen or with close trusted friends or family members. Investigator's impression from interviewing witness suggests the local culture, (Weed, California and near by towns/communities) regard UFOs as an 'open secret'-seen by many but seldom discussed.

Incidentally 'Bigfoot' sighting was reported in this same area just five miles east of UFO landing area on August 4, 2000, in an area north of Weed, between Goosenest and Herd Peak Mountains. BFRO Report # 174 (Class A).

See http://www.bfro.net/GDB/show_report.asp?id=174.

Case completed: January 5, 2009. Case description: Unknown-UAV

Ballester-Guasp (using CMS computer version)=5.75%=0.0575 Steve Reichmuth, Chief Investigator

Assistant State Director MUFON Northern California   (Pictures below)

Above sighting location view: Miller Mountain from Herd Peak Lookout (PSP)

Occurred: December 25th, Christmas Night, 2008; After dark: 1630-2100hrs.

Location: Dunsmuir, CA

Witness: 2
Object: 5 orbs
Duration: Several minutes

Justin Hisey, and his son Blake stepped out onto their deck that faces west, overlooking the Sacramento River on South 1st Street, which runs north to south. Justin a Railroad employee has resided in Dunsmuir his entire life and never witnessed any UFO type sighting. As the father and son duo were warming themselves in front of the outdoor heat source, on the deck, Blake noticed to the east descending down Blackberry Hill, a number 5 dice shape pattern of motorcycle looking round lights moving about 60 yards away right above South 1st Street. The configuration of lights spanned about 15 feet across. There was no seeming structure they were attached to, but moved and shined in unison. Justin and Blake watched the orbs move west, first over the river, then the railroad tracks and headed straight to the Mt. Bradley area, located just north of Castle Crags. As the orbs moved away the lights faded in and out, bright to dim, never going all the way out. Then it was just gone.

Occurred: April 4, 2009; Approx. 2030-2130hrs.

Location: McCloud, CA (Colombero Drive)

Witness: 8+
Object: 2 craft, one being a Jelly-fish like translucent craft
Duration: 3 sightings; 4-6 minutes and around 10 minutes, respectively

In April 2009, Rosemary Cary worked on the McCloud River Railroad's dinner train. Often as was customary, Rosemary's husband, Richard 'Dick' Cary, came to greet her and escort her home upon finishing work. This time, they had two vehicles and as she took off towards the house, Dick followed behind her in his vehicle. While on Colombero Drive, Rosemary noticed Dick was flashing his headlights at her so she stopped to see what he wanted. When Dick showed her what he was seeing in the sky, she said, "It's probably a satellite." Dick said, "No way!" As they were watching the movement, it suddenly came low and to a dead stop, not too far above the tall timber. Dick says, "I thought it was a helicopter on fire and about to crash!" The aircraft quickly left and as it did, Dick said, "We'd better get to our house! It looks like it is headed right that way." They jumped into their respective vehicles, gunning for home and thinking they were about to witness a crash. While driving, they saw the object continue; now clearly seeing it was not a helicopter on fire at all. The craft swiftly arrived over the Haul Road and abruptly stopped again, pausing several seconds, then banked a sharp left, zipping in a flash east on the '89 corridor.

The Carys quickly arrived at their nearby Shasta Avenue home and pulled in. Rosemary says, "I excitedly left the vehicle and headed into the house, but almost immediately, Dick called out to me, saying "'Look, here comes another one!" "The craft came in from the west speedily and was now about 150 feet directly above our house almost skimming the tips of the tree tops. I grabbed my binoculars from the garage," said Dick. Rosemary said, "I also wanted to go back inside and grab my binoculars, but I was afraid I'd miss something. I did grab the phone so we could call the kids in another part of town." Rosemary also said, "The dogs went crazy, either from not being greeted as they were use to or in reaction to the craft." Rosemary did call the kids in another part of town. Dick elaborates, "My daughter and son-in-law leapt outside and they too saw the craft very well from outside their house."

By now, Dick was getting a really clear view of the object. He said, "I could not get the whole craft in the eye viewers, it was too close!" He describes the craft, "It looked like a huge jellyfish. It was a translucent dome, about the size of ½ a house. It was bright with some small highlights of orange and red tones but mostly the lighting was a beautiful light purple hue and glowing. In the middle of it was what looked like an intense blazing fire but with the flames mostly contained or going up, not going sideways or down but like it was powering something, in the manner of a hot air balloon. There was no noise at all. It was quiet, even though it was so close and had all that fire and everything, it was silent."

"It stayed above us at least 5-8 minutes," Rosemary explained. Dick added, "When it left, it slowly meandered off then went and stopped above the Haul Road in the same exact exit pattern the first craft used. It hesitated there about 10 to 20 seconds, then also in the example of the earlier craft, turning east with a sudden motion, then continuing at warp speed, left, heading away towards Pilgrim Creek Road where it just suddenly blinked out right there. "

The next day, first at the bank in McCloud and later at work, Rosemary mentioned the occurrence to some locals. Both at the bank and at work, acquaintances and other employees informed her of others seeing the same thing. Two other older, well-know lifetime locals, a woman and a man separately, had come in speaking of clearly seeing the craft too, as had another adult male and around three or four kids at Hoo-Hoo Park. Most all repeated the same scenario as witnessed by Rosemary, repeating the way the object exited with a fast, hard left, pausing over the Haul Road then zipping east, following the '89 corridor east out towards Pilgrim Creek, then just disappearing while right there in the sky. Dick says, "This was written about in the Mt. Shasta Herald. It may have been a week later but probably it more like two weeks because it took a while for the story to get out and to talk about it.

Occurred: August 2010; Approx. 2200hrs.

Location: Klamath River, Seiad Valley, CA (and nearby mountains)

Witness: 1
Object: Lights/craft
Duration: Minutes

The witness was at Collins-Baldy Fire Lookout in the Klamath National Forest, doing photography of cloud formations. Upon reviewing the photos at a later time, one of the pictures had a UFO in it. The image was not present in the photos taken before or after this shot nor was the craft visible at the time of taking the shots.

Another startling and significant event occurred to the witness's brother and his friend. The two were heading up the Klamath River towards Yreka. As the two men rolled-up on the area around the confluence of the Klamath River and Shasta River at Highway 263, there was construction underway with a closure in effect. While waiting for passage, the two men noticed a dozen or so contractors and state highway employees had all ceased activities and were looking up into the sky. As the two followed the gaze of the workers, they noted a silver cigar-shaped craft hovering in the sky just north east of the bridge, and very close by. The entire group witnessed the UFO hovering in broad daylight.

The witness's family has resided down-river in the Seiad Valley area for about 40 years. At times, the witness has noted odd aerial activity over the Devils, around Lake Mountain, and in the Seiad Valley area proper. Activities have included odd lights performing radical maneuvers while zigzagging and sudden 90 degree turns. The lights have zipped straight up, then straight back down along with sudden disappearances of the lights happening while observing this phenomenon.

Around two years ago, the witness, her spouse and a friend exited their home to observe the Persied Meteor Showers that arrive every August. Upon stepping out, the witness immediately observed an object and pointed it out to the others present, thinking it was a satellite. But the movement of the object suddenly halted then took off again shooting straight up and suddenly disappearing from view. It seems quite apparent that the Seiad Valley and surrounding area holds a secret for UFO activity that many locals are unwilling to disclose. Even with the witness's anonymity, we feel fortunate to have this information.

Occurred: Early Spring 2011; Approx. 2330hrs.

Location: Castle Crags, CA

Witness: 1
Object: 1 disc
Duration: 1 minute

Witness's solar light was triggered on. She got up to investigate and to see what it was that triggered the solar light. A very bright light up by Castle Crags Peak was observed about 5-6,000 feet up with powder blue lights radiating down in a circle from the bottom of the lower dome of disk but not to the ground. The object was a mile in front of her directly to the west. As soon as she went to grab binoculars, the ship took off in a zigzag pattern south to north back south and vanished.

Another incident two decades earlier involved her mother being chased down the road from Castle Crags and followed all the way to Weed by a green iridescent sphere. A white door on the crags that she has also seen on occasion, by the top, which also has been seen by others including the Indians, then disappear. This door allegedly has been said to be another portal into to caverns inside the Crags and possibly a dimensional gate.

Witness is a businesswoman, rancher, very grounded, focused, and competent in her observational skills.

Castle Crags viewed from I-5 in Castella. Looking east at west side as witness saw it. (BDW)

Occurred: Various stories from 1939 to 2009
Location, McCloud, Dunsmuir and Castella, CA
Witness:
Object:
Duration:

A mother of one of our UFO witnesses has stepped forward to share a lifetime of her UFO observations. The witness and her husband have lived in Dunsmuir their whole life and are in their older years. They are both authentic and down to earth people who are credible in every way. She has taken the sightings very seriously and has a crystal clear memory of the events. Some of the dates are unclear but considering the duration of the various sightings and details to environmental factors surrounding the events, it is a miracle. Here are listed her stories, sightings and accounts from family and neighbors she has experienced or has been told.

Interview with (undisclosed) of Dunsmuir, Ca.

1. G & A. The parents of witness's spouse took a trip to Reno, Nevada sometime from 1958 to 1960. She based this estimation on the birth of her first child 1957 and the birth of her second child in 1961. She recalls the Camel Races that year. As the undisclosed couple proceeded out of McCloud towards Reno, a craft was above them and followed them down the '89 corridor and onto the '44 corridor. The craft followed for a lengthy period of time, perhaps an hour or so.

2. The often repeated scenario of lights following vehicles along Highway '89 in the McCloud area, zipping in and out of trees along the road and was a very common scenario that many witnessed. A name mentioned was "T's Mother", L, her husband, and 2 children witnessed it but they all passed away. Many others from this area repeat this story often.

3. In 1966, the witness was to attend a Sara Coventry Jewelry Showing in Weed, Ca. In the South part of Dunsmuir, where she spent most of her life, is where she began her journey this unusual day. Loading her two daughters into the car for the trip North, they left home, and a bright white orb the size of a basketball started chasing the car and followed them virtually all the way to Weed. She varied her speed, drove faster and slower, but could not shake the orb. The little girls were terrified of the unknown object that looked like a bright moon pursing them. The children tried to hide on the floorboards of the car and cover themselves with the coats or the blanket they usually kept, but being either late spring or early fall, the cozies were not in the car. Upon arriving at the event in Weed the witness shared her ordeal with the organizer and a friend.

4. In the mid 1960s, another husband and wife resided on the corner of Crag View and South First Street in the South part of Dunsmuir, and were the owners of a nearby business. One late afternoon or early evening, the woman and her stepdaughter, J., went grocery shopping. Upon returning and pulling in the parking spot, stepdaughter picked up some grocery bags and proceeded into the house. When the woman got out grabbing more bags, her eye caught an odd sight. Directly to her West, gliding over the Railroad Park area was a craft. She stood stunned as the stepdaughter came back out to assist. Mom said, "Do you see this? I don't want you to think I'm crazy." Stepdaughter did, indeed, see 'this'.

5. There was an older lady that lived on South First Street in Dunsmuir. Witness does not remember her name. This woman saw saucer-shaped craft constantly and reported it just as often. Everyone thought she was crazy. People stayed away from her though she often urged neighbors to join her so she could show them what she was seeing. She stated that the craft would come out of Castle Crags heading directly East over Railroad Park, Red's Wrecking Yard and the Pacific Power substation located nearby, then almost always head to the bridge close-by on the Sacramento River, just above the sewage treatment plant, turning South and head downriver. She witnessed these craft, "all the time" and reported the sightings "all the time". No one would ever respond to her requests.

6. Witness's husband was in the Army, serving until sometime in 1952 at Fort Ord, California and Fort Lewis, Washington. While enlisted, he had friends that worked with radar, often calling him in to witness activity on the screen but "there was nothing out there".

7. Approximately three years ago in 2009, witness and her husband were heading south on I-5 in the Mott area after taking a meal at Lalo's Restaurant in Mt. Shasta. The time of year estimate is 'during good weather because we don't go out at night to dinner in bad weather." They looked up as a bright, fluorescent bluish-green ball traversed above the towers above and to the southeast of the Freeway on Girard Ridge, across from the old Shasta Royal Inn. Both of these individuals noted the activity at the very same time. The ball was pretty big but likely larger than it appeared from below due to distance. The 'ball' suddenly zoomed directly into the ground as they watched. They called 911, concerned about the safety of the communications equipment in the direct vicinity. They assured the 911 operator they had not been drinking. No one ever responded to their request nor followed up on the sighting. NOTE: During the 1960s there was a UFO reporting phone number in the police station that now is gone, but sightings were taken much more seriously during those days than they are now. There is erosion in trust that even credible sightings are not taken as genuine, and ignored.

8. G. & M. and Mr. N. lived in Dunsmuir in the past. Mr. N worked at the water company. The N family leased a cabin and land from the R family in the Railroad Park area. The cabin has now been torn down. Mr. N had a sighting and went on the record, trying to relay the story to anyone who would listen. He also had a photograph that had a show of craft though the ship had not been visible to the naked eye. It just showed up in the series of photographs. The witness states she has seen this photo and also says Mr. R. or his sons may have a copy of this item. According to the witness who saw the photograph of the UFO, it is dated on the back and had been professionally tested for authenticity and was confirmed to be genuine. It was taken during daylight hours from the North window of the House of Glass Restaurant in Dunsmuir. The witness claims there may have been two UFOs that showed up in the picture. Family members were contacted to find the whereabouts of the photo. It was looked for and is apparently now missing.

9. Up old Highway 99 by Manfredi's old Market, now South Dunsmuir Avenue, VMDI witnessed two large disks between her car and the Railroad tracks going south. The craft kept up with their car and emitted a low vibration noise. She and her passenger were scared, and afraid that the disks were going to collide with their car. The disks then shot straight up and fast out of sight. Witness can only remember year of 1939. She is now in her late 80s.

10. Same individual also witnessed a dish shaped craft with a light going around on the bottom of craft on many separate occasions coming in and out of Castle Crags in the evening. That also goes along with the sighting of above of woman on South First Street in Dunsmuir. It seems the presence of UFO activity from over five sources confirms their location of coming out of Castle Crags area.

Occurred: October 21, 2011; 1845-1850hrs.
Location: Mount Shasta, CA
Witness-1
Object: -disc/light Duration: 3+ minutes

I thought I would share my recent UFO sighting with you. I have had others but this is the most recent one that took place on October 21, 2011. It was just about dusk between 6:45 and 6:50 pm and I was walking home from my neighbor's house and facing the mountain. I saw an extremely bright light just sitting stationary above and to the right of the mountain. It seemed odd because it wasn't dark enough for stars to be out yet and the sun was still on the mountainside. Coincidentally, I just happened to be carrying my binoculars, which was bizarre timing because I do not usually carry them with me, but that day I had them because I was observing the unmarked jets creating another whiteout of our blue sky with so called 'contrails' to those that are gullible enough to believe crock of bullshit, excuse my language. I'll stick to the subject. Anyway, I stopped in my tracks and began to observe this very bright light, which was actually a reflection from the sun shining on this object. It began to move slowly to the left towards the center of the mountain and it stopped again when it got between Shasta and Shastina. Then it moved forward through the valley between the two mountains and slowly descended downward behind the mountain until it was out of view. I got a very good look at it because, as I said, I had my binoculars on me by some stroke of luck and really strange timing. It's almost like I was meant to see it. Anyway, I checked to see what time it was and determined that it happened between 6:45 and 6:50 pm. My immediate thought was that this thing had to have been captured on the Snowcrest snow cam. So, the next morning I pulled up the video from the previous day and paused it at the 6: 45 pm mark and began viewing it frame by frame. The cam takes a shot every 30 seconds and the incident lasted for 1 to 2 minutes so unless I was hallucinating, there had to be at least one frame with the UFO in it. Sure enough, when I got to the frame at 6:48:31 there it was, barely visible due to the graininess of the video. I knew it was what I saw because it was in the same position and at the same time. I took the frame and filtered out some of the background noise and adjusted the contrast etc. to make it clearer. Attached is a document showing the original frame and the enhanced version. As you can see, the sun is reflecting off of the object in the same direction that it is reflecting off of the mountain. It is disc shaped just like I remember it, except I was able to see more detail through my binoculars. There is actually a rim around the lower part of the object that does not show up on the photo because of the poor quality of the video. The fact that this thing shows up in the video at the time that I noted and in the same position as I saw it confirms to me that the thing that I saw was really there. I have seen others before but this is the first time that I can actually verify a UFO sighting with evidence. Pretty cool huh? I thought so. Keep looking up! There is a lot of things going on in the sky that the general population is oblivious to while happening right in front of our faces without an acceptable explanation. As long as everyone continues looking down at their iPods it will continue to go unnoticed and unchecked. It's time to wake up the masses out of their sleep state and demand some accountability from the corrupt corporate owned puppets that our supposed to looking after our best interests, not against it. I somehow feel that the increasing UFO activity connected in some way to the orchestrated collapse of society that is currently in full swing. Not that ET's are necessarily behind it but who knows. Perhaps we are just being monitored as we self-destruct. Perhaps not. (PHOTOS BELOW)

Mountain 'Snowcam' shows object clearly over Shasta peak.

I-5 and Crag view Drive, where many UFO's have been seen and reported. (BDW)

Occurred: December 2011; 2330hrs.
Location: Mt. Shasta, CA
Witness: 1
Object: 30 feet orb
Duration: 45 seconds

On a clear, cold December night, an individual in early 20s driving home from work driving north on Old Stage Rd. by Deetz Rd. observed a bright teal light coming in at high speed in a controlled flight arc from horizon to tree line, to disappear a half mile to the east between North Old Stage and Summit Roads. From where craft originated in sky, a bright white light flashed and disappeared, like into a wormhole.

The witness also claims her sister, only several days before, observed mountain at approximately 4:00 pm and saw ultraviolet light that lit up mountain from behind it (easterly) lasting for several minutes. Possibly alpine glow but light was far brighter on spectral scale in purple range.

Occurred: Early September 2011; Approx. 2100-2200hrs.

Location: Old Ski Bowl on Mount Shasta, CA

Witness: 2
Object- (2) 1 launching another one
Duration: 3-5 minutes

So on a September night, a friend and myself drove up to the Old Ski Bowl at around 9-10 p.m. to hang out and watch the stars. We parked at the upper parking lot of the old Ski Bowl and we then proceeded to walk up a dirt path until we found a suitable rock for us to chill on and be comfortable checking out the sky. We were looking toward the mountain for about a half hour, chatting and stargazing. When from the far left a solid bright light appeared in the sky right over the mountain. It proceeded from the far left to about 3/4 through the night sky heading to our right. It was not a commercial plane because it had no red or white blinking lights, and we could tell that it wasn't a plane because of its constant speed and the trajectory of its flight path. We watched the object for about 3 to 5 minutes when suddenly it launched a smaller, brightly lit object south of its position. Almost instantly after the launch the main object proceeded to leave our atmosphere and it just disappeared. I say it disappeared because it didn't move in any North, South, East, or West direction, but rather dimmed and disappeared as if it just went straight up away from us.

Note: That was not the only strange occurrence on the mountain tonight that my friend and I experienced. Prior to the object appearing in the sky, my friend and I were looking away from the mountain, down toward town, when suddenly there was just a blinding bright flash of light. It obscured my whole vision for a split second, a bright white light. Then once it disappeared I wasn't blinded, like you sometimes can be when you look at the sun for too long and try looking at stuff after. About 10-15 minutes later the other sighting occurred.

Occurred: May 10, 2012; 1700hrs. (Author's sighting)
Location: Mount Shasta, Ca.
Witness: 1 with camera
Object: 1 Disc
Duration: 5 minutes

Curious disc occupying cloud bank that was almost transparent with lots of blue sky about it. Disc was about 100 feet across overhead 1,000 feet up. It was hovering and stood out within the clouds as a solid metallic, dark object, and with the clouds blowing over it did not move. It was right above my property, in plain view, and if I had a film camera it would have come out totally clear. Took two photos and was amazed that I had time to run in to a get camera as I watched if for a good two minutes before I took the two pictures. Once I got the pictures, it vanished. (photo below)

Center of picture: Saucer (Pine Grove and Old Stage Rd.) 5 p.m. May 2012 (BDW) Behind dense cloud cover.

Occurred: May 12, 2012; 2200hrs. (Author's sighting)
Location: Mount Shasta, CA
Witness: 1
Object: black wedge
Duration: 6 seconds

On a warm spring evening, a Mt. Shasta resident was in hot tub observing the clear, starry western sky, on this quiet night. A dark black triangle quietly sailed by estimated going 45mph with no lights at all entirely silent as it went by and slipped through the night sky. Its skin appeared semi flat and was precise and focused on its northwesterly direction. It came in to my line of vision I registered it as it occupied my full focus for a few seconds and then watched it leave my field of vision as quickly as it came. I thought that it perhaps was one of our new craft with otherworldly appliances but the way in which it behaved, performed, its agility and precision in a lightweight package, 15 to 20 foot equiangular triangle with no fin attachments or appendages countered that thought. If anything its black appearance spoke of concealment, but unlike the SR-71 stealth it was a lot smaller, more agile, it floated, with no noise or heat nor did it put out an aura of heat distortion or vibration. I've seen a few flavors of UFOs but this one was unfamiliar and if it is terrestrial built (seemed like a drone) we have totally reverse engineered others handiwork. It was not a balloon, nor electric as it generated no motion distortions as electric engines do or buzz or whine, (I have many model electric flying devices) It may be a newer D.o.D. drone replacement which takes the place of the once secret DRONE which is being used by law enforcement, Dept. of Agriculture and crashing in Iran when our overconfident fliers forget its not a electric flying toy competition.

Lockheed as well as other aircraft contractors have been making many balloon type prototypes of hovercraft with high-speed capability driven by solar powered quiet motors that have an "otherworldly" appearance, but this was not one of them and much more sophisticated eliciting a suspicious aura about its origins.

Part B: BIGFOOT SIGHTINGS

Bigfoot greets you entering Happy Camp, CA (Shae Wallenstein)

This section originally was to complement the book for the uncommon amount of yeti sightings over the years. It is said that The Karuk Tribe, among several other Tribes, have a kindred affinity with the Bigfoot and have had many reported sightings on their reservation, along with many sightings, in the Happy Camp area. One of the reasons is that the Big foot knows the Indians have a long standing respect for and sacred acceptance of their lives and they invariably feel safe on the sacred ground of the Great Spirit where ever the Indians reside and know they won't be shot at or exploited.

Happy Camp, which is west of Yreka, has a weeklong celebration to the Bigfoot with the town's adornment of a 40-foot tall carving of a Bigfoot welcoming ones to town. The sightings which are available online consists of over 85 pages of sightings from the 1890s to the present and include all American locations including Siskiyou County. With the sheer numbers of sightings, some confirmed, many not, it has been the author's prerogative to list only the ones in the immediate Mt. Shasta area that have good background.

Above foot print given to me by an undisclosed Indian Chief who would not be interviewed, and sent it through a reliable source. He said that this Yeti has frequented the reservation for several generations.

Occurred: 1974; Approx. Evening-dark
Location: Chester, CA
Witness: 1
Object: Yeti
Duration: 1-2 minutes

I read the L.A. Times article about you and your endeavors. When I was a kid (approx 1974, but I'm not sure) I was camping in Nor Cal near Chester, CA and saw what I believe was Bigfoot/Yeti. I was alone in the campsite and my parents were out for supplies & had the only camera. I wish all the time I'd had my own camera! Anyway, I heard what I thought was a bear rummaging in the trash. My mom had told me to stay in the trailer w/the door locked, so of course I opened the door. LOL. The "bear" stood up and walked several feet on its hind legs. Not like a trained bear would, but like a man or ape. It had front "legs" that were too long & were more like arms. It reached around and scratched it's hind end w/the front "leg" and then heard me and walked away quickly on hind legs and eventually on all fours. It stunk like raw sewage. I went out to where it was (I had no fear for some reason...maybe my age!) and looked for fur or droppings or something, but didn't see anything. If you have any questions feel free to write me.

Occurred: Fall 1980; Approx. Morning
Location: Mount Shasta, CA
Witness: 1
Object: Sasquatch
Duration: 3-5 seconds

Witness, Jeff Hough, was in vehicle around Mt. Shasta Sewage Treatment Facility off S. Old Stage, by Lake Siskiyou, hunting. This area once was a small airstrip and a dump but later the surrounding area became The Mt. Shasta Golf Resort. The land appeared leveled and bladed as our witness was driving the dirt roads littered with Manzanita, heading northwest at a very slow pace going away from the sewer pond area. Suddenly, a big foot stepped out of the Manzanita brush directly in front of his truck and had to slam on his brakes and swerve a little to avoid hitting the dark brown coated, over seven foot tall beast. The Yeti, non-chalantly continued to walk as if he did not even notice what occurred nor cared. It walked without breaking his gait in front of the witness and disappeared into the brush on the east side of the road, disappearing from view. Although witness was well armed, he did not even consider taking a shot, as he was entirely surprised.

Occurred: Thanksgiving 1987; Approx. 2200hrs. (Author's sighting)
Location: Mount Shasta, CA (old Ski Bowl Lodge)
Witness: 5
Object: 3 Yetis
Duration: 15 minutes

Arriving up the Everett Memorial Highway, we parked just below the old ski lodge and walked up the spiral narrow road to the where the old ski lodge resided. The outer shell was still left although the inside was charred from a fire a years back. It was partly cloudy and cold. Lots of snow on the ground, some satellites were visible but not the greatest night for stargazing as it turned out, but we had another surprise awaiting us.

Upon approaching the lodge from the south side, all five of us were in a loose line and I on the far left end could see the west side of the building 50 feet away. The solid metal door, which was 7 feet tall swung open and the strongest odiferous, pungent and musky smell blew up our noses being down wind. A dirty, white-haired male Yeti appeared next to the edge of the door, his head as tall as the height of the door and stared at us, as a female came next to him, then followed by a juvenile about 4 feet tall, who held on to its mother's leg in apparent fear. The female was a little over 6 feet 6 inches tall. They all stared at as we stared back at their glowing amber eyes. The smell was bad, and they looked as alarmed as we were, all of us, and a friend said, "Hey, let's rush them" like in football, which I retorted,
" Are you nuts? Let's not, lets just leave them alone." We all began to walk backwards, as he motioned to the Yeti and saying, " We don't mean you any harm and we are going to leave." They all looked as if they understood us on their human looking faces. We all backed up down the hillside and they curiously followed us, keeping the 50-foot distance between them and us. It took about 5 minutes to negotiate the backwards descent, but we made it back to the car, as they kept their distance and let us leave. "Whew," we all exhaled.

That spring, the lodge was torn down and since there had been a little snow on the ground, there must have been tracks of the family of Yetis. Until spring, the snowy road is closed with a gate at Sand Flat warranting my suspicion of the Forest Service knowing the Yeti were holed up there which may have inspired them to take the lodge down. I worried about them being chased, apprehended for many years. Several years later I did inquire with the forest service if they knew the Yeti were there, which they denied, but having close friends with the staff it was discovered later that they in fact knew. Many locals will not admit their existence to protect the Yeti and discourage people from coming up and terrorizing them. They have a hard enough time living in a world dominated by humans that shoot first and dissect later, for either publicity or money. They do in fact live up here but to their whereabouts most of us locals will be tight lipped. It is 26 years since that sighting giving the Yeti a good safe start. If you want to see a Yeti don't look, and perhaps you will see one. Remember they too have a right to live without the fear of being exploited, killed and disrespected. It was about 10 years later near twilight that out my bedroom window that I heard the most intense blood curdling scream coming out of the woods behind the county yard. It was not a bear, a lion, or any animal of which I have heard and I have heard them all around here. It was a Yeti, and I had heard a recording of that exact scream, allegedly of a Yeti sighted in Pennsylvania. It was a huge, large throated animal yell.

Site where old Ski Bowl and lodge resided which avalanche took out. Location of Yeti and where many UFO and other Sightings occur (Snow covers road) (BDW)

Author standing in Yeti sighting area where old Ski Lodge was located atop Everitt Memorial Hwy. Mount Shasta

Occurred: Thanksgiving 1987; Approx. 2200hrs. (Second available witness)
Location: Mount Shasta, Old Ski Lodge
Object: 3 yetis
Witness: 5
Duration: 15 minutes
Second witness: We parked in lower parking lot, and walked up to the old Ski Lodge at the end of Everett Memorial Hwy. We were going to look at UFOs and satellites and one in our group was talking about sightings he had on the mountain in the 1960s. As we approached the old lodge from the south, I did not see the door open but smelled the bad smell of body odor and saw one come out, about 7 foot tall, whisky red, amber eyes, white fur, then another come out, and then a third small one. We cautiously left, walking down the cutbacks between the driveways that spiraled up, till we got back to the car. It was scary.
(Other witness's were not available or locatable)

Occurred: 1994 or 1995; Approx. 0600hrs.
Location: Mount Shasta/Grey Butte
Witness: 1
Object: Bigfoot
Duration:

This happened in 1994 or 1995. I was with other people snowboarding on Mt Shasta having camped out the night before in the snow. I was in the area of Green butte/ Grey butte heading towards the ski park traveling downhill. The sun was not yet up but there was plenty of light due to the reflective properties of snow. Had gotten up early because it was so damn cold. I was cruising through timber when I saw something out of the corner of my eye and stopped my board because I was curious and I needed to adjust a binding anyway. What I saw was standing on its hind legs doing something in/to a smallish tree approximately 80 meters away and give or take 10 meters below me in elevation. It was covered in tan/ash/blonde/grey colored hair. I was downwind of it and noticed no odor/aroma. I watched for a good 10-15 seconds when it started to turn towards me it got far enough around to see its profile before it realized I was standing there and it bolted away and downhill. I was riding in 70-80 cm of good powder and it had no trouble getting up and moving in it. It ran away on two legs. Conclusion= big feet or webbed? The profile most closely resembled artist's renderings of the Neanderthal man. This was a massive creature that I estimated as taller than me (I am 180 cm tall) and probably 1.5 times my weight (I would guess it to be in the neighborhood of 120 kilos). I did not go look at the tracks because I rode in the opposite direction at best speed and got enough speed to break out of that gulch and into the next and put some distance between it and me. I had a healthy fear of the unknown when in the woods alone and still do.

Occurred: 1996; Unknown times of year
Location: Mount Shasta, CA (by hatchery)
Witness: 3 over time
Object: Yeti
Duration: seconds each

I know there is a weird amazing energy down Indian Lane (the old timers called it Lover's Lane) I've felt a lot of different vibes of Energy and it can happen and vary with when you are there. My oldest daughter A.N. came in one night about 9:00-10:00 pm (this was 15 years ago) scared out of her wits, about a big smelly creature she described as Bigfoot. My brother when he lived down the lane in his early 20's described a large "bear like creature" that would shake his Trailer at night so hard it would wake him up.

Occurred: Winter 1997; Approx.1400hrs.
Location: Mt. Shasta, Sand Flat Area
Witness: 1
Object: 1 Yeti
Duration: 20 seconds

In 1997, one witness, (JN), went up to the Old Ski bowl to cross country ski. As a native born in Shasta, locals such as this witness are comfortable in the terrain and are very aware and seasoned in winter sports and outdoor activity. At about 2:00 pm, witness claims he saw a Yeti tearing up some dead wood as looking for something to eat. He was unfortunately downwind from the witness, and in 15 seconds, the witness bolted and dashed down the hill before the Yeti got wind of him. Witness was shocked at his experience. His elder brother had had a similar experience 4 years earlier also in the area of Sand flat.

Occurred: January 2000; Approx. 1300-1500hrs.

Location: Dunsmuir, CA (I-5 on/off ramp, Dunsmuir Ave.)

Witness: 2
Object: Yeti
Duration:

Justin Hisey was driving his son Blake to the Mt. Shasta Elementary basketball Tournament. They were heading north on Dunsmuir Ave to catch the Mott Entrance to Interstate 5 north, just south of Mount Shasta half way down the Dunsmuir grade, about 5 miles. They noticed what looked like a large individual with a huge backpack heading down the steep bank across the freeway just north of the Oak Tree Inn which faces west and is situated above the north bound entrance of the freeway on the east side of the Siskiyou/Dunsmuir Avenue underpass. The individual was moving rapidly and bound onto the freeway surface boldly. Justin stopped his vehicle and stated to his son that " I think were are going to see a drunk person get lit up by oncoming traffic!"

As they watched the individual became clearer in their field of vision. This was not a person with a backpack at all, but a being. The witness described it being between 6 foot 6 inches to 6 foot 8 inches tall, the size of a large football linebacker. It had white hair covering its largely built body. More aptly described as a YETI! It's graceful gait and stride took it with only a few steps half way across the freeway the center divider and with no effort completed its course over the rest of it as quickly. It leapt down the steep off ramp, on the west side of the freeway, and then exit from the Mott Road, just across the street from the entrance to Hedge Creek Falls in its fluid motion, jumped the fence. The Yeti then disappeared behind the north side of the water bottling plant and headed down the bank to the river and vanished.

Neither Justin nor Blake had ever seen anything remotely similar to this.

Yeti scaled freeway from left side all away across divider over other side, down to river. (BDW)

Other side of freeway and view of water bottling plant Yeti disappeared behind. (BDW)

Part C: INNER SIGHTINGS

Several of these articles were sent to me after conversations from individuals who spent time on the mountain. I must thank Ashlyn for these connections. Some had telepathic connections, some were lead here as if a magnet guided them, some had visions and others had visual and audible cues and prompts that appeared as apparitions. The individuals included in this short chapter all are well-educated, experienced and effective people. They all were willing to disclose their identities and show great courage and fortitude. Read them with an open mind. Without prejudice, it would not be fair to negate all the varieties of sightings we are capable of fathoming.

" Seven After Seven" (Author's story)

Condensed chapter from book "If You Wanna Get to Heaven" by Brian Wallenstein
First edition 1976, Second 1997

In 1975 a young man took a vision quest. He gave all his worldly possessions away and hit the road with packpack, sleeping bag, a few bucks and directions to a group forming under the auspices of a UFO cult. Meeting up in Oregon, he got to read the treatise of this group that seemed simple enough, requiring no idol worship but an ascetic type lifestyle, seemed like a good opportunity to do more self-examination, connect with the spirit within and see the West Coast.

Long story short, on the way to Gold Beach, Oregon they drove through Mount Shasta. Unaware of the existence of Mount Shasta, the young man was awe struck to see both the mountain and Black Butte jutting out of nowhere. Enthralled by the sheer size, magnificence and presence of these pyroclastic wonders he had a deep feeling of something for this place. Further down the road, on Hwy. 97 on the way to Klamath Falls they stopped at a rest stop, ate dinner while he pondered the visible beauty of Shasta and the surroundings, and he knew in his bones he would be back. Immediately in thinking that, a large meteor the size of a Mack Truck came crashing through the atmosphere ablaze in a spectrum of colors and hit somewhere to the west of them. It was clearly visible in great detail for it was less than several thousand feet away in its path. Jaws agape they both just looked at each other with wide eye disbelief. To him, it was a sign of providence. To this day it was the largest meteor he has seen. Noting the time it was 7 minutes after 7.

After six months on the road and what most people would consider synchronistic, miraculous and paranormal experiences, this young man finished his tour, replete with a great deal of wisdom and experience and was hitchhiking home from Seattle. He now had the idea to visit Mount Shasta on his way back to the Bay Area, and found several rides, of the last that gave him a ride to Mount Shasta. The driver for some reason gave him a working wristwatch and told him to take note of the times of any significant things that may occur. Still unaware of the myth and lore Mount Shasta held for many, he just wildly appreciated its volcanic propensity and alpine feel.

Being dropped off, the driver offered the young man 20 dollars somehow knowing he may need it for food and supplies. A 10-minute walk took him to the front of what is now the Karate Center on Mount Shasta Blvd. where he met 2 older adults asking them if they knew of a youth hostel or campground he could spend the night. Speaking between each other, the couple offered him a couch but he resisted and said he could sleep outside, but they insisted.

It was now 7 minutes after seven on the 7th of October. After arriving at their home he was greeted by a very attractive young woman sitting at the table, who when seeing him and he seeing her, both dropped their jaws. It was as if looking in the mirror at ones counterpart. They looked like identical twins in male and female bodies. The parents happened to be mentors of sorts for Jungian psychotherapy, ontology and other healing arts, but the real deal. The house the young man lives in today is that very same house he stepped in that day, at 7 after 7. Of course there are lifetimes that occurred, but that is another book.

"Who was that hiking Avalanche Gulch?"
"The Man on the Mountain"
Experiences of a woman in 1970 on the hills of Mt. Shasta
Hi Brian, here's the story. I was about 20 then (now 61), and two of us witnessed this sighting on Mt. Shasta. I was in Redding yesterday, and caught some glimpses of the magnificent peaks after many years, and it brought this memory back. I was Googling today and found out some more about the particulars (including your upcoming book).
It was the sort of short-notice expedition that a college student could appreciate. I'd guess the month was June around 1970. My old friend and ski buddy JoAnne called and asked if I'd like to climb Mt. Shasta on the weekend with some friends. I said yes and rented shoe crampons and an ice axe, and prepared my backpack & down bag. I drove to Tahoe to meet up with the group and seven or eight of us drove up to Shasta in a large van. (Everyone was around the same age.) We arrived at the parking lot in the afternoon, and hiked up to Horse Camp to camp for the night. The plan was to start the hike at 4 a.m. in order to hike to the peak and back in one day. We checked out the lodge, but opted to sleep outside in a row (since it was so cold). Trouble was, I awoke with a migraine headache. I opted not to go, and a girl stayed to look after me. When I got up later in the morning, we spotted our group near the top of Avalanche Gulch making their slow way. We stayed in the beautiful meadow around Shasta Alpine Lodge (pictured below). In the early afternoon, I saw a lone, hatless man all in white near where I had last seen our group. I thought he was a hippie or something. I called the girl's attention to the man on the mountain, and we kept an eye on him off and on for a couple of hours. He followed the zigzag of the steep trail. I remember seeing him walking, and just standing still in different spots.
He didn't seem to be getting much closer, and his presence seemed less 'scary' as time wore on. Finally when we looked up to find him, he wasn't anywhere to be found. When the team got back to camp later that afternoon, we asked them if they'd seen the hiker, but they hadn't, which seemed strange to us.
(There was no one else on the mountain that day.)

Why Are So Many People "CALLED" to Move to Mount Shasta, California?
By
Bob Bordonaro, reprinted here with his permission.

My wife and I, Robin Alexis, moved to Mount Shasta California from Port Angeles Washington in late August 2011. Why were we here? Well, in the middle of August 2010, I woke up and told Robin that I had a dream and that the spirit of El Morya, an Ascended Master who we are told dwells in Mount Shasta, came to me and called me to come to Mount Shasta. It was a Wednesday when Robin and I do our live Radio show, *Mystic Radio® with Robin Alexis*. With Robin's support, we decided to run a recorded show, (which we really don't like doing unless it is absolutely necessary). I packed my bags and left a couple of hours later in my little 1991 Mazda Miata. The west coast (including Washington State) was having a heat wave, so it was a long and hot 660-mile drive, as my air conditioning was not working. I arrived the next day, to view all the glory of Mount Shasta.

I spent the next week exploring, enjoying and falling in love with the mountain. When I left for home that following week, I knew I would be spending much more time here at Mount Shasta. I had such a beautiful experience at the mountain that I was only home for a short while when Robin and I decided to make a second trip back to the Mountain, so she could see first-hand what I had experienced.

We had a beautiful time and then we both knew we were going to be spending more time here at the Mountain.

After we returned home to Washington, we received guidance from many spirits and many people channeling information from spirit for us, saying that we had to go back to the Mount Shasta. Well, I was up for it but Robin was not very happy about leaving our home in Washington.

Fast forward to January 2011. We kept getting more messages to "come to the mountain". I told Robin, we don't want to go in the middle of the winter, let's just see what happens later in the year. The messages didn't stop! Finally, we scheduled another trip to Mount Shasta in early May 2011. I thought this might be a little early in the season to visit the mountain, and boy was I correct. We had a great time but we couldn't even hike up to our favorite spot, *Panther Meadows*, because there was still eight feet of snow up there, so we did our hiking at lower elevations. We had some beautiful weather where we were in shorts and tee shirts, but the day we left, on a May morning, was in a snowstorm. All the while we were there we knew we had to spend more time at the Mountain as opposed to just coming for a few days every so often.

Fast forward to July 2011. We decided to go back to the Mountain and look for a place to live because we just had to be there. Something had called us and we just had to go. It took a year of negotiating with our human selves about coming to Mount Shasta, as both of us fought the messages we had been getting. Nothing about this made sense to our human selves, but our higher selves knew that we were supposed to be here.

In August 2011, after we settled into the house we had rented here at Mount Shasta, we began hiking the beautiful nature trails here. When Robin and I would get deep in the woods we would sit quietly and meditate. Each time Robin would share with me that a Presence known as Saint Germaine, another being assumed to live "in" the mountain, would come to her. He would tell her that she was supposed to live in McCloud, a small town about 15 miles from Mount Shasta. He would show her an image of a house that had big windows looking out the back of the house. Through those windows she would see huge trees growing up through the deck. She would hear and see a body of water running from behind the house. She would tell me about it, saying, "If we decide to buy a house it needs to be in McCloud on the McCloud River."

We decided to start looking at real estate since the prices were very reasonable at this time. We didn't see home in Robin's vision in McCloud so we assumed her visions were not accurate. After a few months of looking for the house in Robin's vision, we decided to stop.

From time to time, I would visit Mike & Tony's, a local Italian restaurant in Mount Shasta. I became friendly with a man named Jim who kept insisting that I take my wife and to look at the house he had let go into foreclosure.

Finally we took a look at Jim's old house. When we entered the kitchen and looked out the back of the house, Robin remarked that the view was like the one she had seen in her vision from St. Germaine, but since it wasn't in town of McCloud it couldn't be, it couldn't be the one she saw, so we dismissed it as well. Then during another meditation St. Germaine appeared to Robin again. He said, "Jim's house is on McCloud Avenue. It is a Portal for the Violet Flame." Then she put it together! The house is located on McCloud Avenue in Mount Shasta, not in the town of McCloud!

Finally my wife recognized that this was the house she was shown in her vision. She had misinterpreted the words and the visions. I was still hesitant, but we made an offer on the house on McCloud Avenue. If our offer was accepted, the house would close escrow in December 2011. We went into meditation together again and asked for a sign from St. Germaine about this house for us.

That afternoon, while driving back from a shopping trip to Yreka California, (a small city about 35 miles from Mount Shasta), without either of us touching the CD player in the car, a CD popped on by itself. It was *"December"*, an album by George Winston. That was our sign, and we bought the house. Within the week after buying the house we discovered it is just down the street from the *"I AM Temple"* for the Saint Germaine Foundation, which just affirmed our decision to buy this house on McCloud Avenue in Mount Shasta California. What other signs did we need to know we made the right decision???

We are now living in Mount Shasta and we are just letting spirit be our guide for the future. We still don't know exactly why we were called here, but what we do know is that we are supposed to be here, and the journey Spirit has planned for us will be reveled in Spirit's time, not in our human time frame. All we know is that we are following the guidance we are receiving and that's just fine for now. We meet people every week who will tell us that they moved here because a Being, or an Ascended Master or a Lemurian, has "told" them to the mountain. There truly is something unique and special about what or who lives in Mount Shasta. Perhaps after you read this, you, too, will be "called" to move here, and we would love to meet you upon your arrival!!!

My Journey to the Heart of Lemuria
by Tina M. Benson, M.A. reprinted with her permission

My partner, Paul and I booked a Sacred Site Trek on June 5, 2011 with Ashalyn of Shasta Vortex Adventures. We started by drawing Medicine Cards for our journey. Mine was porcupine – innocence, trust and faith – and my partner's was bat, which symbolized rebirth. Little did we know the incredible rebirth that was about to occur on our journey.

As we drove up the mountain together, I heard a gentle voice in my head suggest, "Be willing to let go of your story." When we stated our intentions for the day, Paul recalled a dream he had where he and I journeyed to Mt. Shasta together and experienced a very deep and profound ritual inside of Telos, a Lemurian crystal city underneath Mount Shasta. I shared that our relationship was on the verge of total collapse, and I needed to seek clarity of heart, mind and soul about our relationship. Ashalyn gently suggested that I allow myself to move out of the relationship chaos and open myself to the elevated energies of the mountain. As I did that, I was embraced by the exquisite beauty of the mountain and could feel the spirits calling me.

We began our snowshoe journey and about 30 yards in, we stopped to invite the mountain spirits and our higher self to walk with us. Ashalyn asked Paul four separate times to connect with his higher self and each time his response made no sense. Ashalyn looked at me and said, "I can't go on any further until I give this man a healing." This gentle woman had become fierce, intense and purposeful. She turned to face Paul head on and told him, "You have the biggest guru cord in your 7th chakra that I've ever seen, AND, there are eight dark spirits attached to you. They have been controlling you and not allowing you to access your own soul."

Ashalyn called upon the mountain spirits and the healing guides she has been working with for over 30 years to help her remove the dark cords that were wrapped around and through his entire energy field from these unethical 'invaders.' She talked about how Paul (a world-renowned, esteemed philosopher, unbeknownst to her) often blathers on and on, not making any sense – and how those spirits have been confusing and disorienting him. Then I began to channel, saying, "You have never accepted being a human being in a body, and because of that, you can't receive all the goodness that is here." I, who am also a spiritual healer and teacher, shared that I had never met anyone more challenging to heal than Paul. However, when he surrenders himself, in moments such as lovemaking, meditation, or when in nature, he is the singularly most beautiful, luminescent, exquisite human being I have ever known.

Paul was leaning forward with his head resting on the tops of his ski poles, where he stayed for another 20 minutes or so. Both Ashalyn and I simultaneously stepped away from him and out of his aura, to make room for the extraction to continue. Ashalyn asked him once again to open to the golden essence that he is. "Allow that God-like essence to merge with your body and assist with your healing." Paul quietly descended into his own process as I closed my eyes and entered a deep meditation..

I was pulled into the earth by two lavender tunnels of light, into a space where there were other people preparing for a ritual. I was brought into a pool of water with dolphins and was in pure bliss and delight playing with them for quite some time. I ascended back to the earth's surface and felt the mountain and all of the many spirits embracing me with immense love and light, feeling a sense of familiarity with the mountain – a sense of 'returning home.'

Paul slowly awakened from his deep cleansing process. Once we were sure he was back in his body, we continued snowshoeing. I shared my experience with Ashalyn and she laughingly told me that I was taken into Telos before she even had a chance to guide me there! Soon we arrived at the Sacred Circle of Trees where Ashalyn lead a meditation into the Ascended Masters' Retreat and then into Telos.

My guides for the journey were a large man named Hargar, a petite woman named Rose, and a young boy named Wyatt. A swirl of rainbow light lifted us up and out the top of the mountain where we hovered in space above the mountain. I felt myself dissolving into no self, no-body, no-form, until I dissolved into a vapor, along with my three guides. After hovering there in pure unity and bliss, we were guided to a portal that took us into Telos. After several moments I notice that the ground underneath me became a kind of viscous, lavender lava which lifted me up and carried me down, down, down, over a lavender waterfall and to the bottom of a big cavernous area.

A very tall, regal-looking woman carrying a long staff was there to oversee my process. Her name was Aramas. I was told to lie down upon an enormous boulder that held within it a huge, amber-like, pulsating heart-shaped rock. As I lay on my back upon the heart rock, I felt "the heartbeat of the universe – the source of the original OM."

I laid there for quite awhile and then was lifted up vertically and placed back down over an enormous diamond-shaped amethyst crystal, which penetrated my vagina and moved up through my entire being, sending explosions through and out of my body, filling the entire cavern with crystal energy. Then I was laid back down on the heart-shaped rock and told to sleep, dream and stay there all night. I must remember to come retrieve myself in the morning.

With split awareness, I left myself lying on the rock and ascended with Aramas back to where we began the meditation. We then returned to our bodies and snow shoed back in silence to our car. Paul shared his intense meditative journey where he underwent extensive psychic surgery. He, too, had been told to return to that spot the next morning to get his bandages removed.

After dinner, Paul went on a 2-hour walk. He was guided to a forested area near the hotel where he was taken through a shaman's death ritual. For the first time ever, in that particular body, he was able to experience and merge with the golden radiance of his soul. In bed that night, we talked about that beautiful golden essence, which to me is what I fell in love with three years ago. We talked about his identification with his ego and its stubbornness to take a back seat to this beautiful essence. It was the first time since I've known this man that he has been able to glimpse his own magnificent divinity. We both felt humbled by the grace that happened earlier that day on the mountain, as we fell asleep in each other's arms.

(PHOTO CAPTION: Lavender clouds over Mount Shasta, taken Sunday as they returned to finish their journey. Note the large smiling face in the clouds on the left.)

The next morning Paul and I drove back up the mountain to the Sacred Circle of Trees. I took myself back, through meditation into Telos. This time at the portal I was lifted as if I had wings and gently glided down, down, down until I arrived back at the cavernous space where part of myself was still sleeping upon the heart-shaped rock.

I was to sit facing my sleeping body that was on the rock and to enter meditation. I felt my breath attuning itself to the pulsing heartbeat of the rock. Then I was told, "You were kept here overnight so that we could recalibrate your frequency to the Universal Heartbeat. This is Lemurian technology. We must attune human beings to this Universal Heartbeat because they are out of synch with it. Until they are aligned to this frequency once again, there will never be peace upon our planet. Human beings must be recalibrated to this Universal Heartbeat in order for there to be peace upon our planet. We are very honored to share our technology with you."

I was then instructed to return again and again to this place to re-attune my breath to this Universal Heartbeat. I am also to share that process in my healing work with others. As I felt myself rise from sleeping on the rock, Aramas arrived to guide me back to the portal. She and I merged into one being and glided back to the portal and back to my body.

I can only say that what happened for both Paul and I on that mountain, under Ashalyn's gentle guidance, was nothing short of a miracle. I remain profoundly grateful and indebted to her for trusting her inner wisdom and guidance, and I thank the mountain and the many spirit guides there for trusting me enough to lay me down upon that beautiful, infinite, Universal Heartbeat – the Heart of Lemuria. Please know that with the right intention and an open heart, that you, too, can be recalibrated to the Heart of Lemuria.

Tina M. Benson, M.A., Soul Whisperer and Transpersonal/Jungian-Oriented coach has over 25 years experience teaching, leading and facilitating individuals and groups in the U.S. and abroad in consciousness explorations, chakra initiation/meditation retreats, couples' and women's retreats, ecstatic dancing, Voice Dialogue, Enneagram, chanting, ritual and travel to sacred sites around the world. She's founder of the Soul Speaks Project and an ordained minister officiating at weddings, birth, death and other life-passages and celebrations. Contact Tina at: www.soulwhisperer.com, tinabenson1@cs.com

Jung's Spaced Out Client
By Brian Wallenstein

Excerpts used from Dr. C.G Jung's book FLYING SAUCERS A Signet Book, New York 1959,1969.

In this book devoted to the psychological and archetypal significance of UFOs throughout history, Jung makes references to the old school notions of phallic symbolism and the universal urge for union with our higher self and wholeness within a new myth of UFOs. He does however question himself and leaves room in the collective world for UFO phenomena. Jung was instrumental in working with both Dr. James McDonald and J. Allen Hynek in establishing psychological evaluation standards for witnesses. However some chose to use these antagonistically to debunk UFOs entirely as a form of cover up like Edward Condon used. Jung admits that with the convincing testimony of the competent witnesses, pilots and the persuasive photos and group experiences, that yes, they in fact, may very well exist. That was a real stretch for a colleague of Sigmund Freud and the old school boys to admit.

In the final chapter of his book, Dr. Jung brings into focus the interesting story of one of his clients; Orfeo M. Angelucci, and his UFO experiences. Angelucci coincidentally worked for Lockheed Aeronautics in Los Angeles, California. He went through both inner sighting and literal outer sightings and wrote several books upon the subject. Given the similar content to both Fredrick Oliver's book "A Dweller on Two Planets" and Guy Ballard's channeled, "I am Books" it seemed fitting to include a very condensed version of, " The Secret of the Saucers" by Orfeo M. Angelucci Amherst Press 1955 (Jung's client) into focus.

Orfeo wrote this book in 1952, which was a busy year for UFO sightings, and working for Lockheed, it seems he may have had insider information about the sightings, but that is conjecture.

According to Jung, Angelucci had his first genuine UFO sighting August 4th 1946, which at the time he brushed aside but apparently was writing about The Nature of Infinite Entities, of which the content was about atomic evolution, suspension and involution as well as the origin of cosmic rays. Jung further says Angelucci although serious and earnest, "Was a shy and uncultured person having an uncanny working and extensive knowledge of science." (Seems similar to Oliver).

On May 23rd, 1952, Angelucci claims he found his calling, and began to preach the gospel revealed to him by our space brotherhood. He thus was making his living sharing what the saucers revealed to him. As the story goes, he was leaving the night shift from Lockheed at about 11pm, heading home. He experienced the prickly sensation one gets before an electrical storm, and felt unwell and was anxious to get home. He then saw a red saucer ascend from the ground and head west at high speed at about a 40-degree angle. In its wake were two green orbs that spoke to him in plain English, and explained to him to not be afraid and these orbs were like communication organs, and that they were friends from another world. The voice bade him to get out of his already stopped, idling car. As he did, the two pulsating orbs asked him if he remembered his first experience in 1946 and informed him that was his actual first contact with them. Just then Orfeo became very thirsty. The orbs told him to "Drink from the crystal cup you will find on the fender of your car." He drank and it was the most delicious liquid he had ever tasted. He felt refreshed and strengthened.

Between the two orbs, a TV like screen appeared and two figures of the most perfect man and woman appeared. They had magnificently glowing eyes and yet seemed familiar to him. Orfeo realized he was in telepathic communication with these beings as he assessed the situation, and the screen began to fade while the fireballs resumed their former state. "The road will open Orfeo, the voice assured him. We see humanity as they really are and each individual, not by the limits of being human. Your people and planet have been under observation for quite some time, and every person's progress is recorded on our crystal disks in vital statistics as we see each of you as sacred unlike many of your fellow earthlings. We come and reassess your kind at each point of progress. We have a kinship and a sense of brotherhood as we see our own growing pains in our civilization similar to yours."

They continued to inform Orfeo of drone ships, mother ships, that thought travels faster than light and that no one needs UFOs except to appear in the denser material plane. Orfeo seemed from that moment mentally and spiritually quickened and this world lost its confused meaning and he felt aspiring to the ideal of these superior beings.

Again, 1952 was a significant year for UFO flaps on the globe.

On July 23rd, 1952 two months after this experience, Orfeo felt the same condition known to be the precursor to psychic phenomena, where he felt unwell, raw, and electrified. He called in sick and going outside he saw a glow on the ground that became solid, appearing like a small igloo. He stepped inside the door into a gorgeous mother-of-pearl room with smooth walls brightly lit, with18 feet high vaulted ceilings. There was a chair in the room made of the same radiant stuff, which felt like it was hugging him. As he sat, the door disappeared like it was never there. He heard a deep hum that turned into music as the lights dimmed and he fell into a dreamlike state of relaxation. When the room lit up again he found at his feet a bright coin like object. He also felt as if he was being carried away in this craft and he realized he was correct. A nine-foot window appeared and he viewed the earth slipping from view, he was already 1,000 miles away. He began to cry and was comforted by a voice that told him to weep for earth and the inhabitants who are mired in purgatory by hate and self interest, if only they knew they were growing to love universally. He then noticed a huge ship above his own. Orfeo was being instructed further on how this ship worked, on how telepathic responses behaved and how the further working out of souls on the earth towards their good, would occur. They informed him of Jesus Christ's identity of the Lord of the Sun or lord of the flame, and merged with the over soul of humanity in sacrifice to help them realize their real identity, that he was not in the literal sense, the son of God, but allegorically. The story goes on to elucidate Orfeo's further visions into these states, going to other planets, meeting Neptune (who also is another Alien from the Anunaki in Sumerian text), but the coin Orfeo had found on his first journey was stigmatized upon his chest and it was the symbol of the Hydrogen atom, representative of the omnipresence of God, everywhere and nowhere according to Jung. Orfeo's journey is representative of our collective yearning for re-union with our primordial self, the marriage of the confusing opposites into wholeness.

After these spiritual journeys that Orfeo endured, he became a gospel thumper and a martyr true to form, with mockery and disbelief, but several days later on Aug. 2nd, 1952 with a group of eight others, Orfeo witnessed a UFO in the skies that disappeared after a short time.
As earlier mentioned about Carl Jung, he never witnessed the Apollo Landing on the Moon, or the other various space developments and has a very hard time letting go of his deniability that any UFO or spacecraft could physically exist outside his version of the psychic phenomena. It is small-minded as much, to transfer or project impossibility onto something if it pushes out of our comfort zone. Obviously his work with archetypes is invaluable, understanding the unconscious desires we all have to aspire to wholeness, have a union, but as Freud once said, "Sometimes a cigar is just a cigar." I think this aptly applies as well that sometimes a UFO is a UFO! © 2012 Brian Wallenstein
Jung's Book may still be available on Amazon if Orfeo's story interests you.
It is in this book in its entirety. C.J., Jung (1969) Flying Saucers, Signet Press, New York

Part D: ANOMOLOUS SIGHTINGS

Every town has vortex points either real or urban legends. There are haunted houses, gravity or anti-gravity hills, that mysteriously defy the laws of physics, time and space warp portals, clouds that mysteriously appear and disappear, elemental nature entities and benign forces that just seem to occur out of nowhere. Although there are also other types of anomalies that are not so nice, places that reflect injustice and or injury that in reality only want and need to be thought about with love, release and forgiveness, we focus on the flip side, the miraculous and the unusual.

Occurred: Second World War
Location: Bataan / Corregidor
Witness: entire prison camp
Event: physical metamorphosis
Jas's Story (name changed)
Duration: 3 ½ years
This story needs to be told, as it is not just an anomaly but also an anomalous behavior that manifested out of need and compassion by spontaneously mutating an individuals body to accommodate a life in amidst the horrors of war.

There are incidents of spontaneous gender changes when a population is threatened by extinction. Fish, invertebrates and other species have been logged with such behaviors when one of the necessary genders is missing.

This particular situation is of a former resident who happened to be a medic in the army during the Second World War, captured and in a prison camp. Nightly he and several other G.I.'s would sneak over to what was referred to as the zero ward (where dead GI bodies were thrown, even ones who were sick and not dead) and take the still alive soldiers back to their ward and help them. Anyhow the Corregidor/ Bataan Isles occupied by the Japanese, never took prisoners but were forced to. Let alone the miraculous heroism, graciousness and awakening of these soldiers who from a logistical error, were left there in the winds of war in Europe and Hawaii, forgotten for 3.5 years our hero Jas actually with the intervention of a spiritual love for life, saved the day. These soldiers identity will be graciously protected but the incident has been recorded by the medical industry and printed in the Merck Manual decades ago. I know it is something that embarrassed them but states the potential of what can happen when our basic urge to preserve life in the midst of genocide are activated.

One day an infant showed up in the prison camp. It was hungry and there was no milk, no female let alone wet nurse available and the baby's fate seemed sealed to die. Apparently .the infants mother was killed. This individual who we will call Jas the medic, had so much compassion for this child he spontaneously developed active mammary glands replete with milk and fed this child until it could be transported to a village that was days and more likely weeks away by truck. The normal diet of a G.I. in this situation if they were lucky was a piece of moldy, maggot infested bread and water. Is this an anomaly or is it only viewed that way by our lack of understanding of our true potential? 3.5 years later they were finally rescued by the US forces that abandoned them and the island taken over by the US and allied forces. If you're a history buff look up the Bataan Death March. These guys against all odds were supposed to all die, but did not. When Jas told this story he added that, "war is the stupidest invention of all men and it does not matter which side your on, it's a real waste and stupid." He also went on to say that the Japanese treated their own soldiers almost as harshly as the treated the prisoners and there were a few more compassionate Japanese who also realized how stupid war was. He came home with beriberi, malaria, other challenges, emotional trauma and wisdom. He like many soldiers of that vintage who were 19-21 when they were drafted never complained or blamed anyone except the drunken addiction to power and dominance humans seem to need to rise above.

Perhaps these kinds of stories are along the lines of the frail old "grandmother" who finds her grandson pinned under a truck that fell on him while working on it in the driveway, effortlessly picks the vehicle off him and saves his life. Perhaps it is not that far fetched that we have alien kin visiting our planet and us considering we are capable of such amazing feats as it is. The power of care is inherent within us and along with that our own stupidity that covers up this unlived potential. The lesson: wake up; get out of your own way.

Author's Note: Our Hero of this story died on a Thanksgiving Day, many years ago. His daughter had the privilege to spend the last hours of his life with him and recalls that right before he passed on, the house literally shook like a military parade was outside. Both she and he peered out the window from his bed and they actually could see the shadows of his entire platoon, waiting to take him home. He said he could see them all, where all she saw were energy pattern outlines. The Lead man saluted to house, they presented arms, rolled out a red carpet, and while the daughter was peering out distracted, the platoon called to him, secretly and she saw him in full dress joining them in honor. He looked in the window to her, and she could see him back in his vibrant youth smiling in full salute. He died with the greatest honor any person could. Bless you Sir.

Occurred: April 11, 1987; 0430hrs. (Author's sighting)

Location: Mt. Shasta, CA

Witness-1
Object: Elemental entity
Duration:

We were living on North Mount Shasta Blvd. at the time this occurrence commenced. This house built back in the 1940's was a block and cement house, with mostly concrete floors, designed to literally float on the bog they were built on. During the spring thaw the waters from the mountain fill up and saturate the ground, lava tubes, and bring the standing water level to 4-16 inches below the surface making it a marshy mess. These standing waters also carry current, both electrical and hydrostatic and are very active in the spring. No one would know about these phenomena unless they were familiar with physics, but more so Indian mythos and natural elemental spirits. Much folklore has been written about black waters, about simple entities that exist, store motion, store what occurred about them and appear to be apparitions from the standpoint of human understanding. Current science in fact has found that the crystalline nature of water (H_2O) and its plasticity behaves like quartz and stores memories, impressions and holds a tremendous amount of data, much more than memory chips. It also picks up and is sensitive to impressions around it, such as people's thoughts. Water as a rarified current from the standing water state, then in its actions begins to broadcast what it has held in its static state and releases it to what is around it. With this theory in mind, the following experience is of an individual who was affected by the rapid release of black waters from the bog at the base of Mount Shasta in town.

I woke up at 4:30 in the morning unable to breathe, move, talk or function. I was in a kind of what is known as a "Sleep Paralysis" but was choking unable to breathe. My wife to my right in the bed, 4 months pregnant was sound asleep and there was no way for me to nudge, scream, whisper or anything to wake her and alert her to help me. I managed to begin to meditate and with great strides of will, was able to begin to take short, tiny breaths with counting and total focus that seemed like an eternity. My hands had begun to turn blue and with each passing breath, I was able to take more in until I began to breathe normally (approx. 15 min, clock on nightstand) and was able then to get a word out and woke up my spouse. When she awoke I told her the accompanying lucid dreamlike imagery that accompanied this, which was an entity that was coming out of the ground, enveloping the house, and telling me it was taking everything in its flow, including me and it was going to win. Seems it did not, but as I sat up on the edge of the bed, and during this episode there was a strong wind emanating from the floor, (cement) I commanded the entity to leave and it literally blew stronger to the point it puffed the ash in the closed wood stove out the cracks of the door and spun the hanging mobile above it with such velocity it spun fast. I was concerned as we were pregnant and did not like the idea of having our child exposed to this kind of energy. Even though the entity was not evil in its intent but more of an elemental dense nature, we had to find a remedy. The next day I asked several friends about this and it seemed they both knew of the nature of elementals and suggested I water witch the current around the house and use copper rods to change the flow and ebb of the water/electrical current the elemental was following and divert it. I did this and found that yes in fact the pattern and flow of the water current (you could see just by digging and water moving) and I planted 9 rods which moved the current and had no more problems, and it diverted the wind out of the place as well. Very awakening to the life force and how it manifests in different mediums. (Picture shows water now on surface. (Photo below))

Path of black water between houses that accommodated elemental entity, used to be underground. House on right was one affected. (For above sighting) (BDW)

Occurred: October 30, 1997; 2130hrs. (Author and son's sighting)
Location: Mount Shasta, CA Sisson Hatchery
Witness: 2
Object: Time space shift
Duration: 45 seconds

We were driving to Ray's Grocery Store south on Old Stage Road at 9:30 in the evening on a very windy late cold October night. The wind picked up dramatically when we entered the four-way stop sign at Sisson and Old Stage Road. I slowed down, as there seemed to be objects the wind was carrying from the west to the east across the road. They were solid, some like sagebrush, some looked like large flying birds or pterodactyls. It was very spooky. About 50 feet from the stop sign, we both saw something the size of a good size panel truck racing across the road down the slope of the meadow and upon our closer look, it looked like a mastodon. We looked at each other in total disbelief knowing that we in fact, did see it. We both blurted out, "Did you see that, it was a Mastodon." We were in a space-time shift and it was translucent and plasma-like. The energy around us was viscous, as we sat at the stop sign, watching it flow across the road towards the mountain. We could see the outlines of the dire wolf and saber tooth tigers and hear the confusion, howling and growls through the tiny opening in the window. I took off as fast as I could to be sure we did not get stuck in this vortex, but we did see the entirety of the mastodon, its side and its hind going down the hill. The flow of it was obvious and coming from the west from behind the Hatchery. There must be some origin to it to the west, behind the hatchery. Maybe this is part of what happens around unstable magnetic areas, like volcanic areas as Mount Shasta is. It lasted a good 45 seconds. This anomaly continues down Old Stage from around the Historic Berryvale Post Office down to Cantara Loop, which contain many old sites from prior towns with curious history.

4 way stop where procession crossed road, from hatchery (right) down hill (left) (BDW)

Occurred: All year, all the time
Location: Mount Shasta, CA
Object: magnetic anomaly
Part of Everett Memorial Highway and Summit Drive going south on it, by Buddhist Monastery is on a magnetic anomaly. Being a bicycle rider, one is privy to the fit and finish of particular roads in their own region. Both these roads have a gravitational flux that allows a bicycle rider to ride uphill, (not optical illusion) and be pulled for close to a mile. GRAVITY HILLS
There are two magnetic anomalies I found up here. One anomaly is at the base of Black Butte, on the north side of Summit Road, which the Buddhist Monastery is situated upon and the other one, on Everett Memorial Highway, before McBride Springs. Many studies have been done in other places to the optical illusion of the property or being pulled up a hill on your bike or car, and prove that the road in fact is sloping downward when it looks like its is sloping up but both these roads are obvious upward inclines. Get your bicycle and ride it in these places and experience this anomaly.

   The magnetic influence on Summit Road is a larger quadrant and continues west to Old Stage Road, where pedaling downhill half way from graffiti bridge to Deetz Road, a half mile stretch is going down hill and you have to petal like your going up an incline. Be sure to have a speedometer and you will see for yourself these opposing forces at work, on this ride.

* MAP SECTION CHAPTER *

Visual aids significantly improve our understanding and position especially with the subject to UFO study. This has been the most time consuming and difficult task of the entire chronology but worth the effort in helping us all get our bearings straight on what it might look like from the air. I have to thank Google Maps, Google Earth, NOAA, USFS, and The Dept. of Interior for their basic layouts. However, I had to innovate, use, manipulate and generally fashion something that we could convey the flight patterns to you, as well as give you a general idea of where the cites are located in California and Oregon. I am confident that in another six months, I could have flight paths for every UFO craft that has flown in California for the last 60 years laid out on hundreds of trajectory maps, but with the time constraints and limited resources, what has been created will suffice. I am proud of what has been done and you will too be grateful as these maps are tailored in general to the location of the specific sightings chosen for the book. They are located in a convenient place between the local sightings and the regional sightings for your pleasure. Please take note that the connected points do not indicate single trips, or that the UFOs only touched the points once. These represent, in some cases, many visits, and the flight paths do not imply that they traveled that route specifically. They may have taken many routes, but all roads lead to Mount Shasta ☺

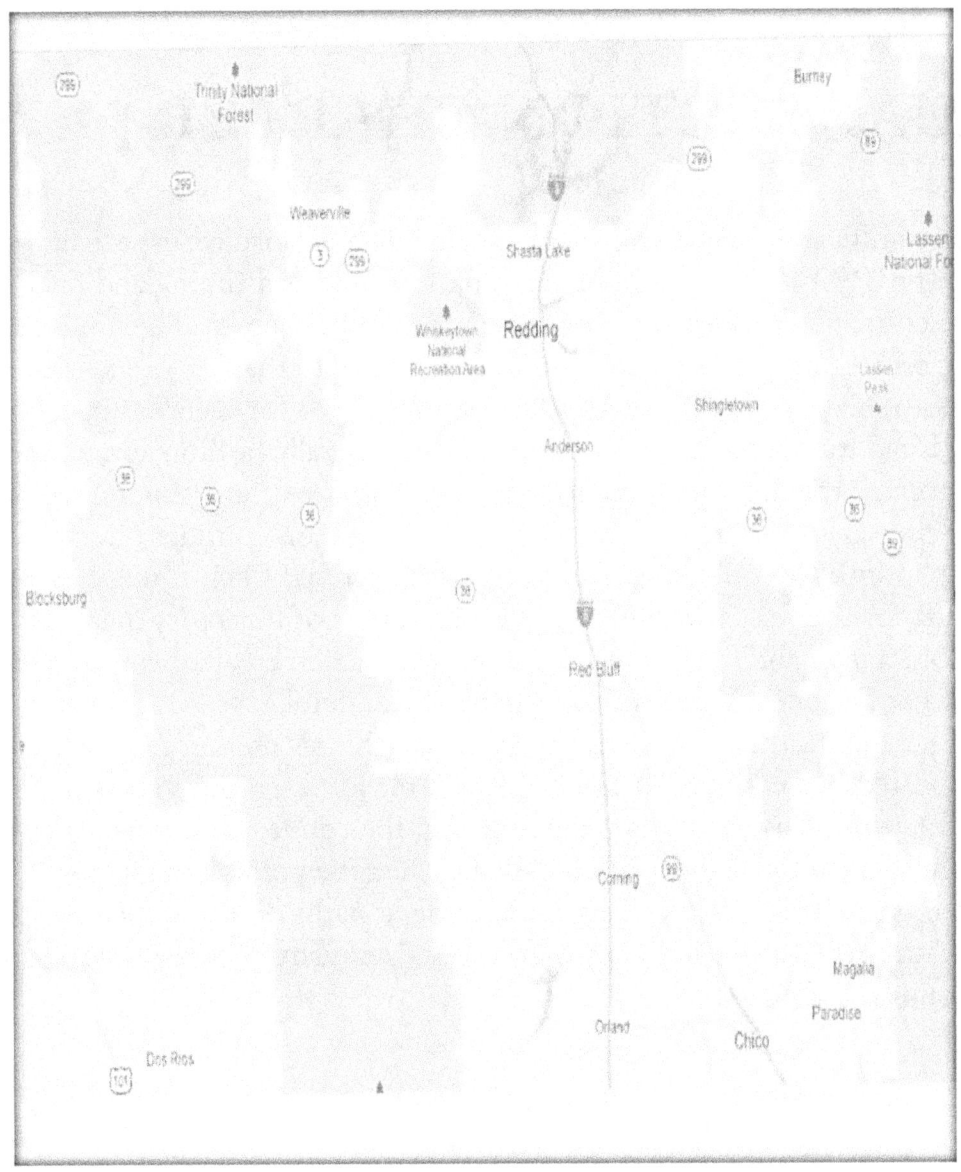

Red Bluff, (cattle mutilations, crop circles) up to Redding, and east to Burney connects back around to McCloud north, on south side of Mount Shasta. *

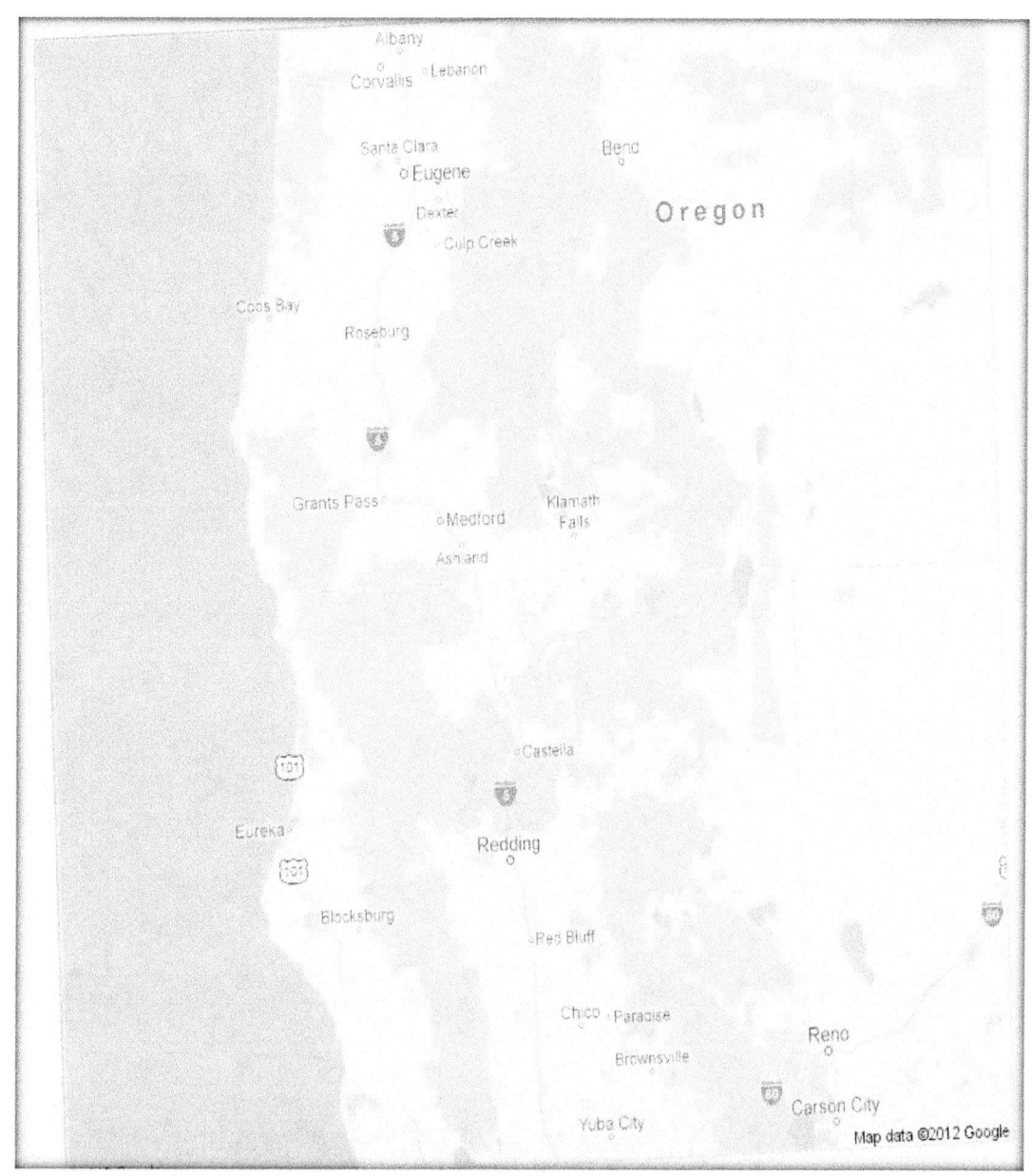

Flowing into Oregon from Flight patterns gives general idea what area looks like.

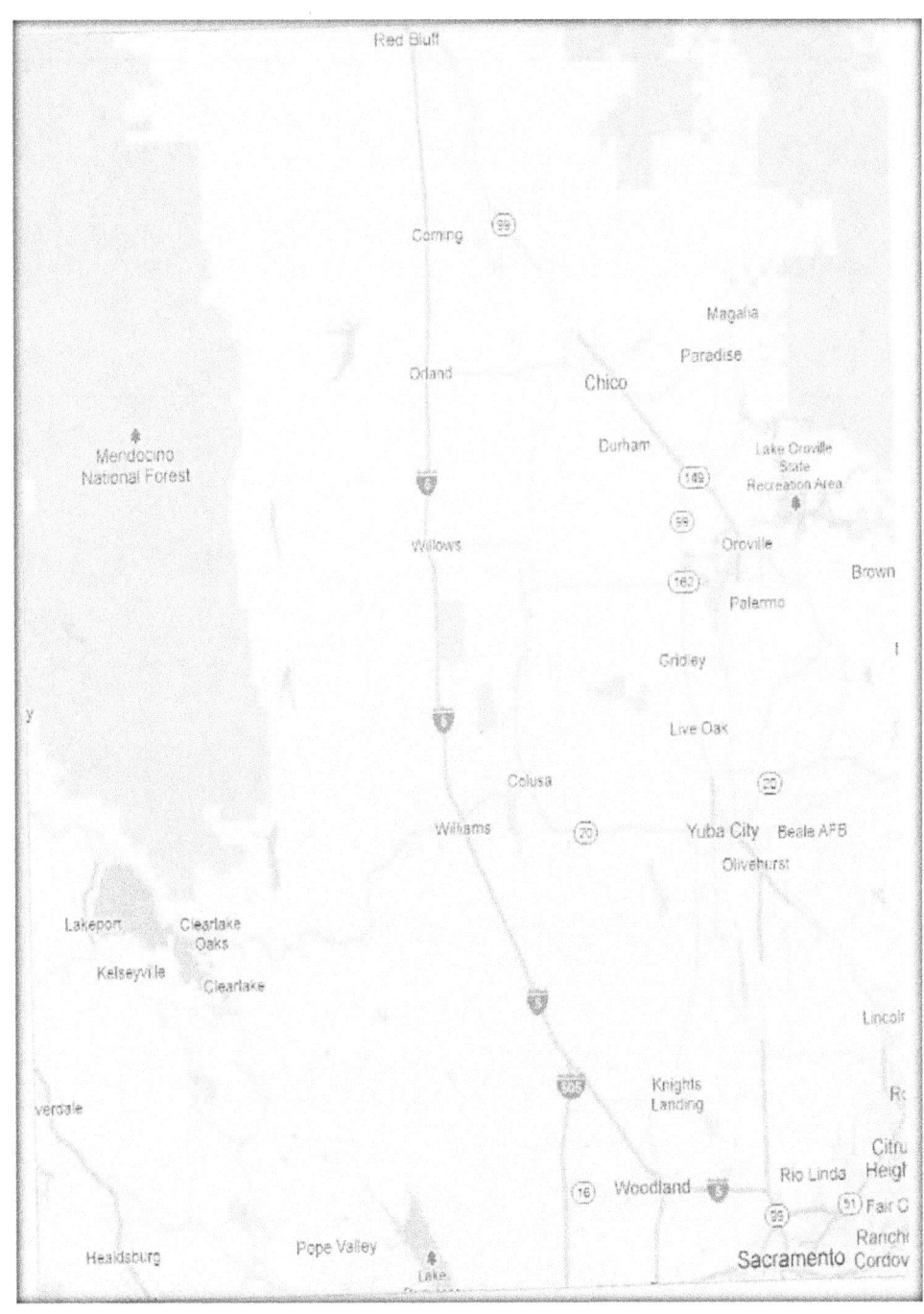

Through Williams, Willows, Orland, Corning, Red Bluff, and Chico/Redding. General travel pattern of UFOs up I-5 Corridor. *

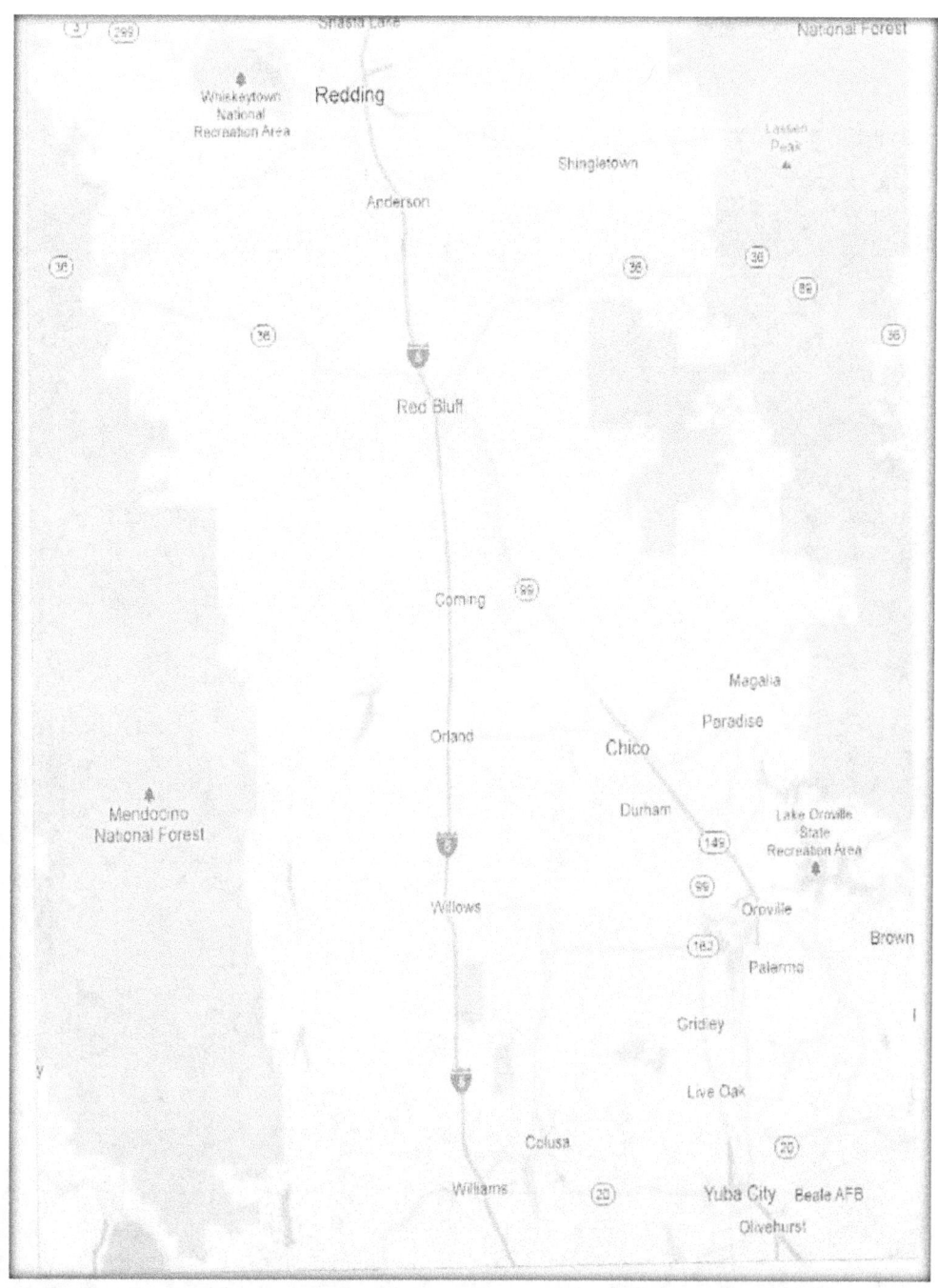

Chico is a hot spot and shows relationship to Red Bluff, Corning and Redding. (Note): Chico, Marysville, Oroville and Redding were all military bases during WW2 that also made them places of interest to the curious ETs.

Palo Cedro is east of Redding and further to Millville, has sightings. *

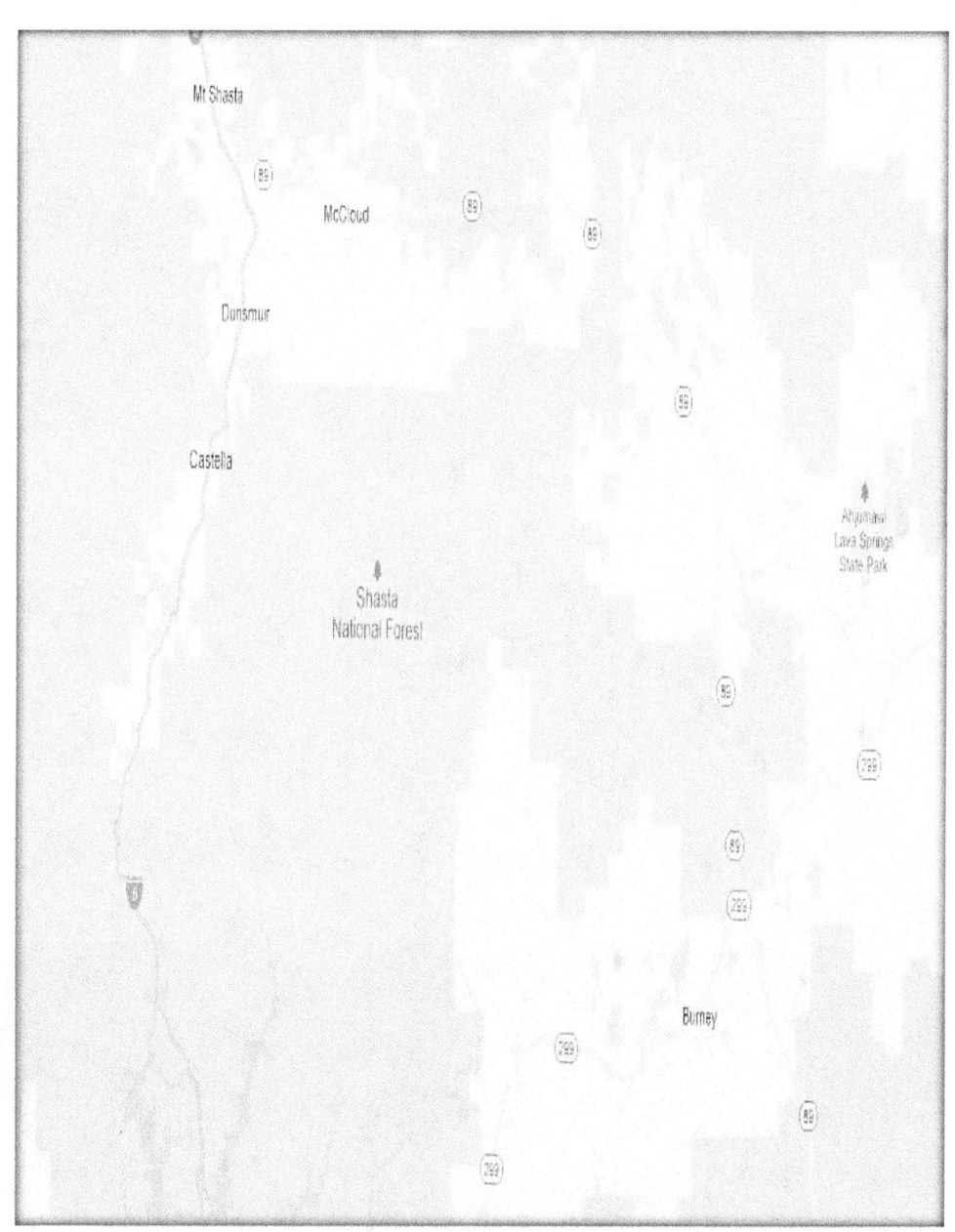

Burney, Castella, in relation to McCloud to Mt. Shasta and Dunsmuir. Chester and Susanville are east of Redding about 100 miles. *

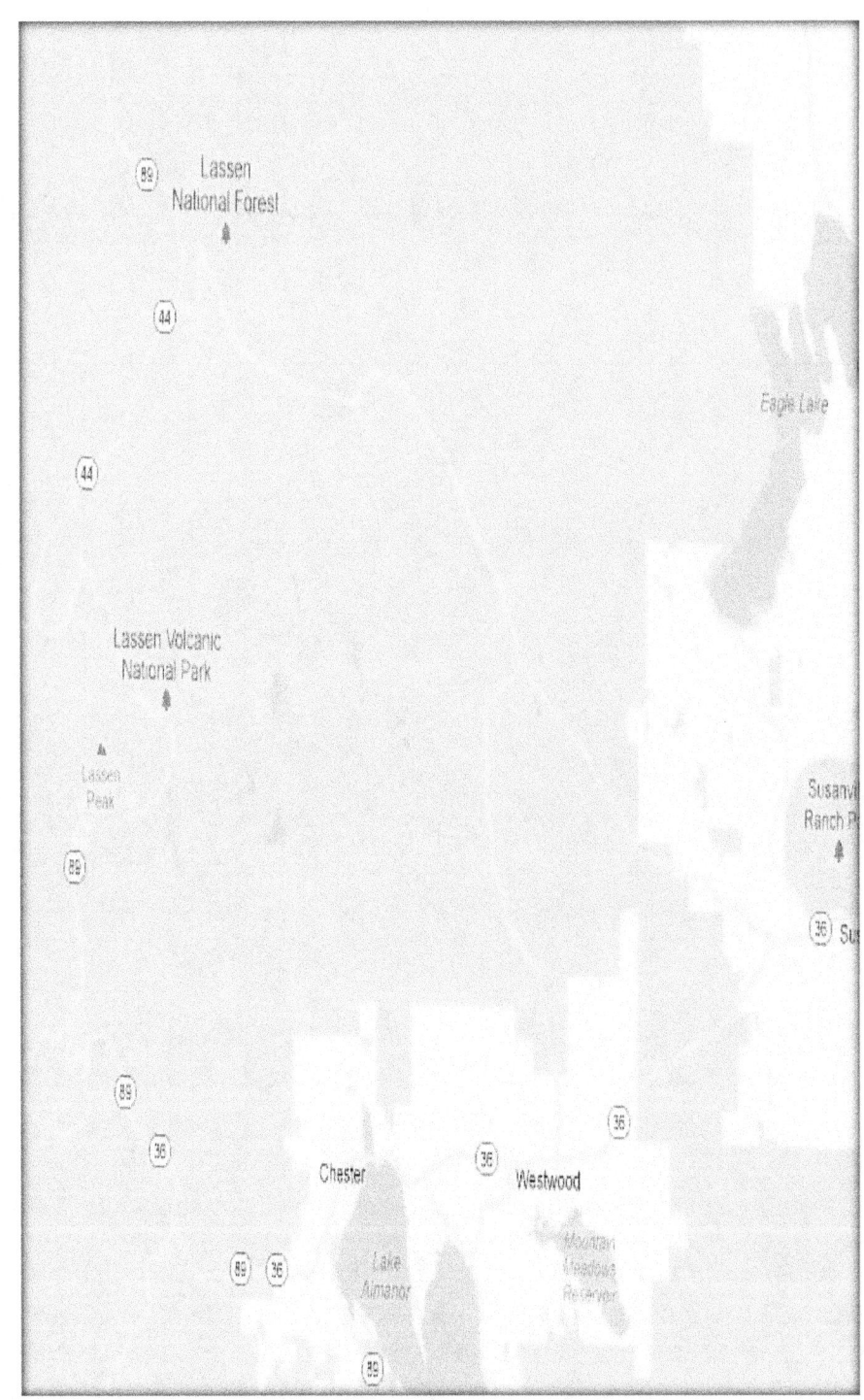

Location of Susanville to Chester and Lake Almanor. *

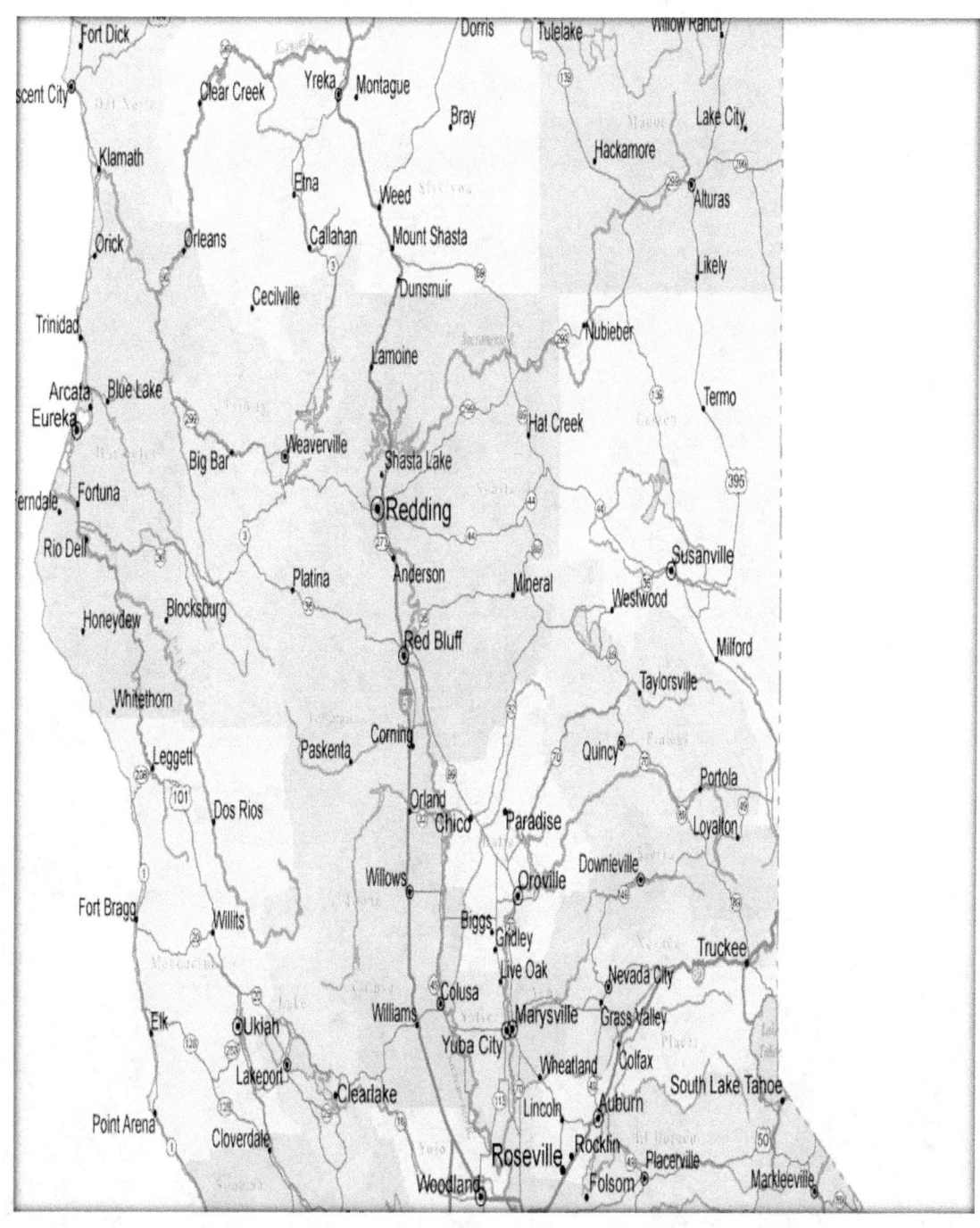

(map unknown)
Note position of Susanville to Mineral, McCloud, Milford, Hat Creek to Redding, Red Bluff, Castella, Mt. Shasta and Corning.

Kelseyville in relation to Clear Lake, another sighting location.

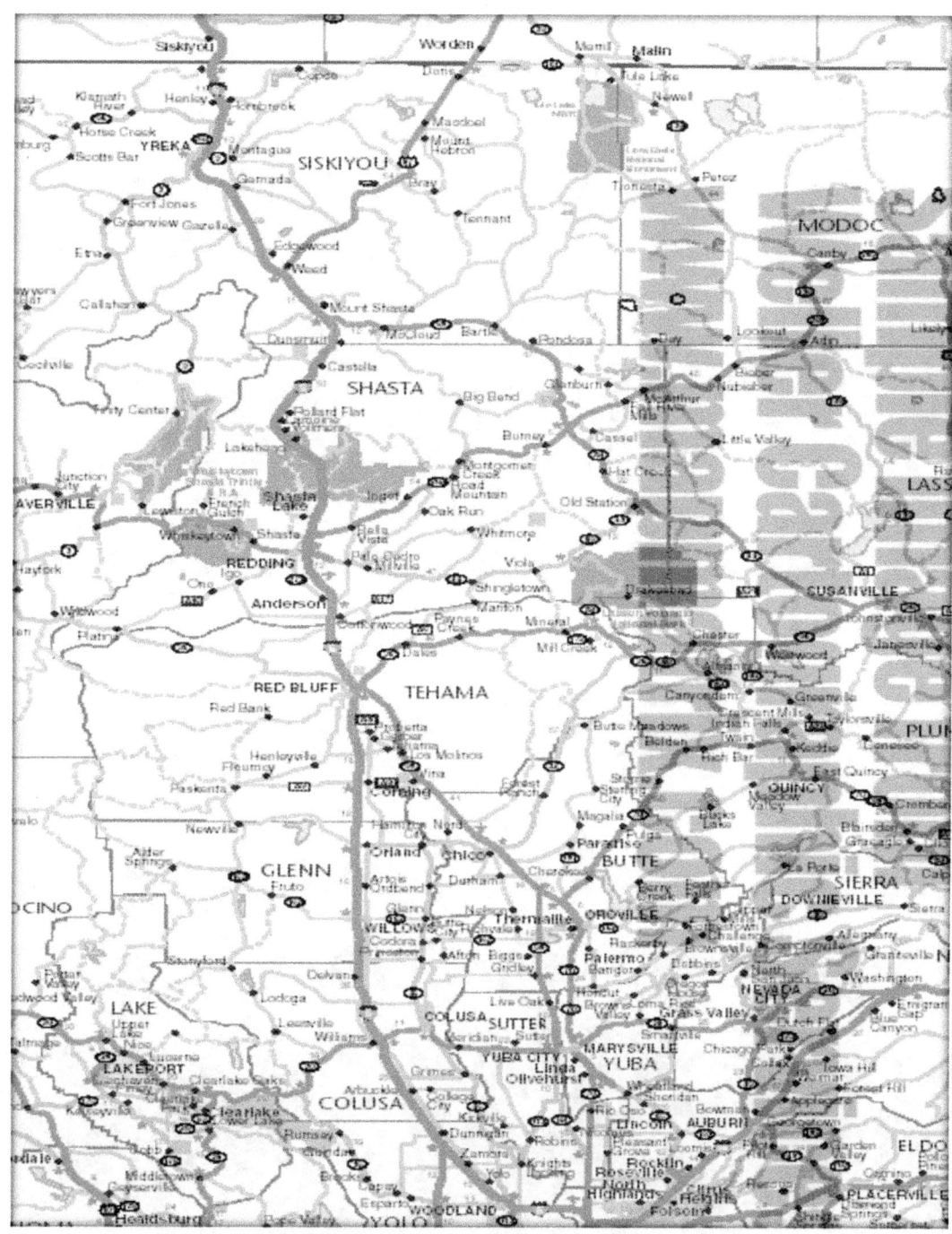

Mill Creek, Palo Cedro, Burney, Big Bend (mapmatrix.com)

Castle Crags view of east side. *

View up Castle Crags from Railroad Park Ave., Dunsmuir Ca. *

Shasta Lake in relationship Redding, Palo Cedro and Millville. *

Happy Camp in relationship to Yreka. *

Herd Peak lookout on Miller Mountain, facing Goosenest. *

Looking north at South side of Mount Shasta. *

Girard Ridge Lookout with Castle Crags and Mt. Eddy behind. *

Burney in Relation to McCloud, Redding, Castella and Mt. Shasta. *

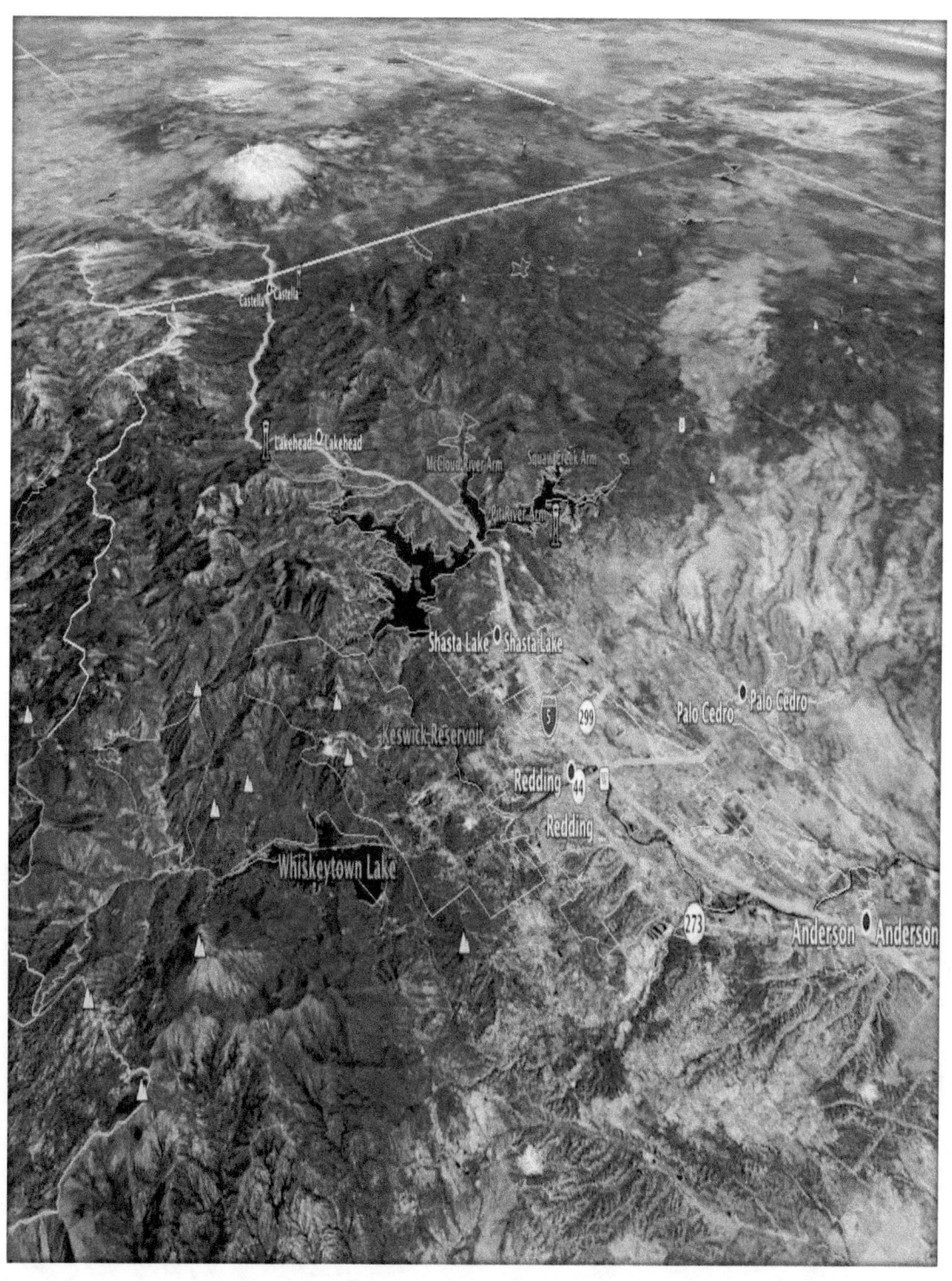

Shasta Lake, Lakehead, Redding, Whiskytown, Palo Cedro and Anderson. *

Hat Creek in relation to McCloud, Burney and Mt. Shasta. *

Topo showing placement of respective cities in Siskiyou County. *

Chester (lower Right under Google) to Anderson. *

Mount Shasta with many identifiers for the enthusiast.

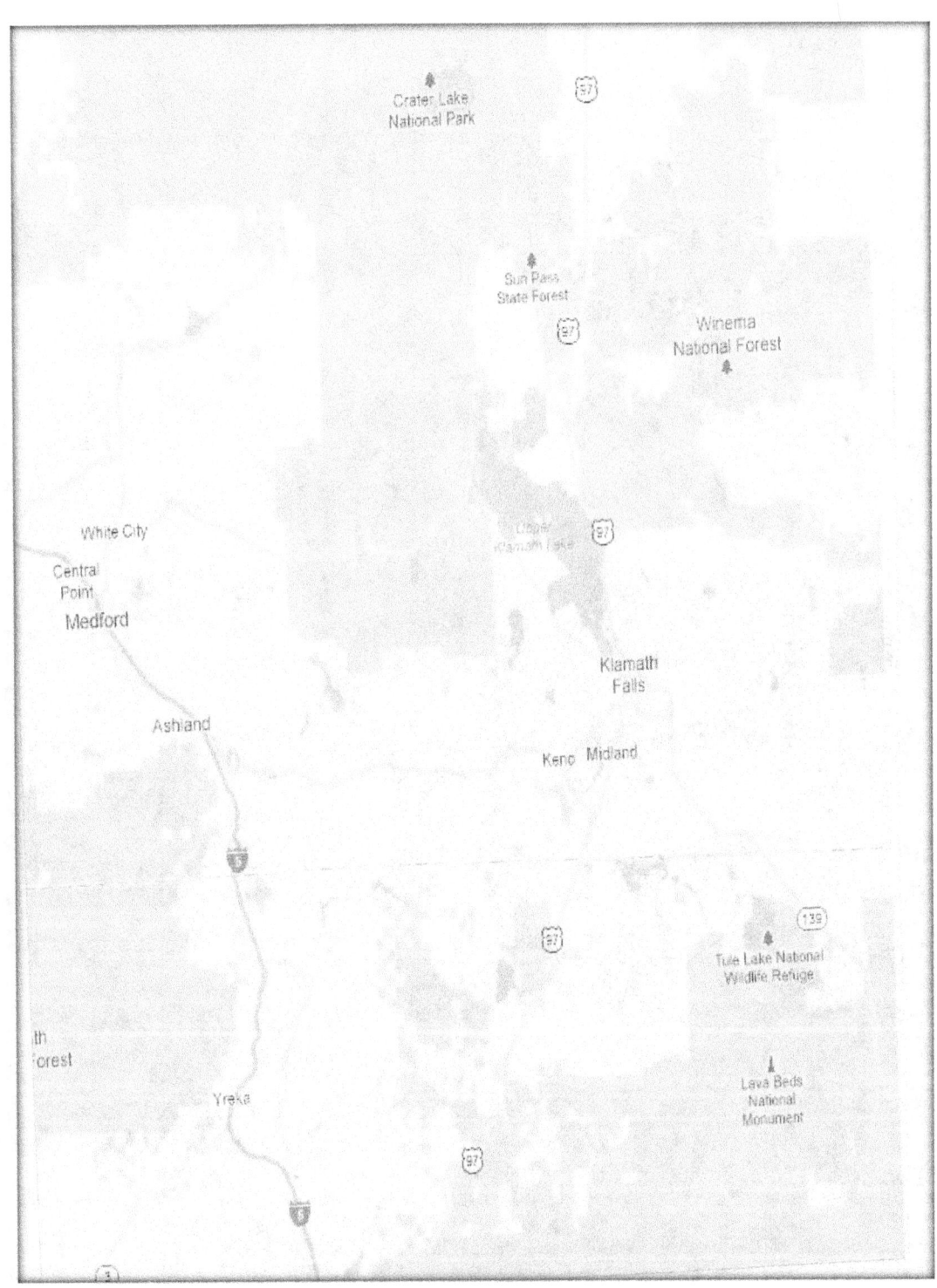

Oregon and California border shows Ashland, Klamath Falls in relation to Yreka and Tulelake. *

McMinnville was a huge historical sighting (In main chronology) that put UFOs on the radar as credible.

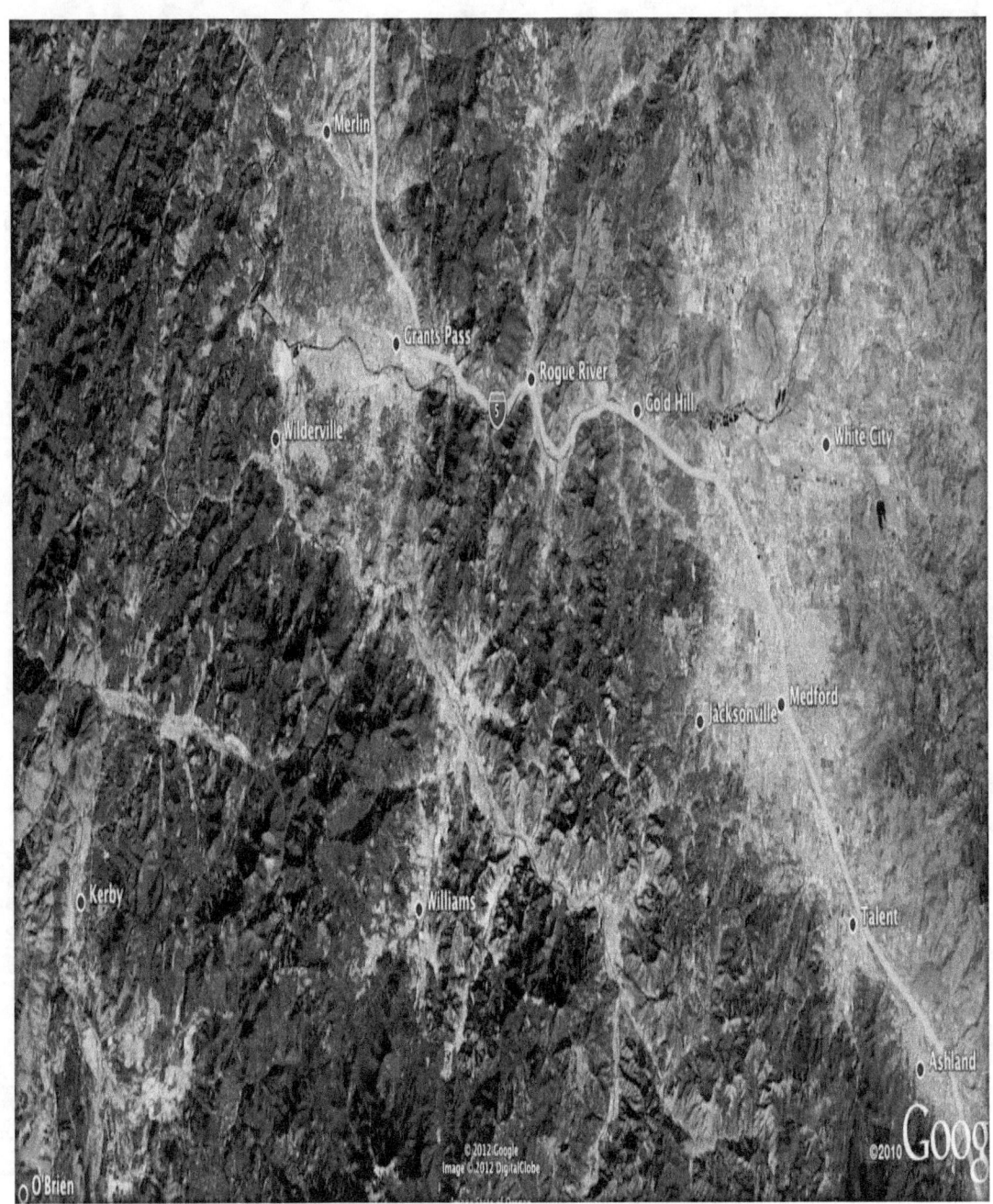

Southern Oregon Topo. *

I-5 Corridor through Oregon with cities visited by UFOs. Oregon Road Map
(origin:unknown)

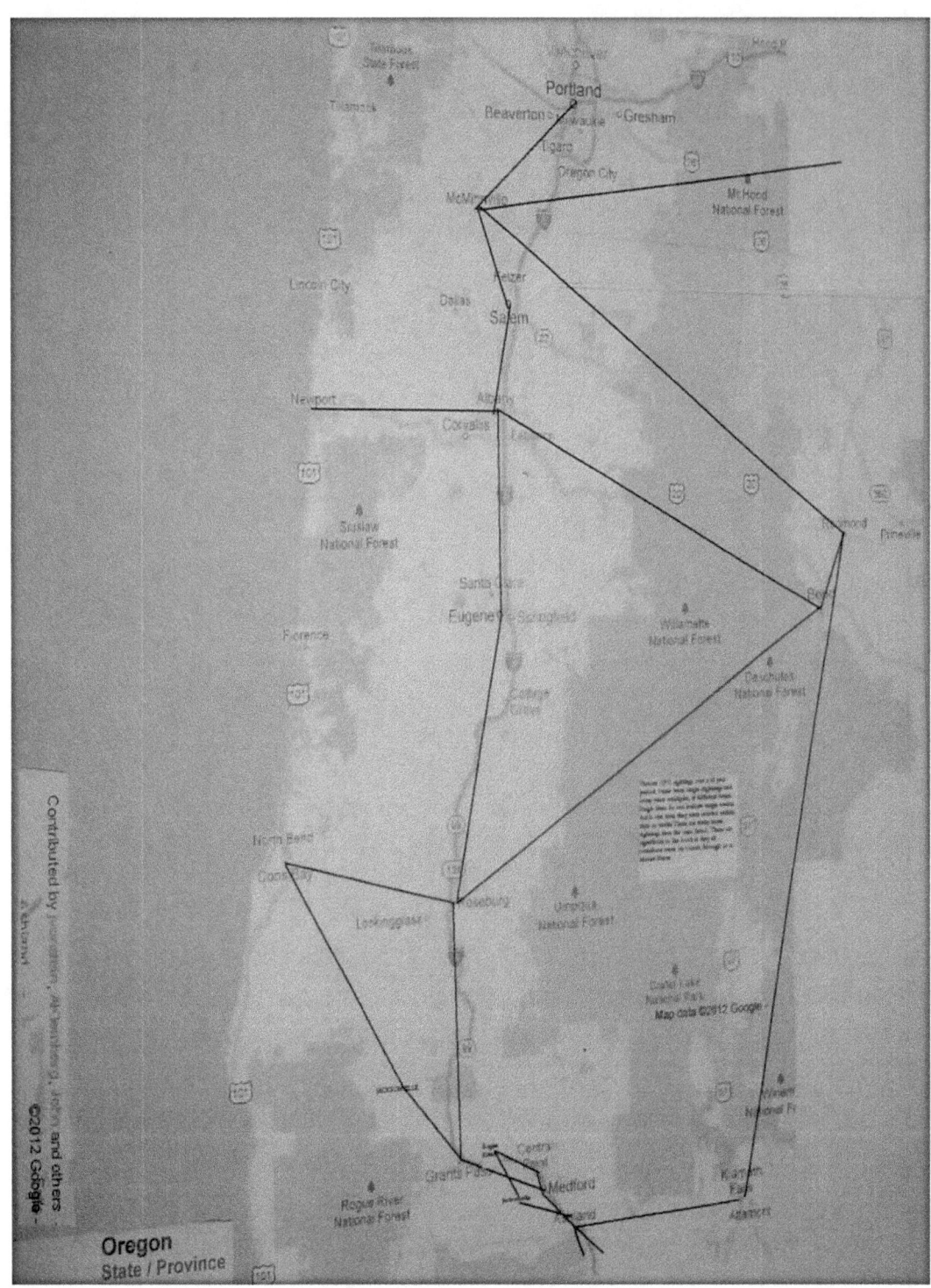

Connected points of UFO sightings in Oregon (Ones relevant to book)

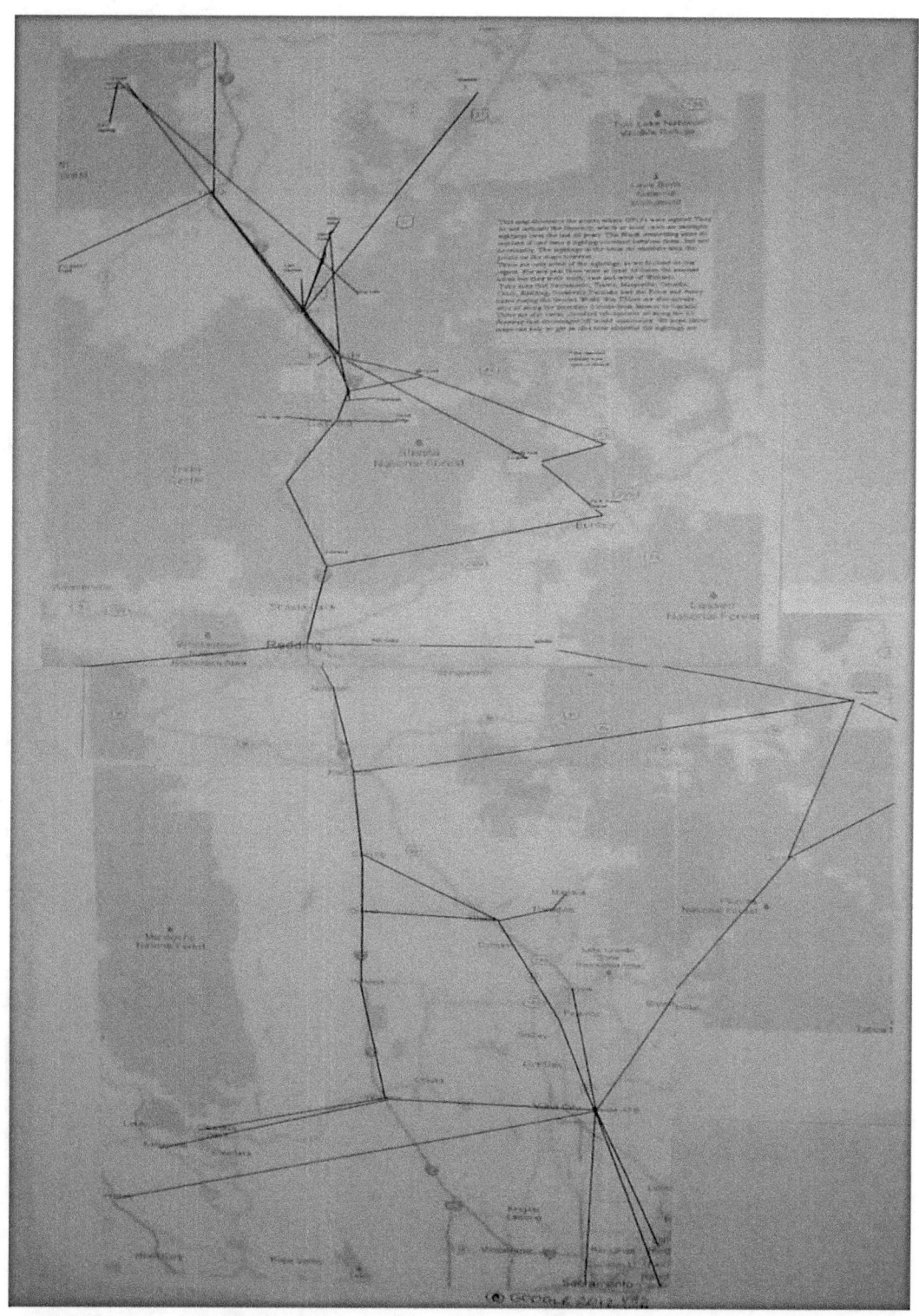

Connected points of sightings occurring over 60 years. Most are multiple sightings and many follow these lines. There are significantly more, but these are relevant to the book.

Template for many sightings in Siskiyou County. *

Siskiyou and Shasta County with their relation to Mount Shasta. *

Southern Oregon Topo * (above) cities frequented by UFOs

Bend, Roseburg and Redmond, OR; all sites of UFO activity. *

Roseburg Oregon is another busy UFO terminal.*

Abandoned Fire Lookout on Grizzly Peak by (douthink) tagged points.jpg

CHAPTER FIVE
Introduction

Chapter five is divided into 10 Sections that span from ancient times to 2012. The inexhaustible amount of research conducted using historic and religious records, graphically reveals the existence of UFOs throughout our history. This lexicon includes cave drawings in South America, the American Southwest, Australia, the Middle and Far East and biblical paintings. Chronicles are included from Benjamin Franklin and George Washington while Thomas Jefferson went on record to disclose his findings.

Ancient Sumerian cuneiform text, cultural and historical notes left in all forms of society, were unearthed in 19th the early 20th centuries and is ubiquitous in the Mayan Culture and the Egyptian culture along with direct connotations to the ETs being responsible for influencing, teaching and helping shape the future of man. Ancient Sumerian, i.e., Mesopotamia, is considered the oldest and original cradle of civilization, predating all other civilizations.

All others borrowed these traditions, beliefs and the records unearthed reveal the most profound origins of humanity in their written history on all fronts. Located in upper Iraq, most of the legends about it have been proven to be accurate and fact. Translated script has been painstakingly cross-referenced with other historical documents from all over the globe.

This is our first sighting and contact with extra-terrestrial life. It is beyond reasonable doubt the authenticity of this history. The ziggurats that held the craft, the advanced mathematics, maps, views of the planet, the story and cosmology of earth, and other planets, the advanced descriptions of every facet of life including carvings, pictures of spacecraft, direct references to these flying machines and the humans who have ridden with them to the techniques they used to do genetic splicing and much more is explained in these ancient texts.

All this exists in great detail in these ancient writings, on stone and ceramic tablets and brass cylinder rollers. The remnants left over from the history, all match exactly. The Annunaki originally landed on earth 450,000 years ago and perhaps even longer than that.

There was also new evidence found in Puma Punko that Virachoco (the one from space) landed 900,000 years ago and helped ready the earth for life. Not only was a royal bowl with cuneiform (Sumerian) writing on it found a quarter mile from this site, the central figure in the middle of this city near Puma Punko, is a Sumerian figure. He looks nothing like the local inhabitants. Zecharia Sitchin brings up that one of the founding brothers, Enki had established a base in South America, and the evidence is being found. To add to the mystery, Puma Punko was an outpost built overnight by the Off-Worlders entirely. No humans were involved or a collaboration of humans and aliens.

Zachariah Sitchin, who recently died, wrote the 12th Planet and the series of books that explain this history in exacting detail with pictures and references. He has a reputation of such high standing, that his work is heralded globally and used by academia and governments alike. This series is a testament and translation about the Sumerian civilization and its history, which is our history.

Sitchin closes his writing of 12 books with End of Days, had translated many of the Sumerian 30,000+ tablets with more still untouched text, discovered that in fact the Anunaki which means, those that from heaven came, the Planet Nibiru which in its elliptical orbit, comes through our solar system every 3,600 years (one of their solar years according to the detailed text and charts), states that the Anunaki landed here to mine gold (not plunder) needed it to revitalize their depleted atmosphere from their own pollution and did not have the manpower to do it all.

Their crew tired and understaffed was ineffectual after 140,000 years of personally mining the gold and planned a mutiny against the leaders if they did not get them the needed help.

There are accurate pictures of the solar system taken from the viewpoint of outside in, accounting for all the planets (before they were discovered) and more. The book contains accurate star maps, which were on these tablets with accurate timetables.

They decided after a great meeting among them, to use their DNA and fuse it with the native peoples to augment them so that they would have an efficient work force intelligent enough to follow directions, since at the time (allegedly 56,000 years or so ago) the native earth man (homo erectus) was not evolved enough or intelligent enough to be of any use. However, both the Anunaki and the earthmen being of the same seed, the DNA would match. So, through both artificial impregnation (the Adam and Eve story) and genetic cloning, a new worker race was created. How fast, it did not say, but it was evolution on the fast track. More and more of this knowledge, if looked at with new eyes in the biblical sense, shows that man perhaps mistook aliens as gods, and yet still reported the stories with some literal accuracy.

The Sumerian Text contains the Epic of Gilgamesh, which is the story of Genesis used in the Bible written thousands of years earlier. The Anunaki were also responsible for educating men in the art of farming, science, language, writing and culture, education, and the first actual high civilization according to the text. Sitchin's new book depicts the return of the Anunaki in our lifetime and depicts most of this history. The point is, what and who created life in the universe? A singe godlike advanced race, a few of them or does all life including our own, possess a single source of being-ness, which is the consciousness of existence? It seems as self-evident as the presence of atoms throughout the universe, and is a source inherent and original within us.

No culture as of yet discovered or claims to have been able to fashion the genes that give attributes to life forms that reflect intelligence regardless if they, or we, can mix and match components to make or remake organisms. It seems we can mix genes, but not create this template or understand the magnetic or atomic tension that gives rise to the momentum, action of life. The interesting aspect about the study of the Sumerian Text about the Anunaki is that one gets the impression that even though they are an advanced ancient race with technology in all matters of science, astronomy, physics, humanities, medicine and ethics, they don't seem very spiritually enlightened. There is not enough data on that, but their preoccupation with being worshipped, satiated with materials, sexually occupied and being overlords, they are competitive, jealous, cocky, a lot like us, and full of themselves.

It seems that no matter how smart a culture can be, if they are short on compassion, or forget how sacred life is, even if one is immortal, what good or of what use are they? It occurs to me that every race that deems themselves superiors, still have a lot to learn about crafting successful civilizations without suffering, maximizing higher emotion with higher intellect to not be cruel, controlling and tyrannical, or lording over slaves.

There are signs all over the planet, the pyramids in Russia, remote battle sentries in Siberia that vaporizes both asteroids and certain UFO craft, (documented and researched a few times in the 20th century) and included in the following sighting section.

It is in the scope of this book to illustrate the amount of sightings that occurred in the Northern California and Southern Oregon Region over the last 60 years. It is in our interest to continue to expand our search to help confirm, and make comfortable all of us to the possibility and fact that we are not alone in this universe, and that we may have allies among the challenges we face. There's innumerable references within every culture that has lived on earth to the Star beings, UFOs and visitations. Petroglyphs left by many cultures such as the Zuni, the Hopi and the Iroquois, all plainly refer to our space brothers who came here in shiny discs out of the skies, and still communicate inter-dimensionally with us. There is a story of an Indian woman falling in love with a star man, who gets her pregnant with star children. It is a common theme and can even be seen in Judeo Christian context with the Immaculate Conception of Mary.

Both people and animals have noticed and noted the presence of UFOs as long as there has been conscious life. Science and scientists conveniently dismiss or call it an outlier, an anomaly if a variable does not fit into their sphere of focus, only until there is enough clamoring for them to let go and observe. If we treated valuable metals and jewels as scientists treat briefly occurring events, and start paying attention, we would not have metals such as gold, platinum, diamonds, and the like. It takes tenacity as well a diffused and wide focus to catch the swiftly flying birds in the periphery of our consciousness that yield the most rewarding findings. Some things don't just bang us or land on our heads to be observed. We must be on the lookout for the unusual, but once we get accustomed to it, then we start to recognize these events more swiftly. That is how we evolve.

What was once invention is now convention. It's not what we're used to, but what we are not accustomed to be which eludes us, but has always been there. Take for example when you decide to buy a new car. You choose the model and then all of a sudden you see a million of them on the road, like they just appeared out of thin air, but it was only until your attention to it was attenuated did you see them and they were always there. Same thing with reviewing history, all of a sudden there has been a mass explosion of reviews of history, archeology and the presence of UFOs is tantalizing, but it was always there. We just were not ready to recognize it. It may seem unnerving and frightening to think our history could include such 'out of this world events' but we know so little about our origins, and the comfort of hiding in possibly inaccurate and misunderstood ideas can be destructive and counter productive. 'Seek the Truth, and the truth will set you free."

At the very least, will give us all a firm foundation of knowledge to base any decisions we make on. If we base our lives on errors then what we reap is unauthentic and thus unreal. This has nothing to do with connecting to a universal source from within but to be able to understand the motivations of any cultures of peoples be they terrestrial or extra-terrestrial and give us a handle on how to go about interacting with all our neighbors, with an open heart, and a clear mind. If, in fact, the concepts we offer are true, then many questions, debates and conflicts scientifically and religiously could be answered.

If we ask ourselves why we are resistant against such notions of our beginnings include extraterrestrials, we may find our reason is not based on logic, but of cultural and religious bias. Maybe deliberate programming and conditioning against it by the cunning use of fear and ridicule at its heart, planted there like a seed by those who do not want us to look any further and find out the facts. It is nice to imagine there are other options beside the Religious, Darwinian, alien interventions or the like. It may be all of them, but we deserve to have all the accurate data that may still be undisclosed.

SECTION 1: NOTES

This first section covers Ancient Sightings up to 1900 using cases files of Mufon, Nicap, Von Daniken and many others. NOTE: all materials are referenced, sources, and cited "in text" with hyperlinks to the originals for your convenience throughout this book. Underlined titles are all hyperlinks www.nicap.org unless otherwise indicated. These links will work if your copy of the book is electronic; otherwise you have the addresses for the full reports on many of these condensed synopses.

You will notice in the following classifications, such as the Hynek or McDonald classifications and degrees there of. We left them so you could see how they prepared their reports. These gentlemen were instrumental researchers the U.S. government hired to run these reports and reporting criteria.

Hynek originally started out disbelieving in UFOs altogether, then realized after trying to debunk them, that they in fact are REAL. He became an advocate of proper research methodology as Dr. McDonald did. These men, along with Major Keyhoe and Chief A. F. Ruppelt, ran NICAP, BLUE BOOK, and APRO and were associated with high governmental offices. You will also notice at the end of these sightings parentheses with initials. They were the persons who took the reports such as: PD= Peter Davenport, FR= Fran Ridge, BB=Blue Book Ted Bloecher=Bloecher

We most graciously THANK RUBEN URIARTE, President of the West Coast MUFON for their allowance and permission to use their sightings, NICAP; NUFORC and PETER DAVENPORT for his allowances for use of their sightings as well. Many other landmark historical sightings are freedom of information or Fair Use 107. We also have given both MUFON and NUFORC all our sightings for their database, as well.

SECTION 1: ANCIENT UFO CASES

329 BC: Alexander the Great records two great 'flying shields'

Date: 329 BC

Location: Central Asia,

Mosaic of Alexander the Great

Alexander the Great records two great silver shields, spitting fire around the rims in the sky that dived repeatedly at his army as they were attempting a river crossing. The action so panicked his elephants, horses, and men they had to abandon the river crossing until the following day. UFO researcher Bruno Mancusi, and Macedonian historian Aleksander Donski discovered these references:

1. Middle East (Reign of Alexander the Great_, 356-323 B.C.) A historian of the reign of Alexander the Great allegedly tells of two strange craft that dived repeatedly at his army, until the war elephants, the men, and all the horses panicked and refused to cross the river where the incident occurred... The historian describes the objects as "great shining silvery shields, spitting fire around the rims... things that came from the skies and returned to the skies." Frank Edwards: 'Stranger Than Science' (Pan Books, London), p. 198.

"Frank Edwards, the noted American UFO reporter, quoting some source unfortunately not disclosed, states 'Intelligent beings from outer Space may already be looking us over.' He exasperates us by claiming 'Alexander the Great was not the first to see them nor was he the first to find them troublesome. What did the things look like? His historian describes them as great shining silvery shields, spitting fire around the rims... things that came from the skies and returned to the skies.'119.

Alexander the Great Siege of Tyre Sighting
Occurred: 332 B.C. Sources: multiple

Location: The Siege of TYRE Macedonia by Alexander the Great

Object: Shiny Discs- 5

Occurred during the Siege of Tyre by Alexander in 332 BC. Quoting Giovanni Gustavo Droysens Storia di Alessandro il Grande, the erudite Italian Alberto Fenoglio, writes in CLYPEUS Anno 111, No 2, a startling revelation which we now translate

'The fortress would not yield; its walls were fifty feet high and constructed so solidly that no siege-engine was able to damage it. The Tyrians disposed of the greatest technicians and builders of war-machines of the time and they intercepted in the air the incendiary arrows and projectiles hurled by the catapults on the city.

One day suddenly there appeared over the Macedonian camp these others were smaller by almost a half. In all there were five. The unknown chronicler narrates that they circled slowly over Tyre while thousands of warriors on both sides stood and watched them in astonishment. Suddenly from the largest "shield" came a lightning-flash that struck the walls, these crumbled, other flashes followed and walls and towers dissolved, as if they had been built of mud, leaving the way open for the besiegers who poured like an avalanche through the breeches. The "flying shields" hovered over the city until it was completely stormed then they very swiftly disappeared aloft, soon melting into the blue sky.'120"

Strangely, Drake interpreted Edwards (ref. 119) and Fenoglio
(120) versions as two different cases! And he didn't understand that Fenoglio DOESN'T quote Droysen, he was quoting the unknown chronicler.

Source: Jim Aho, 'Ezekiel's Wheel'   **ID:** 493

UFO Case Report: UFOs in the Bible: Ezekiel's Wheel, 593 BC

Occurred: 593 BC

Location: Cheber River, Chaldea (in modern-day Iraq),

Of all the UFO accounts in the Bible, the one mentioned most often is Ezekiel's Wheel. One of the passages reads: "This was the appearance and structure of the wheels: They sparkled like chrysolite, and all four looked alike. Each appeared to be made like a wheel intersecting a wheel." (Ezekiel 1:16)

Ezekiel's Encounter with a flying object
http://www.ufoevidence.org/Cases/CaseSubarticle.asp?ID=495
The religious leader Ezekiel writes about an encounter with a flying object year 592 B.C. [sic] near Cheber River Chaldea, in today's Iraq. R The bible is filled with many stories and references to alien gods, not god, or Elohiem being one god but a group of gods and rather than gods, aliens, or perhaps our parent species.

Source: The Bible UFO Connection   **ID:** 497
Case Type: Summary Report

UFO in 1034 depicted in 15th c. book, The Nuremberg Chronicle

Occurred: 1034

Location: Europe

A rare typeset book from 1493 contains what may be the earliest pictorial representation of a UFO. The book Liber Chronicarum, describes a strange fiery sphere, seen in 1034, soaring through the sky in a straight course from south to east and then veering toward the setting sun.

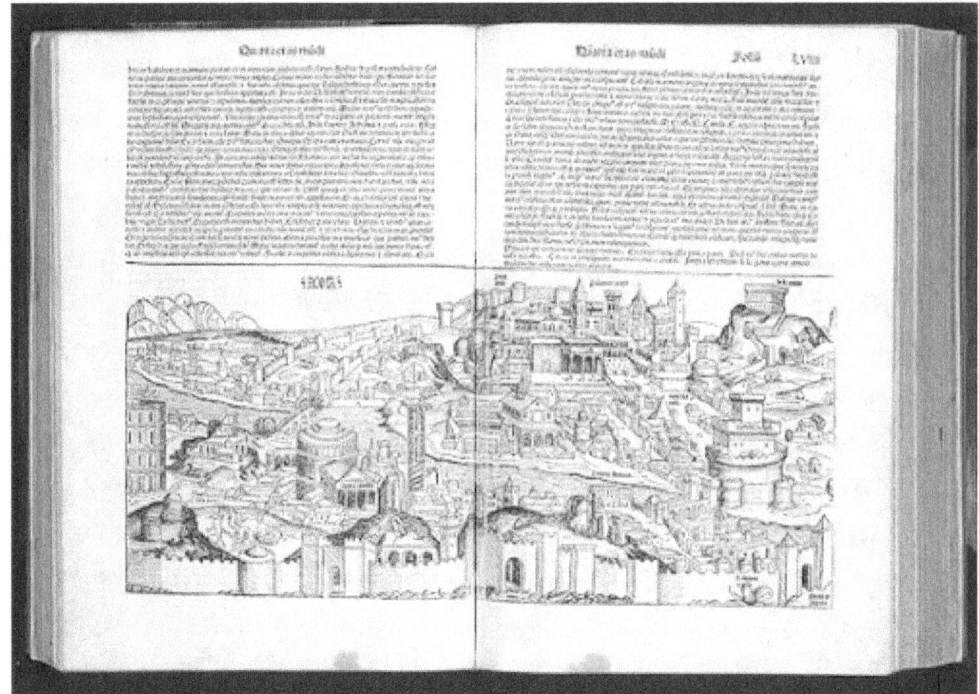

A couple of pages from the The Nuremberg Chronicle, or Liber Chronicarum. (This image does not show the page with the UFO illustration.)

Type of Case/Report: Summary Report
Hynek Classification: Full Report / Article

Source: The Bible UFO Connection http://www.bibleufo.com/ufos2.htm

A rare typeset book from 1493 contains what may be the earliest pictorial representation of a UFO. The book Liber Chronicarum (or commonly known as the Nuremberg Chronicle), describes a strange fiery sphere, seen in 1034, soaring through the sky in a straight course from south to east and then veering toward the setting sun. The illustration accompanying the account shows a cigar-shaped form haloed by flames, sailing through a blue sky over a green, rolling countryside. This may be the first work that actually contains actual illustrations of UFO's. Background information about The Nuremberg Chronicle (Liber Chronicarum)
source info includes,bibleufo.com,nicap and others
http://www.ufoevidence.org/cases/case497.htm

Occurred: June 30, 1400; 0000hrs.

Location: Myers Spring Canyon, TX

Shape: Circle
Duration:
What I have is a picture of Stone Age art painted at Myers spring canyon that looks like a shaman standing next to a UFO. ((NUFORC Note: Date in about 1400 AD is approximate. Photo shows images, which seem mildly reminiscent of a disc-shaped object in the sky of a primitive painting. PD))

Source: Various sources
UFO 'battle' over Nuremberg, Germany in 1561

Occurred: April 14, 1561
Location: Nuremberg, Germany

At sunrise on the 14th April 1561, the citizens of Nuremberg beheld "A very frightful spectacle." The sky appeared to fill with cylindrical objects from which red, black, orange and blue white disks and globes emerged. Crosses and tubes resembling cannon barrels also appeared whereupon the objects promptly "began to fight one another." Hans Glaser depicts this event in his famous 16th century woodcut.

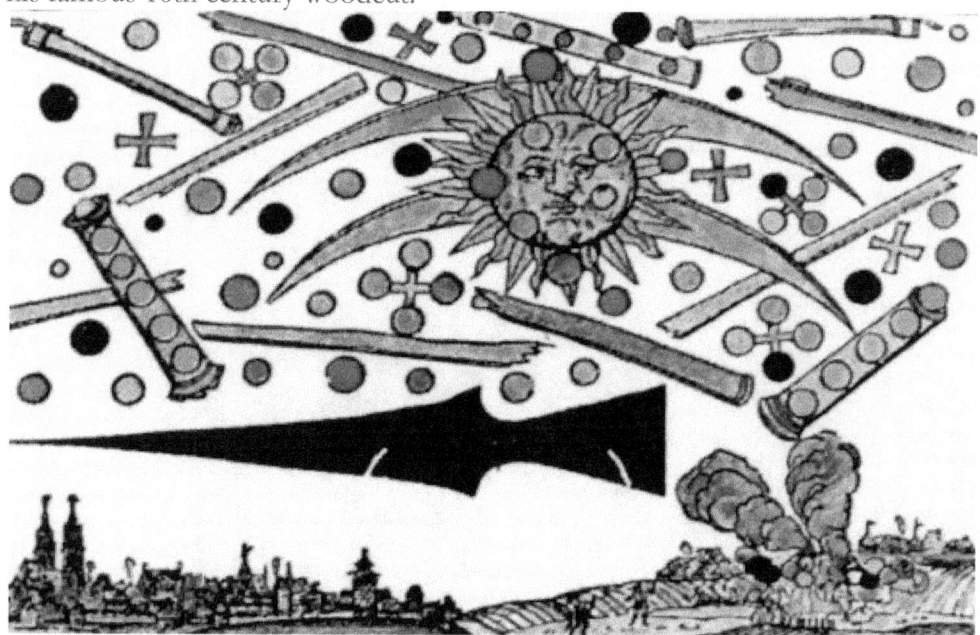

Hans Glaser woodcut from 1566 of the 1561 event over Nuremberg. (Wickiana Collection, Zurich Central Library

Type of Case/Report: Standard Case
Hynek Classification: DD
# of Witnesses: Multiple
Special Features/Characteristics: Multiple UFOs

Below are descriptions of the event from various sources:

At dawn of April 14, in the sky of Nuremberg (Germany), a lot of men and women saw a very alarming spectacle where various objects were involved, including balls "approximately 3 in the length, from time to time, four in a square, much remained insulated, and between these balls, one saw a number of crosses with the color of blood. Then one saw two large pipes, in which small and large pipes, were 3 balls, also four or more. All these elements started to fight one against the other." (Gazette of the town of Nuremberg).

The events lasted one hour and had such repercussions that an artist, Hans Glaser, drew a woodcut of it at the time. It describes two immense black cylinders launching many blue and black spheres, blood red crosses, and flying discs. They seem to fight a battle in the sky, it also seems that some of these spheres and objects have crashed outside the city.

Today in Odd History, an eerie battle raged in the skies above Nuremberg, Germany. It began at dawn, as dozens, if not hundreds, of crosses, globes and tubes fought each other above the city. It ended an hour later, when "the globes in the small and large rods flew into the sun," and several of the other objects crashed to earth and vanished in a thick cloud of the sun," and several of the other objects crashed to earth and vanished in a thick cloud of smoke. Smoke, according to the Nuremberg Gazette, the "dreadful apparition" filled the morning sky with "cylindrical shapes from which emerged black, red, orange and blue-white spheres that "frightful spectacle" was witnessed by "numerous men and women." Afterwards, a "black, spear-like object" appeared. The author of the Gazette warned that "the God-fearing will by no means discard these signs, but will take it to heart as a warning of their merciful Father in heaven, will mend their lives and faithfully beg God, that he avert His wrath, including the well-deserved punishment, on us, so that we may, temporarily here and perpetually there, live as His children." In the same year, a Lutheran clergyman on progress in Nuremberg wrote: "...God the Almighty has ... placed in the heavens many horrible and hitherto unheard of signs... We have seen far more signs now than in any other year. The sun and the moon have been darkened on a number of occasions. A crucifix in the sky was seen, as were biers and coffins with black men beside them. Further, rods and whips and many other signs were seen in a multitude of places... and scarcely a year has passed of late without an eclipse of sun or moon" (News of the Odd)

Occurred: August 7, 1566

Location: Basel, Switzerland

(Wickiana Collection, Zurich Central Library)

Woodcut of spheres seen over Basel, August 7, 1566.

Type of Case/Report: Standard Case
Hynek Classification: DD
# of Witnesses: Multiple
Special Features/ Characteristics: Multiple UFOs

Source: 16th century woodcutting and newspaper

Woodcutting depicts illustrates dark spheres that were witnessed hovering over the town of Basel, Switzerland in 1566. The spheres appeared at sunrise, 'Many became red and fiery, ending by being consumed and vanishing', w Samuel Coccius wrote in the local newspaper on August 7, 1566.

On August 7, 1566, at dawn, many citizens of Basel (Switzerland), frightened, saw during several hours the black spheres involved in a formidable aerial battle, invading the sky of their city: "at the time when the sun rose, one saw many large black balls which moved at high speed in the air towards the sun, then made half-turns, banging one against the others as if they were fighting a battle out a combat, a great number of them became red and igneous, thereafter they were consumed and died out," wrote Samuel Coccius, the student in "crowned writings and liberal arts" who consigned the strange events in the city's gazette.

Occurred: March 1676
Location: United Kingdom

Edmund Halley, the astronomer who discovered Haley's comet, could recall two accounts involving unidentified crafts. His first experience was in March of 1676, when he saw as he said, "Vast body apparently bigger than the moon."

Portrait of Edmund Halley (National Portrait Gallery, London).

Classification & Features

Type of Case/Report: Summary Report
Hynek Classification:
Special Features/Characteristics: Famous Person, Astronomer, Scientist/Engineer, Witness Photo

Source: The Bible UFO Connection
www.bibleufo.com

The astronomer who discovered Halley's comet, Edmund Halley could recall two accounts involving unidentified crafts. His first experience was in March of 1676, when he saw, as he said, "Vast body apparently bigger than the moon." He estimated it at 40 mi. above him. He also stated that it made a noise, "Like the rattling of a great cart over stones." After estimating the distance it traveled in a matter of minutes, he came to the conclusion that it moved at a speed greater than 9,600 m.p.h.

Source: 'The Life and Voyages of Christopher Columbus'

Case Type: Standard Case **Features:** Famous Person, Water-Related, Witness Photo

Christopher Columbus UFO sighting in 1492

Occurred: October 11, 1492

Location: Atlantic Ocean

In 1492, Christopher Columbus and Pedro Gutierrez while on the deck of the Santa Maira, observed, "a light glimmering at a great distance." It vanished and reappeared several times during the night, moving up and down, "in sudden and passing gleams."

Source: 'The Life and Voyages of Christopher Columbus'

From "The Life and Voyages of Christopher Columbus":
Christopher Columbus and Pedro Gutierrez while on the deck of the Santa Maria, observed, "a light glimmering at a great distance." It vanished and reappeared several times during the night, moving up and down, "in sudden and passing gleams." It was sighted 4 hours before land was sighted, and taken by Columbus as a sign they would soon come to land.

From the WaterUFO.net site:
Even Christopher Columbus, it appears, saw a UFO. While patrolling the deck of the Santa Maria at about 10:00 PM on October 11, 1492, Columbus thought he saw "a light glimmering at a great distance." He hurriedly summoned Pedro Gutierrez, "a gentleman of the king's bedchamber," who also saw the light. After a short time it vanished, only to reappear several times during the night, each time dancing up and down "in sudden and passing gleams." The light, first seen four hours before land was sighted, was never explained."

(From Beyond Earth: Man's Contact with UFOs, by Ralph Blum and Judy Blum, which is referring to 'The Life and Voyages of Christopher Columbus' (1850).)

http://www.ufoevidence.org/cases/case487.htm

Painting of the Madonna and Saint Giovannino

Painting in the Palazzo Vecchio in Florence, attributed to the 15th Century school of Filippo Lippi. Look in the upper right section of the painting. A man and dog are apparently looking at something in the sky! Source:
http://www.nicap.org/ancient/painting.htm

UFOs over Hamburg, Germany in 1697

Occurred: November 4, 1697

Location: Hamburg, Germany

There was a UFO sighting over Hamburg, Germany on November 4, 1697, depicted in this artwork. The objects were described as being "two glowing wheels".

Artwork depicting the two 'glowing wheels' over Hamburg in 1697.

Type of Case/Report: Summary Report
Hynek Classification: DD

Source: UFOs at Close Sight / others

There was a UFO sighting over Hamburg, Germany on November 4, 1697, depicted in this artwork. The objects were described as being "two glowing wheels". (Note: if the round object on the far right is the moon, the UFOs were either very low or very large. Notice also the many people pointing -- this was likely a famous event.)
http://www.ufoevidence.org/cases/case488.htm **for full article**

FRESCO IN TURKEY HINTS AT PILOTED CRAFT

The fresco is on the wall behind an altar in a church in Turkey, I believe. The 'pilot' is looking back and around the curved painting to the other side where another craft with a different being in it is looking at the occupant of the craft in the picture. Unfortunately my UFO Library is in storage at the moment as Sue and I separated a couple of months ago and my current space is too small for most of my 'stuff' and I don't have access to the book to get a scan of the other craft.
Errol Bruce-Knapp (As soon as we can get more images and further information, this will be updated. Fran Ridge) source: nicap./org/ancient/fresco.htm

Occurred: April 5, 1800; Approx. 2000hrs.
Location: Baton Rouge, LA
Shape: Light
Duration: 15 seconds

UFO report communicated by Thomas Jefferson

From: http://www.ufodigest.com/article/report-concerning-unidentified-flying-object-communicated-thomas-jefferson

This one page notice appeared in the Transactions of the American Philosophical Society vol. 6 Part 1 (Philadelphia, 1804), p. 25. At the time it was written, Thomas Jefferson was president of the Society and also Vice President of the United States. Apparently it was written and submitted by the naturalist William Dunbar, and communicated or presented to the society by Jefferson. Unfortunately, the plate referred to is missing.
The entire volume (21 MB), which is in the public domain, may be downloaded from:

http://rbedrosian.com/Downloads/TAPS_1804_Jefferson_Ufo.pdf No. III.
Description is of a singular Phenomenon seen at Baton Rouge, by William Dunbar, Esq., communicated by Thomas Jefferson, President A. P. S.

Natchez, June 30th, 1800.
Read 16th January 1801.

A PHENOMENON was seen to pass Baton Rouge on the night of the 5th April 1800, of which the following is the best description I have been able to obtain. It was first seen in the South West, and moved so rapidly, passing over the heads of the spectators, as to disappear in the North East in about a quarter of a minute. It appeared to be of the size of a large house, 70 or 80 feet long and of a form nearly resembling Fig. 5. in Plate, iv. It appeared to be about 200 yards above the surface of the earth, wholly luminous, but not emitting sparks; of a colour resembling the sun near the horizon in a cold frosty evening, which may be called crimson red. When passing right over the heads of the spectators, the light on the surface of the earth, was little short of the effect of sun-beams, though at the same time, looking another way, the stars were visible, which appears to be a confirmation of the opinion formed of its moderate elevation. In passing, a considerable degree of heat was felt but no electric sensation. Immediately after it disappeared in the North East, a violent rushing noise was heard, as if the phenomenon was bearing down the forest before it, and in a few seconds a tremendous crash was heard similar to that of the largest piece of ordnance, causing a very sensible earthquake.

I have been informed, that search has been made in the place where the burning body fell, and that a considerable portion of the surface of the earth was found broken up, and every vegetable body burned or greatly scorched. I have not yet received answers to a number of queries I have sent on, which may perhaps bring to light more particulars.

Occurred: 1864

Location: Cave Springs, GA

In 1864, prior to airplanes, a white object was observed about half a mile high and moving rapidly toward the south. A Confederate soldier's war journal was published as "Blood & Sacrifice," by Blue Acorn Press, Huntington, WV in 1994.On Page 140, the author reports, "Prior to our arrival at Cave Springs, quite early in the forenoon in fact, a white object was observed in the sky to the southeast, apparently about half a mile high and moving rapidly toward the south. We decided that it was a balloon, and that the enemy was endeavouring by that means to ascertain the strength of the reinforcements that were coming to Gen. Johnston." The book editor notes that Union General Sherman's troops did not have balloons with them on the Atlanta Campaign. It should also be noted that observation balloons were tied to the ground and only moved vertically up and down. If a balloon were untethered by accident, it would only move as fast as any winds prevailing at the time, or certainly not "rapidly." The observer (book author) was a Sergeant in Company B, 46th Mississippi Infantry Regiment, and later Sergeant Major, and finally Acting Adjutant of the Regiment. He had been a schoolteacher prior to the War Between the States, and the book is a very factual chronology of his daily experiences during the War. ((NUFORC Note: We express our sincere gratitude to the individual who took the time, and trouble, to submit this report. Indeed, there are many published articles and books, which suggest to us that the UFO phenomenon has been occurring for a long, long time, and that it did not begin in June 1947, as many people believe. PD))

1870-1890

THE APPEARANCE OF THE GREAT AIRSHIPS

A bustling and enterprising endeavour for American and European Inventors starting in 1870 through the 1890s was the invention and development of the great airships. Dirigibles, Zeppelins and Balloons span across the continents. In Europe there were floating palaces for the rich, considered as well, for military applications. The Americans were no different. During the Industrial Revolution, they tested and started manufacturing of these flying machines. Some had wings and some allegedly were powered by motors that ran off harmonic vibration, played by a keyboard, built by an enterprising Californian.

Of course Count Von Zeppelin had been flying these huge, rich passenger airships for fun and profit, and many hoaxes popped up globally, attributing the sighting to ETs. The Spanish American conflict of 1898 added another dimension to the appearances of ships as well.

The Great U.S. Airship Sighting that started in Stockton, California, was well documented in The San Francisco Chronicle and The Sacramento Gazette. After its sighting, dozens of major newspaper articles sprung up on the sighting of this craft that required frequent landings for water to power itself across the continental United States. It turned out there were at least 3-4 patents submitted during this 10-year span and coincidentally both the northern U.S. as well as the southern U.S. had major sightings of airships going at high speeds, hovering over major cities, and landing. The fliers of these ships had to take pit stops to get water or other supplies and stopped at late hours and shocked rural residents of many states with these requests. Meanwhile, only a few days later after these rashes of sightings, a genuine UFO report is witnessed with a live alien aboard. Unfortunately the craft crashed and the small individual died within hours. The town doctor tried to save him but the injuries and anatomy were so different it was futile. There were also multiple airships flying at the same time, so sightings were occurring north and west. The inventors claimed responsibility and their flight paths and landings coincided with the sightings, but also during that time was the genuine UFO event in Aurora, Texas.

It should be noted here that it also seems co-incidental that during this time, Fredrick Oliver's book, "A Dweller on Two Planets" was beginning to make a splash. It appears with the advent of flying on a collective level, the focus and attention on air space occurrences was heightened. It was also during a time when major universities began deeper investigations into archaeology and remains in ancient sites. With the advent of the Industrial Revolution many gears began to turn and if it were not for the love and desire of flight, these airships, which incidentally the Wright Brothers were well aware of, would have not joined the race. Heavier and more maneuverable airplanes were more desirable for military applications even though the U.S. and European Armies enlisted the use of Airships well into the 20s, 30s, and 40s, evidenced during the Second World War. They could stay up for long periods of time at high altitude, but still were slower because of their size and maneuverability.

The Aurora Texas UFO Crash

On April 19th 1897, the town of Aurora, Texas witnessed the approach and crash of an UFO. Judge Proctor was awakened with the explosion of his water-pumping windmill crashing down from the impact of a small fiery UFO crashing on his property. The living occupant, a tiny alien still alive was rescued but died.

The original article, reporting the Aurora, Texas Incident, as written in 1897, in the April 19th edition of the Dallas Morning News reads as follows:

" About 6 o'clock this morning, the early risers of Aurora, Texas were astonished at the sudden appearance of the airship which has been sailing around the country. It was traveling, due north, and much nearer the earth than before. Evidently some of the machinery was out of order, for it was making a speed of only ten or twelve miles an hour, and gradually settling toward the earth. It sailed over the public square and when it reached the north part of town, it collided with the tower of Judge Proctor's windmill and went into pieces with a terrific explosion, scattering debris over several acres of ground, wrecking the windmill and water tank and destroying the judge's flower garden. The pilot of the ship is supposed to have been the only one aboard, and while his remains were badly disfigured, enough of the original has been picked up to show that he was not an inhabitant of this world.
Mr. T. J. Weems, the U. S. Army Signal Services officer, gives it as his opinion that the pilot was a native of the planet Mars. Papers found on his person, evidently the records of his travels are written in some unknown hieroglyphics and cannot be deciphered. The ship was too badly wrecked to form any conclusion to its construction or its motive power. It was built of an unknown metal, resembling somewhat a mixture of aluminum and silver, and it must have weighed several tons. The town, today, is full of people who are viewing the wreckage and gathering specimens of strange metal from the debris. The pilot's funeral will take place tomorrow". E. E. Haydon, who was a part-time reporter for the Morning News, wrote the article. As startling as the news was, no other newspaper in the world ran the story in their pages.

In reference to this event Wikipedia has this listing "In 1897, (The Dallas Morning News) reported that an airship had smashed into a windmill driven pump belonging to a Judge Proctor, then crashed. The occupant was dead and mangled, but the story reported that presumed pilot was clearly "not an inhabitant of this world." (Jacobs, 17) Strange hieroglyphs were seen on the wreckage, which resembled "a mixture of metals it must have weighed several tons.""(ibid.) (In the 20th Century, unusual metallic material recovered from the presumed crash site was shown to contain a percentage of aluminum and iron admixed.) The story ended by noting that the pilot was given a Christian Burial, in the town cemetery. In 1973, MUFON investigators discovered the alleged stone marker used in this burial. Their metal detectors indicated a quantity of foreign material might remain buried there. However, they were not permitted to exhume, and when they returned several years later, the headstone – and whatever metallic material had lay beneath it, was gone.   On May 24, 1973, when the following United Press International account, the event is again in public view.

"Aurora, Tex. -- (UPI) -- A grave in a small north Texas cemetery contains the body of an 1897 astronaut who was not an inhabitant of this world," according to the International UFO Bureau. The group, which investigates unidentified flying objects, has already initiated legal proceedings to exhume the body and will go to court if necessary to open the grave, director Hayden Hewes said Wednesday.

"After checking the grave with metal detectors and gathering facts for three months, we are certain as we can be at this point [that] he was the pilot of a UFO which reportedly exploded atop a well on Judge J.S. Proctor's place, April 19, 1897," Hewes said. He was not an inhabitant of this world." Another UPI account quoted a ninety-one-year-old who had been a girl of fifteen in Aurora, Texas at the time of the reported incident. She said she "had all but forgotten the incident until it appeared in the newspapers recently." Her parents had gone to the sight of the crash, but would not take her. She said the corpse of the pilot, "a small man," had been buried in the Aurora, Texas cemetery.

Then the Associated Press reported, "A North Texas State University professor had found some metal fragments near the Oates gas station (former Proctor farm). One fragment was said to be 'most intriguing' because it consisted primarily of iron which did not seem to exhibit magnetic properties." The professor also said "shiny and malleable instead of dull and brittle like iron."

The Cemetery Association stopped digging up the grounds in search of the "Martian Pilot", this stopped worldwide focus on Aurora until MUFON got wind of the incident.

In 1997, MUFON, conducted a field investigation of Aurora. The results they claimed were "unusual." The entire layout of Aurora is similar to a military base, and it contains small airstrips, like many that were built clandestine and impromptu in the Second World War, for emergency transports. They exist everywhere including in Mount Shasta now with grass growing over them. Staging areas are everywhere but North Texas is a convenient hub for deploying troops, historically as well as aircraft even in the beginning of flying. Other curious facts collected report that the Judge Proctor buried the spacecraft in his well, and it made his terribly ill and disfigured his hands. There was terrible odors from the water the animals and people got sick. Apparently it also was radioactive when some investigators went in, but did not disclose this. The judge filled the well in with cement so no one could dig it out or get poisoned from it. He was uncomfortable with the entire notion of the spacecraft being found as it was and spent hours breaking it up to fit it down the well. Along with the grave of the alien it originally had a headstone which some feel it was removed to protect the body and when MUFON came to investigate it, they used equipment that could identify the skeleton and in fact it was still there, a small figure and strange metal, and were going to return to carefully extract the metal, but to their dismay the entire body had disappeared, and allegedly the military came and took it with no ones permission to further black out the incident. The movie "The Aurora Incident" goes into it in depth as well as " Alien Sightings" 100 years of cover up" as well as a 1986 movie, " The Aurora Encounter," produced by Charles B. Pierce, tells the tale.

In an expose written by James L. Choron in 2005, on the Aurora Crash, says, "This incident has been covered up and ridiculed by the U.S. Government (a standard operating procedure of the MAJESTIC 12 group) and has been widely reported to be a hoax (a weather balloon?). This, to say the least, sounds a lot like Roswell in 1947? The U.S. Government has a long history of cover-ups in regards to such occurrences. It is hoped that the current, renewed interest in the incident will last, and that a new investigation will clear up the Aurora Texas event for good, although much time has passed. It is tragic that most, if not all of the original witnesses are long dead, for, at one time, up until around the early seventies, there were quite a few people still living who had been children at the time and not only remember the crash, but remember a rash of "airship" sightings, all over East and North-Central Texas, as well as the stories which were passed down to them from their "elders". Almost everyone who grew up in those parts of the state have heard stories from their grandparents, or other "old folks" about such events, many of whom were "substantial" citizens, including doctors, clergymen, judges, army personnel, sheriffs and other professionals". Choron 2005

TWENTIETH CENTURY SIGHTINGS
SECTION 2: 1900-1935

Our chronology is primarily devoted to Northern California, however with the magnitude, relevance and depth of the global sightings, we included some of the significant and profound sightings that occurred nationally and internationally. The serious and sobering facts of these sightings had made landmark experiences in our history, forever changing public and government perceptions. Our region's sightings had left an indelible mark in the science of UFOlogy and its evolution. We have included all sources and web addresses (if available) within each sighting for conveniences sake. Many of the sightings in this section to the present are from NICAP, MUFON, CUFON NARCAP, NUFORC and others that were reported to these agencies, in their original form, with typos, grammatical errors and all. With their blessings, we have created this regional chronology and historical document to help inform you of the past sightings.

***** TECHNOLOGICAL AEROSPACE *****

***** MILESTONES*****

1895-1901: Konstanatin Tsiolkovsky, Russian born space and rocketry genius works on idea of rocket propelled spacecraft. 1903 He Russian Scientific review publishes his "Exploration of the Worlds Space with Reactive Instruments. In 1905, he condemns U.S. military's blueprints of rockets for military use.

1902: Then high school student Robert Goddard submits "Navigation of Space" to Popular Science.

1903: Wright Brothers achieve flight in a heavier than air machine.

1905: Einstein publishes paper on theory of relativity.

1906: Russian engineer Karavodine patents the pulsating ramjet. Same year Robert Goddard ponders idea of solar powered electric rocket propulsion in his personal notes.

1907-1908: British biologist A. Wallace, publishes "Is Mars Habitable?" Orville Wright conducts trial aircraft flights for U.S. War Department. Lowell publishes "Mars as the Abode of Life" A falling space object enters over Siberia and explodes known as. Tunguska event

1910: Halley Comet makes a spectacular appearance during its close encounter with the Earth, inspiring an interest in space among many, including future German rocket pioneer Johannes Winkler
Marconnet, pantents a pulsejet engine for use on the aircraft.

1911: The American Aeronautical Society is formed in the U.S.

1912: Nikolai Tikhomirov proposes a project of a solid-propellant rocket to the Russian navy ministry.
Engineer V. V. Ryumin publishes article "By the Rocket into the World's Space" in *Priroda i Lyudi* magazine, which popularizes work by Tsiolkovsky

1914: In the U.S., Robert Goddard registers two patents for a liquid-propellant rocket and a two- and three-stage solid-propellant rocket.
Tsiolkovsky publishes the 3rd part of the "The Exploration of the World's Space with Reactive Instruments."
The novel "Islands of Ephir Ocean" by B. Krasnogosky and D. Svyatsky describes an expedition to Venus.

1915: The U.S. Congress creates Advisory Committee for Aeronautics (NACA), precursor of NASA.

1916: In France, Henri Melot works on rocket engines for aircraft.

1917: The Smithsonian Institution awards a $5,000 grant to Robert Goddard to conduct rocket research in the upper atmosphere

1918: Robert Goddard writes "The Ultimate Migration" describing the exodus of the human civilization from a dying Solar System onboard a nuclear-powered colony. The work would not be published until 1972.

1919: In a letter to the Soviet government, Nikolai Tikhomirov proposes to organize a laboratory for the development of powder rockets.
In the U.S., Robert Goddard submits a progress report entitled "A Method of Attaining Extreme Altitudes," to the Smithsonian Institution.

1920: The Smithsonian Institution publishes Goddard's "A Method of Attaining Extreme Altitudes," which was misinterpreted by the press as a proposal for a rocket flight to the Moon. Hermann Oberth develops a concept of a multi-stage liquid propellant space launcher.
A paper "Riches of the Universe" by Tsiolkovsky is published

1921: A Russian inventor, A. F. Andreev, requests a patent for a portable personal flight vehicle propelled by a liquid engine burning oxygen and methane.

1923: **June:** In Germany, Hermann Oberth publishes work called "The Rocket into Interplanetary Space." **Nov. 1:** In the US, Robert Goddard tests a rocket engine using liquid oxygen and gasoline and supplied by a pump.
Revolutionary Military Council of USSR issues a request to Tikhomirov's laboratory to test the possibility of using jet propulsion for increasing the range of existing munitions.
In Italy, Luigi Gussalli (1885-1950) publishes "Can We Attempt a Space Journey to the Moon?"
In USSR, Aleksei Tolstoy publishes the novel Aelita, describing a civilization on Mars. A year later, Yakov
Protozanov directs a film based on the book.

1924: The Society for Studies of the Interplanetary Travel, OIMS, is founded in Moscow. (386)
Tsiolkovsky's work "Rocket into cosmic space" describes multi-stage rockets.
Tsander publishes "Flight to Other Planets."
Tsiolkovsky, Tsander and Kondratyuk propose the use of the atmosphere as a breaking medium for the spaceships returning to Earth.

1925: Tikhomirov's lab, later known as Gas Dynamics Laboratory, GDL, moves to Leningrad (St Petersburg).
Oberth learns about Tsiolkovsky's work. The first exhibit dedicated to the interplanetary travel is held in Kiev.

1926: In Auburn, Massachusetts, Robert Goddard launches the world's first liquid-propellant rocket. In the US, Hugo Gernsback, an engineer and businessman, publishes first issue of *Amazing Stories*, an early sci-fi magazine, which made huge contribution into popularization of space flight.
1928: Fritz Stamer's Ente (Duck), the world's first aircraft powered by a solid-propellant rocket engine, completes the first 1.2-kilometer flight in Germany after a takeoff from a catapult.

1930: The American Interplanetary Society is formed in New York by G. Edward Pendray, David Lasser and others.

1931: In Germany, the Repulsor rocket reaches 1,006 meters in altitude and lands with a parachute.
 Goddard launches a rocket, which reaches 518 meters in altitude.
1932: Sergei Korolev leads the tests of a rocket-propelled glider. Goddard tests a guided rocket. Wernher von Braun starts his work on rocketry for the German army at Kummersdorf

The above is only part of a much larger list of Milestones. It is included for the early part of the twentieth century, to fill in data that shows the Russian's were further advanced in Space technology than the U.S. for most of the century.

Occurred: February 28, **1904**; Approx. 0610hrs.
Location: Pacific Ocean off San Francisco, California, United States
Witness: **3**
Shape: **Multiple-Oval**

Duration:

Special Features/Characteristics: Multiple UFOs, Water-Related Hynek Classification: **DD**

Three members of the crew of the USS Supply, at 6:10 a.m. local time, sighted an echelon formation of three objects, which appeared near the horizon below clouds, moving directly toward the ship. As they approached, the UFOs began soaring, rose above the cloud layer, and were observed climbing into space, still in echelon. The lead object was egg-shaped and about the size of six suns, and the other two were smaller and round.

Illustration of the sighting. (credit: NICAP / Hall)

Article

Source: NICAP / Richard Hall (1964) citing "Monthly Weather Review" (1904)

1904: Circular UFOs Maneuvered Near Ship

One of the earliest formation cases was reported February 28, 1904, by a ship in the North Pacific off San Francisco. Three members of the crew of the USS Supply, at 6:10 a.m. local time, sighted an echelon formation of three "remarkable meteors" which appeared near the horizon below clouds, moving directly toward the ship. As they approached, the UFOs began soaring, rose above the cloud layer, and were observed climbing into space, still in echelon. The lead object was egg-shaped and about the size of six suns (about 3 degrees of arc). The other two were smaller and appeared to be perfectly round. They remained visible for over two minutes. (Meteors, of course, do not travel in echelon formation, change course and climb, nor remain visible for two minutes).

National UFO Reporting Center (NUFORC)
Curious Phenomenon Reported to the Times of London
Occurred: July 1908

NUFORC Home Page www.nuforc.org

This is a very important event, as it is tied to a global defense system set up by Extra Terrestrials that discharged its contents to protect the planet from an asteroid or hostile forces. It was designed to do this from its location in Tunguska Siberia. The site has been common knowledge to the tribes living in Siberia and studied several times by scientists in Russian and the U.S. in the 1990s. It is confirmed to be of unknown origins.

Note to visitors to our website: The following letter was the first of several letters and articles to appear in "The Times" (London), which describe the highly unusual meteorological events, and magnetic anomaly, that were observed not only over England, but over all of Western Europe, as well.

Some of the readers may quickly grasp a possible explanation for the cause of the strange events. Others will find a statement at the end of the third article, which discusses the possible cause of the effects…

The following article appeared in "The Times" (London) on Wednesday, July 01, 1908.

"Curious Sun Effects At Night"

"To the Editor of the Times."

"Sir,--Struck with the unusual brightness of the heavens, the band of golfers staying here strolled towards the links at 11 o'clock last evening in order that they might obtain an uninterrupted view of the phenomenon. Looking northwards across the sea they found that the sky had the appearance of a dying sunset of exquisite beauty. This not only lasted but actually grew both in extent and intensity till 2:30 this morning, when driving clouds from the East obliterated the gorgeous colouring. I myself was aroused from sleep at 1:15, and so strong was the light at this hour that I could read a book by it in my chamber quite comfortably. At 1:45 the whole sky, N. and N.-E., was a delicate salmon pink, and the birds began their maturational song. No doubt others will have noticed this phenomenon, but as Brancaster holds an almost unique position in facing north to the sea, we who are staying here had the best possible view of it. (pictures missing)

Yours faithfully, Holcombe Ingleby
Dormy House Club, Brancaster, July 1" **(1908)**

The following article was published on the following day, Thursday, July 2, 1908, in "The Times" (London):

"The Aurora Borealis."

The Aurora Borealis was very brilliant again last night. In the higher points in the suburbs from which London can be seen the sight was most unusual. All the outstanding features of the metropolis were silhouetted. Many people were in the suburban roads viewing the sight."

"TO THE EDITOR OF THE TIMES."

"Sir, --I should be interested in hearing whether others of your readers observed the strange light in the sky which was seen here last night by my sister and myself. I do not know when it first appeared; we saw it between 12 o'clock (midnight) and 12:15 a.m. It was in the northeast and of a bright flame-colour like the light of sunrise or sunset. The sky, for some distance above the light, which appeared to be on the horizon, was blue as in the daytime, with bands of light cloud of a pinkish colour floating across it at intervals. Only the brightest stars could be seen in any part of the sky, though it was an almost cloudless night. It was possible to read large print indoors, and the hands of the clock in my room were quite distinct. An hour later, at about 1:30 a.m., the room was quite light, as if it had been day; the light in the sky was then more dispersed and was a fainter yellow. The whole effect was that of a night in Norway at about this time of year. I am in the habit of watching the sky, and have noticed the amount of light indoors at different hours of the night several times in the last fortnight. I have never at any time seen anything the least like this in England, and it would be interesting if any one would explain the cause of so unusual a sight.

Katharine Stephen.
Godmanchester, Huntingdon, July 1."

The third article appeared in The Times (London) on Saturday, July 4, 1908. An interesting aspect of the article is shown in red…

"The Recent Nocturnal Glows"

"The remarkable ruddy glows which have been seen on many nights lately have attracted much attention, and have been seen over an area extending as far as Berlin. There is considerable difference of opinion as to their nature. Some hold that they are auroral; their colour is quite consistent with this view, and there is also the fact that Professor Fowler, of South Kensington, predicted auroral displays at this time from his observations, which showed great disturbances in the sun's prominences. These violent disturbances in the prominences were also described by Mr. Newbegin at the meeting of the British Astronomical Association last Wednesday, the latest disturbance noted being on the morning of that day. There was a slight, but plainly marked disturbance of the magnets on Tuesday night, and this materially strengthened the auroral theory, as the two phenomena are very closely correlated. However, this was shaken on the following night, when the glow was quite as strong, but the magnets were exceptionally quiet. This convinced many, who had before been inclined to the auroral theory, that the phenomenon was simply an abnormal twilight glow; this is supported by the fact that nearly all the observers agree that the glow was vertically above the position of the sun, and moved with it from north-west to north-east during the night; a further argument is that the glow was always near the horizon, whereas aurorae may be seen in any part of the sky.

It is well known that there is some twilight so long as the sun's depression below the horizon does not exceed 18-deg.; in other words, we have no real night in London when the sun is more that 20-deg. North of the equator, or from May 23 to July 21. It is only necessary to suppose that some temporary condition of the atmosphere made this twilight much brighter and redder than usual.

We may recall the circumstances of the wonderful glows which were seen in this country in the autumn of 1883, and which were due to the dust scattered in the upper atmosphere by the terrific outburst at Krakatoa at the end of August. Those glows had many points in common with the recent ones; (1) the deep, lurid colour, suggesting a distant conflagration many were for some time doubtful whether Tuesday's glow was not due to this cause) (sic); (2) both glows were seen at a much longer interval after sunset than ordinary sunset glows, and the latter had already faded before the abnormal glow began. This indicated an extraordinary height for the dust causing the glow, and consequently the extreme fineness of the latter; by charting the places and dates of first visibility of the glows in 1883, it was found that the dust was carried westward by a previously unknown upper current at a speed of 80 miles an hour; it did not reach the British Isles till its third circuit of the globe, each circuit having a wider range in latitude. We thus see that distance is no obstacle in vast cosmical phenomena of this kind, which are absolutely world embracing. No volcanic outburst of abnormal violence has been reported lately; there have, however, been some moderate outbursts in the Pacific during the spring, and it is possible that the dust may have reached us from these, or from some unreported eruption in some little-known region of the world." The curious meteorological effects may have been the result of the mysterious explosion, which occurred over the Tunguska Region of Russia, approximately 600 miles to the northwest of the northern tip of Lake Baikal in Siberia, at 0717 hrs. (Local time) on June 30, 1908. If so, the effects, documented in the news articles, must have been visible over Europe, many thousands of miles to the west of the epicentre of the blast, within a very short period of time.

Question 1: If the meteorological effects were, in fact, caused by the explosion at Tunguska in the heart of Siberia, how could the dust cloud generated by the explosion, and the one presumed by the writers to have been the cause of the meteorological anomalies, have reached Europe, apparently within hours, or perhaps even minutes?

Question 2: Moreover, if the explosion was, in fact, caused by a meteor or comet, is such an event consistent with the magnetic anomaly that apparently was detected throughout Europe, and perhaps even around the world?

In addition, it is intriguing to consider whether the magnetic anomaly occurred at the instant of the actual explosion, or whether it occurred prior, during the possibly 1-2 minutes that the object that exploded was seen by local herdsmen in the area prior to the actual blast.

June 30, 1908; Podkamennaia Tunguska, USSR
Unexplained explosion in the taiga, equivalent to a thermonuclear blast, sometimes interpreted as the crash of an interstellar vehicle. (Magonia #35, Anatomy 18; Challenge 99) (Brad Sparks: was equivalent to an approximately 7 megaton hydrogen bomb blast [larger estimates in the 20-35 MT range are due to incorrect scaling of blast and thermal effects] which occurred above ground, about 3 miles height, resulting in the Hiroshima-like effect of trees left standing near the ground zero with all branches stripped off while trees at an angle because farther from ground zero were flattened outwards. Nobelist Willard Libby found that 1909 tree rings worldwide showed an increase in radiocarbon, suggesting to him that the Tunguska blast released neutrons into the atmosphere like a megaton-range H-bomb, converting nitrogen into radiocarbon. Russian and other scientists have conducted 80+ years of investigations of Tunguska. As early as 1946 Alexander Kazantsev suggested Tunguska was the explosion of a spaceship and compared it to Hiroshima. Mainstream scientists generally hold that it was a comet or asteroid (in actuality a comet could not have survived to such low altitude, but a stony asteroid might have, but the radioactivity remains unexplained).

Author's Note: Serious research has been conducted several times in the 20th century regarding this incident. What was discovered through both science and eyewitness reports, is of an underground laser type anti-asteroid and hostile UFO defense station of alien construction that when alerted, rises from underground, three half-moon shape domes of metal that emit high output energy concentrations, on target and vaporize those targets. The local people of the area have stories passed down from their tribes about those who saw them in use over the ages, including their activation on the asteroid that exploded in pieces. The first group of scientists that researched these domes found seven of these domes and all got deathly ill and never went back, but left research notes that the American scientists studied. They visited the area to confirm the folklore and earlier findings.

The scientists who studied this site in the 1990s, found the domes and became deathly ill around the area too, so whatever is there is extremely toxic. They think it is not only the materials used in the defence projectiles,

but also a way of protecting the site from intruders. It is considered as " no mans land" by the authorities in Siberia. This story is also illustrated in Series 3 of "Ancient Aliens."

Take into consideration while reading the following that there is a global war in progress called "The War to End all Wars," (The first World War), and sightings which undoubtedly occurred in scores were not recorded as people were busy, but as in many wars, the ETs were observing perhaps even rallying. Remember the war ended in 1918 and U.S. involvement started in 1917.

Occurred: **July 5, 1929;** Approx. 1400hrs.

Location: Buchanan, or Burns, OR

Witness: 2+
Shape: Disk and aliens
Duration: 1 minute

We were traveling east of Burns, climbing up thru a cut in the rim rocks when this object very slowly flew over the top of us, some 50 ft above the rim rock. The object stopped, and thru the most left window there was two beings, looked most like our type of people. I could see them pointing with arms and hands.

I stepped out of the car, but my mother demanded that I get back into the car. I stood one the running board the rest of the short time. The craft was two tones of brown. It hovered for about 40 sec. the craft had windows in the middle section. There was a soft hum you could detect any movement, or vibration of any kind.

As it left it moved very slowly one of the persons walk to the next window, about that is when they started to move. The craft must have been some where at 100 yards in length, because you could not quite see total length. Then in a blink it was gone. My mother said at that time, don't say anything about this, because people will think we are crazy ((NUFORC Note: Sighting occurred in the year 1929, irrespective of however it may get "sorted" by a computer. Date is approximate. PD))

Occurred: **June 1, 1930; Approx.** 2200hrs.

Location: Ithaca/Freeville, NY

Witness: 2+
Shape: Very Large Triangle Shaped Object
Duration: 20 minutes
Several other similar reports in the area have prompted me to report this sighting.On Freeville Rd, between Ithaca and Freeville, NY, a dark remote Rd at night, we witnessed a very, VERY large Triangular craft. The weather was clear and the night sky was spectacular for stargazing. There is very little artificial light in this area. At first, I notice a bright star with tiny flashes around it. I thought I was seeing satellite photos being taken. Then I noticed across from it, at a distance of possibly 100 feet, the same thing, and again at a third point, making a triangle shape.

In between the lights, was darkness, no stars could be seen. As we watched, we could see this object, which may have been up to 2000 feet in the sky, was moving very slowly to the east.
((NUFORC Note: Date of sighting is approximate. PD))

Alien captured on film in 1930
Location: Alaska, United States

The picture was taken in the early 1930s by the sender's grandfather, who lived in Alaska. The entity was first seen when the grandfather was on his way to a lake. He chased the entity until he got close enough to take this one picture. It was some four months before the photograph was developed, being in a remote, sparsely populated area. The sender received the picture from his grandfather only last week. His grandfather died the day after giving him the photo, and relating his story. View full report www.ufoevidence.org
Source: UFOCasebook.com (B.J. Booth, editor) ID: 1109

Close-up and cropped picture #1.

Occurred: **February 15, 1931;** Approx. 1400hrs.
Location: Holyoke, CO (6-8 miles southwest of)

Witness: 1
Shape: Oval
Duration: 1minute
In the winter of 1931, I saw a flying machine (UFO) as I was riding on a horse in the draw of a pasture.

I was 17 and I was riding a horse checking on cattle after a snowstorm and I was about 8 miles Southwest of Holyoke, Colorado. I went over a small hill and down to a draw that was about 25-30 feet deep. I was heading south and I got to the middle of the draw. I looked up to the west and I saw an elongated oval shaped flying machine in the air. It was an aluminum color and a door that was transparent like a glass opened. It looked like a doorway and a man appeared that looked human but was not as tall as a normal human. He looked down at me and he had a uniform on that was darker than he was. I looked back down because I was on a horse that stumbled quite a bit. When I looked up again ASAP there was nothing there so they must have left.

((NUFORC Note: Time and date are approximate, although the witness is almost certain the incident occurred in 1931. We spoke with the witness at length, and he seemed to us to be exceptionally reliable and credible. PD))

Occurred: **June 1, 1931;** Approx. 1300hrs.

Location: Abilene, KS

Witness: 2
Shape: Disk
Duration: 30 min. approx.

This is what my mother related to me regarding an experience my grandparents had. They did not tell anyone for 20 or more years because they did not want anyone to think they were crazy. My grandparents were college educated, my grandfather was a dentist and my grandmother was active in civic and church affairs. Neither were prone to being "fanciful"...both children of homesteading parents who had worked their way to actually graduate from college. My grandfather was active in Masons and church activities, and my grandmother was a founding member of the local Daughters of American Revolution and also in church activities (the local Methodist church). They were driving from Abilene to Salina and were on US 40 near the local pump station, known as Sand Springs (this building still exists today). The only information that I was told my mother was that their car engine stopped, their watches stopped (they had to reset them when they arrived in Salina.) The watches were approximately 30 minutes slow. A silver-ish disc like object came down in front of their car and hovered there. They did not speak, too stunned by the event. From their standpoint, it hovered for only a few moments and then zipped off into the sky. However, they always wondered what had actually happened because of the time difference with their watches. That is all the information that I know. Anyone else they might have told is now deceased. However, I was raised with the philosophy that we really can't discount the possibility of other beings in the universe. ((NUFORC Note: Witness indicates that the date of the incident is approximate. We have requested more information about the event, although all the witnesses are no longer living. PD))

1933 UFO stops for repairs in Saskatchewan, Canada

Occurred: 1933

Location: Nipawin, Saskatchewan, Canada

Witness: 1+
Shape: oval craft with 12 occupants

Witnesses saw "a large oval shaped object that was domed at the top and slightly rounded on the bottom. It was supported by legs and from a central door- way, or hatch, about a dozen figures could be seen going up and down a ladder-like stairway. The Occupants appeared to be slightly shorter than the average man, and were all dressed in what appeared to be silver colored suits or uniforms. All appeared to be wearing helmets or ski caps, and all were busy running around ""repairing,, the craft. " View full report www.ufoevidence.org/cases/case1143.htm

Source: John Brent Musgrave (FSR Vol 22 # 6 1976) ID: 1143

Occurred: 1932 or 1933
Location: Oakland, CA

About seven brilliant objects in a group flew from 30 degrees elevation in the east to 45 degrees elevation in the west, in an arc. Witness now college graduate, former USAF pilot. (NICAP UFO Evidence, 1964, Hall)

Occurred: **April 18, 1933; Approx.** 1900hrs.
Location: New York Worlds Fair, NY

Witness: Many
Shape: Light
Duration: 30 minutes

The object seemed to be a lighted amorphous cloud-like shape. At the regular evening music, fountain, light, gas jet exhibition a strange-lighted shape suddenly appeared and hovered for twenty minutes over the crowd. Many people saw it and pointed, including my family and me. It made no noise that could be heard over the music and suddenly darted off straight up and disappeared into the star-lit cloudless sky. I still remember it vividly.(((NUFORC Note: Witness indicates that date of sighting is approximate. PD))

Occurred: September 15, 1934; 1530hrs.
Location: Valley City, ND

Witness: 2
Shape: Disk
Duration: 5 seconds

Two brothers observe, hear, three black discs streak overhead, moving from SE to NW; below high overcast. RE-TYPED REPORT: May 18, 2008
TO: National UFO Reporting Center. Not much to explain but is much different, much earlier, than anything I've ever heard about UFO's. Been trying, at times, to figure out whom to tell this to for many years. Looked to National Enquirer years ago. Asked them for an address to write to explain what we saw. They gave me the address in photocopy, which as you can see was undeliverable. So forgot about it until a UFO presentation on TV tonight. Have a computer, but am not very proficient at it, but found your address. The letter I wrote 1990 was returned and have never been opened. Is still sealed. Here goes! I am 83 and my brother is 88. We saw this in

approximately 1934. We were in the country on our family farm 6 miles east of Valley City, ND. It was in open country ¼ mile from the farm in the afternoon about 3-4 p.m.. The sky had high overcast, no wind, very quiet. We were at an old gravel pit, checking for badger holes. We heard a rushing sound, which got our attention. Looking up, we saw the object moving from southeast to northwest. They were out of sight in seconds. To me, they seemed to be perfectly round, black. I likened them to be like a bowling ball in size, very, very close, very low (had to be low to be able to hear the rushing sound, as they passed by), as low as 50 feet, side-by-side, equidistant apart. Not quite overhead, but off to the west of us. My brother says they disappeared in the clouds. I disagree. I say they disappeared in the distance. I am a private pilot, so I know a little more about cloud conditions. The clouds went much too high. They disappeared in a straight, ascending attitude. Ascending in about a 10-15 degree attitude. Was very slight. That's the story, for what it's worth! Much can be said yet about it because we had nothing like this experimentally. I am quite certain, way back then. Didn't even have much flying, yet. This is not a hoax! This did happen!! If you have any questions, better get at it while our minds are still good! The age is there. Have not told my 88-year-old brother about this writing. My phone number is-- (701) ((number deleted)) my brother's number is (701) ((number deleted)) Listed under his wife's name ((name deleted)) Respectfully Yours, ((name deleted)) ((NUFORC Note: Witness indicates that the date of the incident is approximate. PD))

Occurred: **June 15, 1935;** Approx. P.M.

Location: Crater Lake, CA (actually Oregon)

Witness: 1
Shape: Circle
Duration:
In 1971 I was going thru aunts pictures in attic suite case and found four pictures of her beside old ford overlooking Crater Lake. In background was one small dot to right side in all pictures. DID not seem to be due to camera or dust on negative. My AUNT MOVED IN 1979 AND THRU THE WOODEN SUITCASE AWAY. I wanted to look at pictures in the 80s again and compare the dots and see if same position on all prints. THIS TOOK PLACE IN VANCOUVER BC.
((NUFORC Note: Date is approximate. The photos were disposed of, apparently. The Crater Lake we are familiar with is in Oregon. Perhaps there is one in California, as well. PD))

SECOND WORLD WAR YEARS
SECTION 3: 1936-1945

Einstein's E=MC2 has been around for a while as Nuclear Development is on the rise with uranium enrichment becoming mastered.

In 1936 the Nazi's allegedly found a downed UFO in the Black Forest, which they promptly disassembled and began the task or reverse engineering. Out of it came many discoveries that advanced their war effort. Many UFO Sightings became a matter of national and global security and were not disclosed at all. Nicola Tesla, who by now had introduced 25 years earlier AC current, had developed radar far before Raytheon. Tesla was being watched by the CIA and FBI while he was developing his Free Energy, or 0 point energy transmission as it was stifled by those stealing his patents. Even though that research credit including the radio was given back to him, Tesla's genius scared many, including the government. He developed the 'Death Ray' that would vaporize a battleship or plane at 250 miles, now called a laser weapon. He also developed a waveform generator, or a variable oscillator, capable of oscillating over a broad spectrum, able to cause earthquakes. He aptly called it an, 'Earthquake machine' that now anyone can build with Radio Shack parts for under ten dollars.

Werner Von Braun, the great German Rocket Scientist working for the German's admitted at the end of the war, on film, the he was in fact, " being aided and given ideas by extra-terrestrials."

The German Scientists at the end of the war were all moved over to the U.S. to continue their research for the United States as "Free Citizens"(war criminals working secretly for the U.S.) in secret installations for JPL, Bell Labs, Lockheed, White Sands and more, to keep the U.S. dominant in space and war technology.

The Second World War's accompanying technological breakthroughs of radar, missiles guidance systems, short wave radio, nuclear energy and weapons, the ominous appearance of many UFO sightings increased a thousand fold for good reason. Along with the increasing interest in Theosophy, occultism which Hitler was indulging in, the discoveries being made in Ancient Archeological sites of the near East, South Americas, psychic research being conducted, the research Germany was doing with genetics and cloning, all focused humanity on subjects not normally viewed. It pushed the world into a quantum leap, a technological game for dominance and survival, so the steep climb in breakthroughs, it was incredible.

Hitler had an intense interest in the occult and obtaining as much information and relics as he could. The Indiana Jones movies although fictionalized, did convey the twisted passion of the Nazis for global dominance, purifying the race back to what Hitler believed were our alien roots genetically. Using eugenics and genocide he almost had his way. His group of blond-haired, blue-eyed, pure race was supposed to be offshoots of a group of aliens he called the Brill or Aryans, whom he was preparing to meet with when he achieved world dominance. With the alleged 1936 crash in the Black Forest of an extra terrestrial craft and rumors of his science team reverse engineering these craft, it was of significant importance to the free world and its security which involved the U.S. much more than the noted reason of just saving the Jewish People. It also was a race for the U.S. to keep global financial dominance.

Knowing a well respected leader of the CIA underground (not yet called the CIA) who acted against Hitler, disclosed to me on many insider occurrences of that time, including Hitler's use of visualization and Black Magic to achieve deceptive illusions and the use and progress of not only space/time travel but of past life regressions to find out what actually occurred once a person got beyond the amoeba stage. Allegedly, the spacecraft that landed in Pennsylvania a decade later was the craft one of his (Hitler's) physicists escaped Germany in near the end of the war. That event is still undisclosed. His scientists were on the cutting edge of many breakthroughs in medicine, genetic engineering, nuclear power, alternative energy, zero point energy and anti gravity through use of vaporized Mercury for levitation of aircraft, along with rocket technology, all gleaned from ETs. The U.S. government still has not disclosed vital and significant information, regarding the use of self-hypnosis to genetically alter ones self to the point of undoing evolution, so an individual discorporates entirely. These experiments were carried out for several decades and had their start in the war, using Edgar Cayce's information as a starting point. The U.S. also had to stay neck and neck with Russians in the arena of PSI (psychic research and application). The Americans had successfully transmitted a mouse over fax lines from Washington, D.C. to U.C. Berkeley in the late 1940s up to the 1960s. Later, with the "Philadelphia Experiment" which used massive high electrical energy to disrupt the electromagnetic fabric of time and space, using the Tesla's principles, created portals or wormholes, but it was accidental and yet to be mastered. Unfortunately, the U.S. citizens have no idea that there is a second government above and below the government that poses as the judicious, God fearing government we are supposed to trust. That is a front for much deeper and more accurate powers that control most of the revenue, science and research that poses as a free world when it is not. We have seen the cracking of this façade over the last 47 years and most people realize now they are being duped into believing a total charade exists, that is covering a murky and deceptive, larger than life scenario. This is not a conspiracy theory but actual historical fact.

They have in their minds that the reality of things would plunge the world into chaos, terror, anarchy, and it would but it would be for the truth rather than for some violent scheme.

Chaos simply is the motion between stepping from one age to another as much as it was from the age of the horse and buggy to the age of technology. The excuse is that they think most people would die if their religious beliefs were challenged with fact, when in fact would and could make ones own spiritual connection the more genuine and real. It is also a rationalization to keep the higher-ups real intent obscured, and away from public scrutiny. Beyond world dominance the directive may be more simple and arrogant than that.

You will notice during the actual War years there are not a whole lot of sightings reported, but there were a lot of sightings. From the 'Foo-Fighter' electromagnetic balls of static charges Hitler's engineers created, to scramble and mess up the engines of airplanes, to some of the efforts of the world military, many things remained out of sight and out of the media for obvious reasons of national and global security. Handful of sightings occurred in Los Angeles and other strategic spots, points of conflict in the European arena and Japan. NICAP.org has a plethora of World War Two sightings listed for those with interest in that time. Only from stories from GIs about seeing strange craft, or airmen sworn to secrecy do we get a flavor of the times. We do however have enough sightings available before during and after to fill in the blanks. 1947 and 1952-53 were busy years for sightings. New missile silos, nuclear and microwave applications and technologies were all being experimented and applied and I'm sure we were being observed not only by other countries but other eyes as well.

You will see as you flip through this decades selected chronology, we included some of the political going on about the newly forming UFO study groups, their evolution and connection to Washington, The Air Force and how they were viewed. There were several flaps as they call it, when massive sightings, many craft occur, and over the entire planet in specified places. Take note of this as the West Coast and our area, from Corning on up and into Oregon are very well visited spots, in the 1940s through the 1970s and on…

Author's Note: As previously noted, all the following sightings are how they appeared with no adulteration or alteration from us, including the formatting from the various sources. We tried to preserve the original formatting, but altered some dating layout for convenience.

Occurred: May 1940
Location: Boulder Mountain, Montana
Witness: 1
Shape: disc
At an isolated location miner Udo Wartena saw a large disc shaped object about 35 foot high and over 100 feet across hovering above a meadow. The object resembled two soup plates, one inverted over the other and stainless steel in appearance. Wartena then saw a staircase unfold from the bottom of the craft. Out of it came a man who asked him if the ship could take some water. The man then invited Wartena inside the object. Wartena accepted and met another man inside that told him they had come from a distant planet and were 609 years old. View a full report from www.ufoevidence.org/cases/case1032.htm   www.nicap.org
Source: Warren Aston, UFO Magazine March/April 1998   **ID:** 1032
Case Type: Standard Case   **Features:** Contact, Humming, Water-Related, Physical Trace

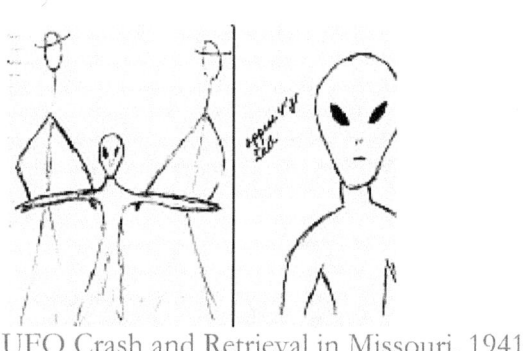
UFO Crash and Retrieval in Missouri, 1941

Occurred: April 1941
Location: Cape Girardeau, Missouri, United States
Witness: 1
Shape: Crashed craft, 3 bodies

One of the most mysterious stories of a crashed UFO with alien bodies preceded the well know Roswell events by some six years. Reverend William Huffman was summoned to pray over alien crash victims outside of Cape Girardeau, Missouri in the spring of 1941. He was shown three victims, not human as expected, but small alien bodies with large eyes, hardly a mouth or ears, and hairless. View full report

www.ufoevidence .org/cases/case858.htm nicap.org

Source: Excerpt from UFO Casebook (BJ Booth) ID: 858

Occurred: July 2, 1941; 1130hrs.
Location: Forest Home, CA
Shape: Circle
Duration: 2 minutes
Small soundless circular craft w/2 beings wearing large goggles hovered watching before streaking away.
In the summer of 1941, for a couple of days, my family & I were the guests of a neighbor who was affiliated w/ a women's lodge in Forest Home, CA. One morning, my father & I went on a walk near the lodge looking for pinecones. I spotted a large one in a tree & my father worked his way into the branches to get it for me. Suddenly & silently a small silvery circular craft appeared hovering above us w/2 beings watching us through large goggles or something like gas masks. I screamed at my father to look, but his glasses had been brushed off in the branches & could not acknowledge my sighting. I continued screaming in terror and suddenly the object streaked away without a sound. Without waiting for my father, I ran as fast as I could back to the lodge where my mother & friend could not console me in my fright.
((NUFORC Note: Witness indicates that date of sighting is approximate. PD))

Photo of UFO over busy street in China, 1942

Occurred: **1942**

Location: Tiensten, China

Young Japanese student Masujiro Kiryu, went through his fathers scrapbook of photographs from the China Campaign, just before World War II, and observed the strange cone-shaped object in the sky above a Tientsien Street. Two people are pointing up at the object while others stare up at it. www. nicap.org

Source: Wendelle Stevens, reposted on UFOArtwork.com **ID:** 259
Case Type: Standard Case **Features:** Photo

The 1942 'Battle of Los Angeles'

Occurred: February 25, **1942**

Location: Los Angeles, California, United States

Imagine a visiting spacecraft from another world, or dimension, hovering over a panicked and blacked-out LA in the middle of the night just weeks after Pearl Harbor at the height of WWII fear and paranoia. Imagine how this huge ship, assumed to be some unknown Japanese aircraft, was then attacked as it hung, nearly stationary, over Culver City and Santa Monica by dozens of Army anti-aircraft batteries in full view of hundreds of thousands of residents. Imagine all of that and you have an idea of what was the Battle of Los Angeles. **Source:** Jeff Rense, Rense.com
ID: 509 Case **Type:** Major Case **Features:** Photo, Mass Sightinghttp://www.ufoevidence.org/Cases/CaseView.asp?section=sortoldest National UFO Reporting Center Case Brief NUFORC

Young Danish boy witnesses UFO in dogfight with three U. S. P-51's. Occurred: May 4, 1945

Subject: German rfz (ufo) May, 4, 1945.

A special event took place on 4th may 1945 at 16.03 when 3 Mustang flying over a very important ((important)) German radar station area Camp "Lindwurm" near Fraer, Jutland Denmark as this event took place:

All three pilots mentioned above ((below)) were on their way back to their base in England from a raid in Danish water "Kattegat" - but they were on their course crossing a big and very important ((important)) German military Radar- and listening Camp- codename "Lindwurm" - with 6 spec. Radar/listening devices/system were placed among other equipments etc. - I think they all three pilots got very surprised on this spot which looks peacefully - here is the event as I remember it:

On that this very last day 4th may 1945, just 14 hours before the second World War (the German surrender) ended both in Denmark and Germany was on a Friday 4th May 1945 and precisely at 16,03 hours did 3 Mustang Jagerfly IV pass by on crossing Jutland and they were passing by on course 280o. - A schoolboy (((witness name deleted)) - 13 year) has stopped on the road, when he heard this special airplane noise which only could come from the allied flights. The boy stopped on the road, on his way home from school, just outside one of the German Wurzburg Reise radar-station - close to the road (30 m) and when he discovered three Mustangs - he signaled to all 3 pilots by waving his arms. The pilot in front tipped with his wings from side to side back to the boy. On the precisely the same time a German flak open fire on these 3 Mustang IV from ((Squadron # deleted)) Sqdn. (Madras Presidency).

All 3 Mustang-flight's now turned 360o around and back again to fire on this creasy German flak, hiding in on a very small forest area - Pilot P/O ((pilot name deleted)) here was hit in his arm, and the aircraft was holed. The P/O ((name deleted)) crash landed few minutes later on an open field, approx 1 Km east of the city Hornum village, where he was taken care of and brought to Løgstør Hospital by civilians (local doctor), just before the German army enters the crashing plane. - The German army now went to Løgstør Hospital to capture ((name deleted)) as prisoner, but the hospital chief/doctor refuse to do it, as he said: "this patient was in treatment." - P/O ((pilot name deleted)) soon recovered and was back in Great Britain within 2 weeks time. The schoolboy shortly after that got in serious trouble from two Germans soldiers, who was stationed at this Wurzburg radar, nearby him - they shout, threaten and running after the boy, they want to give him a beating. The boy jump on his bicycle as fast as he could away home.

A very special event took place as those 3 Mustang turning 360o around to fire back on the flank. Then they aprox. after 90o all meet a colored "grey hat formed object" hanging in the air - max. 100 m of the ground, without no sound! When this "grey hat-formed object" realizes that it now could get in trouble, it started at ones away from those Mustangs in direction SSW. When "the grey hat object" (a German "Haunebu" Vril-7) was on its side marked with a German cross. This object did started with a short heavy noise, a sound as you may imaging coming from many metal dustbins, like rumbling and away was this "grey hat-object" as fast as a now a day Jet-fighter.

This report on this document I hereby sign, testifying as the witness, to that the above report, confirming that this incident did take place as described here. in June 1999.

Please contact: ((name deleted)) e-mail: ((e-address deleted)).tele.dk November 1997 - Signed by my own hand. Note: The special event on 4th may 1945 at 16.13: I would like to get in contact to one or all 3 RAF-pilot who was stationed on Banff Air Base in northern England in may 1945:

Mustang Jagerfly ((a/c # and pilot #2 name deleted))
Mustang Jagerfly ((a/c # and pilot #3 name deleted))
Mustang Jagerfly ((a/c # and pilot #1 name and birth date deleted))
All three Mustang on crossing out over Jutland, some flak positions by Mariager (correct pos.

was Fraer near Skoerping) opened up fire. Mustang Jagerfly ((pilot #1)) was hit in his arm and the aircraft was holed. Bell crash-landed near Løgstør (Aars) and he was taken to Løgstør Hospital. He soon recovered. - Some sources also state that the Banff Strike Wing was "escorted" by their Air Sea Rescue Warwicks as well on 4 may 1945.

It has later been confirmed that P/O ((pilot #1)) has been killed in Kenya by the "Mau-Mau movement" in Kenya sometime between 1952-56.

I'm willing to answer any questions and to your first question I can replay: I did not see this rfz object before the incident tool place because my interest was the well knowing noise from this P-51 a noise which we during the 2. WW have heard many time but only at nights and never have seen, now was possible to see in full daylight and this was a exciding moment for me. After following the P-51 around in the air I here for the first time got the view of this rfz object there was hanging quit was just after the Germans try to shoot the P-51 down and they now turning around very fast- here I got the first look at the rfz -object.

Second question: the rfz did take off from its hanging position after the P-51 was in very close contact to this rfz object and the rfz fly of in the direction SSW south/south/west in high speed possible and my only guess is that it did try to avoid any further contact.

i do hope that my response to your questions be content sincerely ((witness name deleted))

P.S.--My English is not the best but I hope we understand each other.

((NUFORC CORRESPONDENCE WITH WITNESS)

Dear Mr. ((name deleted)),

Thank you very much for the most interesting letter!! The report is remarkably well detailed, and it certainly has the sound of authority to it.

There is little doubt in my mind but what the incident occurred precisely as you have described below.

I have many questions about the gray object, and would like to correspond with you about it. Would you be willing to answer a few questions, please?

Most of all, I would like to know whether you witnessed the object ahead of time, before the P-51's arrived. Also, I would like to know what happened to it, in the final analysis. Did it fly off, in order to escape from the P-51's, or did it stay in the same place?

I will enter your description of the incident as a report, and post it to our website in the near future.

Thank you, again, for sharing the information with our Center!!

Cordially,

Peter Davenport

((END NUFORC LETTER))

Occurred: August 16, 1945

Location: San Antonio, NM-United States

Witness: 2

Shape: Crashed craft

Reme Baca and Jose Padilla were young boys living in San Antonio, New Mexico in August 1945 when, they say, they literally stumbled across the remains of what they believe to have been an alien spacecraft. Their personal account of the case displays many of the key ingredients of crashed UFO lore. View full report www.ufoevidence.org/cases/case852.htm

Source: Ben Moffett, The Mountain Mail, Socorro NM, Nov. 2, 2003   **ID:** 852
Case Type: Standard Case   **Features:** Crash/Retrieval, Humanoid/Occupant, Children, Witness Photo

POST WAR TECHNOLOGY SIGHTINGS, GROUP FORMATIONS
SECTION 4: 1947- ON

With the discovery and developing use of enriched Uranium, the playing field on the planet became quite different. The UFO sightings, as one could guess, became much more abundant and there began emerging patterns of sightings, over Air Force Bases, Military installations of all flavors, and any covert research centers they were probably watching. We can only surmise, from some of the patterns witnessed, that there are still hidden bases and installations the ETs may have an interest in. Sightings began to occur more and more in Northern California, which we will focus on in this section of the book. Take in mind that we include down to Yuba City and up to Central Oregon as part of the Mount Shasta Range of sight. Having included sightings from Europe and the Continental U.S., the reader gets a sense of the "FLAPS" or flurry, of sightings occurring between 1947 and 1952. There were countless sightings, hundreds of witnesses, with many Air Force personnel involved. Missile silos were being built, nuclear proliferation was in full swing and technology on full bore. In this section, many regional sighting, the Roswell Crash and several other crashes most of us don't know about are illustrated. 1947 holds a record for famous sightings of which we listed, so you can get a feel of what it must have been like to pick up the newspaper in that year. All these sightings have significance to Mount Shasta and the general panic that must have accompanied it.

We deliberately left memorandums, letters and other articles of interest regarding the formation of the CIA, NICAP, NUFORC, MUFON, Project Blue Book and the respective relationships they had with the public, the military and each other. The Military had created problems to try and maintain its anonymity in creating black operations outside the confines of a democratic system in the guise of national security that has turned into a backstabbing power struggle, that discredits the honesty and integrity of our very own system, that many still feel today. The art of deception along with ridicule fueled skepticism and intense mistrust flourished, as the public then and now is lied to daily. The government in its infinite wisdom indulges in the McCarthy era tactics as cold war mentality is ramping up. Keep in mind how little information is disclosed today about our geopolitical climate (for real) let alone any credible aerial phenomena or UFOs and alien life. Does it have anything to do with national security or with those who they decide are worthy of the secrets of advanced technology and advanced races, data and the power it possesses? Also during this time Project SIGN turned into Project GRUDGE that evolved into Project BLUE BOOK, which originated NICAP. The CIA formed out of other groups that incidentally had their own UFO operatives that played down the work of the other groups, to cover up, black out and otherwise stifle and choose what they considered classified. The practices of ridiculing the witnesses became a CIA practice. Fortunately many members of the CIA quit and joined forces with the other reporting agencies to help chronicle and make public as much information as is possible. Major Don Keyhoe, NICAP, Dr. James McDonald and J. Allen Hynek with CUFON and BlueBook ran these organizations. We left the historical and noteworthy actions of each committee in the text so you can get a feel of what these brave men contributed to the scientific investigation of UFOs today. As soon as we reach the late 1950s, our focus is primarily Northern California, but as you will see the UFO activity came is waves or flaps in certain times and patterns according to what it seems our engagement in missile building, and high-tech weapons installation locations.

1947

More than any other year, 1947 ushers in massive UFO sightings from the ground, from airborne pilots and on radar. Multiple craft are spotted with dozens of witnesses, the publicity that was created, gave the UFO phenomena the credibility it deserves.

In the following section are famous cases, highlighted in orange (grey in book bound copy) denoting it has more details available from the source given in the summaries. NICAP is the hyperlink to more detail if you do not have electronic copy. Even though we have included cases like the Roswell crash, there were several more crash sightings involving aliens within our Regional chronology that have gone unsung except to Blue Book.

Ted Bloecher who wrote for NICAP and has been highly active over the years in all the classified sightings, wrote the following eloquent introduction to the 1947 UFO Flap, or flurry. In his honor we include it. Thank You, Ted!

"During the summer of 1947 a bizarre and inexplicable situation developed in North America for which, up to the time of the writing of this report (1967), twenty years later, no satisfactory explanation has been forthcoming. Beginning in the latter part of June, people in widely separated places and from all walks of life began to report having seen shining, high-speed, strangely maneuvering objects in the sky. In most of these reports the objects were described as round or disc-shaped. For more than a week, sightings were made in continuously increasing numbers. On July 4th the reports rose sharply and spontaneously, and for five days there was scarcely any part of the United States that had not been visited by these strangely elusive aerial objects. Reports came from many points in Canada as well. The number of sightings crested on July 7th, and during the next few days reports began to diminish until, about a week later, only a handful were being made from scattered sections of the country. Although the objects themselves had all but vanished, interest and speculation about them continued for some time after. A wave of sightings of unidentified flying objects had occurred. Flying saucers had become part of the language and the subject of fickle interest and ever-increasing confusion.
As most people familiar with the history of the UFO phenomenon are aware, the events of 1947 seemed to begin on June 24th, the date of the sighting made by Kenneth Arnold, while flying a plane over the Cascade Mountains of Washington. The date is partly justified, for it was the report made by this Boise, Idaho pilot and businessman, who sold fire-fighting equipment throughout the northwest, that opened the first chapter in the modern record of UFO activity. But Arnold's was not the first sighting of the period. For weeks before that people had been seeing unidentified objects in the sky and keeping the matter to themselves. An important result of Arnold's report was to elicit from these earlier witnesses their accounts of those previously unreported observations.

The 1947 UFO wave is perhaps the most fascinating of any to examine because of its unique position at the very beginning of the contemporary period of UFO activity in this country. There were no attitudes about UFOs in June 1947. There were no preconceptions, no misconceptions, no policies by either press or public, nor by any official agencies, and certainly no pattern existed concerning the phenomenon by which comparisons might be made. Few people recalled the reports of ghost rockets over Sweden during the summer of 1946, and it was only during the crest of the 1947 wave, on July 6th and 7th, that any connection was made with those earlier phenomena. A few World War II veterans, who had observed foo fighters over Germany and in the South Pacific during the war, were now reminded of those earlier incidents by the widespread reports of flying saucers. But for most witnesses, the experience of observing strange aerial manifestations was completely without precedent and profoundly baffling.

We now know that after 1947 it could be expected that a UFO witness might be afraid to report a sighting publicly for fear of ensuing ridicule and intimidation. This is a reaction we have come to expect, one of the many psychological complexities of the UFO phenomenon that has developed out of prevailing public and official attitudes over a long period of time. But in 1947 there were no such precedents to create this type of fear; these witnesses had seen something unaccountable and their fear was of the unknown, a reaction to something totally new and unexpected. There was no place, outside of science fiction, for this kind of inexplicable experience. The appearance of some new phenomenon was not just frightening, it was against all common sense, and if something in someone's experience does not make any sense, it is not likely that this experience is going to be made public - at least not until it is discovered that others have shared the same baffling experience. And so to many, it must have come as something of a relief to read of Kenneth Arnold's sighting, and to discover that they had not taken leave of their senses and were not the only ones to have come face to face with something they were quite unable to explain or understand.

Ted Bloecher

Bloecher:

There was approximately one sighting every other day for the first half of the June. These were scattered over the Midwest and western United States. Then the sighting rate doubled to about 2 per day until June 20. I found 3 sightings for June 20, two for June 21, three on June 22, six on June 23, and then the explosion: I found 20 reports on June 24! " Ted Bloecher

Reminder: <u>All underlined titles or titles in **ORANGE**</u> are a *hypertext links to nicap, unless otherwise indicated for the electronic copy*. Once at the nicap site (or indicated site) put in sighting title in the site's search bar if it does not take you directly to article. Many but not all have the entire link address after the sighting highlight, for the printed book version, which is in gray, not orange. We also left the original formats that Blue Book and NICAP used during the "classified" days of sightings 1940's to early 1960's.

Occurred: June 1, 1947; Approx.1800hrs.

Location: Oroville, CA

Witness: 1
Object:
Duration: 1 minute

I was only six, but I was playing jacks in front of my home, I looked up at the ball, and there they were no noise. I know now, they were traveling in a southeast direction. I can't remember what they looked like, only they were not airplanes or helicopters. It might be associated with the Roswell incident...I can't remember what they looked like and it bothers me today. I couldn't get my mother to come and look. It happened in the evening, it must have been warm weather, because I was playing Jacks outside, if it was daylight savings time, it could have been as late as 8:00 pm. They were in formation and made no noise. The only reason I saw them is because I looked up at the little ball I was playing with. This happened in 1947.   ((NUFORC Note: Date is approximate. PD))

NARA-PBB1-12 - June Sightings
www.bluebookarchive.org/page.aspx?PageCode=NARA-PBB1-12

June 2, 1947; Rehoboth Beach (near Lewes), Delaware (BBU) www.nicap.org
Pilot Forrest Wenyon in aircraft flying N at 1,400 ft saw a silvery jar-shaped object 15 inches [?] in size cross in front of the plane at 1,000-10,000 mph heading E on a straight course at same altitude, with a silver-white fire exhaust. [Daytime meteor?] Sev secs. (Project 1947; McDonald list; FOIA; Bloecher 1967)

June 12, 1947; Weiser, Idaho (BBU)
6:15 p.m. Mrs. H. Erickson saw 2 high speed round objects glistening in the sun at high altitude headed SE in trail formation moving up and down twice and leaving a vapor trail that persisted for over an hour. Sev secs. (McDonald list; FOIA; FUFOR Index; Bloecher 1967)

June 14 [23?], 1947; Bakersfield, Calif. (BBU)
2:15 p.m. Veteran pilot Richard Rankin and a young boy saw 10 "almost round" or Flying Flapjack-shaped objects in formation at 9,000 ft and 300-400 mph headed N on a straight level course, then 7 returned on reverse S course at 2:15 p.m. (McDonald list; FUFOR Index; Bloecher 1967)

June 19, 1947; Webster, Mass.
One of the earliest cases of humanoid reports appeared about two weeks later in, I believe, a Worchester paper -- possibly the Telegram. By an older, unidentified woman who saw an occupant inside an object who looked like "a Navy officer." (Bloecher)

June 20, 1947; Hot Springs, NM
8:00 PM MST. Woman and her daughter observed three revolving groups of three discs in triangular formation, straight course, S-NE (Bloecher,17,II-9)

June 21, 1947; Spokane, Washington (BBU)
11:50 [11:55 PST?] a.m. Civilian woman [Mrs. Guy R.?] Overman saw 8 [shiny silvery and slim-- bodied?] disk-shaped objects the size of a housefly at 600 mph [or slower than a 2-engine army plane?] traveling SSW at 7,000 ft one object below an aircraft, then fall with a dead-leaf motion and land before 10 witnesses on the shore of the St. Joe River, Idaho. Sev mins. (Vallée Magonia 57; cf. FOIA; FUFOR Index; Bloecher 1967)

June 21, 1947; Maury Island, Washington
Cat 6. Maury Island Incident (hoax).

Fran Ridge:
Bloecher found 20 reports on June 24! These were mostly in the far northwestern states of Washington, Oregon and Idaho. Sightings were scattered throughout the day from morning to night. After the 24th, the sighting rate stayed at about 10 per day or higher, with sightings occurring not just in the west but also throughout the country.

June 23, 1947; Cedar Rapids, Iowa
Railroad engineer saw 10 shiny disc-shaped objects, very high, fluttering along in a string toward NW. [UFOE, XII]

Ridge:
To many, the events of 1947 seemed to have begun on June 24th, the date of the sighting made by Kenneth Arnold, while flying a plane over the Cascade Mountains of Washington. It was the report made by this Boise, Idaho pilot and businessman, who sold fire-fighting equipment throughout the northwest, that opened the first chapter in the modern record of UFO activity. But as you can tell, this wasn't the beginning.

June 24, 1947; Mt. Rainier, Washington

At 3:00 P.M., pilot Kenneth Arnold, was flying his airplane near Mt. Rainier and noticed some flashes of light, he then saw the source of the flashes; a string of nine very bright metallic objects.

Kenneth Arnold:
"I spent the next twenty to thirty seconds urgently searching the sky all around - to the sides and above and below me - in an attempt to determine where the flash of light had come from. The only actual plane I saw was a DC-4 far to my left and rear, apparently on its San Francisco to Seattle run. My momentary explanation to myself was that some lieutenant in a P-51 had given me a buzz job across my nose and that it was sun reflecting off his wings as he passed that had caused the flash. Before I had time to collect my thoughts or to find a close aircraft, the flash happened again. This time I caught the direction from which it had come. I observed, far to my left and to the north, a formation of very bright objects coming from the vicinity of Mt. Baker, flying very close to the mountain tops and traveling at tremendous speed.... I observed a chain of nine peculiar looking aircraft flying from north to south at approximately 9,500 ft elevation and going, seemingly, in a definite direction of about 170 degrees."

June 24, 1947; Mt. Adams, Wash. (BBU 12)

This afternoon just about the time that Kenneth Arnold lost sight of his objects, Fred Johnson, listed as a prospector, reported watching five or six disc-shaped craft as they flew over the Cascade Mountains. He said they were round with a slight tail and about 30 feet in diameter. They were not flying in any sort of formation and as they banked in a turn, the sunlight flashed off them. As they approached, Johnson noticed that his compass began to spin wildly. When the objects finally vanished in the distance, the compass returned to normal.

Occurred: June 24, 1947

Location: Pendleton, Oregon

a humanoid report on the same day Arnold had his sighting (Bloecher).

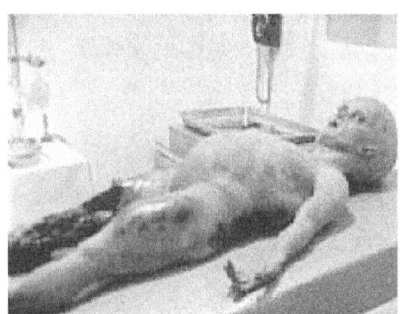

Source, www.nicap.org; www.ufoevidence.org/cases/case809 htm.

255

1947 - Socorro, New Mexico, United States In 1995 London based businessman Ray Santilli caused what has been arguably the biggest controversy in the entire history of UFO research when he launched his 'Alien Autopsy' film across the front pages of magazines and via the TV screen in over 20 different countries. **Source:** Michael Hesemann, Nexus Magazine,1996 **ID:** 809 **Case Type:** Major Case   **Features:** Alien Photograph, Crash/Retrieval, Humanoid/Occupant, Physical Trace

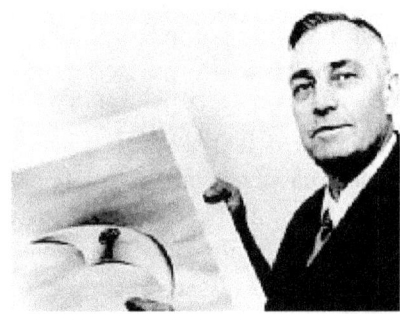

The 1947 Kenneth Arnold UFO Sighting

Occurred: June 24, 1947

Location: Near Mt. Rainier, Washington, United States

June 24, 1947, when Kenneth Arnold spotted nine mysterious, high-speed objects, "flying like a saucer would" along the crest of the Cascade Range near Mount Rainier. His report made international headlines and triggered hundreds of similar accounts of "flying saucers" locally and across the nation. View full report www.ufoevidence.org/cases/case511 Bloecher: According to these (AMC) spokesmen, the investigation at Wright Field was continuing, in spite of what spokesmen in Washington were telling the press. Within twenty-four hours after the release of these official statements, events would begin to take place that would leave everybody, civilians and military personnel alike, in a state closely approximating Ruppelt's description of a flap. Reports of sightings, coming almost simultaneously from hundreds of bewildered citizens, were made to newspapers and police stations all over the country, and adjacent areas as well, from Southern California to New Brunswick, and from Louisiana to North Dakota. People everywhere were experiencing the beginning of one of the most massive waves of UFO sightings on record. Reports came from all kinds of observers: from picnickers and holiday crowds, from policemen and public officials, and from pilots, farmers, professional men, housewives and bus drivers.
Source: Loy Lawhon, www.About.com **ID:** 511
Case Type: Major Case   **Features:** Pilot/Aircrew, Witness Sketch, Witness Photo, Multiple UFOs

Occurred: June 24, 1947

Location: Cascade Mt, Oregon,

compass needle waved wildly as UFO passed overhead.

Fran Ridge:
Bloecher found 20 reports on June 24! These were mostly in the far northwestern states of Washington, Oregon and Idaho. Sightings were scattered throughout the day from morning to night. After the 24th, the sighting rate stayed at about 10 per day or higher, with sightings occurring not just in the west but also throughout the country.

June 28, 1947; 30 miles NW of Lake Mead, Nevada (BBU)

3:15 [1:15 PST?] p.m. AAF pilot Lt. E. B. Armstrong from Brook AAF, San Antonio, Texas, flying F-51 fighter at 6,000 ft saw a tight formation of 5-6 white circular 3 ft objects off his right wing heading 120° [about ESE] at 6,000 ft at 285 mph. (Ruppelt p. 19; FOIA; FUFOR Index; Bloecher 1967).

June 28 [26?], 1947; Maxwell AFB, Montgomery, Alabama (BBU)

9:20-9:45 p.m. 4 AAF officers including 2 pilots and 2 intelligence officers, Capt. W. H. Kayko, Capt. J. H. Cantrell, Capt. Redman, 1st Lt. T. Dewey, saw a bright light just above the SW horizon travel towards them in a zigzag with bursts of high speed, when directly overhead it made a sharp 90° turn and lost to view in the S [SW?]. 25-mins. (Ruppelt p.19; FOIA; FUFOR Index; Bloecher 1967)

June 29, 1947; Des Moines, Iowa (BBU)

3:45 [4:45 CST?] p.m. Bus driver Dale Bays saw a single file line of 4 [18 ?] "Dirty white" round objects between circular and oval in shape, inverted saucer shape about 175-250 ft diameter 12 ft thick, at about 1,200 ft height traveling about 300 mph to the SSE, sound of electric motor or dynamo. Another group of 13 objects were seen heading SSE to NNW [later?]. Few mins. (Battelle/BBSR14?; Mary Castner/CUFOS; FUFOR Index; FOIA; Bloecher 1967)

June 29, 1947; About 20 [15?] miles ENE of Las Cruces, New Mexico (BBU)

About 1:15 [1:20?] p.m. Rocket scientist-engineer Dr. Carl J. Zohn, Admin Asst., Rocket Sonde Section, USN Naval Research Lab (NRL) temporarily assigned to White Sands Proving Ground (WSPG), NRL scientist Curtis C. Rockwood and his wife, and WSPG technician John R. Kauke, were driving in a car from Las Cruces to WSPG headed NE when they saw to their right front [E] a rotating silvery or shiny disc or sphere with no appendages, wings, tail, propellers, reflecting sunlight [pulsating?], crossing the sky at high speed heading N at about 8,000-10,000 ft which suddenly disappeared in mid-air in a clear cloudless sky. Kauke had stopped the car and briefly saw a short vapor trail at one point not reported by the others. Zohn on the passenger side rolled the window for an unobstructed view for nearly 60-secs. (FOIA; cf. Ruppelt, p. 20; FUFOR Index; Randle-Schmitt; Bloecher 1967; etc.)

About June 29, 1947; Jacksonville, Oregon

About 1:00 p.m.. A V-formation of UFOs seen by a group of people on a Sunday either at the end of June or early in July. The date, believed to be June 29, had not been definitely established. The formation was traveling northwest toward Medford, east of the observers. There were nine objects. According to one witness, when first seen the objects were "as white as snow geese"; as they came closer they became blue-white, "like a fluorescent-bulb light." They were sharply outlined and seemed to be solid; "also translucent, like a light, pebbled, frosted bulb." The size of the individual objects was estimated as more than twice the diameter of the full moon -- presumably when the objects were nearest to the witnesses, although this is not stated definitely. (Reference 1, Section II, Page 3)

From: Dan Wilson (Report of the UFO Wave of 1947)
Occurred: about June 29, 1947

Location: Medford, Oregon

To: nicap@insightbb.com form 97BBAbout June 29, 1947, Jacksonville, Oregon: In a sighting report received following the preparation of the Chronology, NICAP obtained information about a V-formation of UFOs seen by a group of people on a Sunday either at the end of June or early in July. The date, believed to be June 29, had not been definitely established at the time of writing. Paul Cerny, of NICAP's Bay Area Subcommittee, sent the report to NICAP.

Early on that Sunday afternoon, a group of people had gathered at Jacksonville, Oregon, a few miles west of Medford, just above the California border. The group included Peter Vogel, M.D., and his wife (now Mrs. Kay G. Kuehnel, of Santa Clara, California, who recently reported the sighting to Mr. Cerny), eight other members of the Vogel family, and about ten others. About 1:00 p.m. PST, a V-formation of oval objects was noticed in the sky above Ashland, 15 to 20 miles southeast of Jacksonville. The formation was traveling northwest toward Medford, east of the observers. There were nine objects.

According to Mrs. Kuehnel, when first seen the objects were "as white as snow geese"; as they came closer they became blue-white, "like a fluorescent-bulb light." They were sharply outlined and seemed to be solid; "also translucent, like a light, pebbled, frosted bulb." The size of the individual objects was estimated as more than twice the diameter of the full moon -- presumably when the objects were nearest to the witnesses, although this is not stated definitely. There was no sound, and no vapor trails were emitted as the formation approached Medford. But when the objects seemed to be over the tower of Medford airport, they each made a spiral ascent, one after the other, and each went behind a cloud that had not been there before and which the objects themselves "seemed to produce."

After the objects had first been noticed in the direction of Ashland, Dr. Vogel went indoors and telephoned the Medford airport tower; Mr. Milligan, Airport Manager, said that he could not see anything. Vogel then ran to his car for his binoculars and camera, but by the time he returned the objects were already out of sight in the cloud, which seemed to be directly over the airport tower.

Both Ashland and Medford were visible from the hillside where the witnesses stood. As the towns are 20 miles apart and the formation had taken about 10 minutes to travel that distance, Mrs. Kuehnel calculated their speed at about 120 miles an hour. Asked to compare one of the objects to a common object that would have appeared similar in the sky, Mrs. Kuehnel answered, "No known object that large and featureless." The objects did not reappear, but the cloud "stayed an oval and stationary shape for over an hour."
Section II, Page 3

June 30, 1947; Near S. Rim of Grand Canyon, Ariz. (BBU) www.nicap.org

9:10 a.m. (MST?). Navy Lt. William G. McGinty flying P-80 from Williams AAF at 30,000 ft heading S saw 2 gray, circular objects about 8 or 10 ft diameter, diving at "unconceivable" speed from about 25,000 ft, which appeared to land 25 miles S of the Grand Canyon. (Vallée Magonia 59; cf. Project 1947; FOIA; FUFOR Index; Bloecher 1967)

Bloecher:

Few of the thirty-nine reports for this period received headline attention when they were printed, and by June 30th, newspaper coverage was not quite as widespread as it had been several days earlier. But UFO sightings would very shortly pick again as the July 4th holiday approached.

In early July the sighting rate climbed beyond 20 per day to 88 sightings on July 4, 76 on July 5, 156 on July 6, 159 on July 7, and a whopping 189 on July 8. After that it dropped quickly back to 20 per day and then only a few per day.

Occurred: July 2, 1947.

Location: Roswell, New Mexico (BBU)

<>9:50 p.m. (MST). Pharmacist Dan Wilmot, wife, and son Paul, at home in downtown Roswell. Fireball came from SE directly or almost directly overhead, heading towards and disappearing over Sixmile Mt. at about azimuth 306° (about NW). (Sparks; Roswell Daily Record, July 8.1947)

July 3, 1947 Roswell, New Mexico

Cat 6 (physical evidence). The Roswell Crash. THIS IS THE FAMOUS ROSWELL CRASH. In order to read it in its entirety go to: **www.nicap.roswelldir.htm**

July 3, 1947 Northern Idaho   www.NICAP.org

This interesting report describes the landing, seen by a family of ten in Northern Idaho, of eight huge objects. This report should have been among those in the Air Force files because it had been reported to intelligence officers from the Spokane Army Air Base, and an intensive air search was carried out by two missions of the National Guards 116th Fighter Group. Local sheriffs deputies also made a ground search, but since no apparent trace of the objects was found, a report was probably never forwarded to Wright Field in Dayton. (Bloecher)

Roswell UFO Incident (Crash and Recovery)
Here is another source of stories about the Roswell Crash. The lower link will take you to a multitude of sources.

Occurred: July 3, 1947
Location: Roswell, New Mexico, United States

In early July 1947 an incident occurred in the desert just outside of Roswell, NM. Many people have heard of the Roswell UFO crash, but very few people know the details of the incident. The following account of the 1947 UFO incident was taken from public records, from information provided by the International UFO Museum and from the press release for UFO Encounter 1997. www.ufoevidence.org/cases/case1134.htm

Source: Roswell Online **ID:** 1134
Case Type: MajorCase **Features:** Crash/Retrieval, Humanoid/Occupant, Military, Witness Sketch, Witness Photo

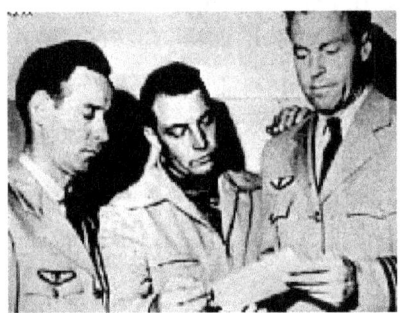
United Airlines Flight 105 pilots witness formation of disc-shaped objects

July, 4, 1947 - Emmet, Idaho, United States

On the evening of July 4th at Boise, Idaho, Captain Smith was walking up the ramp to board his plane, flight 105, for a trip to Seattle when someone mentioned the massive wave of saucers taking place all day over the northwest. Captain Smith joked: "I'll believe in those discs when I see them." Shortly after takeoff five disc-like objects, one larger than the rest, approached Captain Smith's DC-3 head on. View full report www.ufoevidence.org/cases/case723.htm

Source: BB James McDonald, Statement to US Congressional Hearings/UFO Symposium, 1968 **ID:** 723
Case Type: Standard Case **Features:** Interaction/Reaction, Pilot/Aircrew, Witness Photo, Multiple UFOs.

Occurred: July 4, 1947;
Location: Redmond, OR (BBU)
11 a.m. C. J. Bogne of Tigard, Ore., and other witnesses in a car near Redmond saw 4 discs flying past Mt. Jefferson on a straight course at high speed. (McDonald list; FOIA; Ruppelt p. 20; Bloecher 1967).

July 4, 1947; Portland and Milwaukee, Oregon, and Vancouver, Wash. (BBU)

1:05 p.m. Radio newsman Frank Cooley of station KOIN, INS wire service employees in the Portland Oregon Journal Building, Clark County Sheriff's Deputy Fred Krives, Deputy Clarence McKay, Sgt. John Sullivan, Portland Police Officer Kenneth A. McDowell, Harbor Patrol Capt. K. A. Prahn, Harbor Patrolmen A. T. Austad and K. C. Hoff, Portland Police Officers Earl J. Patterson [Paterson?], Walter A. Lissy and Robert Ellis, Oregon Highway Patrol Sgt. Claude Cross, and many others over a wide area saw 5 large discs moving at high speed to the E, 2 flying S and 3 to the E, with oscillating or wobbling motion, sudden 90° turns or zigzagging, radio reports alerted other officers who saw the objects, aluminum or chromium color, disc or hubcap or pie-pan or half-moon shape flashing in the sun, no vapor trail, no noise (except possible humming), some at 10,000-40,000 ft others at about 1,000 ft. McDowell noticed pigeons reacted. Sullivan, McKay and Krives noted low humming sound and reported 20-30 objects. Cooley reported 12 discs at about 20,000 ft. [Further sightings at 2, 4:30, 5 p.m.] Patterson, Lissy and Ellis were pilots. 30-90 secs. (Hynek UFO Rpt pp. 1002; McDonald 1968; FOIA; Bloecher 1967)

July 4, 1947; Seattle, Washington

Photo by Frank Ryman. Coast Guard yeoman took first known photograph of UFO, a circular object which moved across the wind. Photo shows round dot of light. [UFOE, VII, XII]

Occurred: July 4, 1947

Location: Portland, Oregon (BBU)

2 p.m. E. A. Evans saw 3 metallic discs glinting sunlight, 1 moving W to E, followed by 2 others heading N.
4:30 p.m. Mrs. L. J. Hayward saw a silvery disc-shaped object looking like a new dime flipping in an erratic path moving slowly.
July 4, 1947; Portland and Milwaukee, Oregon, and Vancouver, Wash (BBU).
5 p.m. [Other sightings at 1:05, 2, 4:30 p.m.] (Hynek UFO Rpt pp. 100-2; McDonald 1968; FOIA; Bloecher 1967)

July 4, 1947; Near Emmett, Idaho (BBU 34)

9:12 [8:17? 8:12?] p.m. (MST). United Air Lines Flight 105 Capt. Emil J. Smith, First Officer Ralph Stevens, Stewardess Marty Morrow who was called in by Smith as a confirming witness, flying NW on heading 300° from Boise to Seattle at about 7,000 ft, saw 5 disc-shaped objects with flat bottoms and rough tops (possibly 100+ ft size) move at varying speeds, in loose formation [or evenly spaced?] roughly 1,000 ft higher in altitude about 10° left of their heading [or at 290°], with one high and to the right of the others in the distance, all disappearing to the W [NW?] in a gradual climb at about 9:20 p.m. as 5 [4?] additional similar objects came into view slightly higher heading W [or took off to the NW; 3 objects in a line with 1 off to the side]. Smith tried to close on the objects at 185 mph as he climbed from 7,000 to 8,000 ft but could not. 12-15-mins. (Berliner; cf. McDonald 1968; Bloecher 1967; FBI.)

Flying Disc 'Bigger Than Car' Photographed By Youth
July, 9, 1947 - Norfolk, Virginia, United States

The flying disc "was lots bigger then an automobile" and 13-year-old Bill Turrentine of 410 West Fourteenth Street doesn't understand why almost everybody in Norfolk didn't see. He saw the large object, "rocking and spinning like a football" and coming from the southwest. View full report www.ufoeveidence.org/cases/case414.htm

By George Hebert
Norfolk Ledger-Dispatch

July 9th, The flying disc "was lots bigger then an automobile" and 13-year-old Bill Turrentine of 410 West fourteenth Street doesn't understand why almost everybody in Norfolk didn't see.
In fact, when he came to the ledger-dispatch with a photograph he had taken of the object, he wanted to know why the newspaper hadn't already taken a picture of it. "I thought your photographers were fast," he told a reporter.
However, Bill said he had done some fast moving himself shortly between noon Tuesday, when he stood on his front porch and saw the large object, "rocking and spinning like a football" and coming from the southwest.
He had just returned from Summer School classes at Maury, he said, and with all the talk about "flying saucers," had gone out with his camera to see if he could see anything.
I don't see why they call them "flying saucers," he said. "The big one I took a picture of and the two little ones that became behind it a few seconds later didn't look anything like saucers."
Looked Like Football. The boy stuck to his football comparison, explaining that the object was sort of rounded, more oval then disc-like, and though it wobbled in it's northeastward flight it was traveling rapidly, about 600 miles an hour. He guessed that the altitude of the thing was about 5000 feet, just below the clouds. In color it was "gray, almost black," and looked like a "burned crisp," or maybe as rock or stone. The edges glittered, he said, and seemed to be trailing dust. Neither it nor the two that followed made any sound. Bill said he wasn't very surprised when he saw the community, but "almost killed myself" getting shots of it with his camera, set at 1/100th of a second. He called his 18-year-old sister, Josephine, to come and look, but apparently she didn't believe him. As the "flying football" passed over, Bill said he took three shots of it, but when he hurriedly developed the film, after legging it to Olney Road for a vial of developer, only one negative came out well enough for reproduction.
He showed a reporter a contact print he had made himself, and the picture reproduced here is an enlargement made by Photo Craftsmen. Photo experts of this firm are convinced that the boy did a fine job with the old camera he was using. Having closely examined the negative, they said the only flaw was in the kind of film, which perhaps didn't bring out enough detail. They pointed out, that this wasn't really a flaw, in view of the fact that Bill's photograph, with the comparison afforded by the front porch rail and the trees, was certainly the best one taken since the mysterious discs were first reported 1947.

Source: Norfolk Ledger-Dispatch (Norfolk, Virginia), July 9th, 1947   ID: 414
Case Type: Press Report   Features: Photo, Children

Capt. Edward J Ruppelt:
"By the end of July (1947) The UFO security lid was down tight. The few members of the press who did inquire about what the Air Force was doing got the same treatment that you would get today if you inquired about the number of thermonuclear weapons stock-piled in the U.S. atomic arsenal . (At ATIC there was) confusion almost to the point of panic." (Report on Unidentified Flying Objects, p.39)

Michael Swords:
As July wore on into August, (Lt. Col. George D.) Garrett, (General) Schulgen, and (FBI's liaison officer to the AAF/AF) S. Wesley Reynolds became confused by a lack of interest and pressure emanating from the high echelons of the Pentagon. The previous year they had gone through an investigative furor about a subject that they considered to be similar to the flying discs, when hundreds of ghost rocket reports came out of Sweden and other European countries. In 1946, the top brass had exerted continuous pressure to find an answer, but now it had gone completely quiet. This puzzling void has been termed the silence from topside. It was very peculiar to Garrett and the FBI. Their mutual suspicion was that the very highest officials knew what this phenomenon was already (Swords, 1991).

July 30, 1947, "FLYING DISCS"

The following pages are portions of Colonel Garrett's Estimate Report. This is a rough draft of the report. The finalized version has not been located. Page 8 mentions "lack of topside inquiries, when compared to the prompt and demanding inquiries that have originated topside upon former events, give more than ordinary weight to the possibility that this is a domestic project, about which the President, etc. know." Page three text is known but the actual doc page is still being sought.

Disc Zooms Down Snake River Canyon, Idaho

Occurred: August 13, 1947 –

Location: Snake River Canyon, Idaho, United States

A farmer and his two sons were at his fishing camp on August 13, 1947. He noticed an object some 300 feet away, 75 feet above the ground. It was edge hopping, following the contour of the ground, was sky blue, about 20 feet in diameter and 10 feet thick, and had pods on the side from which flames were shooting out. It made a swishing sound. View full report www.ufoevidence.org/cases/case415.htm

Source: Twin Falls Times News, (Twin Falls, Idaho) August 15th, 1947 **ID:** 415
Case Type: Press Report

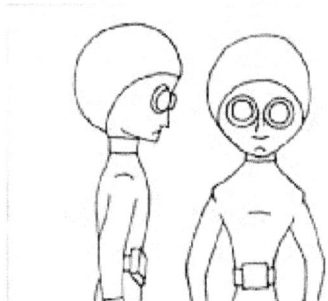

The Villa Santina Case (two humanoids encountered by Italian artist)

Occurred: August, 14, 1947 – Location: Villa Santina, Italy

Italian artist R. L. Johannis was out painting when he noticed a 30 ft. disc-shaped object landing nearby. Next, Signor Johannis saw two child-sized beings standing alongside the object. The artist hailed the creatures. It is possible that this was interpreted as a hostile gesture, for one of the beings touched the centre of its belt and projected a thin vapor that caused the artist to fall dazed onto his back. The creatures approached, picked up the artist's easel, and then returned to the craft. The object then rose, hovered, and disappeared. View full report
www.ufoevidence.org/cases/case751.htm

Source: Charles Bowen (ed.), The Humanoids, Flying Saucer Review Special Issue (1966)

August 14, 1947; 5 miles S. of Placerville, Calif. (BBU)

4 p.m. Insurance adjuster Switzer saw a metallic highly-polished chromium surface object 4-6 ft wide 10-14 inches thick, rounded slightly on top larger in the front, leaving a white trail, at 500-1,000 ft height traveling at high speed. (McDonald list; FOIA; FUFOR Index)
Form 97-BBTime: 1600 hour
Observers: Mr. Ray A. Switzer and Wife
Size: 5 ft. long 12-14 inches in depth Color: Polished chromium
Shape: rectangular
Speed: High rate of speed
Direction: North to South http://www.bluebookarchive.org/page.aspx?PageCode=MAXW-PBB3-85

FBI Memo

August 19, 1947: FBI memo to D.W. Ladd from E.G. Fitch mentioning SAC Reynolds conversation with Lt. Col. George Garrett, re: "there were objects seen which somebody in the Government knows all about.

December 30, 1947; 1 mile W. of Pilot Hill, Calif. (BBU)

7:25 p.m.(PST). Crew of McClellan Field C-47 saw a high speed low altitude object trailing red, green and other colored flames headed E over hills. At 7:58 the crew found a growing ground fire about 7 miles E of Pilot Hill, at 38°50' N, 120°53' W, another C-47 crew sent to investigate found a triangular fire area with 2 points emitting bright blue-green flames, going out at 9:55 p.m. (FOIA) bluebook sighting protocol

Form 97-BB
Date: 24 Oct 2005
From: Francis Ridge
Subject: C-47 Crew Reports Possible Crash of Object, Dec. 30, 1947, Nr. Pilot Hill, California
Category: 11
Distribution: CE, SHG
Pages 373-379Page ID (PID):	NARA-PBB2-373
Collection: NARA Blue Book
Roll Description :   Project Blue Book Roll 2

December 30, 1947; Sawtooth Nat. Forest, Idaho (BBU)

7:26 p.m. (PST). Pilot AAF Lt. Col. W. W. Jones, Hq EPW [Enemy Prisoners of War?], and copilot Major A. A. Andrae, flying a C-54 from Great Falls to Fairfield-Suisun Field at 13,000 ft saw a high speed object trailing green and blue flames descending vertically at their 2:30 o'clock position, but slowing just above the ground. 2-secs. (FOIA)

Pages 242-244, 248, 959-961, 374-375, 377-379 Reference:
1. Report on the UFO Wave of 1947, Ted Bloecher

1948

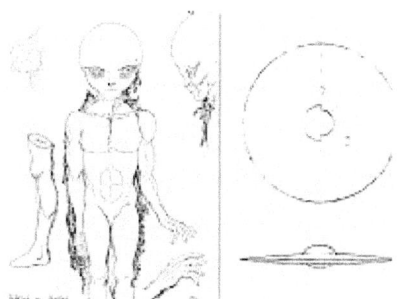

- Aztec, New Mexico, United States Aztec, New Mexico UFO Crash Recovery of 1948

It was the columnist Frank Scully who first alerted the world to sensational stories of recovered flying saucers and little men in his best-selling book Behind the Flying Saucers published in 1950. Scully claimed that up to that time there had been four such recoveries, one of which was alleged to have taken place around Aztec, New Mexico, when sixteen humanoid bodies were recovered together with their undamaged craft. View full report

Source: (Credit: aztecufo.com) ID: 879
Case Type: StandardCase **Features:** Crash/Retrieval, Humanoid/Occupant
http://www.ufoevidence.org/cases/case879.htm

An artist's impression of Capt. Thomas Mantell's fatal encounter with a UFO above the Godman Air Force Base in Kentucky in 1948. (credit: Peter Brookesmith) (Story Below)

Captain Thomas Mantell

Type of Case/Report: MajorCase
Hynek Classification:
Special Features/Characteristics: Military, Pilot/Aircrew, Injury, Witness Photo

The Thomas Mantell Case
January 7, 1948
Fort Knox, Kentucky **Fran Ridge: March 2006, updated 2 Nov 2012**
Air National Guard pilot killed in crash of F-51 during UFO pursuit. Also sighting at Lockburne AFB, Ohio, later same afternoon, UFO maneuvering erratically up and down. [UFOE, V, NICAP site/new evidence] This case was re-investigated in 2006 by the NICAP team and there were major discoveries. The full report in book form and full documentation is produced below.
Brad Sparks:

January 7, 1948. Kentucky-Ohio (BBU)
1-7 p.m. (CST). Mantell case. Kentucky ANG pilot Capt Thomas F. Mantell, Jr., was killed in an F-51D fighter crash at about 3:18 p.m. about 4 miles SSW of Franklin, Ky., (at about 36°40.4' N, 86°35' N) about 92 miles S (more exactly 202° azimuth) from Godman Control Tower, in an attempt to intercept an unidentified object described in one of Mantell's last radio reports to Godman AAF tower, Ft. Knox, Ky., (37°54'23" N, 85°58'00" W, about 725 ft elevation) as "metallic and tremendous in size." USAF 1st Lt. Orner, Commander, 733-5 Detachment, AACS, tracked unidentified white light with red coloration by weather tracking theodolite from Godman Field hangar at 240° azimuth 8° elevation at 5:35 p.m., which disappeared over the horizon at 250° azimuth 0° elevation. See later sightings at Lockbourne AFB and Clinton County AFB, Ohio (below). Expanded & greatly updated version, including new SKYHOOK map

Detailed reports and documents
Preliminary Analysis (Fran Ridge)
The Mantell Incident: Anatomy of an Investigation (2006-2010 A-Team)
Original Case Directory from 2006 with original doc and report links (A-Team)
Mantell Case Analysis (Brad Sparks)
The Mantell Doubters [Oct. 24, 2012] (Brad Sparks)
reports/reports/480108fortknox_carpenter.htm [End Game Map] (Joel Carpenter)
reports/480107mantell_oxygen_thread.htmThe above NICAP links reveal NEW evidence to the Mantel Story, showing definitive UFO involvement in the crash.

Source: Tony Dodd, excerpt from 'Flight to destruction: The strange death of pilot Thomas Mantell'
[go to original source] www.crowdedskies.com_tony_dodd_thomas_mantell.htm

Many stories have been written about the untimely death of Captain Thomas Mantell whose USAF P-51 Mustang aircraft crashed on the 7th. On January 1948, an Unidentified Flying Object was seen hovering in the air close to the United States Army Air Force Base at Godman Field, Kentucky. Like so many incidents at the time, the official Mantell files remain classified, and the truth within them gathers dust in some vault, probably housed at Wright Patterson AFB Dayton, Ohio.

Over the years such UFO related researchers throughout the World have debated stories, but the final proof about such incidents has never been forthcoming. The self-destruct mechanism, which seems to be an inherent part of UFO investigation emerges with uncanny regularity and this coupled with a clever disinformation program, has stifled the startling truth to the present day.

The Thomas Mantell case is no exception to this rule. The tragic death of this brave pilot has officially been put down to pilot error, but who could say otherwise. Mantell could not defend his actions or tell the true story of what confronted him that fatal day. The official Army Air Force verdict, and that shared by many UFO investigators, is that Mantell's aircraft crashed after he blacked out owing to lack of oxygen while attempting to fly too high an altitude in what was later described as a high altitude weather balloon. (It seems that I have heard this weather balloon story before at Roswell)

The case was officially closed and the true circumstances of what had occurred entered the files of Project Saucer, The secret investigation group operating out of Wright Patterson Army Air Field in 1948.

Occurred: August 30, 1948; 1200hrs.
Location: Kelseyville, CA
Shape: Disk
Duration: unknown

In the spring of 1948 in Kelseyville Ca**,** I was laying in a drawer on a pillow out side with mother while she worked, while my 6 year sister played nearby while Mom was hanging clothes on the line a flash of light in the sky caught Moms attention and she watched a silver disk with sunlight glinting on the surface of the craft. The UFO dropped down through the air moving back and forth in a stair pattern, then it moved closer and closer until craft hovered over Mom and she saw a being in the round window, the Being had a sharp shaped face with pointed ears. The middle of the craft stayed still while underneath the bottom of the UFO spun around.The UFO was silent except for the sound of the wind it created which pulled Moms dress and the clothes on the clothesline toward the craft. After a few minutes, the UFO began to rise back up to the sky about 300 ft where it began to wobble in flight then the UFO began to wobble and descend again. The UFO dropped out of site behind the nearby mountain ridge, then there was an explosion and a fire and shortly after that, the forestry dept. made us evacuate. This was the beginning of a life long visits from UFO`s on our family. My Mother brothers report sighting UFO too. ((NUFORC)

Occurred: July 24, 1948
Location: Montgomery, Alabama, United States

Another one of the famous airline sightings of earlier years is the Chiles-Whitted Eastern Airlines case. An Eastern DC-3, en route from Houston to Atlanta, was flying at an altitude of about 5,000 ft.. near Montgomery. ...The object was some kind of vehicle. They saw no wings or empennage, but a pair of rows of windows or some apparent openings from which there came a bright glow "like burning magnesium."

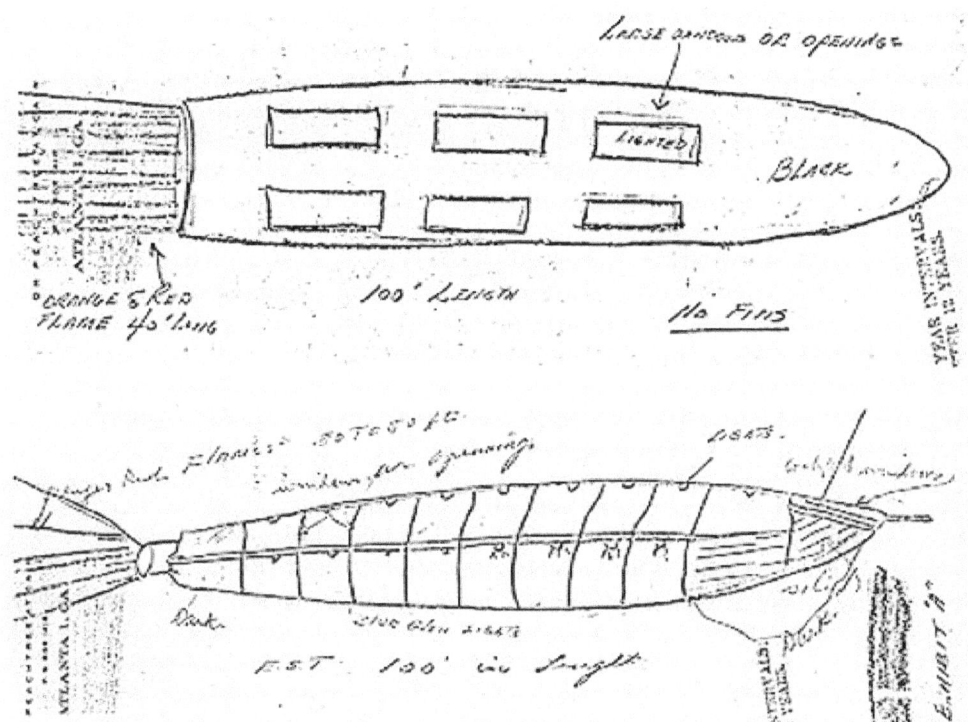

Witness sketches of object.

DC-3 airplane (Eastern Airlines).

Captain Edward J. Ruppelt

On the evening of July 24, 1948, an Eastern Airlines DC-3 took off from Houston, Texas. At about 2:45 A.M., when the flight was 20 miles southwest of Montgomery, the captain, Chiles, saw a light dead ahead and closing fast. R

NICAP

Eastern Airlines pilots, Captain C.S. Chiles and First Officer J.B. Whitted, reported that a cigar-shaped object with lights like portholes approached head-on, accelerated, climbed away. This is a collection of reports and documents from NICAP.

Full Report / Article

Source: Dr. James E. McDonald, U.S. House Hearings, 1968
[go to original source] www.nicap.org

Another one of the famous airline sightings of earlier years is the Chiles-Whitted Eastern Airlines case (Refs. 3, 5, G, 10, 23, 24, 25, 26). An Eastern DC-3, en route from Houston to Atlanta, was flying at an altitude of about 5,000 ft. near Montgomery at 2: 45 a.m. The pilot, Capt. Clarence S. Chiles, and the co-pilot, John B. Whitted, both of whom now fly jets for Eastern, were experienced fliers (for example, Chiles then had 8500 hours in the air, and both had wartime military flying duty behind them.). I interviewed both Chiles and Whitted earlier this year to crosscheck the many points of interests in this case. Space precludes a full account of all relevant details.

Chiles pointed out to me that they first saw the object coming out of a distant squall line area, which they were just reconnoitering. At first, they thought it was a jet, whose exhaust was somehow accounting for the advancing glow that had first caught their eyes. Coming almost directly at them at nearly their flight altitude, it passed off their starboard wing at a distance on which the two men could not closely agree: one felt it was under 1000 ft., the other put it at several times that. But both agreed, then and in my 1968 interview, that the object was some kind of vehicle. They saw no wings or empennage, but a pair of rows of windows struck both or some apparent openings from which there came a bright glow "like burning magnesium." The object had a pointed "nose", and from the nose to the rear along its underside there was a bluish glow. Out of the rear end came an orange-red exhaust or wake that extended back by about the same distance as the object's length. The two men agreed that its size approximated that of a B-29, though perhaps twice as thick. Their uncertainty as to true distance, of course, renders this only a rough impression. There is uncertainty in the record, and in their respective recollections, as to whether their DC-3 was rocked by something like a wake. Perception of such an effect would have been masked by Chiles' spontaneous reaction of turning the DC-3 off to the left as the object came in on their right. Both saw it pass aft of them and do an abrupt pull-up; but only Whitted, on the right side, saw the terminal phase in which the object disappeared after a short but fast vertical ascent. By "disappeared", Whitted made clear to me that he meant just that; earlier interrogations evidently construed this to mean, "disappeared aloft" or into the broken cloud deck that lay above them. Whitted said that was not so; the object vanished instantaneously after its sharp pull-up. (This is not an isolated instance of abrupt disappearance. Obviously I cannot account for such cases.)

Discussion. This case has been the subject of much comment over the years, and rightly so. Menzel (Ref. 24) first proposed that this was a "mirage", but gave no basis for such an unreasonable interpretation. The large azimuth-change of the pilots' line of sight, the lack of any obvious light source to provide a basis for the rather detailed structure of what was seen, the sharp pull-up, and the high flight altitude involved all argue quite strongly against such a casual disposition of the case. In his second book, Menzel (Ref. 25) shifts to the explanation that they had obviously seen a meteor. A horizontally moving fireball under a cloud-deck, at 5000 ft., exhibiting two rows of lights construed by experienced pilots as ports, and finally executing a most non-ballistic 90-degree sharp pull-up, is a strange fireball indeed. Menzels 1963 explanation is even more objectionable, in that he implies, via a page of side-discussion, that the Eastern pilots had seen a fireball from the Delta Aquarid meteor stream. As I have pointed out elsewhere (Ref. 2), the radiant of that stream was well over 90-degrees away from the origin point of the unknown object. Also, bright fireballs are, with only rare exceptions, not typical of meteor streams. The official explanation was shifted recently from "Unidentified" to "Meteor", following publication of Menzel's 1963 discussion (see Ref. 20, p.88).

Other witnesses have reported wingless, cigar-shaped or "rocket-shaped" objects some emitting glowing wakes. Thus, Air Force Capt. Jack Puckett, flying near 4000 ft. over Tampa in a C-47 on August 1, 1946 (Ref. 10, p.23), described seeing "a long, cylindrical shape approximately twice the size of a B-29 with luminous portholes", from the aft end of which there came a stream of fire as it flew near his aircraft. Puckett states that he, his copilot, Lt. H. F. Glass, and the flight engineer also saw it as it came in to within an estimated 1000 yards before veering off. Another somewhat similar airborne sighting, made in January 22, 1956 by TWA Flight Engineer Robert Mueller at night over New Orleans, is on record (Ref. 27). Still another similar sighting is the AAL case cited below (Sperry case). Again over Truk, in the Pacific, Feb. 6, 1953, a mid-day sighting by a weather officer involved a bullet-shaped object without wings or tail (Ref. 7, Rept. No.10). Finally, within an hour's time of the Chiles-Whitted sighting, Air Force ground personnel at Robins AFB, Georgia, saw a rocket-like object shoot overhead in a westerly direction (Refs. 3, 5, 10, 6). In none of these instances does a meteorological or astronomical explanation suffice to explain the sightings.
Source: Dr. James E. McDonald, Prepared Statement on Unidentified Flying Objects, Page 42-43, Hearings, 1968.

August 5, 1948, Top Secret EOTS

The Top Secret "Estimate of the Situation," concluding UFOs were interplanetary space ships, sent to Air Force Chief of Staff. (Brad Sparks: Specifically Aug 5, 1948, the date Keyhoe gave after actually seeing a copy in 1952.)

Ruppelt:
"According to the old timers at ATIC, this report shook them worse than the Mantell Incident. This was the first time two reliable sources had been really close enough to anything resembling a UFO to get a good look and live to tell about it. A quick check on a map showed that the UFO that nearly collided with the airliner would have passed almost over Macon, Georgia, after passing the DC-3. It had been turning toward Macon when last seen. The story of the crew chief at Robins AFB, 200 miles away, seemed to confirm the sighting, not to mention the report from near the Virginia North Carolina state line."

"In intelligence, if you have something to say about some vital problem you write a report that is known as an 'Estimate of the Situation.' A few days after the DC-3 was buzzed, the people at ATIC decided that the time had arrived to make an estimate of the Situation. The situation was the UFO's; the estimate was that they were interplanetary! "

The Gorman Dogfight

October, 1, 1948 - Fargo, North Dakota, United States

On the evening of October 1st, 1948, Lieutenant Gorman was returning from a cross-country flight with his squadron of North Dakota Air National Guard, when he saw an unidentified light source. He closed to within about 1,000 yards to take a good look, later saying, "It was about six to eight inches in diameter, clear white, and completely round without fuzz at the edges." For 27 hair-raising minutes, Gorman pursued the light through a series of intricate maneuvers. View full report

Source: North Dakota Public Radio, Dec. 10. 2003   **ID:** 704
Case Type: MajorCase   **Features:** Pilot/Aircrew

The Gorman Dogfight

Date: October 1, 1948
Location: Fargo, North Dakota, United States

On the evening of October 1st, 1948, Lieutenant Gorman was returning from a cross-country flight with his squadron of North Dakota Air National Guard, when he saw an unidentified light source. He closed to within about 1,000 yards to take a good look, later saying, "It was about six to eight inches in diameter, clear white, and completely round without fuzz at the edges." For 27 hair-raising minutes, Gorman pursued the light through a series of intricate maneuvers.

An F-51 Mustang, like the one flown by Lt. Gorman.

Source: North Dakota Public Radio, Dec. 10, 2003
[go to original source] www.prariepublic.org

Dakota Datebook
December 10, 2003

On this date in 1948, Lieutenant George F. Gorman wrote a letter stating, "...the Air Materiel Command has issued orders classifying the information as Secret. And this makes it a General Court Martial to release any more information. The Command has asked that my commanding officer and myself be court-martialed for releasing what information we did."
The incident the young lieutenant was referring to have since become known as the Gorman Dogfight, one of the early "classics" in UFO history.
On the evening of October 1st, 1948, Lieutenant Gorman was returning from a cross-country flight with his squadron of North Dakota Air National Guard. When the pilots got to Hector airport in Fargo, Gorman decided to log some night-flying time, so he stayed up and circled his F-51 around the city. As he was preparing to land, the control tower advised him that a Piper Cub was in the air. Gorman saw the Piper 500 feet below, but then what appeared to be the taillight of another plane flashed by on the right. The tower insisted there weren't any other planes in the sky, so Gorman told them he wanted to investigate and took off after the moving light. He closed to within about 1,000 yards to take a good look, later saying, "It was about six to eight inches in diameter, clear white, and completely round without fuzz at the edges.
It was blinking on and off. As I approached, however, the light suddenly became steady and pulled into a sharp left bank. I thought it was making a pass at the tower. I dived after it and brought my manifold pressure up to sixty inches, but I couldn't catch up with the thing. It started gaining altitude and again made a left bank," he said. "I put my F-51 into a sharp turn and tried to cut the light off in its turn. By then we were at about 7,000 feet. Suddenly it made a sharp right turn and we headed straight at each other. Just when we were about to collide, I guess I got scared. I went into a dive, and the light passed over my canopy at about 500 feet."
Gorman said he cut sharply toward the light, which was once more coming at him. When it again appeared they'd collide, the object shot straight up in a steep climb-out, disappearing overhead. Gorman again went after it, but his plane went into a power stall, and the object disappeared. The dogfight had lasted 27 minutes. Gorman was so shook up, he had a hard time landing his plane, even though he was a veteran pilot and flight instructor.
The official explanation the Air Force gave was that the light was merely a lit weather balloon. But Gorman's story wouldn't die. In April 1952, LIFE Magazine did a story on UFOs, stating, The Air Force is now ready to concede that many saucer and fireball sightings still defy explanation; here LIFE offers some scientific evidence that there is a real case for interplanetary saucers.
The article went on to describe the Gorman Dogfight: For 27 hair-raising minutes, Gorman pursued the light through a series of intricate maneuvers. He said it was...going faster than his F-51 (300-400 mph). It made no sound and left no exhaust trail. After Gorman landed, the light having suddenly flashed away in the upper air, he found support for his story - the chief of the control tower had followed the fantastic "combat" with binoculars.
That's right. Both men in the control tower saw the whole thing, and so did the two men in the Piper Cub. The Gorman Dogfight has now become one of the most noted UFO encounters in PROJECT BLUE BOOK, the Air Force's official record – and denial – of such things.
http://www.ufoevidence.org/cases/case704.htm www.nicap.org

1949

Occurred: May 24, 1949

Location: Rogue River, Oregon, United States

Employees of aeronautical laboratory see metallic, disc-shaped object

Date: May 24, 1949
Location: Rogue River, Oregon, United States

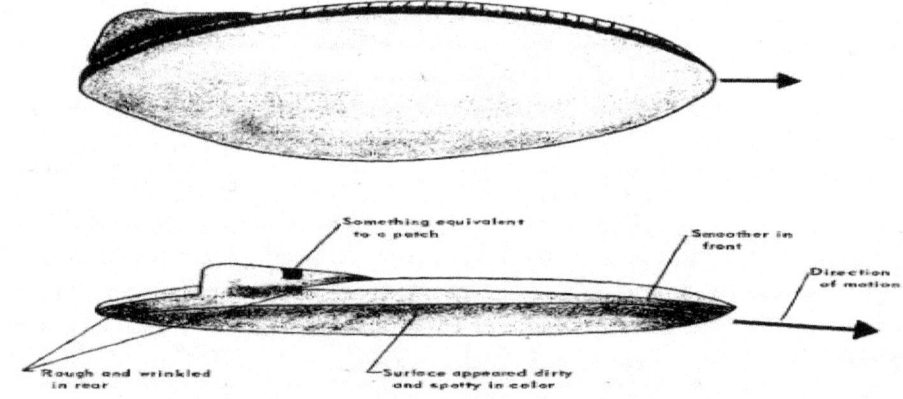

Illustration/diagram of the object from U.S. Air Force Project Blue Book Special Report 14 (Battelle Study) (Credit: UFOs at Close Sight)

Type of Case/Report: StandardCase
Hynek Classification: DD
Shape of Object(s): Disc
# of Witnesses: Multiple **Full Report / Article**

Source: Dr. Bruce Maccabee / U.S. Air Force Project Blue Book Special Report 14 (Battelle Study), 1955

Summary from USAF Project Blue Book Special Report #14:

An employee in the supersonic laboratory of an aeronautical laboratory and some other employees of this lab, were by a river, 2-1/2 miles from its mouth, when they saw an object. The time was about 1700 hours on May 24, 1949. The object was reflecting sunlight when observed by naked eye. However, it then looked at it with 8-power binoculars, at which time there was no glare (Did glasses have filter?) It was of metallic construction and was seen with good enough resolution to show that the skin was dirty. It moved off in horizontal flight at a gradually increasing rate of speed, until it seemed to approach the speed of a jet before it disappeared. No propulsion was apparent. Time of observation was 2-1/2 to 3 minutes.   Other Sources, wwwNICAP.org, Bluebook.org View full report www.ufoevidence.org/cases/case728.htm

The Rogue River Incident
by Bruce Maccabee, from "UFO FBI Connection"

The sighting is important because of the quality of the observations by the several observers, and because the witnesses reported the sighting only to their local security agency, and to the AFOSI.

During the afternoon of May 24, 1950 (sic), five people, three men and two women, were fishing in a boat near the mouth of Oregon's Rogue River. At about 5:00 P.M., they were scanning the river with 8x Navy binoculars looking for signs of jumping fish, when they first noticed a strange circular object approaching from the northeast. They watched it for about two-and-a-half minutes as it hovered east of them before it departed at high speed in a southward direction. The sky was clear and the afternoon sun was at their backs. To the naked eye, it appeared shiny and shaped like a coin with the flat surface parallel to the ground. At its closest it seemed to be only a couple of miles away and about a mile high. They heard no noise.

Two of the witnesses, a draftsman and a wind-tunnel mechanic, were employees of the Ames Research Laboratory at Moffett Field, south of San Francisco. * These two men shared the binoculars. Each man had about a minute to look at the object through the binoculars. The men observed that the object was circular and thin relative to its diameter, with a shape similar to that of a pancake, and with some sort of a vertical fin on the upper surface at the trailing edge. They could see no wings, no antenna, no lights, no propellers, and no jet engines. According to the AFOSI report: The] object appeared round and shiny, something like a fifty-cent piece, viewed from below and to one side. Object's color was silvery and it appeared round in plan view . . . Just before Mn [name censored in the publicly available copy of the original Air Force document] handed the glasses to Mr., the object made a turn on its vertical axis with no tilting or banking ... The trailing edge of the object as it traveled appeared somewhat wrinkled and dirty looking.

* The Ames Research Laboratory carries out research on jet engines, among other things. Employees have Secret or Top Secret security clearances.

The trailing edge looked "wrinkled and dirty;' but there were no exhaust ports. Then they saw it speed off in a southeasterly direction, "accelerating to the approximate speed of a jet plane" in a few seconds without making any noise. About three weeks later the Ames employees reported their sighting to the security office at Moffett Field. The security office then requested that AFOSI agents investigate the sighting.

In 1952, Project Blue Book hired the Battelle Memorial Institute of Columbus, Ohio, to carry out a statistical study of flying saucer reports that occurred between 1947 and the end of 1952. The main intent of the study, called Project Stork, was to determine whether or not there were consistent differences between sighting reports that were explained and those, which could not be explained as mundane phenomena. To carry out this study, the Battelle scientists, working along with the Air Force personnel at ATIC, first analyzed each sighting to determine whether or not it could be identified. The scientists were able to do much better in explaining sightings than the Air Force had done in 1948 and 1949; they were able to explain about seventy percent of the 3,201 sightings they studied. About ten percent of the sightings had too little information for a positive identification. However, about twenty percent had sufficient information for identification, yet resisted explanation. Of these they found a dozen they considered the most descriptive of the sighted phenomena. The Rogue River sighting is the best of that dozen. The two main witnesses were technically trained and, as nearly as can be deter mined from the case file, thoroughly reliable. In the final report of the study, called Project Blue Book Special Report Number 14, the scientists included two rough sketches, based on the verbal descriptions of the Ames employees, showing two views of a circular object with a thickness much less than its diameter, no wings, no tail, and no engine, but a "wrinkled" outer edge.

The object's description by the witnesses is very clear: it was neither an ordinary aircraft nor a hallucination. It either was a hoax or the real thing-a flying saucer. Could it have been a hoax? I would say unequivocally "no" because the two Ames employees reported their sighting to the security officer at Moffett Field. A hoaxer might report a sighting to the press, radio, or TV, but no person of reasonable intelligence whose job depended upon holding a security clearance would try to hoax the security officer at his place of employment. Furthermore, there is no evidence that this sighting was ever mentioned in the popular literature. Apparently these witnesses reported the sighting to no one but the Ames security officer The security officer then requested an investigation by the AFOSI. The AFOSI investigators interviewed all of the witnesses once, and interviewed the Ames employees a second time. The inter views established the high degree of credibility of this sighting.

The Project Blue Book microfilm at the National Archives is supposed to contain all the records collected by the Air Force. In that microfilm two sightings are listed for May 24, 1949. One sighting, case number 402, is explained as "aircraft]' the second sighting, not numbered, is explained as "kites?' The second sighting is not numbered because, according to a handwritten note, the "cards" are missing. This means that someone had removed the copy of the AFOSI investigation report, which originally was sent to Project Grudge, from the sighting file many years before the file was microfilmed and released in 1975. Fortunately, the original AFOSI investigation report was not kept at Blue Book headquarters, but rather at the headquarters of AFOSI, so when the Air Force released the combined AFOSI and Blue Book records to the National Archives, I was able to find the record of this case investigation in the AFOSI section of the microfilm.

How did Project Grudge personnel explain this multiple witness sighting, you may ask? First, accidentally (or intentionally?) they divided it into two parts and treated each separately. The sighting numbered 402 in the master case list contains the interview of one of the two women. According to the AFOSI report, this woman was interviewed on August 8 (the last of the witnesses to be interviewed). The report, in part, read: at approximately 1700 hours, 24 May 1949, she and four other persons, while fishing on the Rogue River near Elephant Rock, approximately 1 '/2 miles above the highway bridge near Gold Beach, Oregon, sighted an object described as being round in shape, silver in color, and about the size of a C-47 aircraft. when first brought to Mrs. [name censored] attention by one of the other witnesses, the object appeared to be three or four miles away. It was coming from the east but later turned to the southwest. It appeared to be traveling at the same rate as a C-47. It made no noise, left no exhaust trail, and made no maneuvers. The interviewee stated that she was not familiar with aircraft; therefore she could not estimate with any accuracy the speed or altitude at which the object was traveling. Mrs. made the comparison between the object and a C-47 because she is familiar with that type of aircraft. Her son has pointed out C-47s to her as they flew over the Gold Beach.

(Note: The names of the witnesses are unknown since they were expunged from the AFOSI records before the records were released to the National Archives in 1975.)

Based on this verbal evidence the Project Grudge personnel identified the object as an "aircraft" because, "No data [were] presented to indicate object could NOT have been an aircraft [capitalization in the original]." The Grudge personnel were able to "get away with" that explanation only because they (1) ignored her claim that the object was circular, and (2) treated her description separately from the descriptions by the other witnesses.

In order to explain the sighting by the other four witnesses, the Grudge personnel used information provided by the AFOSI investigator. This man checked airport records for air traffic or anything else that could conceivably have been in the area at that time. There were no known aircraft in the area. However, the agent learned that radar kites, which are balloons supporting thin metallic radar reflectors, were launched twice a day from military radar installations near San Francisco. He wrote in his report:

These devices are of aluminum sheet, approximately five feet on a side, roughly diamond shaped and containing a double set of triangular fins on the top side. Gas-filled balloons carry these aloft approximately two feet in diameter when they leave the earth. When these devices reach high enough altitude, the expanding gases cause the balloon to burst and the devices known as "kites" fold and drift earth ward. It is possible that one of these devices from one of these radar installations may have been blown as far north as Gold Beach, Oregon, on 24 May 1949.

This is where the official "kites" explanation comes from. (I don't know why the explanation is plural.) Based on this information it is immediately obvious to the most casual observer that what they saw was a radar kite, right?

Think again. The description given above is sufficient to make it clear that a radar kite is not pancake shaped. Furthermore, such a kite, if traveling through the air would be supported by one or more balloons which would have been obvious to the witnesses since the balloons are about the same size as the kite, and it could have moved no faster than the wind. If the kite were falling because the balloons had deflated, the witnesses would have seen it fall rather than accelerate to a high speed and disappear in the distance.

As illogical as it may seem, the Project Grudge and Blue Book analysts, acting on the behalf of the U.S Air Force, have officially accepted "kites" as the explanation for the sighting.

However, the Battelle scientists knew better. They did not accept the Air Force explanation. Neither do I.

I have found another reason to reject the Air Force explanation. An errant radar kite launched from the San Francisco area would have had to be carried north northwestward a distance of about 300 miles. About thirty years after the AFOSI closed their investigation of this case, I reopened it and completed it by obtaining the weather records for the coast of northern California and Oregon. These records show that the winds at all altitudes for the day of the sighting and the day preceding were blowing from the west to the east, and could not have carried a balloon from San Francisco northward to the Rogue River. The AFOSI agent who conducted the investigation of this case could have discovered that for himself in 1949. Apparently he decided to end the investigation once he had found a "possible" explanation.

May 27, 1949; Near Hart Mtn., south-central Oregon (BBU 404)

2:25 p.m. (PST). Oil company vice-president, USNR pilot, former AAF flight instructor, Joseph C. Shell, flew his lightened-load SNJ Navy aircraft from Red Bluff, Calif, to Burns, Oregon, heading NNE at 212 mph ground speed at 9,000 ft above MSL (about 4,000 to 5,000 ft above ground level), saw to his right (about NE) something metallic in the distance [about 20 miles away] at about 42° 38' N, 119° 43' W, which as he approached resolved into 5-8, most likely 6-7, oval or egg-shaped metallic objects, 2:1 length/width ratio, and 1/5 as thick, each the same size less than 20 ft in diameter, fly in trail formation, with an interval equal to 3-4x their length between the lead object and the 2nd object, and only 1/2 to 2/3 object length spacing between the 2nd and all remaining objects, which separation remained constant almost as if being towed by the lead object. He saw the objects "outlined" against the bluffs of Hart Mtn, and could see the dark ground between each object, and noticed they had slightly changed course from a "quartering" path to a path parallel to his course in the opposite direction (heading SSW) at about 230 ±30-40 mph while following the rim of the bluffs (which rim was about 7,000 ft MSL), appeared to be about 1,000 to 1,500 ft below his altitude about 5-1/2 to 7-1/2 miles away at closest approach (to his ESE), but less than the 10-mile distance to the bluffs. Near end of sighting Shell dropped in altitude and then could see the objects at his flight level, at his estimated position 42° 41' N, 119° 49' W, hence his estimate of objects' altitude as about

1,000-1,500 ft below his original 9,000 ft MSL. Objects disappeared on the horizon out of visual range, at estimated position 42° 28' N, 119° 48' W. Visibility being >60 miles.
(Sparks; Berliner; Jan Aldrich; Footnote.com images 6313041 ff.)   5 mins   1 witness   1/30? Full Moon

AVCAT is a special project being conducted by NICAP, with the help and cooperation of the original compiler of AIRCAT, Dr. Richard Haines, and other sources, to create a comprehensive listing of sightings from aircraft with detailed documentation from these sources, including Projects SIGN, GRUDGE & BLUE BOOK.

Occurred: May 27, 1949
Pilot Encounters 5-8 Egg-Shaped Metallic Objects
Detailed reports and documents nicap/docs/490527hartmtn_report.htm (Dan Wilson)
June 5, 1949
Walter Winchell column: "The New York World-Telegram has confirmed this reporter's exclusive report of several weeks before--which newspapermen have denied-- about the flying saucers. Said the front page in the World-Telegram: 'Air Force people are convinced the flying disk is real. The clincher came when the Air Force got a picture recently of three disks flying in formation over Stephensville, Newfoundland. They out-distanced our fastest ships.

Occurred: July 30, 1949
Location: Mt. Hood, Oregon (BBU 496)
9 p.m. Northwest Airlines Capt. Thrush, 2 Portland control tower operators, and a flying instructor (Henry, Penhallegan, Brasford) saw an object with 1 white light and 2 red lights, maneuver and hover. (Berliner; Jan Aldrich) NARA-PBB1-30 - August 1-31 Sightings www.bluebookarchive.org

August 8, 1949, Medford, Oregon
11:20 p.m. to 12:30 a.m. At Medford Municipal Airport, six observers (some military; some civilian CTO's) saw a number of objects, varying from 1 to 7, traveling slow to very fast. the objects were estimated at 30,000 feet and were shiny, and would fly formation for a while and then break off and reform in trail formation. They would disappear for short periods and then reappear in the same general area. Persons from the control tower used binoculars to get a better look at the objects and were able to distinguish wings on the objects but no further identification could be made. Air Force Conclusion: Balls of thistle. (Dan Wilson, BB Archives)

Occurred: August 8, 1949
Location: Medford, Oregon
Report of Sighting of "Flying Discs"
Location: Medford Municipal Airport
Time: 1120 P to 1230 P (Assuming this means Pacific Time)
Observers: Cpl E. ZH. Conger, Medford Airways S/Sgt Arthur Paulson, Medford Airways George E. Milligan, Control Tower Operator Jack Edmonds, Control Tower Operator G. W. White, Communicator CAA Range Station Earl Wescott, Communicator CAA Range Station
Number of objects: Varying from 1 to 7

Speed: Slow to very fast
Altitude: Estimated at 30,000 feet
Color: Shiny
Maneuvers: Objects would fly formation for a while and then break off and reform in trail formation. They would disappear for short periods and then reappear in the same general area.
Comments: Persons from the control tower used binoculars to get a better look at the objects and were able to distinguish wings on the object but no further identification could be made.
Air Force Conclusion: Balls of thistle
http://www.bluebookarchive.org/page.aspx?PageCode=MAXW-PBB7-91
PAGE INFO

Astronomer Clyde Tombaugh UFO Sighting (Astronomer Who Discovered Pluto)

Date: August 20, 1949

Location: Las Cruces, New Mexico, United States

Clyde Tombaugh in 1931 with the telescope he used to discover Pluto (Photo: AP)

Clyde Tombaugh, the American astronomer who discovered Pluto, the ninth planet in our Solar System.

Type of Case/Report: MajorCase
Hynek Classification:
Special Features/Characteristics: Scientist/Engineer, Astronomer, Famous Person, Witness
Photo Source: Wikipedia, the Free Encyclopedia

Clyde Tombaugh was the American astronomer who discovered the planet Pluto. On August 20, 1949, he observed a UFO that appeared as a geometrically arranged group of six-to-eight rectangles of light, window-like in appearance and yellowish-green in color, which moved from northwest to southeast over Las Cruces, New Mexico. He stated:
From Wikipedia: Tombaugh and UFOs
Tombaugh was probably the preeminent astronomer to have reported seeing Unidentified Flying Objects. On August 20, 1949, Tombaugh saw several UFOs near Las Cruces, New Mexico. He described them as six to eight rectangular lights, stating "I doubt that the phenomenon was any terrestrial reflection, because... nothing of the kind has ever appeared before or since... I was so unprepared for such a strange sight that I was really petrified with astonishment." [1] A similar shocked response has been reported by many others, who claim to have seen mysterious aerial objects. Tombaugh was also later to report having seen three of the mysterious Green Fireballs, which suddenly appeared over New Mexico in late 1948 and continued at least through the early 1950s. In 1956 Tombaugh had the following to say about his

various sightings:
"I have seen three objects in the last seven years which defied any explanation of known phenomenon, such as Venus, atmospheric optic, meteors or planes. I am a professional, highly skilled, professional astronomer. In addition I have seen three green fireballs which were unusual in behavior from normal green fireballs...I think that several reputable scientists are being unscientific in refusing to entertain the possibility of extraterrestrial origin and nature." [2]

In 1949, Tombaugh had also told the Naval missile director at White Sands Missile Range, Commander Robert McLaughlin, that he had seen a bright flash on Mars in August 1941, which he now attributed to an atomic blast (mentioned May 12, 1949, in a letter from McLaughlin to Dr. James van Allen). [3] Tombaugh also noted that the first atomic bomb tested in New Mexico would have lit up the dark side of the Earth like a neon sign and that Mars was coincidentally quite close at the time, the implication apparently being that the atomic test would have been visible from Mars.

In June 1952, Dr. J. Allen Hynek, an astronomer acting as a scientific consultant to the Air Force's Project Blue Book UFO study, secretly conducted a survey of fellow astronomers on UFO sightings and attitudes while attending an astronomy convention. Tombaugh and four other astronomers told Hynek about their sightings, including Dr. Lincoln La Paz of the University of New Mexico. Tombaugh also told Hynek that his telescopes were at the Air Force's disposal for taking photos of UFOs, if he was properly alerted.

Some sources, blue book, www.nicap.org
http://www.ufoevidence.org/cases/case355.htm

SECTION 5: 1950-1959

AUTHOR'S NOTE: Starting the 1950s, the magnitude of sightings with use of radar, increased. Sightings over Air Force and Military facilities were also on the rise and created resistance and reluctance of the government to collaborate with civilians with constant denial, cover-ups and ridicule.

We also deliberately left in some international sightings as to illustrate the global nature of the UFO phenomena. This is not some passing fancy to be shrugged off or denied as the military would like, but of great significance to the human race.

January 6, 1950

A new wave of Frank Scully-type hoax stories begin to circulate widely through the media nationwide, including TIME and Newsweek magazines (stories of the AF meeting live aliens, recovering crashed saucers, bodies of little green men). Apparently the new stories were inspired by AFOSI (AF Office of Special Investigations) as a disinformation operation to discredit the Roswell incident in advance, in case Roswell should leak. The AF was fearful that retired Navy-Marine officer-pilot and investigative reporter Maj. Donald Keyhoe, after his blockbuster TRUE article, was hot on the trail of uncovering Roswell, though he was not (the AF had no way of knowing that). AF viewed this as a Navy attack on the AF, exploiting inter-service rivalry and using dirty tricks, and expected more to come (Brad Sparks)

CBS News Anchor Walter Cronkite's UFO Encounter Location: The South Pacific (1950s)

In the 1950s Cronkite was part of a pool of news reporters brought out to a small South Pacific island to watch the test of a new Air Force missile... a large disc-type UFO appeared on the scene. Cronkite guessed that the object was about 50-60 feet in diameter, a dull grey color and had no visible means of propulsion. As Air Force guards ran toward the UFO with their dogs, the disc hovered about 30 feet off of the ground."

Walter Cronkite. Walter Cronkite, anchorman for The CBS Evening News for 19 years (1962–81). He was often cited in viewer opinion polls as "the most trusted man in America,"

Walter Cronkite.

Classification & FeaturesType of Case/Report: StandardCase
Hynek Classification: CE1
Shape of Object(s): Disc
# of Witnesses: Multiple
Special Features/Characteristics: Military, Witness Photo, Famous Person A

Source: Bill Knell, UFO researcher
[go to original source] www.rense.com/general70/crokn.htm

Walter Cronkite's UFO Encounter
By Bill Knell (UFO Researcher)
4-6-2006

Cronkite was very interested in some of the Air Force stories I had collected. He was especially interested in the fact that I had grown-up in an Air Force family as a person interested in researching UFOs. After about 30 minutes of talking, Cronkite said to me, "Let me tell you my UFO story." For the next five minutes I sat in stunned silence as he told me what had happened.

In the 1950s Cronkite was part of a pool of News Reporters brought out to a small South Pacific island to watch the test of a new Air Force missile. After a short inspection of the new system by the reporters, they were lead to an area that was a safe distance from the launch site. The missile was mounted on a specially built launcher that was attached to a cement base. It was obvious that the area had been quickly built just for the test. The details about the missile were going to be given to the reporters in the form of handout sheets and press releases after the test.

Cronkite mentioned that he and the other reporters had been warned that photography of the missile test and any audio transmissions or recordings by the press was forbidden. They would have to give a written account of the event. Just as the test was ready to proceed, everyone was writing as fast as they could. As Air Force Security personnel walked around the perimeter of the test area with guard dogs and the news reporters watched, the missile was fired-up and about to be released. Just then, a large disc-type UFO appeared on the scene.

Cronkite guessed that the object was about 50-60 feet in diameter, a dull grey color and had no visible means of propulsion. Because the noise of activity around him and the missile engine was so loud, he couldn't tell whether the disc made any noise. He did not notice any coming directly from the object.

As Air Force guards ran toward the UFO with their dogs, the disc hovered about 30 feet off of the ground. It suddenly sent out a blue beam of light that struck the missile, a guard and a dog all at the same time. The missile was frozen in mid-air about 70 feet from the launcher, as it had taken off. A guard was frozen in mid-step and a dog frozen in mid-air as it had jumped at the disc. Cronkite reminded me that this all happened within the space of about five minutes or less.

Suddenly, the missile exploded! After that, the disc vanished. The guard and dog looked all right, but were quickly taken away by medical personnel always present at tests in case anyone became injured. At the same time, guards rapidly ushered the reporters into a concrete observation bunker. After about thirty minutes of sitting in that hot box, they were brought out into the air again and addressed by an Air Force Colonel.

The officer told them, "It was all part of the test." Obviously making it up as he went along, the Colonel said that the event was "staged" to test media reaction to UFOs. He reinforced the usual line to the reporters that Flying Saucers were probably not extra-terrestrial, but what people were actually seeing were secret planes being tested by the Air Force. This test was designed to show the media how "shocking" it could be to suddenly view a new technology. Well, Cronkite was certain that what he viewed was a new technology, but he was also sure it was not an Earthly one! He didn't believe the Air Force explanation then, and he still didn't believe it at the time when he told me the story.

After the event, reporters were told that since it was a test of media reaction to new technology, they could not report on it! But, they would be compensated later with exclusive stories on new Air Force projects (a promise that was never kept). Being as private as he was, Cronkite never did share with me his own beliefs about UFOs beyond the story he told me. I was so happy to have heard the story that I was afraid to ask anything further!

The CBS UFO Special was filmed shortly after my meeting with Cronkite and I was included in it for just a few minutes. During the filming, I became aware that Cronkite had not shared his story with most of the other UFO investigators or witnesses. After the special aired I called one of Cronkite's staff members and asked him if he had ever heard the UFO story. He told me he had. Cronkite had only shared it with a few key people and it was NOT covered or even mentioned in the Special.

I wondered why Cronkite had chosen to tell me the story without telling everyone? The staff members told me that most of the others were so busy telling him about themselves that he just never bothered. I guess that the greatest lesson the legendary newsman taught me that day was the art of conversation, knowing when to talk and when to listen!

McMinnville, Oregon UFO Photographs

Occurred: May 11, 1950
Location: McMinnville, Oregon UFO
Case Report: McMinnville, Oregon UFO Photographs

A farmer named Paul A. Trent took two of the most notorious UFO photos of all time on a farm near McMinnville. Beyond many such cases, these two photos have withstood the test of time--through generations of researchers.

McMinnville, Oregon - May 11, 1950: A farmer named Paul A. Trent took Two of the most notorious UFO photos of all time on a farm near McMinnville Oregon. Beyond many such cases, these two photos have withstood the test of time--through generations of researchers. It could be said that this case specified the boiling point for the heated ongoing debate over what qualifies as physical evidence--at least when it comes to the subject of UFOs.

Mrs. Trent was outside getting finished feeding her rabbits, when she noticed the large, disc-shaped object approaching from the northeast. She called out to Mr. Trent who joined her to look at the mystery device. After a few moments, he went back in the house to get his camera. He got back out and picked a spot to stand and took the first picture, then wound the film for the next picture. (That was typically a time consuming process with a 1950 box camera.) Several minutes later, he took the other.

In the meantime, Mrs. Trent called out to her in-laws on their porch, about 400 feet away. They didn't hear her, so she ran into the house to call them on the telephone. Mrs. Trent's mother-in-law didn't get to see it because it disappeared by the time she was off of the phone, but Mrs. Trent's father-in-law did catch a glimpse of it just before it disappeared in the west.

Mr. Trent waited to finish the roll of film before he got the pictures developed. They were displayed nearly a month later in a local bank. (The Trents' banker worked there.) A reporter named Powell, who worked for the McMinnville Telephone Register, saw the photos and convinced the Trents to let him publish their story and photographs. However, it did take a bit of coaxing because the Trent's feared they might get in some kind of trouble with the government.

The photographs and story wound up in the June 26, 1950 issue of Life, who borrowed the negatives from Powell. The Trents didn't get the negatives back for another 17 years. They were requested for examination by the Condon investigation in the late 1960's, and after that analysis was done, they were finally sent back to Mr. Trent.

The first people to examine the negatives found no signs of manipulation of the negative and no visible means of support for the UFO pictured, and so it has remained, for over half a century. No scientific study of this case has ever revealed anything that contradicts the witnesses or their fortunate photos. (NURMUFO) View full report www.ufoevidence.org

Source: STUDIOVNI **ID:** 258

Case Type: MajorCase **Features:** Photo

By Brian Zeiler

Mr. and Mrs. Trent took a classic set of impressive UFO photos in the early part of the evening, just before sunset, on May 11, 1950, near McMinnville, Oregon. According to the Trent's account Ms. Trent, as it appeared over their farm first saw the object, while she was feeding the farm's rabbits. She then quickly called her husband who got the family's camera and Mr. Trent then took two shots from positions only just a few feet apart. The pictures first appeared in a local newspaper and afterwards in Life magazine. Seventeen years later the photos were subjected to a detailed analysis for the University of Colorado UFO Project. William K. Hartmann, an astronomer from the University of Arizona, performed a meticulous photometric and photogrammetric investigation of the original negatives, and set up a scaling system to determine the approximate distance of the UFO. Hartmann used objects in the near foreground, such as a house, tree, metal water tank, and telephone pole, whose images could be compared with that of the UFO. There were also hills, trees, and buildings in the far distance whose contrast and details had been obscured by atmospheric haze.

Hartmann used these known distances of various objects in the photo to calculate an approximate atmospheric attenuation factor. He then measured the relative brightness's of various objects in the photos, and demonstrated that their distances could generally be calculated with an accuracy of about +/- 30%. In the most extreme case, he would be in error by a factor of four. He then wrote:

"It is concluded that by careful consideration of the parameters involved in the case of recognizable objects in the photographs, distances can be measured within a factor-four error ... If such good measure could be made for the UFO, we could distinguish between a distant extraordinary object and a hypothetical small, close model."

Hartmann then noted that his photometric measurements indicated that the UFO was intrinsically brighter than the metallic tank and the white painted surface of the house, consistent with the Trent's description that it was a shiny object. Further, the shadowed surface of the UFO was much brighter than the shadowed region of the water tank, which was best explained by a distant object being illuminated by scattered light from the environment. "it appears significant that the simplest most direct interpretation of the photographs confirms precisely what the witnesses said they saw"

Hartmann further wrote that "to the extent that the photometric analysis is reliable, (and the measurements appear to be consistent), the photographs indicate an object with a bright shiny surface at considerable distance and on the order of tens of meters in diameter. While it would be exaggerating to say that we have positively ruled out a fabrication, it appears significant that the simplest most direct interpretation of the photographs confirms precisely what the witnesses said they saw."

In his conclusion, Hartmann reiterated this, stressing that all the factors he had investigated, both photographic and testimonial, were consistent with the claim that "an extraordinary flying object, silvery, metallic, disc-shaped, tens of metres in diameter, and evidently artificial, flew within sight of [the] two witnesses." [Maccabee op. cit., private communication].
http://www.theblackvault.com/wiki/index.php/McMinnville,_Oregon_(5-11-1950)

Occurred: June 25, 1950

North Korean troops and tanks cross the 38th parallel and invade South Korea in a secret war plan instigated by Soviet ruler Josef Stalin. The Communist aggression opens the floodgates for military funding which increases by 350%, allowing for languishing AF Intelligence and R&D projects to get funded, including UFO investigations (Brad Sparks)

1951

February 1951
Look magazine article: Dr. Urner Liddel, Office of Naval Research, stated "There is not a single reliable report of an observation which is not attributable to the cosmic balloons (plastic "Skyhook" research balloons)."

February 20, 1951 - Air Intelligence Training Bulletin
www.cufon.org
Recognition of Flying Saucers. Document which states, "A flying machine in so featureless a form as a saucer would probably be one of the most difficult things to recognize as such, even in ideal observing conditions."

1952

March 25, 1952. Project BLUE BOOK Named Project Grudge, was upgraded to a separate organization, the Aerial Phenomena Group, and the name was changed to Project Blue Book. According to Ruppelt this change was made because of the steadily increasing number of reports we [the Air Force] were receiving. (Ruppelt, p. 131.)

Turning Point in UFO History - Richard Hall

The summer 1952, UFO sighting wave was one of the largest of all time, and arguably the most significant of all time in terms of the credible reports and hardcore scientific data obtained. Electromagnetic (EM) effects and physical trace evidence were more prominent in other waves, but 1952 (and 1953) featured recurring radar detection of UFOs, often from both ground and airborne radar, visual sightings by jet interceptor pilots sent up to pursue the mysterious objects, and cat-and-mouse chases in which the UFOs seemed to toy with the interceptor.

April 29, 1952, AFL-200-5

Ruppelt:

The number of reports did take a sharp rise a few days later, however. The cause was the distribution of an order that completed the transformation of the UFO from a bastard son to the family heir. The piece of paper that made Project Blue Book legitimate was Air Force Letter 200-5, Subject: Unidentified Flying Objects. The letter, which was duly signed and sealed by the Secretary of the Air Force, in essence stated that UFO's were not a joke, that the Air Force was making a serious study of the problem, and that Project Blue Book was responsible for the study. The letter stated that the commander of every Air Force installation was responsible for forwarding all UFO reports to ATIC by wire, with a copy to the Pentagon. Then a more detailed report would be sent by airmail. Most important of all, it gave Project Blue Book the authority to directly contact any Air Force unit in the United States without going through any chain of command. This was almost unheard of in the Air Force and gave our project a lot of prestige.

May 12, 1952; 40 mi. west of Roswell, NM 9:45 PM UST.

Restricted document shows an unidentified flying object was sighted at 2145 hours UST. The object was blue-green in color and its estimated altitude above the terrain was 30,000 to 40,000 feet. The object traveled three times over approximately the same triangular course. Rate of speed could not be precisely estimated but was faster than that of jet aircraft. Intensity of color brightness varied with the objects altitude. (AF Form 112, Fran Ridge)

Form: 97 Initial Report
Date: Wed, 10 Oct 2007 18:35:43 -0500
From: Francis Ridge <nicap@insightbb.com>
Subject: Sighting nr. Roswell, NM, May 12, 1953, AIIR
Cat: 1
To: CE,SHG, NCP

May 12, 1952; 40 mi. west of Roswell, NM
Restricted document shows an unidentified flying object was sighted at 21h45 hours UST. The object was blue-green in color and its estimated altitude above the terrain was 30,000 to 40,000 feet. The object traveled three times over approximately the same triangular course. Rate of speed could not be precisely estimated but was faster than that of jet aircraft. Intensity of color brightness varied with the objects altitude.

This Air Intelligence Information Report is a follow-up on the May 12, 1952 Roswell sighting. The report notes that the object's "rate of speed could not be precisely estimated but was faster than that of a jet aircraft." It is noted that the object flew three times above the ground each time in a similar triangular course.

May 15, 1952 AIIR: Reporting on Flying Discs (follow-up on May 12, Roswell sighting)

http://www.ufologie.net/

Richard Hall:
Through the first 5 months of 1952, the Air Force Project Blue Book investigators had noticed a build-up of UFO sightings. Then, according to project chief Capt. Edward J. Ruppelt, In June the big flap hit.The objects displayed intelligent control by circling, maneuvering, reacting to pursuit, and otherwise demonstrating extraordinary capabilities unlike any known technology or natural phenomenon, such as sharp turns, rapid vertical motions, and sudden reversals of direction. Radar repeatedly confirmed the presence of unidentified solid objects.

Ruppelt:
The Air Force was taking UFOs seriously because a lot of good reports were coming in from Korea. Pilots were seeing silvery discs and spheres, and radar in Japan, Korea, and Okinawa all had tracked unidentified targets. (Ruppelt, p. 192.)

June 1952 CIA Prepares Secret UFO Report
In the wake of mass public and governmental interest in UFO's kindled by the provocative LIFE magazine article, CIA intelligence experts Sidney N. Graybeal (Chief, Guided Missiles Branch, Weapons & Equipment Division, Office of Scientific Intelligence OSI) and Irl D'Arcy Brent (Chief, Ground Branch, W&E Division, OSI) prepare a summary of the UFO subject for the CIA/OSI hierarchy based on the past several years of OSI intelligence (and OSI predecessor documents going back to ghost rockets of 1946) and mentioning sightings go back to the Bible. The possibility of swamp gas in Michigan as an explanation for UFO's is suggested by Brent. (foreshadowing the Hynek swamp-gas fiasco in Michigan in 1966). (Report has never been acknowledged or released by the CIA despite FOIA litigation. Report's existence and contents was revealed in Sparks interviews with Brent and Graybeal and other OSI officials in 1975-6.) (Brad Sparks)

June 5, 1952 AF Intelligence Initiates Staff Studies on UFO's
AF Intelligence initiates a series of internal Staff Studies on UFO's, inspired by Gen. Garland's new policy emphasizing instrumentation, which are circulated within AFOIN and its field element ATIC. Staff Studies lead to policy and project plan approved by Director of Intelligence, Gen. Samford, on July 28. (Brad Sparks)

Richard Hall:
The summer 1952 UFO sighting wave was one of the largest of all time, and arguably the most significant of all time in terms of the credible reports and hardcore scientific data obtained. Electromagnetic (EM) effects and physical trace evidence were more prominent in other waves, but 1952 (and 1953) featured recurring radar detection of UFOs, often from both ground and airborne radar, visual sightings by jet interceptor pilots sent up to pursue the mysterious objects, and cat-and-mouse chases in which the UFOs seemed to toy with the interceptors. Further, Air Force investigators who plotted the sightings noticed that they were concentrated around strategic military bases, and this clearly posed a threat to national security since their origin was unknown. Senior generals in the Air Force concluded that UFOs were interplanetary in origin, and broadly hinted this belief in LIFE magazine for April 1952.

Early July 1952 Mysterious Dr. "X" Predicts UFO Flap
A mysterious government scientist visits Ruppelt at Project BLUE BOOK and predicts the UFO flap, as hitting New York City or Washington, D.C. I have identified this Dr. "X" as Dr. Stefan T. Possony, Acting Chief of the AFOIN Special Study Group and top scientific adviser to AFOIN Director Maj. Gen. John A. Samford, who was also a leading military strategist and psychological warfare expert. Possony evidently studied the plans for the continental joint SAC-ADC operation Exercise SIGN POST planned for late July and deduced that the planned simulated SAC "attack" on either NY or Washington to test ADC air defenses would trigger false UFO sightings (and in fact SAC did "attack" Washington, but the simulated air raid was on July 23 not on the July 19-20 or 26-27 dates of Washington National UFO incidents). (Brad Sparks)

Occurred: June 28, 1952
Location: Pacific between Hawaii and Calif. (BBU)
10:50 p.m. USAF C-47 pilot saw a very bright light pass across the flight path from left to right. (Project 1947)

Dan Wilson:
June 28, 1952; Pacific between Hawaii and California (29 N 145.20 W)
At 2:50 a.m. local time, Capt. Thomas L. Teate aircraft commander of a C-97 aircraft observed a bright white light that traveled very fast on a horizontal plane on an estimated track of 130 degrees left to right across the flight path of the C-97. The C-97 was on a course of 54 degrees at 9,000 feet. The object seemed to narrow down and completely disappear after 6 seconds. Date-Time 28/2050Z. (Project 1947)

July 10-17, 1952 Dr. Kaplan Visits ATIC Project Blue Book
UCLA Geophysics Prof. Joseph Kaplan, a member of the AF Scientific Advisory Board previously involved with a highly secret compartmented UFO tracking project in 1949 leading to Project TWINKLE, visits ATIC and Project BLUE BOOK, advising on plans for a top scientific panel to establish the importance and credibility of the UFO problem within the scientific community (a later distorted version of the plan is forced on the CIA by the AF as the Robertson Panel and intentionally designed by the AF to fail spectacularly). The Battelle Memorial Institute scientists are deemed not prominent enough to secure support within the scientific community, but will continue with statistical studies of BLUE Books' case files (ordered by Gen. Samford in Dec 1951 to specifically verify Ruppelt's sighting pattern analysis, showing UFO concentrations around atomic weapons bases, after his briefing disturbed Samford). Battelle also continues special lab analyses of alleged UFO physical evidence from time to time. (Brad Sparks)

July 14, 1952 Ruppelt-Maj. Herman Briefing of CSI
Ruppelt and ATIC Maj. Isidore H. Herman present the second ATIC briefing on UFO's for the private CSI group in Los Angeles. (Brad Sparks)

July 15, 1952; Pendleton, Oregon
2055 Zulu. Many civilians in 5 cars and an Oregon State Trooper observed a spherical-shaped object, 35-100 feet in diameter, silver in color, flying very fast at an estimated altitude of 4,000 feet. Object was moving eastbound and appeared to be dipping in flight. Object upon last visual contact at 2100Z, was heading due north. At 2155Z a strike force of six B-36 aircraft were over Pendleton, Oregon, heading north.
Form: 97 BB

Occurred: July 15, 1952, 2055Z,
Location: Pendleton, Oregon, UFO report
Many civilians in 5 cars and an Oregon State Trooper observed a spherical-shaped object, 35-100 feet in diameter, silver in color, flying very fast at an estimated altitude of 4,000 feet. Object was moving eastbound and appeared to be dipping in flight. Object upon last visual contact at 2100Z, was heading due north.

At 2155Z a strike force of six B-36 aircraft were over Pendlton, Oregon, heading north.
http://www.bluebookarchive.org/page.aspx?PageCode=MAXW-PBB12-364

Page ID (PID)   MAXW-PBB12-364
Collection
NARA-Maxwell
Roll Description
Maxwell Blue Book 12
Document Code
N/A
Top of Form

The New York Times
USA, July 22, 1952 "Strange objects in US sky - observed by radar".
 The Washington Post
USA, July 28, 1952   ""Saucer" outran jet, pilot reveals Investigation on in secret after chase over capital - Radar spot blips like aircraft for nearly six hours - only 1.700 feet up."
 The Bridgeport Post
USA, July 28, 1952   "Jet interceptors fail to contact ghostly 'saucers' over capital".
 The Alexandria Gazette
USA, July 28, 1952   "Jet fighters outdistanced by "Flying Saucers" over Mt. Vernon and Potomac" observations in North America around the famous 1952 Washington flap.
 Grand Haven Tribune
USA, July 28, 1952   "See odd objects in Michigan skies".
 Observer-Dispatch
USA, July 28, 1952   "New reports of saucers above state and Washington".
 Wilkes Barre Record
USA, July 28, 1952   "Air Force jet fighters fail to catch objects flying over Washington".
 Ypsilanti Press
USA, July 28, 1952   "'Things' in sky here Sunday appear like flying saucers".
 The Alexandria Gazette
USA, July 29, 1952   "Flying saucers circled the Northern Virginia area again this morning:" observations in North America around the famous 1952 Washington flap.
 The Sheboygan Press
USA, August 4, 1952   "This time it's hard to brush off those mystery "saucers""
 The Washington Post
USA, July 21, 2002   "50 Years Ago, Unidentified Flying Objects From Way Beyond the Beltway Seized the Capital's Imagination:" 50 years later, the Washington Posts remembers the events.
 Athens News
USA, October 17, 2002 "Strange happenings: Similar UFO sightings occur exactly 50 years apart".

The Press, July 22 to Aug 4

Newspaper Articles in BB Files for the period (pdf file)

AIRMAN'S INTELLIGENCE DIGEST

Sometime in August the article by Captain Edward J. Ruppelt for the AIR INTELLIGENCE DIGEST was slated for release. So far the evidence indicates the article was never printed or appeared on another date. In any case all we have is the first draft found in Project Blue Book files above. The spectacular radar-visual sightings at Washington, D.C., on the weekend of July 19/20 were repeated with some new twists on the following weekend.

The Washington Invasion, July 26/27, 1952 - Richard Hall

July 24, 1952. Carson Sink, Nevada. (BBU 1584)
Brad Sparks:
3:40 p.m. (MST). USAF HQ Directorate of Operations Lt. Cols. John L. McGinn (Deputyof Ops, Fighter Br) and John R. Barton (AFOOP-OP-D) flying E in a B-25 bomber at 11,000ft and 185 knots airspeed saw 3 silver white, delta-shaped or arrowhead-shaped objects at their1 o'clock position slightly larger than the size of F-86's (40 ft), each with a ridge along the top, in V-¬formation, cross in front of and above the B-25 from right to left (S to N) at about 1,200 to 2,400 ft away at about 1,800+ mph. (Berliner; NARCAP; cf. Ruppelt pp. 11; NICAP)

The Carson Sink Case

One such case had to be believed because it came from extremely reliable witnesses. Their sighting occurred over Carson Sink, Nevada. It involved two competent observers who were both Air Force officers. These witnesses, Lieutenant Colonel John L. McGinn and Lieutenant Colonel John R. Barton, were also intimately familiar with every type of aircraft or missile in the world at that time. For that reason their sighting drew great attention at both ATIC and the Pentagon when it surfaced. The incident took place while on board a twin engine B-25 bomber that they had requisitioned for a cross-country flight beginning from Hamilton AFB (formerly Hamilton Field). After take off they were headed for Colorado Springs on a very clear day with unlimited visibility. Such perfect flying conditions are appreciated by any veteran flyer, especially the two lieutenant colonels that were looking forward to a smooth ride and the chance to take in some beautiful scenery. While over the Sierra Nevada, McGinn and Barton did see some amazing terrain but soon saw something even more spectacular.
Between Sacramento and Reno they entered "Green 3," the airway's version of a highway into Salt Lake City. At 3:40 P.M. MST while at 11,000 feet over the Carson Sink area of Nevada, the pilots spotted three aircraft ahead of them and to their right. At first the lieutenant colonels assumed these must be F-86 jet fighters. The "bogies" were moving much like the new jets - although something just didn't add up. If they were F-86s, they should be lower in accordance with civil air regulations, and it also appeared odd to see military jets fly in what appeared to be a perfect V formation.
   In short order their B-25 closed in on the objects, close enough for a better look. The pilots then immediately realized they weren't F-86s at all, but what were they? Each craft appeared very bright silver in color with a delta wing-like airfoil. They thought these couldn't be a new type of delta jet because they had no tails or pilot's canopies. The craft all displayed a clean upper triangular wing with a definite ridge running from nose to tail. Before McGinn and Barton fully grasped the fantastic sight before them, the strange objects made a left bank and zoomed within 400 to 800 yards of their B-25, an uncomfortably short amount of space in the air. Their speed estimated by the men was the very least three times that of any conventional jet then flying! Yet after four short seconds the hair raising maneuver was over and the UFOs were gone.

As soon as McGinn and Barton landed at Colorado Springs they were on the phone to Air Defense Command Headquarters. When they learned that no civilian or military aircraft were anywhere near them at 3:40 P.M., the magnitude of their sighting finally sunk in. McGinn and Barton were both command pilots with very distinguished service careers, having logged several thousand hours flying time each. They were assigned to the Pentagon with highly classified assignments and were perfectly familiar with even the most secret foreign and domestic aircraft designs. Neither had seen anything remotely like those objects before, but they indicated that they had probably witnessed what friends of theirs had observed, "flying saucers."

Blue Book made their own study and located all delta wing jets, and then exclusively flown by the Navy, yet none were in the Green 3 area. They also checked other sources, which had no records of aircraft, balloons, or anything of any kind over Carson Sink at the time.36 Although only one of 22 reports which made it into ATIC that day, and just one of about 100 worldwide sightings for the 24th, the Carson Sink Sighting, case number 1584, is the best of those in the files marked unidentified.

Edward J. Ruppelt version:

Here is a "good" UFO report with an "unknown" conclusion:

On July 24, 1952, two Air Force colonels, flying a B-25, took off from Hamilton Air Force Base, near San Francisco, for Colorado Springs. The day was clear; not a cloud in the sky. The colonels had crossed the Sierra Nevada between Sacramento and Reno and were flying east at 11,000 feet on "Green 3," the aerial highway. At 3:40 P.M., they were over the Carson Sink area of Nevada when one of the colonels noticed three objects ahead of them and a little to the right. The objects looked like three F-86's flying a tight V formation. If they were F-86's they should have been lower, according to civil air regulations, but on a clear day some pilots don't watch their altitude too closely.

In a matter of seconds the three aircraft were close enough to the B-25 to be clearly seen. They were not F-86's. They were three bright silver, delta wing craft with no tails and no pilot's canopies. The only thing that broke the sharply defined, clean upper surface of the triangular wing was a definite ridge that ran from the nose to the tail.

In another second the three deltas made a slight left bank and shot by the B-25 at terrific speed. The colonels estimated that the speed was at least three times that of an F-86. They got a good look at the three deltas as the unusual craft passed within 400 to 800 yards of the B-25.

When they landed at Colorado Springs, the two colonels called the intelligence people at Air Defense Command Headquarters to make a UFO report. The suggestion was offered that they might have seen three F-86's. The colonels promptly replied that if the objects had been F-86's they would have been able to recognize them as such. They colonels knew what F-86's looked like.

Air Defense Command relayed the report to Project Blue Book. An investigation was started at once.

Flight Service, which clears all military aircraft flights, was contacted and asked about the location of aircraft near the Carson Sink area at 3:40 PM. They had no record of the presence of aircraft in that area.

Since the colonels had mentioned delta wing aircraft, and both the Air Force and the Navy had a few of this type, we double-checked. The Navy's deltas were all on the east coast, at least all of the silver ones were. A few deltas painted the traditional navy blue were on the west coast, but not near Carson Sink. The Air Force's one delta was temporarily grounded.

Since balloons once in a while can appear to have an odd shape, all balloon flights were checked for both standard weather balloons and the big 100'-diameter research balloons. Nothing was found.

A quick check on the two colonels revealed that both of them were command pilots and that each had several thousand hours of flying time. They were stationed at the Pentagon. Their highly classified assignments were such that they would be in a position to recognize anything that the United States known to be flying anywhere in the world.

Both men had friends who had "seen flying saucers" at some time, but both had openly voiced their skepticism. Now, from what the colonels said when they were interviewed after landing at Colorado Springs,they had changed their opinions.
(Source: Michael David Hall, "UFOs, A Century of Sightings", Page 187 & Captain Edward J. Ruppelt's, Report on Unidentified Flying Objects, Page 10).

July 26, 1952; Williams, Calif. (BBU Missing)

5:15 p.m. (PST). [N Calif. F-94C intercept case involving large orange yellow object moving fast and slow, tracked by airborne and ground radars?? (Weinstein)] Air Defense Command radar detected a UFO, F-94 jet interceptor scrambled, locked onto the object with its radar, crew saw a yellow-orange light. As confirmed by ground and airborne radar, the UFO played tag with the F-94, alternately accelerating away when it got close, then slowing down until it caught up again. (Ruppelt, pp. 222-223.)

RADCAT is a revitalized special project now being conducted jointly by NICAP & Project 1947 with the help and cooperation of the original compiler of RADCAT, Martin Shough, to create a comprehensive listing of radar cases with detailed documentation from all
previous catalogues, including UFOCAT and original RADCAT.

F-94 Intercept With ADC Detection
July 26, 1952
Williams, California
Brad Sparks:

As the interim week came to a close, UFOs were still much on the minds of the people and the military. Ruppelt wrote that on the same night as the second round of the Washington Nationals began, Blue Book received a really good report from California. The ADC [Air Defense Command] radar had picked up an unidentified target and an F-94C had been scrambled. The radar vectored the jet interceptor into the target, the radar operator in the '94 locked on to it, and as the airplane closed in the pilot and RO [radar operator] saw that they were headed directly toward a large, yellowish-orange light. For several minutes they played tag with the UFO. Both the radar on the ground and the radar in the F-94 showed that as soon as the airplane would get almost within gunnery range of the UFO it would suddenly pull away at a terrific speed. Then in a minute or two it would slow down enough to let the F-94 catch it again.

Ruppelt did interview the pilot himself over the telephone. The pilot told Ruppelt that he felt as if they were involved in a big aerial game of cat and mouse with the strange object. The pilot said that he hadn't liked it and was afraid that at any moment, the cat would pounce, possibly destroying his aircraft in the process.

Ruppelt noted in his book, "Needless to say, this was an unknown."

As the California fighter pilot was chasing his single UFO, the formations of strange lights had returned to Washington National. Focus changed as Air Force personnel, including a Naval radar expert assigned to assist them, watched the UFOs dance through the skies over the nation's capital from inside the radar room at Washington National. Before long, everyone, including President Truman, was demanding answers

Source: WASHINGTON INVASION, Page 67.

Please read about the unexpected Washington Invasion and theone staged by the Air Force to get funding to protect our nations capitol. The UFOs did show up for real unlike the staged ones in the covert event.

Joel Carpenter:
JOINT SAC/ADC EXERCISE. The next flying exercise took place on 27 July, as the wing launched 21 B-36s (7-9th, 7-436th and 7-492nd Bomb Squadron) from Carswell, as part of a joint SAC/ADC attack on Detroit, Michigan. En route to Detroit, the bombers were intercepted by Air Defense Command North American F-86 and Lockheed F-94 fighters. The North American F-86 Sabre was the Air Forces first swept-wing fighter, entering operational service in February 1949. The Lockheed F-94 Starfire was the first jet-powered all-weather fighter to enter service with the Air Force and first to feature a speed-boosting afterburner. It became operational in May 1950 with the Continental Air Command. Fighter opposition was considered ineffective as all bombers attacked the target then returned to Carswell the same day.

Occurred: August 19, 1952
Location: Red Bluff, Calif. (BBU 1928)
2:38 p.m. GOC observer Albert Lathrop saw 2 objects; shaped like fat bullets, fly straight and level, very fast. 25 secs. (Berliner)

Brad Sparks:
Blue book listed both these cases under the same number even though five hours apart.

December 2, 1952; CIA Memo
Chadwell Gives Director of CIA His Opinion. CIA knows what UFOs are NOT and is concerned.

Brad Sparks:
Top CIA officials (Chadwell, Robertson, Durant) visited ATIC Project BLUE BOOK to obtain the withheld UFO investigation reports that Ruppelt indicated in phone conversation with CIA missile intelligence officer Frederick C. Durant III on Dec. 9 were being held back from CIA by orders of his boss ATIC Technical Analysis Division Chief, Col. Donald L. Bower, evidently acting at the behest of the AF Intelligence leaders, Gen. Garland and Dr. Stefan Possony. In other words an AF cover-up to help conceal evidence of UFO reality from the CIA.

Col. Bower was blocking Ruppelt's planned visit to CIA in Washington, DC, to prevent him from delivering these reports showing them to be sensational cases (movie film, theodolite triangulation, landing case with burn injuries) but IFO's and not UFO Unknowns or best of the best, as the AF had falsely claimed in the briefing given to CIA on Nov. 25. Ruppelt's investigative reports would have undone too soon the false pro-UFO impression the AF had given to CIA -- the false "UFO" reports were intended to be revealed as IFO's at the CIA Robertson Panel to embarrass the CIA to stay out of AF business, and not sooner. Col. Bower himself had given the deliberately misleading AF briefing to CIA on Nov. 25, falsely promising CIA the AF's "full cooperation," and bringing along the lower-ranking pro-ETH advocate Maj. Fournet whose participation was calculated to reinforce pro-ETH conclusions on the CIA. The AF briefing convinced the leaders of CIA/OSI (Office of Scientific Intelligence) that UFO's were extraterrestrial spacecraft.

Ruppelt gave the CIA team led by Dr. H. Marshall Chadwell (director of CIA/OSI and now convinced of the ET origin of UFO's) dozens of additional "best UFO" reports to study but in fact they were all IFO cases designed to blow up in CIA faces at the Robertson Panel. Ruppelt completely withheld from CIA, and concealed the existence of, his special file of more than 63 Best Unexplained UFO cases, no doubt by direct orders of Col. Bower, whose name keeps popping up in the story of devious AF cover-ups on UFO's in 1952. (Brad Sparks)

Occurred: December 28, 1952
Marysville, Calif. (BBU 2302)
Civilian witness, (es). Case missing. (NARA)

Turning Point in UFO History - Richard Hall

The summer 1952 UFO sighting wave was one of the largest of all time, and arguably the most significant of all time in terms of the credible reports and hardcore scientific data obtained. Electromagnetic (EM) effects and physical trace evidence were more prominent in other waves, but 1952 (and 1953) featured recurring radar detection of UFOs, often from both ground and airborne radar, visual sightings by jet interceptor pilots sent up to pursue the mysterious objects, and cat-and-mouse chases in which the UFOs seemed to toy with the interceptors.

Major Turning Point in UFO History
Almost as incredible as the UFO phenomenon itself was the fact that Gen. Samfords temperature inversion/radar mirage explanation was not strongly challenged by anyone and the explanation stuck. The scientific communities shameful neglect of serious UFO data was largely responsible for this outcome. Apparently Air Force leaders decided at this point that they had tried to send a message, but since the message was ignored and even ridiculed by scientists, that they would change their entire approach to dealing with the subject.

<>Some scientists within the Air Force, not to mention numerous high-ranking operational officers, were not convinced by the temperature inversion theory, but they were outranked. Since some senior general officers in the Pentagon at this time thought that UFOs were probably extraterrestrial, one has to assume that Maj. Gen. Samford was speaking on behalf of higher authority and stating policy decided at the top.

In Project Blue Book Status Report No. 8 dated December 31, 1952, it was stated that several widely publicized theories about UFO phenomena in recent months had been discussed with atmospheric physicists. They have agreed that none of the theories so far proposed would account for more than a very small percentage of the reports, if any. By this time a separate review of UFOs by the Central Intelligence Agency was underway, leading to the so-called Robertson Panel study that was published in January 1953, essentially debunking UFOs as any serious matter for science to address.

Dr. James E. McDonald, University of Arizona atmospheric physicist, later analyzed the weather data for the July 1952 Washington sightings. He said:

The suggestion that an inversion of the sort exhibited by radiosonde data...caused the reported effects is absolutely absurd [his emphasis].... The optics of mirages and the optics of radar ground returns are significantly different in several respects, so that false targets would not seen to lie in the same place in the sky to a visual observer and a radar observer.... Mirage effects are confined to lines of sight that do not depart from the horizontal by much more than a few tens of minutes of arc.

In other words, simultaneous radar-visual sightings and those involving airborne radar lock-on militate strongly against the validity of the theory announced by Samford on July 29. An internal Air Force analysis endorsed McDonalds conclusions, adding:

Our results clearly show that the temperatures and temperature gradients needed to produce mirages, which occur at an angle of one degree, or more from the horizontal are extraordinarily large. It is also clear that these temperatures and temperature gradients are not found in our atmosphere.

The activities, conclusions, and aftereffects of the CIA Robertson Panel are a highly significant story in them. Just as was the case with the later University of Colorado study in 1996-1967, it appears that political considerations overrode objective truth-seeking efforts and nothing remotely resembling real science was done. Instead, scientists were used to advance a political objective in ways that were very unflattering to science and the scientific method.

BIBLIOGRAPHY

Borden, R.C., and T.K. Vickers. Preliminary Study of Unidentified Targets Observed on Air Traffic Control Radars. (Civilian Aviation Administration, 1952.)

Durant, F.C. Report of Meetings of Scientific Advisory Panel on UFOs [Robertson Panel Report]. (Central Intelligence Agency, 1953.)

Hall, Michael David, and Wendy Ann Connors. Captain Edward J. Ruppelt: Summer of the Saucers 1952 (Albuquerque, NM: Rose Press International, 2000.)

Hall, Richard. Radar-Visual UFO Cases in 1952: The UFO Sightings That Shook the Government. (Fund for UFO Research, 1994.)

Keyhoe, Donald E. Flying Saucers From Outer Space. (New York: Henry Holt & Co., 1953.)

McDonald, James E. UFOs: Greatest Scientific Problem of Our Times? Address to annual meeting of the American Society of Newspaper Editors, Washington, D.C., April 22, 1967.

McDonald, James E. Symposium on Unidentified Flying Objects, House Science & Astronautics Committee, U.S. Congress, July 29, 1968. Comments on radar-UFOs, pp. 18-86.

Menkello, Frederick V. Quantitative Aspects of Mirages. (Air Force Foreign Technology Division, April 1969.)

National Investigations Committee on Aerial Phenomena, U.S. Air Force Projects Grudge and Blue Book Reports 1-12 (1951-1953); (Washington, D.C.: NICAP, 1968).

Project Blue Book Unknowns. (Original BB Unknowns are listed as (BBU + the number), cases added by Brad Sparks in his "Comprehensive Catalog of Project Blue Book Unknowns" are simply marked (BBU).

Randle, Kevin. Invasion Washington: UFOs Over the Capitol. (New York: HarperTorch, 2001.)

Ruppelt, Edward J. The Report on Unidentified Flying Objects. (New York: Doubleday, 1956.) Also available online at www.nicap.org
Swords, Michael D. UFOs,

1953

January 10, 1953; 8 miles N.W. of Sonoma, Calif. (BBU 2326)

Jan. 10, 1953; 8 miles NW of Sonoma, Calif. (BBU 2326)_3:45 or 4 p.m. [4:45 p.m. PST?] Retired AF Col. Robert McNab, and Mr. Hunter of the Federal Security Agency saw a flat object to the NW at 45° elevation traveling about 2,400 mph make three 360° right turns in 2-3 secs each in about 1/8 radius required for jets [i.e., about 1/4 mile radius and 300 g's], two abrupt 90° turns to the right and left, each turn 5 seconds apart, almost stop, accelerate to original high speed, almost stop again, speed up again and finally fly out of sight vertically. Sound similar to F-86 at high altitude. (Hynek UFO Rpt pp. 115-6)__Detailed reports and documents
http://www.nicap.org/docs/530110sonoma_docs.pdf

The Miners UFO Incident- Red Bluff/Brush Creek 1953

Red Bluff has been a hotbed of UFO sightings, beginning in early April 1953. John Black and John Van Allen, two titanium miners, were working just outside of their mine near Bluff Creek, when they both observed a "metallic saucer" hovering over the mine area. For the entire story please visit the site, and leave your thoughts. www.1.bp.blogspot.com

The "saucer" returned on four other occasions, during the next few weeks.
On April 20, 1953, Black watched the same craft, which was less than a quarter mile distant. It became quite clear to the two miners that the strange object had a keen interest in their mine.

January 27,1953 CIA letter to Julius Stratton from Marshall Chadwell
Page 8 was part of Dewey Fournet's famous but not seen "motions study" of UFOs as presented to the Robertson Panel. According to what one can reasonably assume by looking at the page, Dewey presented 17 cases [from which he deduced that UFOs were guided by intelligence and the flight characteristics indicated that the intelligence was beyond "us".] This number rings true as Ed Ruppelt says that Dewey sifted his cases down to between ten and twenty. The page shows arrangements of UFOs in the chosen cases, and given the strong likelihood that most if not all of them were 1952 cases on Dewey's watch, they might be specifically identifiable. (Mike Swords)

January 30, 1953; CIA Memo on ONE Briefing
Briefing of the Office of Naval Estimates Board by CIA on Unidentified Flying Objects, included the showing of the Utah and Montana films.

May 1953; AFM 200-3 Promulgated
Handbook for Intelligence Officers. The manual's theme is the importance of having high grade air technical intelligence in order to avoid "technological surprise" from a foreign power. Although UFOs are not specifically mentioned, the illustration of "flying saucers" above an Air Force bomber appears on page 9-3 of the manual, and this chronology shows the timing/context of its publication. (Richard Hall)

May 1953 - CAA Report on Radar Targets
CAA Technical Development Report 180, A Preliminary Study of Unidentified Targets Observed on Air Traffic Control. This flawed report states the July 1952 objects were weather targets. (Courtesy of CUFON)

May 9, 1953; ADDR 200-1
34th Air Defense Division Regulation No. 200-1, Reporting of Information on Unidentified Flying Objects. Including: 34ADD Form 127: Unidentified Flying Object Report (FLYOBRPT)

Occurred: September 11-13, 1953
Location: Chiloquin, Ore
Police Chief, others, watched top-like UFOs three consecutive nights. [UFOE, VII]

December 28, 1953; Marysville, Calif. (BBU 2844)
11:55 a.m. Yuba County Airport Manager Dick Brandt saw a saucer, with a brilliant blue light, reflecting on a nearby building, hovering briefly at one point. (Berliner) Form 97 Original Report
http://www.bluebookarchive.org/page.aspx?PageCode=MAXW-PBB19-1744
Frames 1744 – 1748
This was the year of the Walesville crash, the BOAC case, the Sea Fury incident, and the famous Bermuda radar case, and many others. A significant number of unusual sightings occurred in Europe during what has become known as the "Great Wave of 1952" Much of this wave has already been documented in UFO literature, but what was not known until the release of the CIA files is the fact that the CIA collected many of the sightings through normal intelligence channels, something which they had supposedly fought to block with the Robertson Panel inquiry. (Clear Intent, 132) Examples of reports excerpted from official CIA information sheets are provided below in their chronological sequence. The CIA reports are mostly translated press clippings; newspaper and dates should have been included but were not available.

1954

January 6, 1954; Cleveland Press Headline
"Brass Curtain Hides Flying Saucers." Reporters seeking information were banned from Wright-Patterson AFB, Ohio.

Occurred: 1954 or 1955
Location: Coos Bay, Oregon

"I was idly gazing at the blue sky and scattered clouds to the south, or maybe a little west of south. The sky was very blue and the air very clear, except for the scattered clouds, which were practically motionless. My attention was directed to two white irregular roundish clouds and the sky beyond.

"Suddenly, what appeared to be a huge aluminium discus appeared coming on a decline from above and beyond the cloud to my left and when it appeared to be about midway between and beyond the clouds and about even with the bottom of each cloud it suddenly turned a little to the left (my right) and soared upward and backward at a terrific speed... (Cf., April 26, 1954 report above; Professional Men.)

"As it reversed and started up and back it flattened again so that it was travelling with its perimeter longitudinal to its diameter in my line of sight. The sun was to the right of the clouds and as I remember they may have been slightly pinkish on their western sides, but the object was remarkably clear and well defined - no fuzzy edges or vapour streaks, and it appeared to have ridged or terraced sides. An ordinary track and field discus describes it perfectly as to shape, as I saw it. "I am not capable of judging how far away nor how high it was, but as I remember it appeared to be about two-thirds or three-fourths the area of the usual appearance of a full moon." [41] Arnold W. Spencer Former Town Selectman (12 years)

1955

Occurred: January 19, 1955; Pacific Ocean (BBU) 8:10 a.m. U.S. military pilot saw a white-reddish globular object flying level with the aircraft. (Project 1947)

http://www.nicap.org/550207knightslanding_dir.htm
4:30p.m.local. During observations of conventional jet aircraft the witness observed the antics of a winged UFO that reportedly executed a vertical climb and later, a steep dive. This incident was discovered in an OSI document to the FBI, found in the Blue Book file collection. If this actually occurred it would be an interesting, but distant encounter, worth posting in the Cat. 1 file. (Dan Wilson, BB files)

NOTE: A Hynek Classification of Distant Encounter is usually an incident involving an object more than 500 feet from the witness. At night it is classified as a "nocturnal light" (NL) and during the day as a "daylight disc" (DD). The size of the object or the viewing conditions may render the object in greater detail but yet not qualify the sighting as a Close Encounter, which is an object within 500'.

Winged Unidentified Executes Vertical Climb Near Jets
Occurred: February 7, 1955

Location: Knight's Landing, California

Fran Ridge:
Feb. 7, 1955; Knight's Landing, California (BB)
4:30p.m local. During observations of conventional jet aircraft the witness observed the antics of a winged UFO that reportedly executed a vertical climb and later, a steep dive. This incident was discovered in an OSI document to the FBI, found in the Blue Book file collection. If this actually occurred it would be an interesting, but distant encounter, worth posting in the Cat. 1 file.
Dan Wilson: This case is listed as Kings Landing, Calif. in the PBB Master Index. There is no Kings Landing, Calif. Knights Landing; Calif. is mentioned in the letter to the FBI
Detailed reports and documents
Here are the 3 pages from the OSI files and the master index that have to do with the UFO sighting. The actual PBB file on this case is over 85 pages long and much of that is about complaints against the missile flight test center at Cape Canaveral, RCA, etc. which I have not completed as yet. I am not sure if we are going to do this case.
http://www.bluebookarchive.org/page.aspx?PageCode=NARA-PBB91-1064
Details of UFO sighting at Knights Landing, California

http://www.nicap.org/docs/550207knightslanding_docs.pdf

SUBJECT: _____ DOB: 26 October 1919

1 _____ nia, called at the Edwards AFB OSI Detachment Office on 8 March 1955 and advised that he wished to report an instance of having seen a flying saucer on 7 February 1955.

2. Mr. ¬¬¬¬¬ furnished for identification purposes an Honorable Discharge Certificate which indicated him to be a former First Sergeant Serial No. ¬¬¬¬¬ ¬¬¬ ¬¬_____ and discharged November 1955 from the U. S. Army. Additionally _____ displayed a temporary Student's Permit No._____ issued 29 June 1952 by Civil Aeronautics Administration. Allender stated at the present time he had approximately 145 hours pilot time.

3. _____ advised that on 7 February 1955, while at _____ ranch at Knights Landing, California, at approximately 1630 hours, be observed, while looking north, a jet aircraft at altitude over 20,000 feet flying in a southerly direction. He advised that the sky was very clear and that the afore-mentioned jet was leaving a very heavy vapor trail. Almost immediately thereafter he noticed a second and third jet flying parallel to the first jet and headed in the same direction. He then looked down toward the horizon and noted an unidentified aircraft approximately three miles distance, from 2,500 to 4,000 feet altitude, with an approximate size of one and one-half times a B-36. He stated that this unidentified aircraft was in a vertical bank flying at an extremely high rate of speed. He opined that the speed of the unidentified aircraft was at least equivalent to the speed of the jets proceeding overhead. He looked above this aircraft and noted four more jets headed in the same southeasterly direction and almost parallel to the course taken by the three jets already sighted. A short time later all the jets converged and then assumed their parallel course continuing to fly in a southeasterly direction. At this moment he again noted the unidentified aircraft, which could be distinctly seen exhibiting a sweptback wing configuration as it rolled out of a steep bank to a horizontal altitude. Thereafter the unidentified object commenced a vertical climb disappearing from view in an extremely short time. Stated that he heard absolutely no sound, which was distinguishable. He further advised that witnessed this incident. Approximately 12 seconds later he noticed a saucer like object appear above and slightly ahead of the jets. This object was at least 3,000 or 4,000 feet above the jets and giving a shiny aluminum like reflection and moving in approximately the same direction of the jets. Thereafter the object commenced a dive at approximately a 50° angle, at which time _____ lost sight of the object for a few seconds. While looking in a southerly direction a short time later he noticed the unidentified object materialize before his eyes at an altitude of 2,500 to 3,000 feet. He advised that this object resembled a huge wing in a complete slip as it descended.

4. _____ stated that this object then slowed down and in a hovering position horizontal to the ground commenced to move slowly away from him toward the south. As he watched the object move away from him it then began a vertical bank and rolled out to the horizontal and then commenced a vertical climb at which time the object took on its saucer like appearance again. _____ stated that this object went out of sight and then he noticed what he thought was the same object slightly east and above the jets, which were beginning to go out of sight. _____ stated the total surveillance time of the unidentified object was four or five minutes and that the object was in view in its various positions for a total time of two minutes. In conclusion _____ stated that this object had absolutely no markings and no distinguishable power source, although the area around the trailing edge of the wing seemed hazy and several slotted areas appeared therein.

NARA-PBB1-125 - April Sightings

During the week of April 3-9, five green fireballs were reported in New Mexico and two in northern California. After a number of sightings reported about mid-morning April 5, LaPaz said: "This is a record. We believe we have it narrowed from the many reports to three. But they were seen within a very few minutes of each other."

Air Force (Project Blue Book) "Fact Sheet, issued in October 1955

National UFO Reporting Center
Looking Back - by Bob Gribble
October 1955
NUFORC Home Page

One of the most powerful U.S. senators in modern history actually eye-witnessed two UFO's while on a fact-finding trip through Russia in 1955—and the U.S. government kept the sightings a secret for more than three decades. The incredible encounter is detailed in 12 TOP SECRET CIA, FBI, and Air Force reports—and declassified in 1985. Those startling reports reveal that Senator Richard B. Russell, Jr. (D-GA)—then chairman of the Armed Services Committee—was on a Soviet train when he spotted a disc-shaped craft taking off near the tracks. He hurriedly called his military aide and interpreter to the window—and they saw the UFO, plus another one that appeared a minute later. The astonished trio reported the sightings to the U.S. Air Force as soon as they were out of Russia.

"The three observers were firmly convinced that they saw a genuine flying disc," says an Air Force Intelligence report, dated October 14, 1955, and classified TOP SECRET at the time. Senator Russell served 38 years in the Senate. He was its senior, and one of the most influential, senators at the time of his death in 1971. He was chairman of the Armed Services Committee from 1951 to 1969, and unsuccessfully sought the Democratic Presidential nomination in 1952. The mind-boggling documents detailing his UFO encounter were made available by the Fund for UFO Research and its chairman, Dr. Bruce Maccabee. The group through the Freedom of Information Act obtained several key documents. "These long secret documents are of major importance because they show for the first time that one of the most powerful U.S. Senators witnessed and reported a UFO," said Dr. Maccabee.

The Air Force Intelligence report says Russell and his two traveling companions spotted the UFO's on October 4, 1955, while traveling by rail across Russia's Transcaucasus region. "One disc ascended almost vertically, at a relatively slow speed, with its outer surface revolving slowly to the right, to an altitude of about 6000 feet, where its speed then increased sharply as it headed north," the report states. "The second flying disc was seen performing the same actions about one minute later. The take-off area was about 1-2 miles south of the rail line…"

Russell "saw the first flying disc ascend and pass over the train," and went "rushing in to get Mr Efron (Ruben Efron, his interpreter) and Col. Hathaway (Col. E. U. Hathaway, his aide) to see it," the report said. "Col. Hathaway stated that he got to the window with the Senator in time to see the first (UFO), while Mr. Efron said that he got only a short glimpse of the first. However, all three saw the second disc and all agreed that they saw the same round, disc-shaped craft…as the first." Lieutenant Col. Thomas Ryan, who interviewed Senator Russell's companions in Prague, Czechoslovakia, on October 13, after they arrived there from Russia shortly after the sighting, wrote the Air Force report.

In his report, Col. Ryan called the sightings "an eyewitness account of the ascent and flight of an unconventional craft…by three highly reliable United States observers. He added that Col. Hathaway led off his account of the sightings by saying: "I doubt if you're going to believe this, but we all saw it. Senator Russell was the first to see this flying disc…we've been told for years that there isn't such a thing, but all of us saw it…"

CIA documents show that the agency later interviewed the three eyewitnesses in the Russell party—and also a fourth person, unidentified in the reports, who had seen the UFO's. An eyewitness—whose name was blacked out on the CIA report prior to its declassification—said one of the UFO's "had a slight dome on top" and also a "white light on top." The edge of the disc was glowing pinkish-white, he added. The UFO rose "vertically with the glow moving slowly around the perimeter in a clockwise direction, giving the appearance of a pinwheel."

Interpreter Ruben Efron told the CIA that visibility was excellent. As one UFO approached the train, he said, "The object gave the impression of gliding. No noise was heard and no exhaust was heard, and no exhaust glow or trail was seen by me." After the encounter, Senator Russell told the men with him: "We saw a flying disc. I wanted you boys to see it so that I would have witnesses," according to the CIA documents. And an FBI memo, dated November 4, 1955, also discusses the sighting—and admitted Col. Hathaway's testimony "would support existence of a flying disc…" Dr. Maccabee, of the Fund for UFO Research, believes that Senator Russell and his group never publicly revealed their incredible sightings "because they were no doubt advised not to talk. These documents provide startling new evidence that UFO's exist."

Mr. Tom Towers, in his January 20, 1957, column, "Aviation News," for the Los Angeles, CA, Examiner, printed the contents of a letter from Senator Russell, which was in response to a request for information about the sightings in Russia. Mr. Towers had originally contacted Senator Russell's office by letter with the request that he be given permission to "break" the story. The Senator wrote: "Permit me to acknowledge your letters relative to reports that have come to you regarding aerial objects seen in Europe last year. I received your letter, but I have discussed this matter with the affected agencies of the government, and they are of the opinion that it is not wise to publicize this matter at this time. I regret very much that I am unable to be of assistance to you." The letter was dated 17 January 1956.

1956-1957

Jan. 28: Ruppelt's Book Published

Sign Historical Group UFO History Workshop Proceedings Table of Contents Proceedings Online

The landmark publication by Doubleday of the revealing book by former Project Blue Book Chief, Capt. Edward J. Ruppelt, marked a new phase in understanding the inside workings of the AF on UFO's, a term Ruppelt helped popularize within the government as well as outside with this book. With frequent references and allusions to high-level AF policy debates and numerous detailed accounts of inexplicable UFO cases many people found Ruppelt's book to make a sober and disturbing case for the existence of the UFO phenomenon by someone who officially investigated the subject for the US Government. Ruppelt's book was an enlargement of his article in TRUE magazine of May 1954, which was in effect a first draft outline of the book which he worked on with ghost coauthor-editor and Long Beach newspaper reporter James Phelan through 1955 (see numerous notes in the unpublished Ruppelt papers). Ruppelt's Book Ruppelt Coverups paper by Brad Sparks in the SHG 1999 Proceedings online

FORBIDDEN PLANET: Starring Walter Pidgeon, Anne Margret and Leslie Nielsen, debuts in American Theaters. This was the first Science Fiction Movie featuring, "Robbie the Robot," which changes the thoughts of space exploration robotics and artificial intelligence forever. Arriving from earth in a large nuclear powered flying saucer, the crew lands on an abandoned advanced civilization's planet and discover the former residents were eliminated by, "monsters from the ID," especially when sex and power are involved. Based on Shakespeare's Play, 'The Tempest', the undisciplined, uncontrolled power of the unconscious mind destroys the ideals of a civilization. Shows how denied emotions throw a monkey wrench into the best-laid plans of utopia.

UFO Landing at Edwards AFB

Occurred: 1957 Gordon Cooper Astronaut/Witness
Location: Edwards AFB, CA
Type of Case/Report: Standard Case
Hynek Classification:
Special Features/Characteristics: Military, Famous Person, Witness Photo
Full Report / Article Source: John Cooke
In 1957, Cooper was one of an elite band of test pilots at Edwards Air Force Base in California, in charge of several advanced projects, including the installation of a precision landing system.
"I had a camera crew filming the installation when they spotted a saucer. They filmed it as it flew overhead, then hovered, extended three legs as landing gear, and slowly came down to land on a dry lakebed! "These guys were all pro cameramen, so the picture quality was very good.
"The camera crew managed to get within 20 or 30 yards of it, filming all the time. It was a classic saucer, shiny silver and smooth, about 30 feet across. It was pretty clear it was an alien craft.
"As they approached closer it took off."
When his camera crew handed over the film, Cooper followed standard procedure and contacted Washington to report the UFO and "all heck broke loose," he said.
"After a while a high-ranking officer said when the film was developed I was to put it in a pouch and send it to Washington. "He didn't say anything about me not looking at the film. That's what I did when it came back from the lab and it was all there just like the camera crew reported." When the Air Force later started Operation Blue Book to collate UFO evidence and reports, Cooper says he mentioned the film evidence. "But the film was never found supposedly. Blue Book was strictly a cover-up anyway." Cooper revealed he's convinced an alien craft crashed at Roswell, N. Mex., in 1947 and aliens were discovered in the wreckage. "I had a good friend at Roswell, a fellow officer. He had to be careful about what he said. But it sure wasn't a weather balloon, like the Air Force cover story. He made it clear to me what crashed was a craft of alien origin, and members of the crew were recovered."
Why has the government kept its UFO secrets for so many years?
"It started in World War 2, when the government didn't want people to know about UFO reports in case they panicked," said Cooper. "They would have been fearful it was superior enemy technology that we had no defense against.
"Then it got worse in the Cold War for the same reason.
"So they told one untruth, they had to tell another to cover that one, then another, then another . . . it just snowballed.
"And right now I'm convinced a lot of very embarrassed government officials are sitting there in Washington trying to figure a way to bring the truth out. They know it's got to come out one day, and I'm sure it will.
"America has a right to know!"   Case ID: 357 edit: 357
<http://www.ufoevidence.org/admin/casereportedit.asp?id=357>

August 3, 1957; About 175 SW of San Francisco, California (BBU)

7:45-8:24 a.m. (PDT). USAF 965th Aircraft Early Warning & Control Sq (552nd AEW&C Wing), pilot 1st Lt. Robert J. Springer, Jr., Tech. Sgt. Herman L. Giles, and 16 other air crewmen, while on routine Airborne Operations Center radar early warning patrol over the Pacific aboard RC-121D aircraft (s/n 53-3400) detected a target on IFF Mode 2 transponder only. At 7:56 the IFF target became direct radar "skin paint," at 8:02 the IFF equipment APX-6/APX-7 was turned off but target was still tracked on airborne radar. At 8:15 target was at 2 o'clock position 10 miles range when aircraft started a right turn to reverse course putting target at dead ahead and target "suddenly" took off to the NW at "very high" speed, disappearing at 58 miles range (within 1-2 mins? at 1,800-3,600 mph?). Regained radar contact at 8:18 at 1 o'clock position 22 miles range moving right to left, crossed in front of aircraft again, closing distance to 8 miles at 11 o'clock position at 8:20 when target turned to head on parallel path. We lost contact at 8:24 at 7 o'clock position behind the plane at 15 miles, IFF remaining off, no visuals. (Jan Aldrich) 37-39 minutes

Occurred: October 4, 1957
Soviet Union launches the world's first satellite, Sputnik 1, at 2:28:34 p.m. (EST).

Occurred: November 2, 1957
Soviet Union launches the second earth satellite, Sputnik 2, at 9:30:42 p.m. (EST), this time with a dog on board

November 7, 1957
El Paso, Texas, Times: "Some of the nation's top scientists are 'pretty shook up' about the mysterious flying objects sighted in New Mexico and West Texas skies this week, said Charles Capen (a scientist at White Sands). 'This is something that hasn't happened before,' (he said)." (UFOE)

December 1957; Pacific Ocean
Photograph of alleged disc-shaped UFO. [UFOE, VIII]

1958

January 22, 1958
NICAP Director Major Donald Keyhoe is cut off the air in mid-sentence on a heavily controlled major CBS television program "Armstrong Circle Theatre." Keyhoe was about to mention a secret U.S. Senate investigation of the AF's secrecy policies and the TOP SECRET Estimate of the Situation by the AF at Wright Field, which concluded that UFO's were interplanetary (see Aug. 5, 1948, entry in the UFO Chronology). A public controversy erupted over the blatant censorship, which shocked many viewers, some of who were able to hear some of Keyhoe's censored words that came through faintly on another guest's microphone. CBS admitted Keyhoe was deliberately cut off, not as an accident or technical difficulty, by a network producer to satisfy what they thought the government would want, though not part of any official orders to do so. A few months later the IRS by letter would explicitly deny NICAP tax-exempt status on the grounds of NICAP's public opposition to AF policies. (Brad Sparks) (See actual BB docs)

Armstrong Circle Theater
AIR FORCE CENSORSHIP OF TV BROADCAST ABOUT UFOs STIRRED CONTROVERSY IN 1958

By Richard Hall
In one of the more bizarre incidents of UFO history, a major CBS Television broadcast about UFOs, sponsored by a large corporation, was cut off the air when Maj. Donald E. Keyhoe, USMC (Ret.) departed from the script and started to ad lib.
The "Armstrong Circle Theater," sponsored by the Armstrong Cork Company of Lancaster, Pennsylvania, was a popular program at that time. The UFO discussion, broadcast on Jan. 22, 1958, was titled UFOs: Enigma of the Skies. It had been carefully scripted under the strong influence of the Air Force.
Frustrated by the continued efforts of the Air Force to control what he wanted to say, Maj. Keyhoe started to announce that NICAP (the National Investigations Committee on Aerial Phenomena, which he directed) had been working with a Senate committee to investigate UFO secrecy, when the sound level of the audio - very obviously -- was abruptly cut.
When CBS-TV was deluged with phone calls, letters, and telegrams protesting what appeared to be overt censorship, CBS justified the action by stating:
"This program had been carefully cleared for security reasons.... public interest was served by the action taken by CBS."
CBS Television letter of January 31, 1958
This strange explanation was given at a time when the Air Force regularly claimed that there was nothing at all to UFOs, that the Air Force had nothing to hide, and there were no UFO-related national security implications. This being the case, what exactly did they (and CBS) fear that Major Keyhoe might say that could possibly justify their action? Why did the Air Force quite literally demand complete control over the script? Major Keyhoe's ad lib comments actually were innocuous (see following text).

The entire episode is very revealing about official attitudes and beliefs at that time. As the program neared the scheduled broadcast date, some of the prominent participants dropped out, including Kenneth Arnold (famous UFO sighting witness) and Edward J. Ruppelt (former chief of the Air Force UFO project), rebelling against the (from their viewpoint) emasculation of the script, and the rigid controls being placed by the Air Force and the complicit program producer on what they could say.

On Jan. 28, 1958, Major Keyhoe sent a form letter to the NICAP membership list informing them that what he was about to say when he was cut off the air was:

In the last six months, we (NICAP) have been working with a Senate committee investigating official secrecy on Unidentified Flying Objects. If open hearings are held, I feel it will prove beyond doubt that the flying saucers are real machines under intelligent control.

One can only conclude that Major Keyhoe's credibility was such that someone feared a panic. Certainly the notion that one man's opinion would somehow violate national security or cause any sort of panic among a public eager for more information about UFOs seemed very strange at the time, and makes no more sense in retrospect.

In February, deluged by hundreds of letters and telegrams of complaint, the public relations director of Armstrong sent a form letter to inquirers, including many NICAP members. Clyde 0. Hess, manager of public information, said:

It was most regrettable that a portion of Major Donald C. [sic] Keyhoe's statement had to be deleted because he departed from the prepared script. The producer and director, in observance of strict network rules, had no alternative but to order the audio cut off since they had no idea what the statement might contain. This network policy is designed to prevent any unfortunate incidents that might be embarrassing to the participants, the sponsor, the public, or the network.

In an amusing sidelight, the producer of the program, Robert E. Costello, had written a letter to Major Keyhoe the day after the broadcast, apparently before the storm of controversy had broken. It said, in part:

The response to the show has been very good. Any inquiries on your work we have referred to your office. You may be interested to know that you have thousands of fans as evidenced by the CBS switchboard which had calls backed up for forty-five minutes following the show.

Chagrined about the uproar he had created, Major Keyhoe promptly sent a letter to the Armstrong Cork Company and CBS taking personal responsibility for the incident, which he blamed on a "misunderstanding." His statement said:

Due to a misunderstanding on my part about rules of approval on script changes, it was necessary for Armstrong Circle Theater and CBS to interrupt a statement I was about to make...While I mentioned it to one or two persons connected with the program, I had not discussed it with the director or producer or any representative of CBS.

Certain minor ad lib changes, which I made, had been allowed and on that basis I had assumed that the deleted statement would not be contrary to the program rules. Since then I have been told that CBS Continuity has to approve extreme departures from scripts. Therefore, the producer and director had no alternative but to order audio cut-off since they had no idea of what I was about to say.

I regret the misunderstanding and wish to make it plain that this was not an attempt at censorship by CBS or ; Armstrong Circle Theater.

Left unsaid was the fact that it was an attempt at censorship by the U.S. Air Force, whose representatives applied constant pressure on the program to prevent Major Keyhoe from stating his case in his own words. Their heavy-handed influence also caused Ruppelt and Arnold to bow out at the last minute.

A very wrong-headed official policy, which NICAP was fighting to change by exposing it to public scrutiny, sought to totally control a public discussion of the facts and issues. Thanks to Major Keyhoe's courage and determination, the policy backfired badly in this case.

Typical Reactions to the Incident

"What was so important in Major Keyhoe's speech that we were not allowed hearing? I say this is an outrage." - Letter to Portland Oregonian 1/24/58

"I take no stand either way on UFOs, but I am going to fight for my right to know whatever it was [Major Keyhoe] said." -Letter to CBS-TV 1/23/58

"What makes me boil is that the powers that be consider the American public too stupid and childish to take this thing in stride ...Just what were your last words that we weren't allowed to hear?" - Letter to Maj. Keyhoe 1/23/58

"Call it anything you like, but it appeared to be a very shocking display of censorship; and certainly offensive to the intelligence of the American public." -Letter to CBS-TV 1/23/58

Source: Journal of UFO History, Jan-Feb 2005, Vol. 1, No. 6, Richard Hall

(Note: On March 1, 2006, researcher, Francis Ridge, located most of the script for the show in the Project Blue Book Archives. Files can be located by going to NARA-PBB87-746 and navigating to page NARA page 821.)

May (middle), 1958; Malmstrom AFB, Montana

Nighttime. UFO approached from the north and hovered over the alert hanger, at about 1,000 feet, appeared as a round metallic looking object (called a flying saucer by the guard), no estimate of size is recalled, nor any other details on the object itself. The base radar 1-1/2 miles distant picked up the object as did the FAA radar about five miles away across the city of Great Falls. Object was apparently hovering over the alert hanger and atomic missile and bomb storage building right nearby. Object then slowly moved down the length of the runway, then moved across town (about 3 miles) to the Municipal Airport at Great Falls and hovered over the National Guard (F-89) parking ramp, then flew off. (Jan Aldrich, Bay Area Subcommittee NICAP)

June 1958.

Richard Hall arrives at the National Investigations Committee on Aerial Phenomena, 1536 Connecticut Ave., N.W., Washington, D.C., where he served for ten years.

August 4, 1958; Malmstrom AFB, Montana (BBU)

11:15 p.m. (MDT). <>A/2C Bejnard G. Bell, 29th FIS roving aircraft guard, observed a delta wing-shaped object, flying directly overhead. The object was silver in color and appeared as large as volleyball at arm's length, was solid in shape with no openings, no props, no exhaust, no tail section, and no conventional means of motivation visible. The object made a varied pitch whistling sound, but was much quieter than a jet. The object moved gradually to the north and was observed for approximately 30 seconds. The object was picked up on the approach control radar for four sweeps at Malmstrom AFB. No scramble of aircraft was ordered. Officials at ATIC were not happy about no scramble being made and apparent non-compliance with AF Reg. 200-2, specifically Par 1, (Requirement for rapid identification). (McDonald list, Dan Wilson)

November 20, 1958; W. of Calif. coast (BBU)

At 6:15 a.m. PST, a man in the right seat of a C-118 transport flying at 11,000 feet on a heading of 60 degrees observed a round silver object traveling straight at high speed at high altitude heading toward the sun. The object was observed for 15 seconds. The C-118 was at approximately 840 miles west of Los Angeles [3447 N 133.07 W]. (Dan Wilson, BB files, Project 1947)

1959

February 1, 1959 - JANAP 146(D) Joint Army-Navy-Air Publication 146, CIRVIS Communication Instructions for Reporting Vital Intelligence Sightings. Feb.1 1959Revisedversion of JANAP 146(C) (March 10, 19

September 24, 1959. Near Redmond, Oregon.(BBU)

www.nicap.org
About 4:55 a.m. (PST). Redmond Police officer Robert Dickerson saw a strange bright light [white ball shaped?] rapidly descending north of the airport then stopped and hovered several hundred [200?] feet above ground for several mins where it lit up the juniper trees below. He drove toward it on the Prineville Hwy then turned toward the airport, when the object turned orange [reddish-orange?] and moved rapidly to [dive and hover?] about 10 miles NE of the airport at about 3,000 ft [height? altitude? Redmond is at 3,000 ft elevation MSL]. Dickerson arrived at the airport to report sighting in person at 4:59 a.m. at Redmond FAA Air Traffic Communication Station. FAA Flight Service Specialist Laverne Wertz, Dickerson and others viewed object through binoculars. FAA station reported UFO to Seattle Air Route Control Center at 5:10 a.m., which in turn reported it to Hamilton AFB, Calif., which scrambled 6 F-102 jets from Portland [?] to intercept UFO. FAA station observers saw object hover and emit long tongues of red, yellow and green light that extended and retracted at irregular intervals. As F-102's approached the object from the SE [?] it turned into mushroom shape, emitted red and yellow flames from lower side and ascended rapidly, disappearing above scattered clouds at about 14,000 ft [altitude? height?]. [Object's departure forced one F-102 to swerve to avoid collision, another nearly lost control from UFO's turbulent wake; tracked on F-102 airborne radars but jets unable to intercept.] Object reappeared about 20 miles S of Redmond at about 25,000 ft. Seattle Center reported at 6:20 a.m. radar contact with object about 25 miles S of Redmond at 52,000 ft was made by USAF ADC radar site at Klamath Falls, Ore., which tracked a large 300-400 ft[?] target and vectored B-47 and F-89 aircraft to identify. Redmond FAA controllers lost sight of object. Seattle FAA reported at 7:11 a.m. that Klamath Falls radar still tracked object at 25 miles S of Redmond but varying altitude from 6,000 to 52,000 ft. (Fran Ridge/NICAP UFOE, V)

October 1 1959; Telephone Ridge, Oregon (BBU 6534)
9:15 p.m. Department store manager C. A. Cissman saw a bright light approach, hover about 30 minutes, then take off and disappear in 2 seconds. (Berliner)

Huge Disc Sparks Scramble

Occurred: September 24, 1959

Location: Redmond, Oregon

Category 9, RADAR
Preliminary Rating: 5

RADCAT is a revitalized special project now being conducted jointly by NICAP & Project 1947 with the help and cooperation of the original compiler of RADCAT, Martin Shough, to create a comprehensive listing of radar cases with detailed documentation from all previous catalogues, including UFOCAT and original RADCAT.

Fran Ridge:

A cat 9,11 incident. A huge disc hovered over the city. Six F-102 fighters were scrambled from Portland. As they took off, the Air Force radioed the pilots of a B-47 bomber and a F-89 fighter on routine flights nearby, ordering them to join the chase. The Air Force radar at Klamath Falls tracked the disc. Given the Air Force "explanations" and some incorrect data in 1959 and published in the UFO Evidence (1964), the validity and seriousness of this encounter is much clearer in the MUFON report published much later in 1989. Dan Wilson located 55 documents on this case in the Blue Book files.

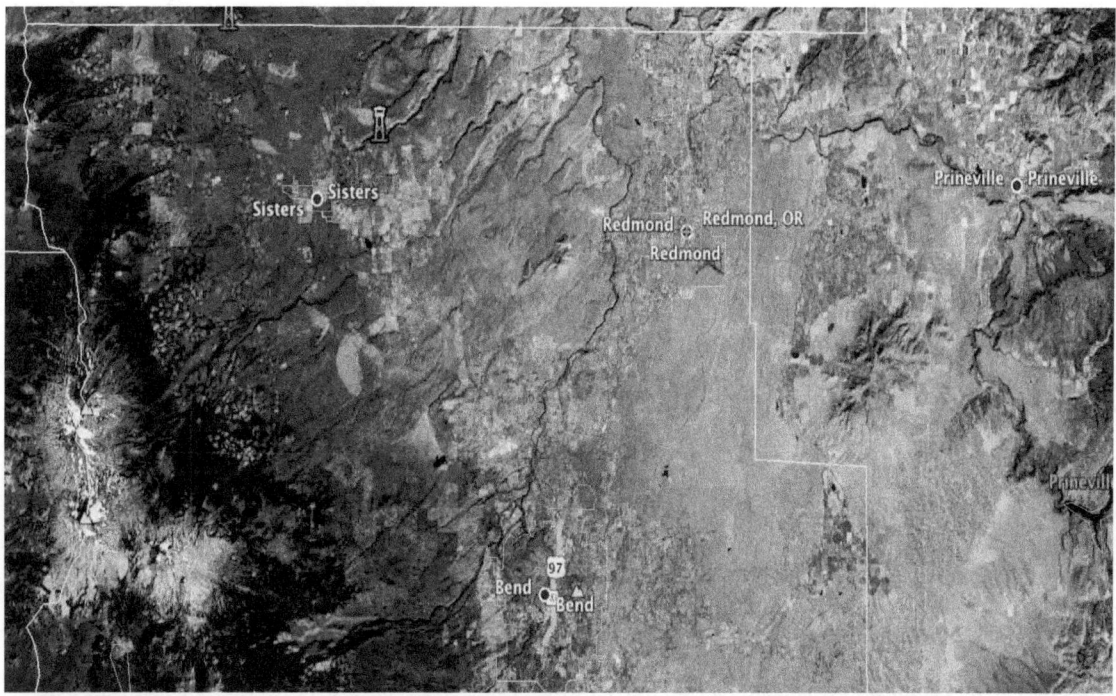

Redmond Oregon in relation to Bend and Sisters. Google Earth, NOAA 2012

Brad Sparks:
September 24, 1959; near Redmond Oregon (BBU)
About 4:55 a.m. (PST), Redmond Police officer Robert Dickerson saw a strange bright light [white ball shaped?] rapidly descending north of the airport then stopped and hovered several hundred [200?] feet above ground for several mins where it lit up the juniper trees below. He drove toward it on the Prineville Hwy then turned toward the airport, when the object turned orange [reddish-orange?] and moved rapidly to [dive and hover?] about 10 miles NE of the airport at about 3,000 ft [height? altitude? Redmond is at 3,000 ft elevation MSL]. Dickerson arrived at the airport to report sighting in person at 4:59 a.m. at Redmond FAA Air Traffic Communication Station. FAA Flight Service Specialist Laverne Wertz, Dickerson and others viewed object through binoculars. FAA station reported UFO to Seattle Air Route Control Center at 5:10 a.m., which in turn reported it to Hamilton AFB, Calif., which scrambled 6 F-102 jets from Portland [?] to intercept UFO. FAA station observers saw object hover

and emit long tongues of red, yellow and green light which extended and retracted at irregular intervals. As F-102's approached the object from the SE [?] it turned into mushroom shape, emitted red and yellow flames from lower side and ascended rapidly, disappearing above scattered clouds at about 14,000 ft [altitude? height?]. [Object's departure forced one F-102 to swerve to avoid collision, another nearly lost control from UFO's turbulent wake; tracked on F-102 airborne radars but jets unable to intercept.] Object reappeared about 20 miles S of Redmond at about 25,000 ft. Seattle Center reported at 6:20 a.m. radar contact with object about 25 miles S of Redmond at 52,000 ft was made by USAF ADC radar site at Klamath Falls, Ore., which tracked a large 300-400 ft[?] target and vectored B-47 and F-89 aircraft to identify. Redmond FAA controllers lost sight of object. Seattle FAA reported at 7:11 a.m. that Klamath Falls radar still tracked object at 25 miles S of Redmond but varying altitude from 6,000 to 52,000 ft. (Fran Ridge/NICAP UFOE, V)

Detailed reports and documents
The Redmond Oregon Case - UFO HISTORY - Loren Gross
The Redmond Report (MUFON UFO Journal #257)
Original report from NICAP files (1964) - UFOE, 113 (Hall)
Copy of Letter to NICAP - Jan 15, 1960 (FAA)
Aviation Experts Othan Than Pilots Report UFO At Redmond - UFOE, 44 (Hall)
reports/590924redmond_report.htm (Dan Wilson)
reports/590924redmond_report2.htm (Vicente-Juan Ballester Olmos)

The Redmond Oregon Case
UFO History
Loren Gross

The Air Force considers this case to be two separate incidents. The first incident took place at the city of Redmond, Oregon, which is about 150 miles southeast of Portland. The most detailed account of the "visual" incident is the interview of officer Robert Dickerson of the Redmond police department by Dr. James McDonald. The questioning took place in 1966. (158.) Writing to Richard Hall, Dr. McDonald says:

"I asked him to describe the episode as he recalled it. He said he'd never seen anything at all like it, before or since, and isn't sure he'd report it to anyone if he did see another! He was working the 'graveyard shift," before dawn, and was driving east on the highway to Princeville (Hwy. 126, he said). Suddenly he saw what, at first, he took to be a 'falling star.' He thought it was going to go all the way to the horizon, but it stopped instantaneously, at a point which he estimates at perhaps 200 feet above the ground, at a distance he puts at not over 1/4 mile. It illuminated all the juniper trees and the ground, and looked like a glowing object rather than an airliner. Longer across that it was high, but he saw no structure, just a uniform bright glow, comparable to a mercury-vapor light, he stated. When it stopped in this abrupt manner, he did not know whether to run, or what to do, he said with a chuckle. But he stayed, and it was stationary for perhaps a few minutes, as nearly as he could now recall. Then the light went out, and he saw it

climb off eastward at an angle, giving off a dull red glow and looking more like a red streak at this time. Since he was heading east and it went off to the east, he decided to drive out to the airport and see if he could find anyone there who had seen it and see if he could get any information out of them."
(159.)

When Officer Dickerson arrived at Roberts field, he went to the air traffic communication station, which was upstairs in the administration building. According to the Redmond station log Dickerson reported in at 1259Z or 4:59 a.m. (160.) McDonald continues:
"FAA employee Laverne Wertz was on duty and Dickerson got him outside and the two watched with binoculars. At that juncture it was just a glowing white light, oval, with longer axis horizontal. It lay off to their east. It made small oscillations, but did not change its general location. There seemed, he said, to be something like 'heat radiation waves' emanating from it. He puts its range at perhaps 7-8 miles then, but said this was a guess influenced by subsequent reports that its brightness awakened several persons in the town of Powell Butte, which lies at that 7-8 mile distance. They watched it for about 30 minutes then he decided to drive towards Powell Butte to get a better view of it. Before he left, Wertz contacted Seattle FAA ARTC radar and they said it was showing on their radar. A request to scramble USAF jets originated somewhere, Dickerson doesn't know where."
(161.)

The Redmond FAA log states Seattle ARTC was contacted at 1310Z or 5:10 a.m. This is interesting because the Seattle log says the Remond station contacted them at 1400Z or 6:00 a.m. (162.) Even more interesting is an Air Force document that states two F-102s were scrambled from Portland International Airport to identify an unknown flying object at 1300Z. or 5:00 a.m. That would mean that the jets took off ten minutes before Redmond reported in? The same Air Force document states the jets were turned over to the control of Mt. Hebo radar site at 1311Z or 5:11 a.m. If these times have any meaning, it would seem the scramble was for a radar return and not the visual report from Redmond. In any case, the jets got to Redmond quite fast. According to the Redmond FAA log, observers (Wertz and someone else? Mr.Davis?) saw the UFO do something extraordinary:

"We continued to observe UFO. Stayed very steady and projected long tongues of red, yellow and green light. These tongues of light varied in length and extended and retracted at irregular times. Observed high-speed aircraft approaching from southeast [Portland is northwest from Redmond. Were the jets redirected?]. As aircraft approached UFO took shape of mushroom, observed long yellow and red flame from lower side as UFO rose rapidly and disappeared above clouds estimated 15,000 feet, scattered layer." (163.)

According to what Dickerson told McDonald, he wasn't at Roberts airport when the UFO shot upward. Instead, officer Dickerson was in his patrol car speeding toward Powell Butte. McDonald writes:

"As Dickerson drove east towards Powell Butte, he had gone only about 2-3 miles [along Highway 126], watching the stationary luminous object through his windshield, when suddenly it shot straight upwards, with almost instantaneous acceleration. He emphasized the way in which it lit up the broken cloud deck as it passed through it, spreading a momentary whitish bright glow over the deck [These clouds could have hidden Venus. This is an important point that came up later]. He, himself, never saw jets, but Wertz informed him later that Wertz was monitoring traffic and heard the communications indicating that the jets were just approaching the area when the object shot up. Dickerson states that Wertz also saw the rapid ascent, from the Control Tower. He said Seattle FAA radar confirmed its rapid ascent. Wertz told Dickerson that Seattle had it at 3,000 feet prior to ascent, and that it went to 54,000 feet. He said all this could be found on the FAA logs, but he had not seen them, himself." (164.)

McDonald wondered if Seattle had HRI capability but the Redmond log, according to chief Davis, confirms, to an extent, Dickerson's claim. Prior to the object's rise the Redmond log says the UFO was: "...approximately 10 miles northeast of the station at an estimated 3,000 feet [note this is an estimate]. (165.) After the jump, the Redmond log says: "UFO reappeared south of Redmond approximately 20 miles and at an estimated [must have been a visual guess] 25,000 feet." (166.) The next figures specify radar confirmation: "Seattle Air Route Center advised radar contacted UFO at 1420Z [6:20 a.m.] located 25 miles south of Redmond at 52,000 feet. No further sightings made at this station." (167.)

There is a definite possibility altitude information was being fed to Seattle by the Klamath Falls GCI site which was controlling the jet interceptors.

Was the UFO Venus?

McDonald adds:

"The next AM, at the same time, he and Wertz checked the eastern skies. Although they had not noticed Venus the preceding morning, they saw it on the AM of the 25th, in the eastern skies, very bright [a bit of cloud could have cloaked it]. However, he said they were looking down the direction of [Highway] No. 126, which provided a definite reference line, and while Venus lay to the left (north) of Highway 126, they had seen the bright, hovering, oval object to the right (south) of 126. Also, he indicated that the object was not a circular (!) light like Venus; it was oblong, he said. He could not account for the fact that they had not been aware of Venus the preceding morning, and indicated they got some ridicule over this, though none in the press, apparently." (168.)

The radar incident. Part two of the Redmond case. The Keyhoe puzzle.

The director of NICAP was not known to invent information, but the lack of footnotes in his books can lessen the credibility of his writings. He was protecting his sources in every case one assumes. The Redmond "jet chase," as related by Keyhoe, is sensational stuff, but from where did he get his material? He gives no hint? The earliest Redmond account appeared in NICAP's UFO investigator (March 1960), a second very brief reference appeared in Flying Saucers: Top Secret (1960), and an extensive article was published in the UFO Evidence edited by Richard Hall (1964). In none of these accounts is there the sensational version of events Keyhoe mentions in his 1973 book, Aliens from Space. (169.) In a Chapter titled: "The Hidden Gamble," Keyhoe asserts that the jets scrambled out of Portland were on a "secret mission," the purpose of which was to: "...capture the UFO--and its crew, if one was aboard. This was the goal--the powerful driving purpose behind the Redmond mission." (170.)

As the UFO shot upward, in Keyhoe's Aliens From Space version the:

"...nearest pilot frantically banked to avoid a collision. As the UFO shot up past him another jet, caught in the churning air from the machines exhaust, almost went out of control. Three other pilots pulled out of their dives and climbed after the fleeing disc. But even with extra speed from their afterburners they were quickly left behind.

"As the UFO disappeared in the clouds at 14,000 feet, one AF pilot, guided by gunsight radar, climbed after the unseen craft. His approach apparently was registered aboard the disc, for it instantly changed course, tracked by height-finder radar at Klamath Falls. Even after the AF pilots gave up the hopeless chase, the radar operators were still tracking the UFO in high-speed maneuvers between 6,000 and 54,000 feet.

"When the pilots landed, still tense from their frightening experience, they were hurried into an Intelligence debriefing session. After describing the UFO encounter they were ordered not to discuss the pursuit, even among themselves." (171.)

Keyhoe even adds more sensational claims:

"Within hours, a new development upset the AF. When HQ learned of the discs' exhaust blast, it feared that the UFO might be using nuclear power. Through the FAA at Seattle, Flight Specialist Wertz was ordered to make a flying check for abnormal radioactivity. Using a Geiger counter, Wertz and the pilot of a TriPacer circled at various altitudes in the area where the UFO had hovered. The results, teletyped to the AF, were never released." (172.)

"Track JB-129."

The Air Force admits scrambling jets to investigate a unidentified blip which was given the number "JB-129." In a letter to Senator Magnuson, the Air Force said the Seattle and Redmond FAA logs were "misleading." The scramble was in response to the notification of a UFO sighting, but that there was no: "...evidence of radar tracking of the UFO or any success of the attempted intercept." (173.) Moreover: "The radar return bearing track No. JB-129 on the Klamath Falls ground control interception station was not an unidentified flying object. It was determined by the four senior controllers on duty, during the period of search that this radar return on the ground station scope was a radar echo from a gap filler antenna located on a mountain at the 8,010-foot level. This radar return did not move during the entire period of the search." (174.)

The way the letter to the Senator describes it, there didn't seem much to be excited about. So why did the Air Force go to so much trouble to check out a "stationary" target?

The 25th Air Division compiled a report on the radar/jet case in early 1960. Fortunately this was more than three months after the incident, because after three months recorder logs and intercept action reports were routinely destroyed, thus the official records that would have showed the speed and altitude performance of track JB-129 were not available and never would be. The Senior Controller on duty, Captain Gordon Poland, was also unavailable, having been transferred to Iceland on February 1, 1960. What we do know is in the 25th Air Division summary. The two F-102s scrambled at 1300Z (5:00 a.m.) returned to base at 1420Z (6:20 a.m.) with "negative results." (175.) We assume no radar or visual contact was made in contrast to Keyhoe's "wild" story. We are then told two more jets were sent up at 1452Z (6:52 a.m.). (176.) It should be remembered that 1420Z Seattle reported the UFO was about 25 miles south of Redmond at 52,000 feet. At 1511Z (7:11 a.m.) Seattle notified Redmond that the UFO was 25 miles south but that the UFO was at: "various altitudes from 6,000 to 52,000 feet." (177.) Orders came from McChord AFB, Washington to have two more F-102s held in readiness. (178.) At 1529Z (7:29 a.m.) the airborne jets returned to base "without results." <179.) At 1608Z (8:08 a.m.) a single F-102 was scrambled. At 1613Z (8:13 a.m.) the Air Division summary says: "One civilian Tri-Pacer aircraft took off from Redmond for low altitude search. One of the crew members was equipped with a Geiger counter." (180.) This confirms some of what Keyhoe claims. At 1655Z (8:55 a.m.) the single F-102 aloft reports in from the area of Pauline Mountain.

The pilot is making a visual search of the ground apparently. He reports seeing boats on a lake and a radio relay station. He also reports weather conditions: "scattered clouds and bumpy air." (181.) At 1718Z (9:18 a.m.) another single F-102 was scrambled. This pilot reports clouds from 20,000 to 12,000 feet. He sights a ranger station near Pauline Lake. (182.) At 1821Z (10:21 a.m.) the Tri-Pacer aircraft returned to Redmond. No object sighted and no radiation detected. (183.) At 1823Z (10:23 a.m.) yet another aircraft, an F-89, was scrambled from Portland. The F-89 returned to base with negative results. At 1949Z (11:49 a.m.) the senior controller at 25th Air Division ordered the scrambling of an H-19 helicopter from Kingsley Field. The craft conducted a search until 2154Z (1:54 p.m.) with negative results. At 2155Z (1:55 p.m.) track JB-129 was scrubbed from plotting board. (184.) The summary noted that beginning at 1700Z, or 9:00 a.m., the 25th Air Division scrambles were made part of an Air Defense exercise. (185.) It had to be full daylight by then and would explain the single plane scrambles and the helicopter. Apparently the pilots of these flights were in no hurry because they were requested to check for any ground features or weather that might be affecting the radar beam. The Pauline Mountain /lake area was where the gap filler radar antenna was located. The Air Force record states: "Repeated passes were made by the interceptors at levels from 40,000 down
to 12,000 feet. The interception being directed through [My emphasis-L. Gross] our radar return JB-129. No interceptor received a radar return on his airborne radar. At altitudes below 12,000 feet visual search was conducted under VFR conditions with negative results." (186.)

With the information the Air Force provides, there isn't much of a case for a Redmond radar object, but there is some indirect evidence the military felt there may have been a strange craft in the sky over Oregon on the 24th. A press account states: "Wertz also said the Seattle FAA office ordered him to search the Redmond area by plane. He and private pilot, Ben Jacques, canvassed a large area southeast of here [Corvallis] using a device to check for radioactivity, but nothing turned up." (187.) This confirms some of what Keyhoe claims and tells us it wasn't Wertz's idea, nor could it have been the FAA's idea. The agency isn't in the nuclear business. The request had to come from the military.

F-102,F-89 At 5:00 AM on September 24th, 1959, a huge disc hovered over the city of Redmond. Six F-102 fighters were scrambled from Portland. As they took off, the Air Force radioed the pilots of a B-47 bomber and a F-89 fighter on routine flights nearby, ordering them to join the chase. The Air Force radar at Klamath Falls tracked the disc. At Redmond, the FAA observers watched the disc when they heard the jets. When the fighters dived toward the object, it accelerated at terrific speed and shot straight up in the jets ' path. One pilot banked to avoid a collision. Another fighter caught in the churning air caused by the UFO, almost went out of control. The three other pilots pulled out of their dives and climber after the disc, but were quickly left behind. During the chase, the fighter's radar operators were still tracking the disc. The FAA radar tracked also the disc for some times at altitude varying from 6,000 to 52,000 ft. Later the same day at 16:00, two large shiny disc-shaped objects were seen maneuvering over Portland, Oregon.

AVIATION PERSONNEL OTHER THAN PILOTS SEE REDMOND UFO
September 24, 1959
Aviation personnel other than Pilots -- Federal Aviation Agency (FAA) control tower operators and flight controllers, flight crewmembers, ground crews, airport supervisors, etc. -- have made regular reports of UFOs. The FAA often has cooperated with NICAP, in some cases furnishing logs, Teletype reports, and other documentary material. Some of the information has come from NICAP members employed by the FAA, other from public servants (not NICAP members) who apparently have no prejudices about UFOs and merely believe that the subject should be treated frankly and openly.

September 24, 1959: Redmond Airport, Oregon, is situated southeast of the city. Just before dawn, policeman Robert Dickerson was cruising the city streets when he noticed a bright falling object like a meteor. Instead of "burning out" the object took on a larger, ball-like appearance, stopped abruptly, and hovered about 200 feet above the ground. Is glow lit up juniper trees below it.

The patrolman watched the UFO for several minutes, and then drove toward it on Prineville Highway, turning in at the airport. The UFO, meanwhile changed color from bright white to a duller reddish-orange color, and dived rapidly to a new position NE of the airport.

At the FAA office, Flight Service Specialist Laverne Wertz had just completed making weather observations minutes before, and had seen nothing unusual. Now Patrolman Dickerson, Wertz, and others studied the hovering object through binoculars. The UFO was round and flat, with tongues of "flame" periodically extending from the rim.

At 1310Z (5:10 am. PST) official logs show, the UFO was reported to Seattle Air Route Control Center. Logs of the Seattle center show that the report was relayed to Hamilton AFB. The Seattle log continues: "UFO also seen on the radar at Klamath Falls GCI [Ground Control Intercept] site. F-102's scrambled from Portland."

As the Redmond Observers studied the UFO, they noticed a high-speed aircraft approaching from the southwest. The log continues: "As aircraft approached, UFO took shape of mushroom, observed long yellow and red flame from lower side as UFO rose rapidly and disappeared above-clouds."

The UFO was seen again briefly, hovering about 25 miles south of the airport. Radar continued to show the UFO south of Redmond for about two hours.

December 24, 1959; "UFOs Serious Business"

Air Force Inspector General's Brief issued to Operations and Training Commands: "UFOs SERIOUS BUSINESS"; UFO investigating officers to be equipped with Geiger counters, camera, binoculars, and other equipment.

SECTION 6: 1960-1969

1960 brings incredible sightings: RED BLUFF, CORNING and MOUNT SHASTA. It establishes a precedence of UFO credibility not matched. Some of the information is still unavailable and classified but we have included half dozen reports from all official sources, including MUFON, NUFORC, B.B., local, McDonald's testimonies with witnesses and Keyhoe comments. There are many references but the sightings themselves are of incredible value. We have also included Capitan Jack Brown and Ray Chinca's sighting on the tail end of this flap that is still classified and undisclosed.

Hat Creek Radio Observatory, the Allen Telescope Array

Lassen Area Photo Tour

Allen Telescope Array

© 2008 Betsy Malloy Photography. Used by Permission.

This collection of satellite antennae form the Hat Creek Radio Observatory, also called the Allen Telescope Array. The observatory has been in this location for over 50 years, run by the UC Berkeley Radio Astronomy Lab and SETI Institute (Search for Extraterrestrial Intelligence).

When completed, the array will have 350 individual units. It's a working facility that offers a self-guided tour off-season and guided tours during busy times, but is only open weekdays.

1960

April 13, 1960
Red Bluff, CA
State Police encounter with highly maneuverable elliptical object, red light beams swept ground (NICAP UFOE, V).
June 4, 1960; Pacific Ocean (BBU)
(McDonald list)

Occurred: June 13, 1960; Approx. 0200hrs.
Location: Quincy, CA
Shape: Disk
Duration: several minutes
A lighted object was silently descending vertically into the canyon, then stopped and hovered

A few days ago, while researching UFO California sightings, I came across the Red Bluff Incident, and it absolutely blew my mind! I am convinced without a shadow of doubt that I was a witness to this event.

From the latter part of June until the 24th of August 1960, I along with a friend, prospected for gold on the middle fork of the Feather River. We gained access to the river through the little California mine, between Oroville and Quincy Ca.

For many years, I've wanted to collaborate this event of August 1960, but have been unable to do so, until just a few days ago. I'll try to be brief and precise in this. On an evening in mid August 1960, I awoke on an open-air cot on a sandbar, and was watching the stars and SAC planes when my attention was diverted to movement down the canyon. The river elevation was 1,700' and the ridge elevation was 6-7000'. A lighted object was silently descending vertically into the canyon. It stopped and hovered. The craft seemed to be rotating with internal rotating lights. As I was watching, the vehicle sent very bright beams of light on the canyon walls, and down to the river. From my vantage point, and because of the extreme terrain, I could not see the floor of the canyon or the river. I may have been 200 - 300 yards, but it is impossible to estimate the actual distance. After a minute or so, the bottom lights disappeared and the vehicle silently ascended and was gone very quickly.

I have discussed this event with my wife and very close friends, but have never joined any groups, or sought information about what I had witnessed. I would welcome some feed back on this if there is still anyone around.

AUTHOR'S NOTE
******ATTENTION********ATTENTION******

***THE FOLLOWING IS MUST READ CASE. IT ESTABLISHES A PRECEDENCE THAT ALL OTHER CASES FOLLOW. ***

If you live in Northern California, then THE RED BLUFF CASE is absolutely necessary to learn about. Since we included this sighting in the Local Sighting Chronology, we have left the general article and the following links to all the details on various websites, such as Blue Book, Nicap and others. The information itself would take hundreds of pages of text. However, all the ESSENTIAL FACTS and DETAILS are conveyed below.

http://www.nicap.org/600813dir.htm   - The Red Bluff Case

August 13-14, 1960; Red Bluff, CA. (BBU)

11:50 p.m.-2:05 a.m. Officers Charles A. Carson and Stanley B. Scott plus 3 others observe maneuvering silent red light with 5 white lights to the E descending to 100-200 ft height, reversed course, lifted to 500 ft, hovered, swept ground with red beam, aerial gymnastics, then headed E chased by police car, joined by similar object from S, disappearing in the E. (Hynek UFO Rpt pp. 92-94; NICAP UFOE, VII)

August 13-18, 1960; CA
A concentration of UFO sightings, mostly in north, included many police witnesses. (NICAP UFOE, XII)

August 15, 1960
Policy letter to Commanders, from office of Secretary of Air Force: The USAF maintains a "continuous surveillance of the atmosphere near Earth for unidentified flying objects -- UFOs."

Report : August 13, 1960; Red Bluff, California. Police, vehicle encounter, light beam, E-M, maneuvers, cat-and-mouse pursuit.   (EM Cases)

April 13, 1960; Red Bluff, CA
State Police encounter with highly maneuverable elliptical object, red light beams swept ground (NICAP UFOE, V). Report : August 13, 1960; Red Bluff, California. Police, vehicle encounter, light beam, E-M, maneuvers, cat-and-mouse pursuit. http://www.nicap.org/cat3-to1964.htm

UFO Case Report: Red Bluff Incident

Date: August 13, 1960
Location: Red Bluff, California, United States

California Highway Patrol Officers Charles A. Carson and Stanley Scott were on patrol when they sighted what they thought was an airliner about to crash. When the UFO had descended to about 100 or 200 feet altitude it suddenly reversed direction and climbed to 500 ft. Description: round or oblong surrounded by a glow (color not mentioned) and having definite red lights at each end. They continued to watch the UFO as it performed "unbelievable" aerial feats. (NICAP, 1964)

Classification & Features

Type of Case/Report: Major Case
Hynek Classification:
Special Features/Characteristics: E-M Effects, Radar, Police

Dr. James E. McDonald, UFOs: Greatest Scientific Problem Of Our Times?

August 13-14, 1960; Red Bluff, CA. (BBU)

www.nicap.org

11:50 p.m.-2:05 a.m. Officers Charles A. Carson and Stanley B. Scott plus 3 others observe maneuvering silent red light with 5 white lights to the E descending to 100-200 ft height, reversed course, lifted to 500 ft, hovered, swept ground with red beam, aerial gymnastics, then headed E chased by police car, joined by similar object from S, disappearing in the E. (Hynek UFO Rpt pp. 92-94; NICAP UFOE, VII)
August 13-18,1960; CA
A concentration of UFO sightings, mostly in north, including many police witnesses were present. (NICAP UFOE, XII)

www.nicap.org/600813.htm

(RED BLUFF INCIDENT)

August 15, 1960
Policy letter to Commanders, from office of Secretary of Air Force: The USAF maintains a "continuous surveillance of the atmosphere near Earth for unidentified flying objects -- UFOs."

Dr. James E. McDonald:
A rather detailed account of this sighting can be found in reference 3 (see pp. 61, 112, and 170). I have interviewed one of the two California Highway Patrolmen who were the principal witnesses and have spoken with two other persons who were involved in the incident. CHP officers C.A. Carson and S. Scott, driving east at 2300 hours on a back road south of Red Bluff, suddenly sighted what they first took to be an aircraft about to crash just ahead of them. Pulling their patrol car to a rapid stop and jumping out to be ready to render whatever assistance they could, they were astonished to see the long metallic looking object abruptly reverse its initial steep descent, climb back up to several hundred feet altitude, and then hover motionless. Next, it came silently towards them until, as Officer Carson put it to me, "it was within easy pistol range". They had their pistols ready and were debating whether to fire when it stopped. Attempts to radio back to the nearest dispatcher failed due to strong radio interference, which

recurred each time the object came close to them during the remainder of this 2-hour long sighting.

Huge bright lights at either end of the object swept the area. Carson stated to me that one light was about six feet in diameter; other smaller lights were also discernible on the object. After some initial minutes of hovering only 100 to 200 feet away from them and at about the same distance above the ground, the object started moving eastward away from them. They then contacted the Tehama County Sheriff's office that handled their night dispatching work, and asked for additional cars and for a check with Red Bluff Air Force Radar Station. Then they began to follow the object.

The full account is too involved to relate here (see reference 3), but it is important to point out that a number of witnesses confirmed the object from various viewing points in the county. A call to the AF radar unit brought confirmation that they were tracking an unknown moving in the manner reported by Carson and Scott.

However, when Carson and Scott went the following day to talk with personnel at the Red Bluff radar base, they were informed that no such radar sighting had been made. Their request to the officer in charge to speak with the radarman on duty at the time of the incident was denied. The Bluebook explanation that came out after a few days attributed this very detailed, close range sighting of a large object, seen by two experienced officers to "refraction of the planet Mars and the two bright stars Aldebaran and Betelgeuse".

NICAP referred the question to one of their astronomical advisors, who found that none of these three celestial objects were even in the California skies at that time. Bluebook then changed the explanation to read Mars and Capella. Capella, the only one of those celestial bodies that was even in the sky at 2300, was nowhere near the location of the sighted object, and could not, of course, give the impression of the various movements clearly described by the officers.

Carson subsequently stated, "No one will ever convince us that we were witnessing a refraction of light." Carson remarked wryly to me on the Bluebook explanations, "I'd sure hate to take one of my cases into court with such weak arguments." Dr. Menzel (reference 9, p. 254) concurs with the Air Force explanations and speaks of this being a night of "fantastic multiple inversions of temperature and humidity", such that he would have expected many more reports of UFOs. I should like to know what radiosonde data Dr. Menzel is citing, since the data I obtained does not fit that description.

Any such casual putting aside of the details of the basic report has no scientific justification in the first place. If Menzel and Bluebook think California Highway Patrolmen draw their .44s in uneasiness over looking at a refracted image of Capella, and misinterpret it as a 100-foot object with huge bright lights, I am afraid I cannot share their readiness to so easily discredit and discount reliable witnesses. When I spoke with Carson a few months ago, I found him still deeply impressed by this incident six years after it occurred. "I've never seen anything like it before or since."

Dr. James E. McDonald
Source: Greatest Scientific Problem of Our Time? (April 22, 1967)

SECTION VII

Officials & Citizens

The reports of technically trained observers, military and civilian pilots, in themselves are sufficient to make a strong case for UFOs. However, when we also realize that a broad cross-section of reputable citizens have described identical phenomena, it seems incredible that UFOs are not an acknowledged fact. The disc-shaped, elliptical and other main types of UFOs observed by pilots and such responsible persons as judges, civil defence officials, professors, lawyers and clergymen have reported scientists with great frequency.

Some of these individual observer categories could fill another complete section of this report. From the hundreds of cases on file, the following have been selected to provide a survey of what has been seen by officials and private citizens of various backgrounds.

LAW ENFORCEMENT OFFICERS

The Police switchboards normally and logically are the first to be swamped with calls during concentrations of sightings, since there is no established procedure for citizens to follow when they see a UFO. Examples abound of cases in which police responded to citizens' reports of UFOs, and saw the objects for themselves. Police Officers on patrol duty, too, have observed unexplainable objects maneuvering overhead.

During a six-day concentration of UFO sightings in northern California, August 13-18, 1960, at least 14 police officers were among the numerous witnesses. At 11:50 p.m. (PDT) August 13, State Policeman Charles A. Carson and Stanley Scott were patrolling near Red Bluff when they noticed an object low in the sky directly ahead of them. (Their report of the sighting was put on the police Teletype, a copy of which was submitted to NICAP confidentially by a police source. Later, NICAP Adviser Walter N. Webb contacted Officer Carson and was sent another copy of the teletype report, a sketch of the UFO, and a letter giving additional information.)

Verbatim text of the police Teletype report to the Area Commander:

"STATEMENT MADE BY OFFICER CHARLES A. CARSON CONCERNING OBJECT OBSERVED ON THE NIGHT OF AUGUST 13,1960.

Officer Scott and I were E/B on Hoag Road, east of Corning, looking for a speeding motorcycle when we saw what at first appeared to be a huge airliner dropping from the sky. The object was very low and directly in front of us. We stopped and leaped from the patrol vehicle in order to get a position on what we were sure was going to be an airplane crash. From our position outside the car, the first thing we noticed was an absolute silence. Still assuming it to be an aircraft with power off, we continued to watch until the object was probably within 100 feet to 200 feet off the ground, when it suddenly reversed completely, at high speed, and gained approximately 500 feet altitude. There the object stopped. At this time it was clearly visible to both of us. It was surrounded by a glow making the round or oblong object visible. At each end, or each side of the object, there were definite red lights. At times about five white lights were visible between the red lights. As we watched the object moved again and performed aerial feats that were actually unbelievable. At this time we radioed Tehama County Sheriff's Office requesting they contact local radar base. The radar base confirmed the UFO - completely unidentified. Officer Scott and myself, after our verification, continued to watch the object. On two occasions the object came directly towards the patrol vehicle; each time it approached, the object turned, swept the area with a huge red light. Officer Scott turned the red light on the patrol vehicle towards the object, and it immediately went away from us. We observed the object use the red beam approximately 6 or 7 times, sweeping the sky and ground areas. The object began moving slowly in an easterly direction and we followed. We proceeded to the Vina Plains Fire Station where it was approached by a similar object from the south. It moved near the first object and both stopped, remaining in that position for some time, occasionally emitting the red beam. Finally, both objects disappeared below the eastern horizon. We returned to the Tehama County Sheriff's Office and met Deputy Fry and Deputy Montgomery, who had gone to Los Molinos after contacting the radar base. Both had seen the UFO clearly, and described to us what we saw. The night jailer also was able to see the object for a short time; each described the object and its manoeuvres exactly as we saw them. We first saw the object at 2350 hours and observed it for approximately two hours and 15 minutes. Each time the object neared us we experienced radio interference.

We submit this report in confidence for your information. We were calm after our initial shock, and decided to observe and record all we could of the object.

Stanley Scott 1851
Charles A. Carson 2358."

Extracts from Officer Carson's letter of November 14, 1960, in answer to Adviser Webb's questions:

"We made several attempts to follow it, or I should say get closer to it, but the object seemed aware of us and we were more successful remaining motionless and allow it to approach us, which it did on several occasions.

"There were no clouds or aircraft visible. The object was shaped somewhat like a football, the edges (here I am confused as to what you mean by edges, referring to the outside visible edges of the object as opposed to a thin, sharp edge, no thin sharp edges were visible) or I should say outside of the object were clear to us . . . [the] glow was emitted by the object, was not a reflection of other lights. The object was solid, definitely not transparent. At no time did we hear any type of sound except radio interference.

"The object was capable of moving in any direction. Up and down, back and forth. At times the movement was very slow. At times it was completely motionless. It moved at high (extremely) speeds and several times we watched it change directions or reverse itself while moving at unbelievable speeds.

"When first observed the object was moving from north to south [patrol car moving almost due east]. Our pursuit led in an easterly direction and object disappeared on eastern horizon. It was approximately 500 feet above the horizon when first observed, seemingly falling at approximate 45-degree angle to the south.

"As to the official explanation [See Section IX.], I have been told we saw Northern lights, a weather balloon, and now refractions.

"I served 4 years with the Air Force, I believe I am familiar with the Northern lights, also weather balloons. Officer Scott served as a paratrooper during the Korean Conflict. Both of us are aware of the tricks light can play on the eyes during darkness. We were aware of this at the time. Our observations and estimations of speed, size, etc. came from aligning the object with fixed objects on the horizon. I agree we find it difficult to believe what we were watching, but no one will ever convince us that we were witnessing a refraction of light.

/s/ Charles A. Carson
Calif. Highway Patrol." [1]

August 13, 1960, UFO Sighting

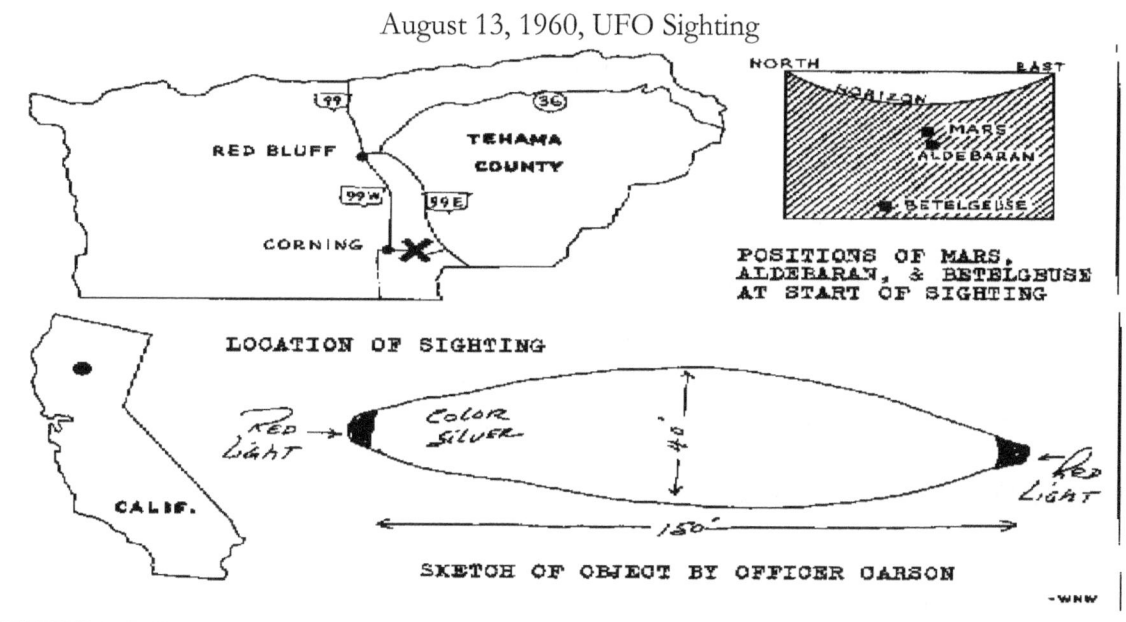

Occurred: September 5, 1960
Location: Sonoma County, CA
Sheriffs observed six varicolored UFOs flying in V-formation. [NICAP UFOE, VII]
Sept. 19, 1960; Susanville, CA (BBU)
(McDonald list)

Occurred: September 15, 1960; 2100hrs.
Location: Susanville, CA
Shape:
Duration: 5-6 minutes
Three craft in formation, high altitude, lights constant across sky - unlike aircraft running lights which blink and "hide"

Observed three orange lights in formation, no sound, high altitude probably 15,000 est. travelling roughly north to south - notable because I have watched aircraft night and day since childhood as son of Navy aircraft electrician.

These "lights" were not red or green or white, nor did they ever blink. Aircraft running lights are sometimes hidden by a wing or fin, but even at high altitude the different colors of running lights are distinguishable one from another. Because of the altitude and darkness, there was no way for me to describe shape of objects.

These objects were constant - unblinking, color unchanging as they traversed from north to south. Not satellites, which were visible at that time, but were single (no formation) and would "wink out" as they passed into earth shadow.

While I do not remember exact date, sighting was within an hour after dark - not once but three consecutive nights. The only difference is on the third night, one of the objects, which from such distance I could only describe as an "orange light" broke off from the "formation" and shot away to the east, while the other two lights continued southward.

Within days I heard news reports of law enforcement officials and other reporting cigar shaped objects in the area of Redding, California I am submitting this report because Peter Davenport asked for anyone who has ever seen anything unidentified to file a report.

Posted 21 December 2010 - 06:46 PM

Hope everyone has a jolly fine Xmas - below are some interesting statements about the UFO/OVNI subject.

Moon Dust Report near SAC Base 28 Sept. 1960

Occurred: August 14, 1957; 2100hrs.

Location: California (airborne sighting)
Shape: Fireball
Duration: 5 seconds
Red Fireball Over California

From 1954-58 I was a flight engineer (mechanic) on USAF C-46 twin-engine cargo aircraft. I was assigned to the 2343rd Air Force Reserve Flying Center (AFRFC) at Portland Oregon. Even though we were regular Air Force, we had to go to Chico, California with the reserve unit for two weeks summer encampment during the first part of August each year .I do not remember the exact time, date or year, of this event, as we went to Chico for 4-5 years.

I was in the cockpit of a C-46 in a four-plane formation that had been out over the Pacific Ocean for nighttime navigation training. (Long before GPS) The navigators used a sextant through the astrodome (glass bubble) on top of the aircraft to determine our location.

We had completed the over water training and were headed southeast back toward Chico which was 35-50 miles distance, when we noticed a bright cherry red fireball coming from the south at a very high rate of speed. It had a vibrating, quivering motion as it headed north. We were flying at 8,000-9,000 ft. and the fireball passed across in front of us from the 12:00 o'clock to the 9:00 o'clock positions. It was level to slightly below our altitude. It was about the size of a dime at arms length at its closest point. As the fireball got in our 10:00 o'clock position it split into 3-4 pieces.

These pieces all continued north, more or less parallel to each other.

When we landed at the base in Chico most of the ground crew that were waiting to park and service the aircraft after we landed had seen the fire ball pass at about a 30 degree angle to the west of them.

Over the next few days we heard that the fireball was first seen over southern California. After it passed in front of us, the pieces supposedly joined back together over Klamath Falls, Oregon where it hovered for a short while, then flew off to the north.

About that time we had a C-46 transporting another flight engineer, who had injured his back, to our home base at Portland, Oregon. The injured man and the flight engineer of the aircraft were both close friends of mine. As they were on the final approach to the base, the control tower called them and told them to make a "go-around," as a UFO was following them in for landing. They applied power, raised the landing gear and the flaps and made a "go -around" to the left. The UFO followed them down across the runway, and then it also headed north. From what I remember the UFO was last seen near the Canadian border.

Total elapsed time from Southern California to the Canadian Border was about 5 minutes.

I have tried (to no avail) to go into the archives of some of the newspapers in that area to find and record of this incident.

((NUFORC Note: Date is approximate, although we have invited the witness to investigate whether he can determine the date unambiguously, as well as contact other members of the flight and have them submit independent reports about their recollections of the event. PD

December 14, 1960 (50-page Summary in pdf file format)
Full Report (219 pages in pdf format)
The Brookings Research Institute in Washington releases a report prepared during 1960 for NASA entitled "Proposed Studies on the Implications of Peaceful Space Activities for Human Affairs", including a section entitled "Implications of a Discovery of Extraterrestrial Life". (Commonly referred to as "the Brookings Institute Report".) The report discusses effects of meeting extraterrestrial life: "It is possible that if the intelligence of these creatures were sufficiently superior to ours, they would choose to have little if any contact with us . . . " (New York Times, Dec. 15, 1960)

Author's Note: The above is "The Brookings Report" in which NASA is allowed to keep all its findings under wraps. The 216 page report may have missing pieces of articles of law that expose it for what it is, a type of Intergalactic Marshall Law Document.

1961

<u>April 25, 1961</u>
U.S. Air Force Guidance Collection Letter No.4, originally classified Confidential, describes and provides guidance for project Moon Dust reporting. Several items of interest appear in the document: classification level of Moon Dust Alerts and reports, focus of Moon Dust on "foreign earth satellite vehicles," and destination agencies for Moon Dust reports among them. (CUFON)

May 1961: Joint statement by 21 American Scientists released by NICAP. Calls for open investigation of UFOs without secrecy and the need for a more thorough investigation protocol for circumstantial evidence. It states the Air Force should have a more straightforward information policy, specifically to give out all facts on major UFO sightings.

1962

Date: Thu, 05 October 2006 13:06:48 +0100 (BST)
From: Daniel Wilson <daniejon2000@yahoo.co.uk>
Subject: April 18, 1962; New York to Eureka, Utah, to Nellis, Las Vegas, Nevada (BBU)
To: Francis Ridge <nicap@insightbb.com>

The following pdf file contains all of the resized Project Blue Book documents below and is now housed on the NICAP site for security reasons.
http://www.nicap.org/docs/620418nellis_rep3.pdf

April 18, 1962; New York to Eureka, Utah, to Nellis, Las Vegas, Nevada (BBU)
Reported crash. High-speed brilliant maneuverable object is tracked by radars and sighted visually across the continent by numerous military and civilian witnesses. (Don Berliner, Dan Wilson)
Category 9, RADAR
 Preliminary Rating: 5

RADCAT is a revitalized special project now being conducted jointly by NICAP & Project 1947 with the help and cooperation of the original compiler of RADCAT, Martin Shough, to create a comprehensive listing of radar
cases with detailed documentation from all previous catalogues, including UFOCAT and original RADCAT.

National Air Defense Alert
April 18, 1962

From New York to Las Vegas - A Ten-State Incident

October 8, 2006; updated: 19 Feb 2010 (* 78 new documents by Dan Wilson); 11 June 2010 Holloway paper

Fran Ridge:

The Air Defense Command was alerted after the object was tracked by NORAD and ground witnesses as it traveled for 32-minutes from Oneida, New York across the Midwest, through Kansas & Colorado, Utah, and the disappeared from Nellis AFB radar at 10,000 feet. Jets had been scrambled from two locations. Power outages were reported and the object made a turn toward the east and landed. There is quite a bit of information from numerous sources concerning this major incident, including Project Blue Book documents, and now possible confirmation by a radar man at ATIC. The case is also NOT explained in a BB monthly sighting listing for April 1962. It is very interesting that every one of these states except Utah, has or was in the process of obtaining ICBM Bases: New York, Plattsburg AFB; Kansas, Forbes AFB, McConnell AFB; Utah, Minuteman production at Air Force Plant 77 at Hill AFB; Idaho, Mountain Home AFB; Montana, Malmstrom AFB; New Mexico, Walker AFB; Wyoming, F. E. Warren AFB; Arizona, Davis Monthan AFB; California, Beale AFB. (Don Berliner, Brad Sparks, Dan Wilson)

Detailed Reports:

The Las Vegas UFO Crash (Kevin Randle)
reports/nevada_fireball.doc (Scott Holloway)Project Blue Book Documents AF Confirms Object No Meteor & Unidentified; MISC-PBB2-1051-1053 (Fran Ridge) rep1.htm
Front Page: Myst Obj Trkd ADC Over 9 ICBM States, MISC-PBB2-1058 (D. Wilson)_rep2.htm
NY to Eureka, UT, to Nellis, Las Vegas, NV, MISC-PBB2-1071-1074, 1105 (D. Wilson)_rep3.htm
Fireball Lights Skies In Ten States, MISC-PBB2-1057 (Dan Wilson)_rep4.htm
Hynek & Lt. Col. Friend Say UFO Was a Bolide, MISC-PBB2-1113-1114 (Fran Ridge)_rep5.htm
C-119 Pilot, Hill AFB, MISC-PBB2-1064-1067 (Fran Ridge)_rep6.htm
Object Tracked on Radar - No Meteor, NARA-PBB1-268 (Fran Ridge)_rep7.htm
Radar at Nellis Confirms Obj Chnged Course, MISC-PBB2-1048-1050, 1058 (F. Ridge)_rep8.htm
Memo for Congressional Inq Div Says Object A Bolide, MISC-PBB2-1112 (Fran Ridge)_rep9.htm
* Eureka, Utah, Rocky Mountain Area (78 Documents in 4 pdf files !!!) (Dan Wilson)_rep10.htm

Other Sources

April 18, 1962, Power Out for 42 Minutes (Dan Wilson)
First Titan 1 Wing Became Operational at Lowry AFB (CO) on April 18, 1962 (Dan Wilson)
xxxApril 18, 1962, NSC Meeting, Kennedy Approved Atmospheric Test Series (Dan Wilson)

At 0319Z on April 19, 1962, a brilliant midair illumination appeared over central Utah, seen by many witnesses. As many as 20-30 explosions were heard after that, and an object was seen falling toward the ground. Observers at Jericho, Utah, stated that the object was emitting a gasping sound, retarding forward movement, and surging ahead three or four times. The object was thought to have landed in the Rush Valley in the vicinity of the McIntire Ranch (39* 29' N, 112* 23' W). (Wilson)

The original over-sized docs from the footnotes.com site are listed below:
Page ID (PID) MISC-PBB2-1071-1074, 1105
Collection Other Official Microfilm
http://www.bluebookarchive.org/page.aspx?PageCode=MISC-PBB2-1105

October 1962; During the Cuban Missile Crisis; Palermo AFB, NY

A "hot bird" (an F-106) was scrambled when radar tracked an unknown on the East Coast. (Fran Ridge)

1963-1964

Occurred: June 15, 1963; 1815hrs.

Location: Sacramento, CA

Witness: 2
Shape: Cylinder
Duration: unknown

My niece and I were setting on the front porch when I felt some one watching us. We looked straight above us and seen a large craft hovering above us.

As we watched, the craft started coming down, as it did so it was turning slowing at the same time. It was the shape of a cylinder and silver in color, and it was all lit up with a light that came from all over the craft. I couldn't see where the light was coming from, as the whole craft was light.

We were so frightened that we jumped up from the step we were setting on and turned to run in the front door. We were frozen in place and couldn't move for a short time it seemed. I could move my eyes, but that was all. I had seen my niece beside me frozen too.

Then it released us and we ran into the house. When we came in my sister made the comment that we had been out there a long time. To us it seemed we had only been out there a few minutes.

Occurred: June 26, 1963
Location: Pine Crest, Calif.
A technician and many other witnesses observed four glowing greenish objects with halos. A similar object approached three more objects moving westerly from the west. The fourth object stopped and hovered as the three approached, split formation, and continued west. Then the fourth object continued east. [Report via Bay Area NICAP Subcommittee.]

Occurred: September 14. 1963
Location: Susanville, Calif. (BBU 8548)
Witness: 1
Shape: 2 objects
3:15 p.m. E. A. Grant, veteran of 37 years training forest fire lookouts for the U.S. Forest Service, saw a round object intercept a long object then either attach itself to the latter or disappear. (Air Force Project Blue Book 1963 Summary, Berliner)

May 22, 1964
The Inspector General Brief, "Reporting Unidentified Flying Objects",
Operations and Training updated from previous edition.

Occurred: July 12, 1964; Approx. 1200hrs.
Location: Drew, OR
Shape: Circle
Duration: 20 minutes
Date/Time: 07/1964
Shape: Silver (Metallic) Disc
Duration: 30 minutes
Summary object hovered above work location.
The summer of 1964 I was working with a geophysicist checking the ground for mineral deposits by sending electro magnetic transmissions through the ground from a power generator.. Myself and a co-worker was taking our lunch break when i noticed this silver looking object coming towards us i thought at first it was a weather balloon.
The object hovered approximately 500ft above us it was a silver (metallic) in color it also had a dome. We yelled for help but no one heard us..
Finally the object started moving north from our location at a high rate of speed I have never seen anything move as fast as this object moved..
But this sighting was not over we no longer settled down and we saw the UFO again this time it was flying above the cockpit of a airliner I don't think the pilot even new this object was above him. The UFO followed the airliner to about the same location where it left the first time and it was gone at a high rate of speed.
I really believe we were making contact with this UFO through our electro magnetic transmissions.
I would take a polygraph test to prove what I witnessed that day...
((NUFORC Note: Witness indicates that the date of the sighting is approximate. PD))

September 15, 1964; Big Sur, Cal.
UFO reportedly filmed disabling Atlas warhead.
The Big Sur Filming / UFO Disables Dummy Warhead?
September 1964

Fran Ridge: USAF Lt. Robert Jacobs was Officer-in-charge of Photo-Optical Instrumentation for the 1369th Photographic Squadron at Vandenberg AFB. He disclosed his crew's filming of an Atlas missile launch where a UFO allegedly caused the ICBM's warhead malfunction. At the time of the filming apparently no one knew anything about a UFO. But the next morning Jacobs was ordered to report to the office of Maj. Florenz Mansmann, First STRAT AD, his commanding officer where he was shown the film and was told to forget it ever happened. Kingston George, the project engineer for the experiments, and who probably never saw the film, "identified" the object as "nothing to do with UFOs" in an article by the Skeptical Enquirer (Winter of 1993). Before Mansmann's death, and 40 years after the actual event, the major confirmed a UFO incident in writing. The opinion of some serious researchers on our team, however, suggests that the telescope imaging system was not adequate enough to produce the results described by Jacobs and Mannsman. At least four researchers have shown that with the seeing conditions at the height of the equipment used, and the imaging systems operating at that shoot, what was reported could have actually occurred. And four others who were involved in this type imaging attest to it being "remarkably successful". So, the question remains, however, was this a UFO incident or a cover-up about an important warhead test? Detailed reports and documentsUFO Filmed Circling Atlas Rocket [dated Sept 1988] (T. Scott Crain)

December 4, 1964; Baker, Oregon (BBU)
At 4:50 a.m. local time, a glowing round object that had the apparent size ranging from a baseball to a basketball at arm's length and was seen by at least 4 observers, Harold H. Eves TSGT Radar Operations Crew Chief, a city policeman, Donald H. Stinett 821st Radar Squadron, and Vernon W. Meador 821st Radar Squadron AC&W Operator. Observations were made with binoculars and a surveyors scope, and there were also radar returns observed on an AN/FPS-35 Search Radar. Length of observation was approximately 2 hours. Some observers saw from 3 to 4 objects. (Dan Wilson, McDonald list) Object Observed & Tracked on Radar For Two Hours December 4, 196 Baker, Oregon Dec. 4, 1964; Baker, Oregon (BBU) (McDonald list)

Brad Sparks: At 4:50 a.m. local time, a glowing round object that had the apparent size ranging from a baseball to a basketball at arm's length and was seen by at least 4 observers, Harold H. Eves TSGT Radar Operations Crew Chief, a city policeman, Donald H. Stinett 821st Radar Squadron, and Vernon W. Meador 821st Radar Squadron AC&W Operator. Observations were made with binoculars and a surveyors scope, and there were also radar returns observed on an AN/FPS-35 Search Radar. Length of observation was approximately 2 hours. Some observers saw from 3 to 4 objects. (Dan Wilson) Detailed reports and documents Reports/641204baker_rep.htm (Dan Wilson)

December 1964-January 1965 sighting concentration around Washington, D.C., led to CIA contact with NICAP to obtain information.

1965

Lost in Space (TV Series 1965–1968) - IMDb

Created by Irwin Allen. . Actors: June Lockhart: Maureen Robinson · Mark Goddard: Maj. Don West · Marta Kristen: Judy Robinson · Bill Mumy: Will Robinson · Angela Cartright.

Occurred: 1964-1965; 1400hrs.

Location: Mount Shasta, CA

Witness: many
Shape: Cigar
Duration: 30 minutes +

Metallic cigar shaped object, windowless, soundless, motionless, unlighted, daytime/clear blue sky sighting over old Drive-in movie theater at the junction of Hwy 89 at Interstate 5. Object was very low in the air, 500? It just hung there, sort of tilted to one side. Many cars and people stopped and watched as this object. All NB/SB traffic on I-5/Hwy 89, people of Dunsmuir and Mt. Shasta MUST HAVE SEEN THIS! I was a 5 or 6-year-old child at the time. We were returning to McCloud from Mt. Shasta via Hwy 89 at the I-5 interchange. My mother pulled over at the entrance to the Drive-In Movie theater and I saw other cars stopped and people outside of them looking up into the sky. I remember my mother being awed, and all the others who were watching. This cigar shaped, metallic, windowless, soundless, unlighted object was hanging, slightly tilted, in the clear blue afternoon sky. I know this was the time frame because I remember the sun was behind us as I looked up at it. I was not overly impressed (at that age) but I could tell the adults were. My dad and I always watched the space shots, I remember him telling me the news said it was some type of satellite or rocket part falling from the sky. But it never fell, or moved at all. I remember even as a small child I knew this was not true and wondered why this explanation was given.

I do not know the duration of the event. It was actually rather boring after you got over the shock of this strange thing in the sky. Eventually we quit watching and drove on home. I do not know the duration of the event but estimate it as 30 minutes +, (about that time is all we watched but it was there when we went on, still hanging. I have always wanted to speak to others who witnessed this, especially those older than myself (I'm 39 now) I wrote this because of seeing other, similar descriptions in other locations during 1964/65. I will send a map, by mail. I have not looked for microfiche newspaper coverage. Many, many people had to have observed this object as it was in a very visible, high traffic area being over the interchange of Hwy 89 at I-5 on a very clear, sunny day. The sun was not in the same part of the sky, it was in the west, behind anyone looking at the object. It was most directly over the Drive-In Movie theater/Mott Airstrip in the triangle shaped area NE of the Hwy 89 at I-5 junction.

Occurred: May 31, 1965; Approx. 1300hrs.

Location: Shasta Lake, CA

Witness: Many
Shape: Disk
Duration: 1-hour approx.
Silver disk craft approx 40ft diam. flies over school kids and teachers outside for school meeting and lands in back of school.
At Deer Creek Elementary, now Deer Creek Middle School, in Central Valley California, now known as the city of Shasta Lake, the school was for 5th and 6th grades only. This was in 64/65. As students, we were outside for a meeting with the principal, re: something for all students to attend, such as a student body meeting. The weather was clear. A silver disk shaped craft flew over the school grounds with no noise. Its movement was north to South, very low, just above the tall pine trees. We all watched, teachers, secretaries, students...Over 45 of us... as it went over us and appeared to land in back of the school grounds, behind some trees. The area was woods and no houses. We, as students were immediately sent to our classrooms, where we sat silent, due to teacher's orders, until 2 pm. I observed a man walk by the class window. He was not a sheriff, as we had been told the sheriff had been called. He was in a black suit, with a black briefcase. I did not see any others. He was walking towards the craft, to the south, and not on the side of the principal's office, which was on the east side of the buildings. At 2 pm. The alarm to go home sounded, and we were told we could go. A girl and I headed towards the direction of the craft where it had landed. The tall pine trees had covered its descent, but we did not see it leave. My friend and I met other students coming back from behind the school from the landing site; some had went to find the ship. We were told it was gone, but we still went to the site, where we found a large approx. 40ft diameter ring. It appeared to be burned at the ring all around the edge.(Later on when older, I would return, and find for years nothing grew there.) My friend and I saw nothing but the ring, so we left. Something scared us, and we began to run, but I do not recall what. I went home and told my mom, who came back to see the ring. She saw it also. That night something scared me again by my home, which was not far from the school. Later on in the 70's there would be many sightings in the county. I would know people who had major sightings, and as far as I know besides a report in the papers occasionally, few people reported the many sightings.   ((NUFORC Note: Date is approximate. PD))

Minuteman II Operational in August 1965
The LGM-30F Minuteman II featured many improvements. It had a completely new Aero jet General SR19-AJ-1 second stage motor, which increased range by about 1600 km (1000 miles). Its guidance unit used solid-state circuitry, and could store up to eight sets of target coordinates. The first LGM-30F launch occurred in September 1964, and the first missiles became operational in August 1965.

During the first days of August 1965, major portions of the American Midwest and west were the scenes of numerous sightings of mysterious flying craft. The night of August 2-3, saw a large number of sightings by a large number of witnesses and Santa Ana, California, was the setting the next morning for Rex Heflin's famous Polaroid images of a dome-topped UFO zooming across the highway.

August 3, 1965
Denver Post editorial: "Maybe it's time for more people to get serious about the UFO question. ... If we still choose to be skeptical, we nevertheless are not nearly so ready as we once were to dismiss all reports of variously shaped but elusive flying objects as products of midsummer night dreams."   (NICAP chronology)

August 4, 1965
Fort Worth Star Telegram (TX) editorial: "They can stop kidding us now about there being no such thing as 'flying saucers.' Too many people of obviously sound mind saw and reported them independently. Their descriptions of what they saw were too similar to one another, and too unlike any familiar object".

August 6, 1965
Cascade (ID) News editorial: "An objective observer is about forced to the conclusion that there are objects of some sort appearing in the skies that cannot be explained by any conventional circumstances. There is absolutely no reason to deny the UFOs' existence because we don't understand them."

August 13, 1965
Portsmouth (NH) Herald editorial: "Perhaps we really do have neighbors from somewhere beyond our present kin."

August 16, 1965
Christian Science Monitor editorial: "[UFOs] sighted early this month over Texas may give scientists something to think about for a long time. . . . They give the clearest evidence of all that something strange actually was in the sky. ... It makes the clearest case yet for a thorough look at the saucer mystery."

August 21, 1965

Science editor, Christian Science Monitor: "Flying saucers are all but literally knocking on the laboratory door. Something definitely is going on that cannot yet be explained."

GIANT EAST COAST BLACKOUT DUE to UFOs

I distinctly remember this blackout growing up on the east coast. It affected all of us and it was so broad that there was no way of denying it. We have included it to rekindle your memories or educate you to how widespread the UFO activity was and the newspaper reports that highlight the witnesses.

November 9, 1965

Massive power blackout occurred in northeastern United States at about 5:25 P.M. (EST). Some reports of UFOs coincided, with resulting speculation about a possible relationship (section VII).

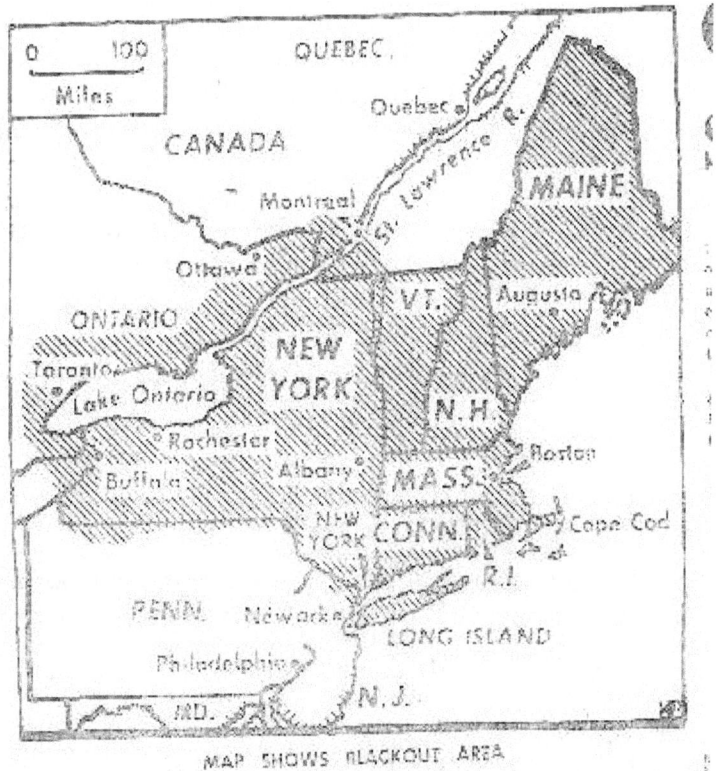

MAP SHOWS BLACKOUT AREA

On November 9th, 1965 New York State as well as portions of six neighboring states and eastern Canada were plunged in to darkness for several hours. Besides the loss of power, the blackout triggered sensors that placed the Mt. Weather facility (to house the president in time of nuclear attack) on red alert. There were also a number of reports of anomalous lights, and speculation that the blackout may have been the related to UFO activity in some way.

UFO REPORTS: Prior to and coincident with the blackout, there were a number of reports of unusual lights and one report from NYC of communication with alien beings:

Tidioute, PA. Two UFOs pace men in a light plane, until jets pursue. Then they shoot away some forth-five minutes prior to the blackout.

Larry Hatch's 'U' database

At 5:20 P.M., enroute between Syracuse and Rochester, Renato Pacibi, conductor of the Indianapolis Symphony Orchestra sees a bright light in the west rapidly descend then head towards Syracuse. Moments later, word comes on the radio that the blackout had occurred.

UfO the secret history pg 145

Camillus (Near Syracuse) A housewife and three of her children reported a 'huge dome-shaped' fireball just prior to the blackout. It was seen for five minutes and as it rose over the moon and moved forward, the lights began to dim. It then moved back, disappeared in a flash, when the electricity went off.
Syracuse Herald-Journal 11-13-65

Reporting on a subsequent sighting over the Sir Adam Beck Power Plant, a newspaper reported:

"The sighting of four strange lights over the Sir Adam Beck power plant of the Ontario Hydro Electric Power commission early today revived memories of the big power blackout that hit the northeast Nov. 9, 1965.

"Moments before the lights flickered and failed all the way to New York City, people reported seeing a strange ball hanging over the Beck plant. A pilot landing at Niagara Falls International Airport saw a weird object hovering 'over the Niagara Falls power station." (UFO's had also been reported in the area some six weeks prior.)"

At the end of the article it mentions: "After the big blackout, spokesmen for the power firms denied a strange light was spotted over the power plant on the night of Nov. 9. Since then, however, they have admitted that sightings were reported by hundreds of people."

Niagara Gazette 4-2-68

During the blackout there were numerous 'fireballs' reported around the Syracuse area, originally attributed to dump fires, barns burning and other such explanations. Some of the reports that made the papers included: Cicero Swamp (Near Syracuse). Pilot and passenger in a small plane report seeing a huge fireball where high voltage lines cross the Mohawk River. They had been approaching Hancock field when the lights went out and saw a 10 second flash they thought may be a barn full of hay.

Syracuse. The Deputy City Aviation Commissioner was also airborne when the blackout began. After landing he was looking down the runway and saw a ball of light towards Tompson road that appeared 100 in the air and 50feet wide.

Along with several more reports of 'fireball' like displays one person reported seeing a kite-shaped object with a 'giant white light' over Port Leyden, near Lowville (to the far north of Syracuse near Canada.)

Herald-Journal 11-15-65

Numerous additional reports of anomalous lights near the Syracuse area during the blackout were subsequently reported in the paper, along with speculation that some thought they may be UFO related.

Syracuse Herald-American 11-14-65

As mentioned in the excerpts below, there were also reported sightings in Manhattan and Sea Cliff (Nassau County, L.I) during the black out.

Finally, there was the odd testimony of actor Stu Whitman. Mr. Whitman claimed that a glowing UFO had hovered outside his hotel room and communicated to him telepathically, warning that the blackout had been a demonstration of their power.

The Philadelphia Daily News 12-20-65

The Great Northeast Blackout November 9, 1965

On November 9, 1995 the north eastern region of the United States and Canada was abruptly plunged into blackness. The worst blackout on record came to be known as the 'Big Blackout'.

The facts are well known. At 5:16 pm, at the height of the evening rush hour, electrical power to one-sixth of the continent's population was suddenly cut off, trapping millions of people on expressways, in elevators and in office buildings. Altogether, thirty million people in eight U.S. states and in the province of Ontario were affected by the disruption (1)

In Ontario the blackout was confined to the eastern portion of the province - from Timmins in the north, across to Cornwall in the east and south toward Sarnia. Windsor, Ottawa and Sudbury were the only eastern centres to escape the blackout.(2) Yet within three hours power was restored to most parts of the province.

Mass media coverage naturally focused on the human aspect of the blackout and to a lesser extent, on the delay in determining the cause of the breakdown.

There was, however, an even more dramatic story.

UFOs had been reported in the vicinity of strategic hydro installations at the time of the blackout. The impressive number of credible sightings led many researchers to consider the possible role these craft may have played in the power collapse.

The researchers included the late Dr. James E. MacDonald,(3) a physicist at the University of Arizona; former NICAP director Major Donald E. Keyhoe; and astronomer Dr. J. Allen Hynek, the current director of the Center for UFO Studies.

Immediately following the breakdown, the U.S. Federal Power Commission and the Ontario Hydro-Electric Power Commission launched a full-scale investigation into the cause. At first, it was reported that the trouble originated with a mechanical breakdown in a high voltage line between Buffalo and Niagara Falls.

According to the [Toronto] 'Globe and Mail':

The report turned out to be false. Then a sub-station near Syracuse was reported to be the cause of the failure, but repairmen found it in perfect condition. (4)

Finally, six days after the blackout, Ontario Hydro engineers traced the trouble to the mammoth Sir Adam Beck No.2 Generating Station at Queenston, Ontario north of Niagara Falls.

It seems that just prior to the blackout, power was flowing from Sir Adam Beck No.2. into Ontario, then across the border via Cornwall into New York State. In graphic terms, power was flowing clockwise in a loop around Lake Ontario.

At 5:16pm, a backup relay on one of the six lines linking Sir Adam Beck to the rest of the province mysteriously tripped the line's circuit breaker, which acts much like a household fuse.

In quick succession the cut-off power jumped to the other five lines, causing an overload that tripped the circuit breakers on these lines as well.

A veritable tidal wave of electricity - 1.1 million kilowatts - flowed in the opposite direction into New York State. (5) Inexplicably, the relays on the New York lines failed to isolate and contain the overload. Within seconds, the entire grid of thirty-one interconnected power utilities of CANUSE (Canada-United States Eastern Grid) had broken down.

Although experts could pinpoint the origin of the blackout, they were baffled by the cause of the relay malfunction and the failure of the protective systems to contain the overload.

In the words of Ontario Hydro's system supervising engineer, Jim Harris: "It's incredible! I would have said this was impossible if I hadn't seen the evidence." (6)

The mystery deepened when it was discovered that the relay had not in fact malfunctioned, but had merely reacted to a sudden surge of power from an unknown source.

As stated in the final report of the U.S. Federal Power Commission:

"The precise cause of the backup relay energization is now known." (7) Where did the unexplained surge of power come from? To this day that question has remained unanswered. Or has it?

Although inconclusive, one answer might lie in the findings of the late Dr. James McDonald who contended that the magnetic fields accompanying UFOs can create sudden power surges in transmission lines as the craft flies overhead.(8) In theory, these power surges could produce blackouts of massive proportions.

Since the 'Big Blackout', McDonald's theory has gained considerable support in the light of strong evidence confirming widespread UFO activity on that fateful evening.

The Syracuse Herald-Journal was inundated with calls reporting more than one hundred sightings in the Syracuse area. One of the first came from Syracuse Deputy Aviation Commissioner, Robert C. Walsh, who was flying over Syracuse at the time of the blackout.(9) Despite the darkness, he managed to land safely at
Hancock Airport.

Standing on the runway, with some airport officials he suddenly noticed an enormous circular ball of light, drifting overhead. "It appeared to be one hundred feet in the air and fifty feet in diameter.(10) It rose for several seconds, then suddenly disappeared. Moments later, a bewildered Walsh and his companions watched an identical device ascending over the airfield before mysteriously 'blinking out', as did its predecessor. Unlike the known high-speed plunges of fireballs, these craft moved upward at moderate speed clearly under some form of intelligent control. At the same time, the mysterious craft were also being observed overhead. Veteran flight instructor Weldon Ross and his student, James Brooking, were approaching the darkened airport when they spotted a second fiery object below.

The Giant craft, estimated at well over one hundred feet in diameter, appeared to be positioned directly over the Clay sub-station, a strategic installation that channels power from Niagara Falls to New York City.(11)

It was the same sub-station where hydro-investigating teams had initially pinpointed the origin of the blackout. In a relentless pursuit of a possible UFO-blackout relationship, Herald-Journal reporters succeeded in uncovering even more explosive evidence. In a front-page story seven days after the blackout, the paper carried photographs of the mysterious red craft taken by Mr. William Stillwell, a sexton at St. Paul's Episcopal Church. He described what he had observed through a 117-power telescope: The center was rotating, around and around and around. It came from the direction of DeWitt and shot off at an angle and then went back the way it came. (12) He had watched the glowing object for as long as two hours before it streaked away.

While investigating teams continued to dig for the mysterious cause of the power failure, press coverage of a possible UFO connection gained momentum.

In a strongly worded editorial, the Indianapolis Star urged: The answer is fairly obvious - unidentified flying objects! It is one angle the multi-pronged investigation should not overlook. (13)

Support for the UFO possibility intensified, as news of other sightings became known. In New York City, twenty minutes into the blackout, witnesses in the Time-Life Building spotted a peculiar glow in the sky over darkened Manhattan. According to Major Donald Keyhoe: It appeared to come from a round object hovering over the city. This was twenty minutes after the lights began to go out. A Time Magazine photographer took several photographs, and one of which appeared in the November 19th issue. (14)

Although clearly visible in the photograph reproduced here, Time editors failed to make any reference in their photo-caption to the spindle-shaped craft. Was it Journalistic oversight or deliberate omission? The only hit of any unusual aerial activity came in a facetious reference to a Soviet satellite:

Some New Yorkers, claiming that they had seen a satellite pass over at the moment the lights failed, argued that the Russians had done it again. (15)

But UFO investigator and author, the late Frank Edwards disagrees with both the UFO and the Soviet satellite explanations: The spindle-shaped thing could have been a UFO--but it certainly wasn't. It was nothing more than an optical ghost, the result of reflections between the elements of an air-spaced lens.(16)

While disputing the validity of the Time photo, Edwards strongly supported the contention the UFOs were somehow involved in activating the blackout. In fact, while conducting his own investigation into the
cause of the blackout he discovered that U.S. military authorities had been well aware of the UFO presence, at least forty-five minutes prior to the power failure. (17)

This startling disclosure came from two commercial pilots, Jerry Whitaker and George Croninger, who were flying over Tidioute, Pennsylvania, when they spotted two disc-shaped 'shiny objects' overhead.

Even more surprising was the sight of two military jets chasing the mysterious craft.

Moments later, one of the discs put on a 'burst of speed' and quickly outdistanced its pursuers. While watching the fast-disappearing UFO, the dazed pilots lost sight of the other object, which had presumably departed in the same manner.

The most spectacular UFO revelation, however, came one day prior to the release of the 'official' explanation when, speaking before a nationwide television audience, NBC commentator Frank McGee announced that a private pilot had spotted a "round, glowing object near the Niagara Falls power plant".(18)

Associated Press picked up the story and numerous newspapers subsequently carried it. The following morning, a well-documented article appeared in the New York Journal American blaming UFOs for the
disastrous power-grid breakdown.

Any further media focus on the UFO connection was brought to an abrupt halt, however, with the release of the 'broken relay' explanation.

Despite mounting evidence, the Federal Power Commission had predictably chosen to side step the possible UFO connection. Dr. James E. McDonald who, as a respected scientist, was allowed to interview certain FOCI officials eventually confirmed this omission. They admitted they had the Syracuse and Niagara Falls reports, also most of the others on that night. But they wouldn't discuss the UFO possibility.... No matter what they believed, I think they were convinced the facts shouldn't be given to the public, and that's why they agreed to the 'broken relay' story. At any rate it was obvious they were covering up_. (19)

Under the circumstances there seems to be a strong possibility that Canadian authorities were also involved in the cover-up. Ontario Hydro-Electric Power Commission investigators, having become aware of the UFO reports, collaborated with the FPC by exchanging information that eventually led to the 'broken relay' explanation. (20)

Furthermore, this explanation had apparently been pre-arranged and was released simultaneously in both countries. (21)

The Ontario Hydro press statement similarly neglected to include UFOs as the possible cause for the blackout.

One reputable American ufologist went so far as to point an accusing finger at the late Lester B. Pearson, then prime minister. Major Donald Keyhoe contends that: To shift attention from the UFO explanation, the 'broken relay' story was invented. Since this could be construed as blaming Canada, the Prime Minister must have been convinced it was best for both countries not to disclose the true situation. (22)

It that was the case, then it represents one of the most shocking deceptions ever perpetrated - leaving the heads of thirty-one utility companies and thirty million people to grope around in the dark in more ways than one!

(1) Time Magazine (November 19, 1965) Canadian Edition, p.24. (22) Donald E. Keyhoe, op. cit., p. 180.

1966

'Star Trek' appears for the first time on Television, September 8, 1966. The public now goes where "No man has gone before, the Final Frontier." It still is inspiring generations to explore our galactic neighborhood.

January 1966; Popular Science Article

"Why I Believe in Flying Saucers", by MacKinlay Kantor, Pulitzer Prize winning author of "Andersonville". The noted writer, co-author with Gen. Curtis E. Lemay of "Mission with LeMay. My Story", tells of the strange personal sighting that convinced him that UFOs are real. (Copy provided by Ole Jonny Brænne)
Date: Wed, 04 Oct 2006 23:42:02 +0100 (BST)
From: daniel wilson <daniejon2000@yahoo.co.uk>
Subject: ICBM Deployement : To Defend & Deter
To: Francis Ridge <nicap@insightbb.com> Just a starter...

Occurred: January 3, 1966; Night
Location: Los Altos, CA
Witness: 1
Shape: Speeding Lights
Night. Army colonel expert on rockets and missiles sighted a group of bright light sources speeding overhead at an estimated 1,000-1,200 M.P.H., maneuvering back and forth. Soon afterwards he observed several searchlights sweeping the area as if trying to spot the flying objects. (NICAP report; U.F.O. Investigator, Vol. Ill, No. 6, Jan.-Feb. 1966, p. 2.)

January 12, 1966; Sagan Requests Materials

Dr. Carl Sagan requests information from Blue Book on sightings. The Tridade Island case is referred to and the Kelley/Hopkinsville case was requested also.

February 1966
Dr Brian O'Brien heads a panel of the AF Scientific Advisory Board that completes a review of Project Blue Book and recommends that the AF contract with several universities to conduct UFO studies. (Sparks)

March 31, 1966; JANAP 146(E)
Joint Army Navy Air Publication 146(E), changed from 146(D) of Feb. 1, 1959. Added that photographs should be sent to the Director of Naval Intelligence. Also added special reporting instruction for unidentifiable objects. (See JANAP 146 History and Evolution)

June 6, 1966 AF Foreign Technology Division (FTD) officials decide to leak the AF-manipulated CIA Robertson Panel Report with Minutes and Comments report (Durant Memo) to visiting Navy scientist Dr James McDonald, who was working under an ONR (Office of Naval Research) contract. The AF evidently wanted to make the CIA look bad when the report was inevitably hushed up by the CIA but complained that AF did not clear this declassification with the CIA. Fearing that McDonald was part of an ulterior Navy plot to make the AF look bad, this big show of AF openness and pretense of trying to overcome excessive CIA secretiveness would serve to make the AF look good instead. As an added bonus this sugar-coating of feigned AF openness in trying to release the Robertson Panel report actually covered a poison pill inside, since the Robertson Panel report was itself an AF-manipulated, virulently debunking anti-UFO document which set forth all of the AF's desired anti-UFO PR positions which had been systematically planted on the CIA and the Panel back in 1952 and 1953. (Sparks)

Occurred: June 15, 1966 Approx.1100hrs.

Location: Roseburg, OR

Witness:

Object: golden figure

Duration: about an hour. I saw YHVH and afterwards there appeared a UFO in the following week that took me from a tent and tattooed me, project blue book there. I had just finished my 7th grade class and had been given the advice to use the summer to think about what vocation I might have an interest in for the rest of my life. I came home and talked about it with my favorite person, my foster dad, who worked for the veterans' administration in town. We were living in the Melrose Eden bower addition and were in the foothills next to the Callahan Mtn. on Doerner Rd. we lived as devout Roman Catholics, and he said to pray about it and smiled. So i fasted and prayed in my bedroom for seven days. I felt a breakthrough in the spirit on the seventh day; a mind image of a path that was unmarked but available and I then quit and resumed household duties. I expected I'd just be shown something, but i was unprepared as to what was about to happen.

It was a summer morning on the farm; everyone went outside after breakfast, leaving me alone to clean the living room. I was idly thinking while I cleaned about a career in either anthropology or Egyptology or in veterinary science while I was cleaning. I heard a man calling my name, each time of the three times getting louder and louder! The last time it felt like i was being defragged like a computer program. I was bewildered! I thought that it was my two playful practical joker brothers and a microphone, but there was neither of them at the house at that time, and when asked who it was, I smelled the fragrance of frankincense and sandalwood and myrrh, like old style Roman Catholic high church incense and as I looked around the room my attention was directed to the ceiling and heard a voice say, "this is my beloved daughter in whom I am well pleased"...I did a double take, amazed because I was adopted and didn't know who or where (not much) about where I was from)and sensed but at this point I did not see another dimension which contained a large kings throne room and many people assembled all around a mighty King. They were in a previously not seen or felt space above my head and I was below their feet - about four feet. Then I said weakly, who are you? Tell me who you are? I thought I was hallucinating, and was scared! I was also only about 13 years old! Then I looked up above the top of an opening "arch" doorway into the dining room, and the wall, the ceiling above as well, disappeared and I was looking upon a large object in the sky! It was far away of about at least 10 miles from me but it was huge! It was a waterfall of what looked like molten gold, and it was surrounded on both sides by thunder clouds flashing thunder and lightning and surmounting the waterfall and its pool which was outrageous with detail of molten gold for water and golden flames, but also the powdered golden "mist" that arose from it and a huge rainbow over the top, with a white crystal throne and two persons radiating immense light that made me feel like I was dying every time I looked up at them- a man standing to the right in long white robe, red robe that went over one shoulder and he looked just like the popular Jesus Christ pictures. The one sitting down had a long grey beard and wore a long flowing white robe. One voice very powerfully vibrated through the air and me said, "I AM JEHOVAH, the Father Almighty, the Creator Almighty." (Note: this means the tetragrammaton. I was standing there amazed, with a broom in my hand and I noticed I was still holding it and I immediately thought of the Moses movie and took off my shoes and put the broom on the ground and did the same thing, got on the floor on my knees, and started praising God.

I thought to ask HIM. HE asked me to be HIS Wayshower, and showed me a vision of a misty white large road in the sky with viscous black walls on either side, and that the light of the Creator fell on this road. HE said, "If anyone argues with you about what you think, do say and testify to leave them by the wayside." I was shown a image of me grown up as an adult showing a Indian man into the viscous wayside, where GOD looked down on his entrapped form crying and turned away from him with finality. I then said "I let everything I think, do say and testify to be of you and not of myself so that others will know you have sent Me." then I worshipped THEM and I looked up from where I was bowed and they were gone. In the afterward, my foster mom and two child siblings entered the house and said to me "what happened? You're glowing very bright." I was lit up like a neon sign, and they put a sheer curtain over my head so they could look at me.

Afterwards, there was a UFO silver disk, with a bulging ring of portholes, about 15' across that landed in the next week while I was up at the next-door neighbors camping out and I was "abducted" from the tent, given a physical examination and tattooed with an Egyptian hieroglyphic that is the name for NIT- a weaving shuttle- on my right groin area over the ovary. My stomach bulged and felt like I had radiation sickness for the next 2 weeks, causing my family to be tightlipped and very worried, but they refused to tell anyone because my dad worked for the government.

The blue book came out and white curtained the area of the landing, because the craft burnt depressions into the asphalt road at the bend right past the house where I lived. I peeked over the edge of the hill and was spotted and photographed (glowing) because I wanted to see what was going on behind the white curtains. That fall, the family doctor gave everyone a complete physical examination and photographed my tattoo for the FBI (I saw the markings card marked FBI which the doctor with a lot of exclamations filled out and took a one step Polaroid picture of the mark. I didn't tell him the truth.

And that began to be the first of countless OFU contacts with the good guys of Heaven. Unfortunately, years later, the two other girls with me in the tent have succumbed to cancer, although that didn't happen till they hit their adult years. I began to be very talented and did amazing things that my foster family couldn't get over, like levitation and fire shooting from my hands and body. I was a pretty good little girl, and I was known as a mind reader and very pronounced psi talented. There were other events, but this is where it first started, although I remember instances in Milpitas California as a baby where shadowy "white" people like ghosts used to come into our house and visit me, telling me not to forget what I was here for, to be with me when I was scared at night, etc. a lot about don't forget the promise, the plan. That's all for now. These are not all the details of what happened the first time for me on Doerner road in Oregon. Oh, and the guys that got off the craft were humanlike, with golden blonde short, curly hair and gold and purple close fitting jumpsuits and white silvery boots. They didn't seem to have seams in the clothes that piloted the craft that burnt the holes in the road.

((NUFORC Note: Witness indicates that the date of the incident is approximate. This is one of two reports from same source. PD))

Occurred: June 30, 1966; Approx.1600hrs.

Location: Corning, CA

Witness: 2
Shape: Disk
Duration: 5 minutes
2 small (2 ft) & 1 large saucers seen at close range by multiple witnesses, Northern CA about 1966

When I was about 12 years old 2 rather small UFOs approached me at close range. They were classic saucers, but only about 2 feet in diameter, and with what looked like "antennae" sprouting from the top center. They came within about 30-40 feet from my cousin and me, and were also observed by several other family members within about 100 feet. One was my aunt, about 30 years old at the time and standing about 30 feet behind me - perhaps 60 feet from the craft. She also saw them clearly as did two other, older cousins. This occurred on my uncle's farm in northern California roughly 1965-66. There were a number of animals nearby that reacted violently to the objects - especially one calf that was tethered and unable to run away. There was also a 3rd UFO nearby - but that one was roughly 30 feet in diameter and was seen only as the smaller ones flew directly away from us and the large craft joined in front of the smaller two. It came from the side, from several hundred yards away when first noticed, and was seen only in silhouette against the afternoon sky. The small ones, however, were so close we could see them _very_ well.

They were gray and black, spinning about 30 RPM, exhibited the characteristic "falling leaf" motion when descending, and one discharged "angel hair" at it's lowest point (about 15-20 feet altitude). No residue was seen later, and the material seemed to "dissolve" in the air while falling. The objects made no sound.

The small objects hovered in and around some eucalyptus trees about 30 feet from us and perhaps 20 feet in altitude - so we saw them both behind and in front of the trees giving a good visual "fix" of their size and position. They never remained completely still, but made small movements in a jerky way constantly. After about 3-4 minutes of watching them, they rose vertically to about 100 ft altitude and flew slowly directly away from us slowly ascending. The larger craft flew in front of the smaller two, then all 3 flew rather slowly out of sight in a straight line, continuously ascending until they flew out of sight. The total sighting lasted about 5 minutes - which seemed like a _very_ long time under the circumstances. Unfortunately, no camera was available, for there was plenty of time to take photos.

Other family members remembered a report being made to the Air Force and, perhaps, a small article in the local newspaper, but I do not remember the latter. No one involved has had another sighting before or since, to the best of my knowledge.

((NUFORC Note: Witness indicates that the date of the incident is approximate. PD))

Mufon Report: Historical Mt. Shasta Case

Occurred: July 1966; 0300hrs.

Location: Mt. Shasta, CA.

Shape: disc

In an interview on May 27, 1967, Paul Cerny, then chairman of the Bay Area subcommittee of NICAP, learned that Captain Jack Brown of the Mount Shasta, California, Police Department, had encountered a peculiar craft-like object in the early morning hours of a July morning in 1966. The following article is based on Mr. Cerny's report.

At 3:00 a.m., while on duty and patrolling the northwest city limits of Mount Shasta, Captain Brown first observed the object through his car's windshield. The UFO closed with him rapidly, apparently on a collision course with the patrol car. To avoid a crash, Brown swerved violently, barely missing a ditch, and came to an abrupt halt. Fear and plans to escape by foot in case the UFO exhibited any further hostilities prompted the police officer to get out of the car and lean on the open door, one foot planted on the door frame.

The alleged craft, a smooth disc, approximately 30 to 40 feet in diameter and 10 to 12 feet thick, hovered silently 20 to 30 feet above the car and emitted a blue-white glow, which seemed particularly bright at the disc's rim. The intensity of the light hurt Brown's eyes as he gazed upward at the disc. The object had no visible windows or openings; however, on its underside were two thick tubular protrusions, curved and tapered, about eight inches around at the ends, skid-like in appearance, but with no perceived function. Also, Captain Brown commented that he detected no exhaust trails coming from these structures.

Brown stated that after the apparent craft lingered for 3 or 4 minutes, it drifted off, moving erratically across the road and adjacent fields "as if nonchalantly surveying the immediate area." Suddenly, the object shot off toward the slopes of Mt. Shasta where it stopped and hovered again before accelerating a second time. It climbed eastward over the mountain and disappeared.

The patrol car's headlights and two-way radio functioned normally throughout the encounter. In fact, Brown contacted his dispatcher, David Vacari, who after dashing to a window sighted the glowing object in Brown's vicinity. Furthermore, Officer Mazzeri, 8 miles north of Mount Shasta, in the town of Weed, California, spotted the UFO heading for the mountain but did not observe the object hovering over the car because the terrain blocked his view. (Mr. Cerny saw fit to point out the absence of electro-magnetic effects on Brown's body and automobile; the Captain experienced no paralysis or tingling sensations, and as previously stated, the car performed as usual during the incident.)

According to Mr. Cerny, Captain Brown, 42 and a family man, seemed well above average intelligence, physically fit, in good mental health, and, in short, a competent observer.

Captain Brown claimed that he would be reluctant to report any future sightings because of the way the Air Force had handled his first report. Evidently, the Air Force insisted on his completing many tedious and irrelevant forms and report sheets before informing him that what he and his fellow officers had seen was something other than what they had described.
THIS IS THE SECOND TIME CAPTIAN BROWN HAD A SIGHTING. LOOK AT AUGUST 1960 FOR THE RECONSTRUCTION OF HIS CASE. At one time the sightings was reported to the Mt.Shasta Police department, in their logs, it is acknowledged by the Nicap in their testimony of the Red Bluff Flap of 1960 but his testimony is missing, classified, blacked out in all the findings, the McDonald tapes, the logs in the Mt.Shasta PD are MISSING now, and a witness who was read the report in 1980 by a the dispatcher, relative, and collaborating stories from kids all coincide with the story in the local sighting and 1960 chronology.

Occurred: November 29, 1966

Location: Above Mount Shasta, CA

Witness: Pilot Jack Brown
Shape: oblong craft
Duration: seconds
Sighting on Mt.Shasta.

National UFO Reporting Center
November 29, 1966: Just after noon on the 29th, Jack Brown was flying his small aircraft northbound over California near Mt. Shasta when he saw a bright object streaking across the sky. "There framed against the blue sky was an object that failed to meet any mental description of an aircraft carried in my think box. It looked every bit like a big butane tank (the oblong type you see sitting by the roads with 'Butane Sold Here' signs on them) cutting across the sky at something in excess of three or four hundred miles per hour. It was on a southwest heading which would take it out over the California coast. I blinked my eyes as I watched what appeared to be a white oblong tank flash thru the sky and disappear. With each blink I thought my eyes might focus on a tail assembly or some projection on the object to give me a clue as to its identity. All this had taken just a few seconds and yet I know I saw something and as well know exactly what it looked like. The sixty-four dollar question was - whatinell was it?" **THIS IS NOT Police Officer JACK BROWN FROM MOUNT SHASTA NOR A RELATIVE.**

1967

Author's Note; E. U. Condon is responsible for the circumvention of the public freedom of information, and discrediting of prior research done by professional scientists for the purpose of burying the UFOology in black operations and classified status, so no one would know what is going on. It is the end of the era of open conversation and inquiry for the public and the government to cooperate as a unit. Meanwhile Corning, Anderson and the lower valley are getting their fair share of UFO activity. Corning and the Red Bluff area again are on the radar for sightings by officers.

January 25, 1967
Dr. E. U. Condon, scientific director of the Colorado UFO Project, spoke to the Corning, NY, and section of the American Chemical Society, stating that the government should get out of the UFO business, since there was apparently nothing to it.

January 26, 1967 newspaper article, "Most UFOs Are Explainable", Edward Condon
Authors note: "Read it and puke."

Occurred: February 13, 1967; Approx. 1915hrs.
Location: Davis, CA
Witness: 2

Shape: Triangle lights, disc base
Duration:
7:15 p.m. PST. Two women driving home saw an object that came within 100 yards of their car. The initial light resolved itself into three huge lights in a triangular formation. The object tipped up to display a disc base and a central red light and 5-6 dimmer white lights (body lights). (NICAP report form; McDonald, 1967, case 7; reprinted in Vaughan, 1995, pp. 205-6.)

Occurred: February 13, 1967; Approx. 2030hrs.
Location: Woodland, CA
Witness: 1
Shape: Bright light
Duration: 25 minutes
8:30 p.m. PST. An ex-Air Force man with pilot training saw a bright light for 25 minutes, which maneuvered and seemed to pace his car. (NICAP report form, Los Angeles NICAP Subcommittee report, NICAP files.) [Note: There are two towns named Woodland in California. One, in Yolo County, is 12 miles north of Davis. The other, in Tulare County, is 245 miles to the southeast of Davis.]

February 23, 1967 - AF to Declassify Reports

Lt. Col. Robert Hippler of AFRSTA said that the Secretary of the Air Force had established the policy that all USAF information on UFO's classified up to and including Secret is to be provided to the University of Colorado. 2

One of the Launch Control Facilities at Malmstrom AFB in Montana

Missileers report major Minuteman Missile shutdown in 1967

March 16, 1967; Nr Lewistown), Montana

Form: 97 NCPDIR
Nuclear Connection Project
NCPSubject: "Echo Flight" / Malmstrom AFB Missile Incident; Lewiston, Montana; March 16, 1967

Francis Ridge:
What could be more interesting and important than an incident involving UFOs and our strategic defenses, especially where there appears to have been a direct relationship between the incident and the failure of vital missile targeting systems, ten missiles at one site and ten at another, 35 miles apart! "Some signal had been sent to the missiles which caused them to go off alert status." What most people do not understand is, this was not just an unfounded rumor. The incident has been confirmed. Although it is normal for authorities to "explain away" almost all UFO incidents, this one is not that easy. This is the story of extraordinary events that happened in 1967 to Strategic Air Command Missile Combat Officers; Missileers assigned to operate the Minuteman Intercontinental Ballistic Missile, an essential part of America's Cold War strategic nuclear deterrent. Our thanks to Joel Carpenter for locating the image of what he suspects is the LCC capsule that was apparently targeted by the UFO. Most importantly, we give our thanks to Jim Klotz and Robert Salas for the official reporting of this incident to the world. And now, their book, "Faded Giant", tells the full story. (See link below).

Brad Sparks:
In the 1967 Malmstrom incident 10 Minuteman DID get knocked out, but the investigation was conducted not by NSA but by the missile contractor Boeing, which issued a SECRET classified "Report of Engineering Investigation of Echo Flight Incident, Malmstrom, Mont - 16 Mar 1967." The trick was that the UFO "rumours" were denied so that the missile disabling could be investigated and reported as if unrelated to the UFO's (which WERE sighted, contrary to the flimsy denials). No satisfactory explanation for the missile disabling was found. (See below) Presumably whenever these kinds of national security UFO incidents occurred that affected military equipment there were investigations by contractors who built or maintained the equipment, and the UFO aspects were compartmented off. At a higher level someone cleared to know about the UFO details could read these reports, and they could ignore the perfunctory denials of UFO activity written into the contractor studies.

The Malmstrom AFB UFO/Missile Incident (1967) - Jim Klotz and Robert Salas (CUFON)
Minuteman Missiles Shutdown - Robert L. Salas (MUJ-345, 15)
Incident at Malmstrom Base - ufocom

Occurred: April 5, 1967
Location: Davis, CA
Witness: 1
Shape: Cone
Duration: 10 minutes
1:50 a.m. PST (4:40 p.m. EST) A woman saw a basket-shaped (truncated cone) object with flashing lights for 10 minutes. The object moved slowly across the sky at low altitude oscillating up-and-down, and also seemed to oscillate laterally while moving. The body lights were red when the object moved upwards and white when it moved downwards (color/motion correlation). The object was described as tumbling end over end. (NICAP report form.)

April 22, 1967
American Society of Newspaper Editors sponsors panel discussion on UFOs at its annual meeting in Washington, DC. Panelists included Dr. James E. McDonald, Dr. Donald Menzel, and Maj. Hector Quintanilla.

Occurred: June 1, 1967; 2030hrs.
Location: Myrtle Creek, OR
Witness: Multiple
Shape: Circle
Duration: initially 15 minutes last
10 PM August 1967 sphere observed by multiple witnesses moving along ridgeline 1/2 mile distance 500 Lights On Object: Yes
Summer 1967 family observed a large (size of the full moon) bright round object moving very slowly along the north-to-south ridgeline of hills ½ mile from our house. Object did not change elevation…just cruised slowly along the crest line. Intermittent red and blue flashes of light occurred. The sighting lasted 15 minutes and object just disappeared.
2 weeks later…8:30 pm past sunset but still light outside my mom walked with me next door to Gram's house where I slept …Mom said goodnight to Gram while I walked into my bedroom…I had walked 3 steps into bedroom when out the window appeared rotating flashes of white-red-blue-green light. I initially thought it was the rotating lights of a piece of road equipment next door at the county road crew shop, but became alarmed when the colors started rotating far too fast and multicolored for any piece of road equipment I had ever seen. I stepped back one large step to reach the door handle of the closed bedroom door (which I had not closed). Gram was on the other side of the door yelling my name. We could not open the door even though we tried hard.
The next morning I "awoke" late at 8:30 AM to hear my Mom talking to Gram in the kitchen…I was standing fully dressed in exactly the…3 steps into the room position that I had been in the night before. I noted that the bed was fully made-up and uncrumpled. Mom asked if I was ok and if I remembered what happened last night. I said "yes". Later she said that as she was walking back home she was alarmed to see the large round object hovering over the part of house I was in and a voice in her head told her "just go home and go to bed" she protested and the voice again gave the same reply. She said she felt compelled to go home and go straight to bed.
When I told the story to a relative many years later – 1993- the relative promptly reported my "descent into delusions" to my Mom…who thankfully confirmed the story to the relative stating that she regretted being unable to do anything about it.
I had my 14th birthday between the initial sighting and the "bedroom events".

Occurred: June 15, 1967; Approx. 2300hrs.
Location: Baker, OR
Shape: Disk
Duration: 30 minutes
Disk shaped object with different colored revolving lights

I was in the Air Force at Baker, Oregon, in 1965 until 1968. And it had to be 1966 or 1967 that our radar site identified a UFO from the 35 tower and 2 6 towers. There were approximately 15 Air Force personnel that saw this object. It was very large with red green white lights and a disc this should be in the Air Force files.

The only other name that I can give you that actually saw it, is the guard at the AP gate. There was only 3 35 towers on the west coast at that time. This is no hoax, I'm 63 now and have to get this out ((NUFORC Note: Witness indicates that date of incident is approximate. PD))

Occurred: June 15, 1967; Approx. 2000hrs.
Location: Arbuckle, CA
Witness: 2
Shape: Triangle
Duration: 3-4 minutes

At dusk in 1966 or 67, three lights in formation or triangular craft with no visible substance floated silently into the West. This took place at our ranch, 7 miles out of our tiny town of Arbuckle, CA. in the Sacramento Valley near the Foothills to the West. Dad and I were outside on a warm summer evening after dusk.

It was 1966 or 1967. We saw an enormous, low flying, silent, very slow moving triangle of, I think, three lights, no real interior substance, more like dim lights (the brightness of small stars) in triangular formation moving as one uni, giving the feel of a "solid" object. There was no movement of the lights, any blinking or rotation, just total "stillness", even though they floated across the sky, slower than anything we had ever seen in the sky. And then an enormous and silent presence we felt upon us.
Stunned, we acknowledged to each other what we were seeing, Dad telling me to run to the Jeep and get the binoculars, QUICK! I took off and fumbled recklessly, not wanting to miss any of this show. I could not find the glasses and so ran back quickly. All that was left in the sky was off into the distance, over the Western Foothills. A bright Turquoise streak, like a wisp of cloud, as if the speed of the thing was such that it left this turquoise vapor behind, in the early twilight. All the reading I've done throughout the years of others' sightings, I've never heard of anyone else seeing this Turquoise light streak. Anyone? Please let me know. I'd be grateful!
((NUFORC Note: Witness indicates that the date of the sighting is approximate. PD))

Occurred: July 1, 1967; Approx. 0000hrs.
Location: Maury Mountains, OR
Witness: 3
Shape: Fireball
Duration: Unknown
Possible Crashed UFO viewed in Oregon's Maury Mountains In The Summer Of 1967.
During the summer of 1967 my parents and myself went on a camping trip in the Maury Mountains of Central Oregon. The campsite was around 30-35 miles east of Prineville, Oregon past a small village called Post.
One night during the trip I was awakened by a streaking lime green fireball moving very quickly across the sky. It was emitting sparks (also lime green in color) and I could hear a crackling sound from above. This whole memory is only remembered in a dream-like state.
In the brush across from the campsite about 30 yards away I saw a creature in a spacesuit-like outfit. It was short and rotund, like the Michelin Tire Man. It was in a form-fitting light brown color jumpsuit/spacesuit outfit. It also wore a dark brown space-helmet with some sort of night vision/binocular type device in front of where its eyes would be. It stood about 4' tall. It was just observing our observations of the fireball streak.

However the next morning a group of military personnel in a jeep and two US Army trucks questioned my parents about the event. I think they were working on the idea that this was a crashed UFO!

I remember the creature probing my emotional state at the time. Were we being tested in some way? My parents are no longer around, I'm 47 years old now, and I was 9 years old at the time. ((NUFORC Note: Date of incident may be approximate, although the witness does not indicate that fact. PD))

Occurred: July 4, 1967; 0515hrs.

Location: Corning, CA

Witness: 5+
Shape: Oblong metallic with light
Duration: 10 minutes

5:15 a.m. PDT. At least five witnesses from two independent locations saw an odd oblong, metallic-appearing object with a brilliant light on top and a smaller light on the bottom near the front (body lights). Two policemen and a businessman described it as a dark gray flattened sphere with a brilliant light beam on top directed upward, and a smaller and dimmer light on the bottom directed downward. A dark band circled the midsection. Two men north of Corning independently saw the object. The witnesses estimated a diameter of 50-100 feet. At first the object appeared to be hovering, then it moved slowly at a few hundred feet above the ground, finally picking up speed and disappearing from view to the south after being visible for about 10 minutes. (Corning Daily Observer, 7/6/67, UFO NICAP files; McDonald, 1968b, p. 74, reprinted in Vaughan, 1995, pp. 129-30.)

A third report is taken from a paper Prof. McDonald presented at the 12 March 1968 Canadian Aeronautics and Space Institute Astronautics Symposium, Montreal. "At about 5:15 am., PDT, on the morning of July 4, 1967, at least five witnesses (and reportedly others not yet locatable) saw an object of unconventional nature moving over Highway 5 on the edge of Corning, California. Hearing of the event from NICAP, I began searching for the witnesses and eventually telephone-interviewed four. Press accounts from the Corning Daily Observer and Oakland Tribune afforded further corroboration. "Jay Munger, operator of an all-night bowling alley, was drinking coffee with two police officers, James Overton of the Corning force and Frank Rakes of the Orland force, when Munger suddenly spotted the object out the front windows of his bowling alley. In a moment all three were outside observing what they each described as a dark gray oval or disc-shaped object with a bright light shining upwards on its top and a dimmer light shining downward from the underside. A dark gray or black band encircled the mid-section of the object. When first sighted, it lay almost due west, at a distance that they estimated at a quarter of a mile (later substantiated by independent witnesses viewing it at right angles to the line of sight of the trio at the bowling alley). It was barely moving, and seemed to be only a few hundred feet above terrain. The dawn light illuminated the object, but not so brightly as to obscure the two lights on top and bottom, they stated. "Munger, thinking to get an independent observation from a different part of Corning, returned almost immediately to telephone his wife; but she never saw it for reasons of tree-obscuration. At my request, Munger re-enacted the telephoning process to form a rough estimate of elapsed time. He obtained a time of 1-1.5 minutes. This time is of interest because, when he completed the call and rejoined Overton and Rakes, the object had still moved only a short distance south on Highway 5 (about a quarter of a mile: perhaps), but then quickly accelerated and passed off to the south, going out of their sight in only about 10 seconds, far to their south.

Occurred: August 1, 1967 Approx. 1900hrs.

Location: Roseburg, OR

Witness: 3
Shape: Sphere
Duration: 15 minutes
3 round craft, with red-blue-green swirling lights, (no sound) , and sparks emitting from beneath them.
During the summer of 1967, I was about 5 years of age but have been having recurring memories of a 3-day event that happened.
My mother and sister are still alive and have validated my memories of the event. I went back home to Roseburg Oregon and interviewed the neighbor lady and her husband who still lived there who also remembered the event.
In August of 1967, approx 7:00pm in the evening, 3 saucer aircraft appeared from the west traveling toward the east, and hovered above our neighborhood (we lived in a valley), approx 500 feet off the ground. I remember my sister ran in the house telling us her car was enveloped in a blue light as she was heading to work in a local movie theater. I recall being scared as my mother turned all the lights off in the house.
We watched them from inside our home for about 5 minutes, and then slowly went outside to look. There was no sound. Silent. In the air above our home approx 500 feet up were 3 disc shaped objects, with red-blue-green swirling lights rotating around the craft in a clockwise circular pattern. They were about the size of a house, quite large.
I remember the colors being very striking, almost a neon glow only more intense. On top was a large orange glowing dome, which reminded me of a lantern.
After 10 minutes, sparks began to fall from beneath the crafts, at the same time. Long showers of sparks were falling with NO SOUND at all.
Suddenly, they accelerated so fast in a north direction that it almost looked as if they disappeared. There was NO WIND either, as the sparks were all that remained falling in the air where the craft were. The sparks did not blow around as the craft left. It was as if the craft "disappeared".
This event occurred precisely at 7:00pm for the next 2 evenings. Exactly the same things happened, with the sparks falling after 5 minutes of them approaching from the west.
On the 3rd day, my sister went to her wealthy friends house on top of a nearby hilltop. They had a telescope for star viewing. My sister said when they appeared; she plainly saw what appeared to be round black windows where the orange glow dome was. The craft was metallic. The lights swirling around the craft were something she said she has never seen before. Beautiful colors just swirling around the outside edge of the saucer shaped craft. There were flames and sparks emitting from the bottom/center of the craft.

The airport was called during the 3-day event. They said they saw it, as the airport was only 5 miles away. They said there was nothing on the radar screen, yet there they were. The local news station had a report on the objects on the evening news the next day. But right after that, no one spoke about it. There were military officers I remember in the news, but not a word about any craft. My mother, sister, friends all talked about it amongst ourselves, but never to anyone else as people would think we were crazy, so we never talked about it.

I began recalling this whole thing about 15 years ago, and it has been getting more vivid, as if it just occurred yesterday. If I could take a photo of them from my mind, you would also agree this was not a helicopter or airplane that we know of.

I am sure there were pictures taken by townsfolk. There were FAR too many people who witnessed this. But what was weird was that the media NEVER talked it about again.

At any rate, I felt I needed to report this. ((NUFORC Note: Witness indicates that the date of the event is approximate. PD))

Occurred: September 15, 1967; Approx. 1930hrs.

Location: Chiloquin, OR

Witness: 1
Shape: Disk
Duration: 2 minutes

It was steel gray, semi gloss exterior. It was a disk. The sun was behind the horizon.

It was dusk. The craft was hovering over the river, about a football and a half away from us. Its energy source appeared to be like a blue flame. It hovered for a minute or so. Then it elevated itself to about 40 or 50 feet above the river. It was there for just a few seconds, then hung a left, and in a puff of smoke it was gone. There were no lights. ((NUFORC Note: Date may be approximate, although the witness does not indicate that fact. PD))

Occurred: December 11, 1967

Location: Chico, CA

Witness: 2
Shape: Domed Disc
Duration:

Night. Two women saw a hat-shaped object with a rounded top (domed disc), a glowing light in the center of the underside, and a series of blinking white and green lights along the rim (body lights). The object maneuvered at high altitude, circling, starting and stopping, and bobbing up and down like a yo-yo. Reportedly other witnesses saw a similar object 2 hours earlier at low altitude. (Unidentified newspaper clipping, NICAP files.)

1968

Special Issue No. 2 June 1969

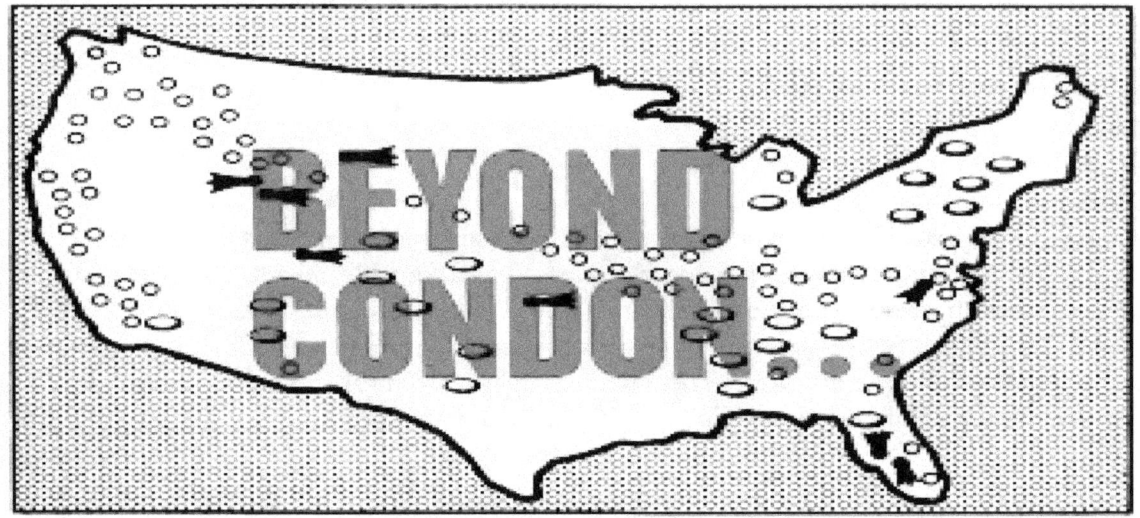

NORTH AMERICAN REPORT ON RECENT UFO CASES AND RESEARCH

Speaking at the July 29, 1968, House of Representatives symposium on UFOs, Dr. James E. McDonald supported the extraterrestrial hypothesis, but added a proviso: "... if the UFOs are not of extramundane origin, then I suspect that they will prove to be something very much more bizarre, something of perhaps even greater scientific interest than extraterrestrial devices."

Dr. James McDonald: "Earlier last year, on September 18, 1967, Condon, Low and Saunders had met for the first time in many weeks. As a result of his reading of the memo, Saunders was deeply concerned about the negative approach to the UFO problem. Saunders was led to believe that if by chance the Extra Terrestrial Intelligence (ETI) hypothesis was substantiated, the announcement would be sent by Condon directly to the Air Force and the President, and never be allowed to go to the public. This troubled him, because Saunders had been given a clear understanding that the report would go first to the National Academy of Sciences, then to the public and Air Force simultaneously. Saunders felt he could not let the problem drop. Another meeting was agreed to. At this point, Keyhoe suddenly sent word that NICAP was going to take a strong stand against the Condon committee and no longer would supply material and reports. The reason, Keyhoe said, was a new speech made by Condon at the Atomic Spectroscopy Symposium at Gaithersburg, Md., on September 13, 1967. A report of the new Condon speech had already reached Dr. McDonald in a letter from a colleague at the University of Arizona, William S. Bickel, assistant professor of physics on the campus. The rest was all down hill, but the Condon Report later tried to say they couldn't get cooperation, as excerpts inserted in the chronology will demonstrate."

The USAF-Sponsored Colorado Project for the Study of UFOs - Michael Swords

As the Colorado UFO Project prepares to announce the failure to find anything of substance to UFOs and the need to close the Air Force Project Blue Book, other studies, such as the secret NSA document below, and later the RAND document (Nov. 27), give a more accurate report.
Sign Historical Group
THE USAF-SPONSORED COLORADO PROJECT
FOR THE SCIENTIFIC STUDY OF UFOS
1995 MUFON Symposium Proceedings Presentation by
Michael D. Swords, Ph.D.
Professor Natural Sciences
Western Michigan University
(Reproduced by SHG with permission of the author)
ABSTRACT
One of the most significant elements in the history of UFOlogy was the so-called Condon Project, centered at the University of Colorado in 1967-1968. This paper discusses the origin, methodological philosophy and overview of the research problem, the activities, results, and external impacts of this work. The paper finds a complex mix of personalities, attitudes, and theories enmeshed in political and social forces, which predestined the project's conclusions and crippled its ability to make any scientific contribution toward the solution of the UFO mystery. Its resultant impacts were nevertheless formidable, both negatively and positively.
THE ORIGINS OF THE COLORADO PROJECT

When telling a story one is told to begin at the beginning, but, time and life being continuous rivers stretching back into the past, where does one really begin? Although starting with the Big Bang and working forward to 1966 might be scientifically most defensible, perhaps beginning with one of my favorite people, J. Allen Hynek, would be preferable. Dr. Hynek, in his famous role as Project Bluebook scientific advisor, had been around the idea of transferring responsibility for UFO research to academia (or some more dedicated non-military research institutions for over a decade. General Thomas D. White, USAF chief of staff, had suggested as early as 1955 that Air Force Intelligence turn over the UFO problem to an outside contractor, such as Battelle or Rand (Watson, 1955). Hynek, and the military personnel at Bluebook, had in the interim toyed with the idea of enlisting NASA, the National Science Foundation, and the Brookings Institution for aid. In the summer of 1965, the Pentagon asked Hynek for his views on involving the National Academy of Sciences. Hynek replied in August of 1965 (Hynek, 1965). Hynek's letter to Colonel John Spaulding agreed that NAS involvement would strengthen the potential for solving both the scientific and the sociological problems, which the Air Force currently faced. And, the structure, a working panel of committed experts, should include both physical and social scientists, and involve itself over a several month period.

SECRET NSA Document on UFOs - 1968 Draft

Document released by the National Security Agency, prepared in 1968, entitled UFO HYPOTHESIS AND SURVIVAL QUESTIONS. "It is the purpose of this monograph to consider briefly some of the human survival implications suggested by the various principal hypothesis concerning the nature of the phenomena loosely categorized as U F O."

Condon Report Plays Down Russian Cooperation

Condon wrote to (Prof. Feliks) Zigel (Moscow Aviation Institute) to explore the possibility of cooperation between the reported Soviet and Colorado projects. Condon's letter was transmitted to Prof. Zigel as an enclosure with a letter from Dr. Frederick Seitz, President of the U. S. National Academy of Sciences, to Academician M. V. Keldysh, President of the Soviet Academy of Sciences for subsequent transmittal to Zigel. The letter was mailed on 16 January 1968; as of 31 October 1968, no answer had been received. (Condon Report)

Occurred: January 1, 1968; Approx. 0000hrs.

Location: Lawndale, CA

Witness: 1
Shape: Circle
Duration: 1 minute
A craft with a portal
NOV. 28, 2009 I have been wanting to tell someone about my sighting for a long time, not because I saw a UFO but because of pieces of information that may have some significant value.

This was in the late 60's in California. I was maybe 15 at the time, living with my parents. I was sleeping, in my own room at the rear of the house, when I was awakened by a noise. It was the sound that a jet would make, however it was of a slightly different pitch, I knelt on my bed looking out of my window and was frozen in place by the sight of a UFO hovering approx. 100' above our neighbors and our house. This UFO was the stereotypical UFO for the time and I am not sure that I could even describe it accurately now. This craft hovered maybe 15 seconds in this spot (which seemed like 15 minutes) and then it departed in a vertical direction until it was pretty high, before moving horizontally. The next thing that happened was pretty cool. It was a clear night with a lot of stars visible and I was watching this craft depart out of frozen, excited, curiosity. It was traveling horizontally when it came to a spot in the sky that was pure black, no stars, and darker in! color than the night sky, as it got to this spot it turned into it an vanished instantly.

I was disappointed that it was gone because I wanted to continue to watch and follow this craft, but I was excited because I felt I knew a little something that others didn't, and that was an explanation of how they are able to travel such great distances. I have told very few people of this sighting because they tend to not believe you, and I have never really known whom I could give this information to. If this information helps in any way than my job is done. Thanks. ((NUFORC Note: Witness indicates that date is approximate. PD))

January 20, 1968; Russian Documents "Fantasy or Reality? (Unidentified Flying Objects)"

January 31, 1968; Hypothesis About Flying Plates
Hypothesis About Unidentified Flying Objects found in newspaper account, Russian document.

February 9, 1968
Dr. David Saunders and Dr. Norman Levine, members of Air Force-sponsored Colorado UFO project, fired by Dr. E. U. Condon for alleged "incompetence" in controversy over project management (Section XV, Vol. II, The UFO Evidence).

Second Week of February 1968 - Analysis Ran on Ubatuba Fragments
Dr. David Saunders: "During 'the second week of February, 1968, Roy Craig [a physical chemist on Dr. Condon's UFO Investigation staff flew to Washington to run the Neutron Activation Analysis [of the magnesium fragment] at the FBI Laboratory. We learned that he had some interesting results. The sample actually was not pure magnesium, but the pattern of impurities was very odd. "

Early March, 1968 - Klass' Book Released
"UFOs -- Identified!", written by Phillip Klass was published.

April 11, 1968 - Condon Committee Seeks Help From Other Nations?
According to the Condon Report, the cooperation of the Department of State was enlisted to seek information about UFO programs of the governments of other nations. On 11 April 1968 the cited air gram (see link above) was sent to various American embassies over the signature of Secretary of State Dean Rusk: (Condon Report)

Occurred: August 9, 1968; Approx.2100hrs.

Location: Eugene, OR

Witness: 1
Shape: Cigar
Duration: 3 to 5 minutes
A cigar shaped large object that blended colors from red to white. Stayed about 3 min.
I was 18 years old when this was seen. There were no reporting sites then. If you told someone, you took a big risk.
I was looking out my upstairs bedroom window and over the tree top, which was about a half a city block away, and about 60 feet high, appeared a large cigar shaped object. It made no noise, but changed colors from red to white. Not blinking, but blending.
It stayed for about 3 to 5 min. then shot straight up and was gone. I heard later that a policeman in New Mexico had seen something that was described the same way.
I have never forgotten this, although it's been 37 years ago. ((NUFORC Note: Witness indicates that date of incident is approximate. PD))

May 14, 1968 LOOK Magazine Article

"Flying Saucer Fiasco" by John Fuller \Reporting internal controversy in the Colorado UFO

Project and the Low "trick" memo.

The extraordinary story of the half-million-dollar "trick" to make Americans believe the Condon committee was conducting an objective investigation
FLYING SAUCER FIASCO
By John G. Fuller
A statement from the director of NICAP
LOOK   5-14-68
A STRANGE SERIES of incidents in the University of Colorado Unidentified Flying Objects study has led to a near-mutiny by several of the staff scientists, the dismissal of two PhD's on the staff and the resignation of the project's administrative assistant.
The study, announced as a totally objective scientific investigation of one of the most puzzling phenomena of modern times, has already cost the taxpayer over half a million dollars. The committee is scheduled to release its report by the end of the year.
The announcement by the Secretary of Defense in October 1966, that the Air Force had selected Dr. Edward U. Condon and the University of Colorado for the UFO research contract was welcomed both by skeptical observers and those convinced of the existence of flying saucers.
Maj. Donald Keyhoe and his National Investigations Committee on Aerial Phenomena, who were among the severest critics of the Air Force's study, publicly announced cautious support and offered NICAP's nation-wide UFO reporting system to the new research group.

Condon, then 64, a distinguished physicist, former president of both the American Association for the Advancement of Science and the American Physical Society, had grappled with and subdued the House Un-American Activities Committee, and served as director of the U.S. Government's National Bureau of Standards from 1945 to 1951. His leadership appeared to promise pure scientific objectivity in the study. Only two details seemed to disturb some observers. Four out of the first five investigators appointed were psychologists. And Robert J. Low, project coordinator and key operations man in the study, held a master's degree in business administration (although his bachelor's degree was in electrical engineering). Some critics felt that more physical scientists were needed. Condon assured them that the staff would become more balanced, and later, it was.

The project staff received a minor jolt early in October of 1966, when the Denver Post published a story: CU AIDE SLAPS UFO STUDY. Low was quoted as saying that the UFO projects "comes pretty close to the criteria of non acceptability" as a university function.

But the massive problems of getting the project started left little time for debate over that statement. Briefings were held in which Dr. J. Allen Hynek, chairman of the Department of Astronomy of Northwestern University and one of the few scientists in the country who had given UFO's serious study, gave the staff the background information he had acquired in his 20 years as scientific consultant for the Air Force. Later, such authorities as Major Keyhoe and Richard Hall 'from NICAP, Maj. Hector Quintanilla, of the Air Force UFO study, and Dr. James McDonald, physicist at the Institute of Atmospheric Physics and professor in the Department of Meteorology at the University of Arizona, addressed the group. McDonald had carried out an extensive investigation on his own.

After examining the hundreds of well-documented reports of sightings by military and airline pilots, radar operators, police, technical observers and articulate, rational laymen, McDonald rejected as highly unlikely such conventional explanations for UFOs as ball lightning (plasma), hallucinations, hoaxes and misinterpretations of natural phenomena. He concluded "only abysmally limited scientific competence has been brought to the study of UFO's within Air Force circles in the past 15 years. Unfortunately, during all this time, the scientific community and the public were repeatedly assured that substantial scientific talent was being used...."

From the beginning, the relationship between Dr. McDonald and Robert Low, the project coordinator, was abrasive. Low, who speaks softly, smoothly and guardedly, contrasts sharply with McDonald, who is intense and bluntly articulate.

The relationship between the Colorado group and NICAP was especially important. NICAP was large and well organized, and could supply information on UFO sightings on a nationwide scale. NICAP hoped that the Colorado group would retain its scientific objectivity by concentrating on the estimated ten percent of "high credibility" cases; such as those Dr. McDonald was investigating.

The first major turbulence in the new project came early in February 1967. Condon, burdened by heavy responsibilities in many public and educational projects, could not spend much time in the project offices. Low assumed the responsibilities for most of the decision-making. But on January 25, Condon, known for his breezy, anecdotal style, spoke before a chapter of Sigma Xi, the honorary scientific fraternity. The Elmira, N.Y., Star-Gazette reported:

"Unidentified flying objects 'are not the business of the Air Force,' ...Dr. Edward U. Condon said here Wednesday night.... Dr. Condon left no doubt as to his personal sentiments on the matter: 'It is my inclination right now to recommend that the Government get out of this business. My attitude right now is that there's nothing to it.' With a smile, he added, 'but I'm not supposed to reach a conclusion for another year...'"

The story also quoted Condon as saying: "What we're always reduced to is interviewing persons who claim they've had some kind of experience. I don't know of any cases where the phenomenon was still there after the person reports it... and it seems odd, but these people always seem to wait until they get home before they report what they saw."

Keyhoe knew of cases where "the phenomenon was still there after the person reported it," and where the observer didn't wait to get home before he reported it. He bristled. He knew that Condon had not yet investigated any field cases personally, nor had any members of the staff completed any meaningful research. The project was only three months old. "I have to admit," Keyhoe told David Saunders, a key staff member, "that I'm shocked by these statements. Is this a scientific investigation or isn't it?"

Condon wrote Keyhoe that some of his remarks had been taken out of context. NICAP then issued this statement: "Although we retain some reservations about the impressions of Dr. Condon's attitudes conveyed through some press accounts, we find no reason to go along with the skeptics who interpret the project merely as the latest gambit in an Air Force propaganda campaign. Having met most of the scientists involved, we are generally satisfied with their fair-mindedness and their thorough plans..."

The NICAP cooperation made it possible to establish an Early Warning System, and staff investigators were now being dispatched for field reports. Saunders gave particular attention to field surveys, as well as to the development of a master casebook and staff discussions of major cases. Low was giving the staff members' considerable leeway in the approach they were taking. Condon, with his office some distance away, did not appear frequently, and some of the staff felt that it was often frustrating to try to reach him. During this time, it seemed to some of the staff that Low turned down several potentially interesting cases for investigation for what were apparently specious reasons.

Another scientific investigator, Dr. Norman Levine, joined the project and immediately became aware of the strained atmosphere developing between Low and several members of the staff. Condon himself was heard to say that he wished the project could give the money back.

A senior member of the staff who was asked to make a speech before a teachers association began looking for specific details on the origin of the project. He was told that he might find some information in the open- files folder under the heading AIR FORCE CONTRACT AND BACKGROUND. The relaxed open-file system was part of a general overall policy to keep the project out of the cloak-and-dagger category. (In a later memo, Low said:

"The key point to keep in mind, it seems to me, is that our own files are not secure, they are not confidential, they can't be kept confidential, nor should they be. It is inconsistent with the purposes of a university to keep confidential any records of research activity, or any other records for that matter.")

The staff member found most of the material about the contract rather dull going, but one memo, written by Low to university officials on August 9, 1966, contained a few fresh details. The memo, labeled "Some Thoughts on the UFO Project," had been written before the contract was signed. In it, Low said, "Our study would be conducted almost exclusively by non-believers who, although they couldn't possibly prove a negative result, could and probably would add an impressive body of evidence that there is no reality to the observations. The trick would be, I think, to describe the project so that, to the public, it would appear a totally objective study but, to the scientific community, would present the image of a group of nonbelievers trying their best to be objective, but having an almost zero expectation of finding a saucer. One way to do this would be to stress investigation, not of the physical phenomena, but rather of the people who do the observing - the psychology and sociology of persons and groups who report seeing UFO's. If the emphasis were put here, rather than on examination of the old question of the physical reality of the saucer, I think the scientific community would quickly get the message. I'm inclined to feel at this early stage that, if we set up the thing right and take pains to get the proper people involved and have success in presenting the image we want to present to the scientific community, we could carry the job off to our benefit...."

When Levine read the memo, he was disturbed by the word "trick" and the phrase about making the investigation "appear a totally objective study" to the public. Others on the staff had a similar reaction.

Many staff members were also disturbed by the news that Condon had decided to attend the June Congress of "Ufologists" in New York. This was a convention of far-out supporters of undocumented and highly colorful UFO sightings.

On September 18, Condon, Low and Saunders met for the first time in many weeks. As a result of his reading of the memo, Saunders was deeply concerned about the negative approach to the UFO problem. It would be easy, he felt, to concentrate on the nut-and-kook cases and persuasively eliminate any serious consideration of the real problem.

The meeting went on for three hours. Low did most of the talking. Condon seemed tired. Low's position was that Saunders was sticking his nose into something that was none of his business. Condon's position was that he didn't understand what Saunders was talking about.

Saunders was led to believe that if by chance the Extra Terrestrial Intelligence (ETI) hypothesis was substantiated, the announcement would be sent by Condon directly to the Air Force and the President, and never be allowed to go to the public. This troubled him, because Saunders had been given a clear understanding that the report would go first to the National Academy of Sciences, then to the public and Air Force simultaneously. Saunders felt he could not let the problem drop. Another meeting was agreed to.

At this point, Keyhoe suddenly sent word that NICAP was going to take a strong stand against the Condon committee and no longer would supply material and reports. The reason, Keyhoe said, was a new speech made by Condon at the Atomic Spectroscopy Symposium at Gaithersburg, Md., on September 13, 1967. A report of the new Condon speech had already reached Dr. McDonald in a letter from a colleague at the University of Arizona, William S. Bickel, assistant professor of physics on the campus.

"Dr. Condon's speech was funny and entertaining," Bickel wrote. "But to me, it was also disappointing and surprising. Dr. Condon emphasized mostly funny things. He told of an offer made to him by a contactee, who, for a sizable sum deposited in the right bank, would introduce him to a UFO crew. ... He told how he tracked the case down and concluded that it was very likely a hoax. My feelings about UFO's are similar to those of many people - I don't know what they are, I believe people are seeing real things, and I believe a scientific attack on the problem will solve the mystery - whatever they are. The net effect of Dr. Condon's talk was zero, if not negative...."

In reply to Bickel, McDonald wrote, " The crackpots are so immediately recognizable that one need not waste any time at all on them.... I fail to understand why a scientific group should be given an address by any member of the Colorado team on the topic of the crackpot fringe...." Word came from Keyhoe that he was drafting a long letter to the Colorado study group, and NICAP would reconsider its cooperation only if the answers to a list of questions were satisfactory.

On September 27, the Rocky Mountain News (Denver, Cob.) published this headline: UFO RESEARCH CHIEF AT CU DISENCHANTED. Condon was quoted as saying: "I'm almost inclined to think such studies ought to be discontinued unless someone comes up with a new idea on how to approach the problem.... The 21st century may die laughing when it looks back on many things we have done. This [the UFO study] may be one."

The majority of the staff began exploring several proposals, including the possibility of the entire staff resigning enmasse or issuing a press release or a minority report. Another proposal was the establishment of an independent scientific group to explore the rational sighting reports and eliminate the crackpot-fringe static. There was general agreement that an objective study of the UFO problem should be made and that accurate and unbiased findings should reach the National Academy of Sciences, the public and the Air Force. A confrontation with Low and Condon was arranged. Condon expressed regret that his statements had appeared in the press. Several members of the staff told of their concern that the content and form of the final report would reflect what they now felt was Condon's and Low's prejudice and would be unjustifiably negative. Staff members speculated that Condon was tired as well as disenchanted. He remained an enigma because the staff saw so little of him.

At an informal meeting in Denver on December 12, 1967, Saunders, Levine, McDonald and Hynek agreed that a new organization might be formed consisting only of professional-level members, designed to assure the continuation of intelligent UFO study regardless of whether the Condon report were negative or positive. After Hynek left, McDonald first became aware of Low's memo, and expressed his shock.

On January 19, 1968, Low phoned McDonald at the University of Arizona. McDonald reminded Low of the clearly negative tone of Condon's public statements over a period of time, including Condon's disturbing preoccupation with the crackpot elements. He also brought up Condon's failure to investigate personally significant field cases or to question any of the working staff that had been making a serious UFO study. McDonald stressed that he was not opposed to negative findings. What bothered him was that negative findings were already being clearly expressed by both Low and Condon. Low hung up in anger. McDonald prepared a long letter to Low to review his complaints. Low did not get around to reading the letter until February 6. In it, McDonald mentioned for the first time his concern about the memo, quoting to Low the phrases about "the trick." "I am rather puzzled by the viewpoints expressed there," McDonald wrote, "but I gather that they seem entirely straightforward to you, else this part of the record would, presumably, not be available for inspection in the open Project files...."

Mrs. Mary Louise Armstrong, who had worked directly with Low as his administrative assistant, was in the office as Low finished reading the letter. Low exploded. He said that whoever gave the memo to McDonald should be fired immediately. Then he seemed to cool down.

On Wednesday, February 7, Saunders was summoned to Condon's office. Low and Condon were present. The questioning focused on the memo. Did Saunders know of it and know where it was kept? Saunders said that the memo was only part of the whole problem. It alone did not seem especially important, he felt. The broader issues of scientific integrity were at stake. Condon, furious that he had not immediately been informed that McDonald knew of the memo, told Saunders, "For an act like that, you ought to be ruined professionally."

Saunders countered by saying that Condon and Low seemed to be treating the symptoms rather than the disease. He reminded them of the efforts of the entire staff to get Low and Condon to modify their intractable stance. He reviewed the long sequence of events and reminded Low that he had blocked the investigation of one particularly startling UFO case. Low protested that the investigation on this was completed. No mention was made of any dissatisfaction with Saunders's work.

Dr. Levine was summoned while Saunders was still in Condon's office. Saunders offered to stay. Low rose from his chair and physically ushered him out the door. Levine was unnerved by the forcible ejection of Saunders. Again, the questioning went straight to the memo. Levine said that he was at the Denver meeting when the memo was given to McDonald. He understood there was nothing whatever confidential about the memo, and did not see anything wrong with the action. Condon asked why Levine had not brought the memo to him, and Levine said that Condon's public and private statements had indicated that there was little likelihood of effective communication. He told Condon that Low had slammed the door in his face when he brought up the handling by Low of an Edwards Air Force Base case, and recalled that Condon himself had suggested that Levine call in sick when he was scheduled to make a talk at Colorado's High Altitude Observatory.

Condon accused him of being disloyal and treacherous, and Levine replied that loyalty to a scientific goal might take precedence over personal loyalty. Condon asked why Levine didn't invite him to come over and investigate the important cases. Levine indicated that he did not feel it was his place to invite the chief scientist of the project over. The questioning lasted about an hour. Condon dismissed Levine abruptly.

Mrs. Armstrong had joined the project at its inception with no convictions whatever about UFO's. By February, 1967, she was convinced that the study was being gravely misdirected. When, on February 7, 1968, Condon told her that he was going to fire Saunders and Levine the next day, Mrs. Armstrong's first impulse was to resign immediately. But she then decided first to confront Condon with what she regarded as clear, unassailable documentation of the factors behind the disagreement and low morale of the staff.

She talked to Condon on February 22, 1968, at his office. She told him frankly that there appeared to be an almost unanimous lack of confidence in the project coordinator and his scientific direction of the project. She pointed out that Low had indicated little interest in talking to those who carried out the investigations or in reading their reports. She said that her long, close association with Low gave strong evidence that he was trying very hard to say as little as possible in the final report, and to say that in the most negative way possible. At Condon's request, she wrote a follow- up letter in which she added that the tone of the memo indicated that Low was not unbiased from the beginning. Condon then wrote her: "My position is that that letter is a confidential matter between the two of us and that for you to disclose it to anyone else would be gravely unethical." But after long consideration, Mrs. Armstrong felt that it was more important to the public interest to state her feelings clearly.

The others who left the project also felt they had an obligation to speak out, and when Condon failed to respond positively to his outspoken letter of criticism, McDonald brought the matter before the executive officers of the National Academy of Sciences in a vigorous written protest. Saunders and Levine cleared their desks at Woodbury Hall and left.

Asked about the near-mutiny in the investigating staff, Condon said that he would make no comment. Low stated that he had absolutely "zero comment" to make about the dismissals. Thurston Manning, vice president and dean of the faculties of the University of Colorado, delivered word through his secretary that he had nothing to say. Scott Tyler, in charge of public relations for the university, said that he had no comment.

The hope that the establishment of the Colorado study brought with it has dimmed. All that seems to be left is the $500,000 trick. END

1969

Woodstock, The Vietnam War, the Nixon era, the love generation, The Fillmore theaters and the best music this world had heard in a millennium came to pass. This year brought in a genuine sense of enlightenment that people are still trying to imitate. The age of the flower child, the beginning of the Age of Aquarius and the realization that we do in fact have the power, the power of love to affect good change and to push to the side, impractical, dishonest, dirty polluting ideas and make businesses responsible.

July 18, 1969; Apollo 11 Sighting

Object apparently not the S4B and Buzz Aldrin recounts encounter. Brad Sparks is convinced this is debris, a piece of Mylar covering for the LM, that came off when attitude control rockets blew it off when the controls kept misfiring. It's trajectory and timing and size and shape all fit.

July 20, 1969; Apollo 11 Lands on Moon

The Lunascan Project Moon Shot Series

The first manned spacecraft landing on the Moon was at 3:17 p.m. E.S.T. on July 20, 1969, when the Apollo 11 Lunar Module, Eagle, landed in Mare Tranquillitatis, located at 0°4'5"N latitude, 23°42'28"E longitude. The Eagle landed approximately 50 kilometers from the closest highland material and approximately 400 meters west of a sharp rimmed blocky crater about 180 meters in diameter. Astronaut Aldrin immediately began describing the view from the window:

". . . it looks like a collection of just about every variety of shapes, angularities, and granularities, every variety of rock you could find . . . it looks as though they're going to have some interesting colours to them."

Before the Apollo 11 landing took place there were some reconnaissance flights performed by the Lunar Orbiters, in particular LO-5.

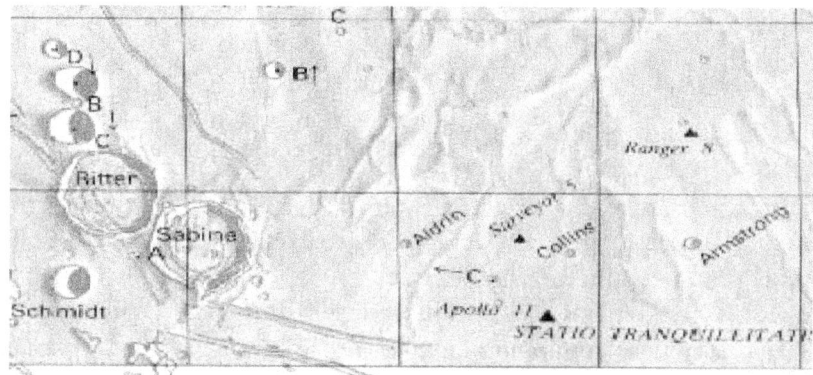

Used in two other previous web pages in the Moon Shot Series, the above graphics shows locations of all three NASA missions, which culminated in the Apollo 11 landing. The chart above is the bottom portion of Rukl Section 35. For comparison purposes the crater Sabine is 30 km wide.

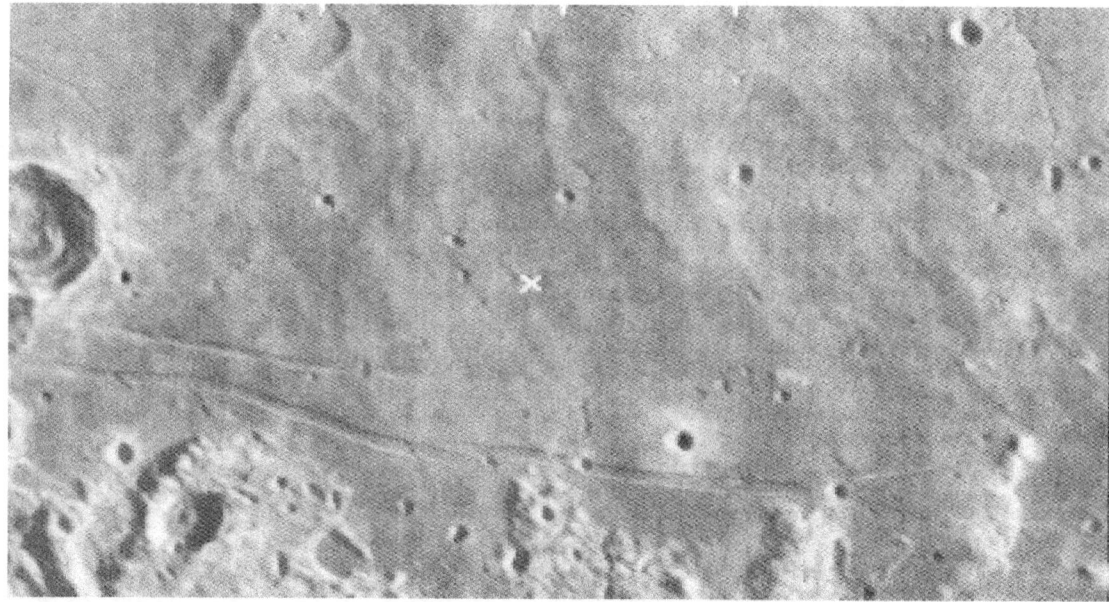

June 15, 1969; "Erase Ridicule", Dr. James E. McDonald

McDonald asks U.N. to begin study of UFOs. Tucson, Arizona Daily Citizen, by John Riddick, Citizen Staff Writer.

'ERASE RIDICULE'

McDonald Asks U.N. To Begin Study of UFOs

Tucson, Arizona Daily Citizen, June 15, 1969

By JOHN RIDDICK

Citizen Staff Writer

University of Arizona physicist James E. McDonald is trying to persuade the United Nations to begin a global study of Unidentified Flying Objects.

At the request of Secretary-General U Thant, McDonald briefed the U.N. Outer Space Affairs Group in New York last week. An appointment with U Thant was canceled at the last minute because the war in the Mideast reached a crisis point in the Security Council.

"Apparently U Thant is very much interested in the problem," said McDonald, senior physicist of the UA Institute of Atmospheric Physics.

McDonald began correspondence with U Thant on UFOs following reports that delegates of some smaller nations, particularly African, have become concerned about increased sightings.

A leading atmospheric physicist, McDonald has been studying UFOs the last year and has come to the conclusion that they are the No. 1 scientific problem of the day.

McDonald told the U.N. committee last Wednesday that the most "acceptable hypothesis for the quite astonishing number of creditably-reported low-level close-range sightings of machine-like objects is that they are some form of extraterrestrial probes."

He added: "I know of no other current scientific problem that is more intrinsically international in character than the problem of the nature and origin of UFOs."

"Scientists all over the world have been misled by over-confidence in the quality of the research on UFOs by the U.S. Air Force," he said. "First among the jobs is to erase the ridicule which has placed the UFOs on the shelf as a 'nonsense problem'."

McDonald said: "Because of the current official, journalistic and scientific ridicule, there has been almost no scientific attention given to the problem."

If it should be proved untrue that UFOs represent reconnaissance of the earth from outer space, alternative hypotheses will be "even more bizarre," he said.

As one technique where worldwide cooperation can be effective, McDonald suggested coordinating the search of the skies by radar on the different continents.

McDonald said there have been unexplained radar sightings, as well as a large number of observations, where hovering UFOs have caused electromagnetic disturbances.

Above all, he said, it is necessary to find if there is a global pattern to the sightings that have accelerated greatly in number during the last two decades.

A Russian member of the U.N. group, taking may notes, told McDonald: "We are seriously concerned with the problem." The group is a secretarial organization for the committee on Peaceful Uses of Outer Space.

In his efforts to persuade his own countrymen to take UFOs seriously and to begin a large-scale scientific study of them, McDonald has talked to numerous groups, most recently including technical people at the headquarters of the Federal Aviation Agency and a regional meeting of the Civil Air Patrol.

Next week he leaves for Australia to study meteorological problems. During his visit he will interview persons who have reported seeing UFOs. There is considerable interest in the problem in Australia.

On. December 17, 1969, Secretary of the Air Force Robert C. Seamans, Jr., announced the termination of the two decades of operations of the highly visible AF investigation of UFO's, Project Blue Book. This was only the announcement date, not the actual termination date, but the AF release was worded in such a way as to suggest immediate termination of BB. In fact BB did not terminate until Jan. 30, 1970, at 3:30 p.m. EST, as NICAP found out and published in the May 1970 UFO Investigator (p. 3a).

In the last year of its official existence, Blue Book received 146 UFO reports of which only one received the unidentified classification. For the 22 years that the Air Force investigated UFO's it received nearly 15,000 reports of which some 587 were classified as unidentified. (Air Force press releases listed the total number of 701 unidentified in the statistical summaries of yearly totals. But UFO sighting date and location in the declassified monthly indexes lists today only about 587.) Due to diligent research, the number of "unknowns" has doubled from that 701 figure to more than 1,600 in Brad Sparks' revised catalog, and may reach as high as possibly 3,000 to 5,000, based on estimates of the late Dr. James McDonald and Sparks.

Towards the end the BB files received fewer and fewer military cases. The Air Force's position was that UFO's were no longer seen by the military simply because they were trained observers who cannot be fooled by such things. Historically, however, that was not true and did not explain why so many military observers in the past saw and even instrument-tracked UFO's. In reality, the trend in the BB files reflected the changes in UFO reporting channels. The Air Force had started shifting military reporting of UFO's into operational reporting channels such as those set up under AF Manual 55-11 of 1965 (now AF Instruction 10-206), and many classified regulations, which bypassed BB.
All seemed dead on the UFO front, but major events were just a few years away. The UFO debate was rapidly dying out in 1969 in the wake of the Condon Report and the closure of BB. NICAP and APRO catastrophically lost members, down from roughly 14,000 for NICAP and 8,000 for APRO to just a few thousand.
Francis Ridge
NICAP Site Coordinator

SECTION 7: 1970-1979

Edward CONDON put the screws to us all by having PROJECT BLUEBOOK ABANDONED. He accomplished his goal to discredit all UFO related research to keep all sightings classified. Perhaps since they all have their toys, reverse engineering functional, made contact, found artifacts, have what they want. There is no need anymore to cite their goings on for the publics benefit. These mavericks are caught in GROUPTHINK, and won't disclose what they know to the public. PLEASE note; since Project Blue Book had closed, (January 30th, 1970) there are no Project Blue Book reports listed on this chronology or ones in the ensuing years. However, since UFO reports were now being handled by other agencies See, Bolender Memo, October 20, 1969, ("reports of unidentified flying objects which could affect national security are made in accordance with JANAP 146 or Air Force Manual 55-1 and are not part of the Blue Book system.") any such report will be noted as to source in this or any future chronologies." The problem is that all relevant and serious contact, technologies both beneficial and dangerous are kept away from public awareness. Problems with national security in regards to weaponry only can obviously afford secrecy, but all other secrecy is insane. The exposure of this secret government, afforded by the privileged that run the military industrial complex, including old Scientists from the Nazi era, is a sad thing.

Early 70's, possibly sometime in 1970; Kingsley, OR

Keno Mountain Radar incident RADCAT Case Directory
Category 9, RADAR
Preliminary Rating: 5
Radar Alert on Keno Mountain
Early 1970's
Kingsley Field, Oregon
Fran Ridge
Sometime in the 1970s, Keno Mountain, Oregon
The incident began with a civilian Close Encounter of the Third Kind, followed by an official alert involving defense radar and F-106's. At that time, Beale AFB, California, was carrying on operational flights with the fantastic SR-71 Blackbird and the hover site of these objects was very near that location.

At the time of the incident, Harold Hartig was a Staff Sergeant, Air Force supervisor and radar operator (rated, 27370) in the United States Air Force. His security clearance was SECRET throughout his entire career. In 1989 Mr. Hartig found out that he had only 2-3 years to live. He decided that several events that he had taken part in should be recorded for historical and scientific reasons. The following report was filed with me on September 7, 1989. Mr. Hartig surprisingly lived another seven years, but passed away in January of 1996.
Detailed reports and documents
The Keno Mountain Alert - Harold Hartig
http://www.nicap.org/70XXXXkeno_dir.htm
WORTH READING. Please visit the NICAP archives

Keno Mountain Alert
UFO Watching SR-71?
Early 1970's
Kingsley Field, Oregon

January 30, 1970; Project Blue Book Officially Closed, at 3:30 PM.

Secretary of the Air Force Robert C. Seamans, Jr., had announced termination of Project Blue Book UFO study on Dec. 17th of 1969. This was only the announcement date, not the actual termination date, but the AF release was worded in such a way as to suggest immediate termination of BB. In fact BB did not terminate until Jan. 30, 1970, as NICAP found out and published in the May 1970 UFO Investigator.

April 28, 1970 Clandestine CIA Meeting with McDonald
Leading UFO scientist, atmospheric physicist Dr. James E. McDonald of the University of Arizona had a clandestine 1-1/2 hour bag-lunch meeting at the outdoor Lincoln Memorial in Washington, DC, with top-ranking CIA official Arthur C. Lundahl, Jr., Director of CIA NPIC (National Photographic Interpretation Center), who was in charge of 4,000 classified CIA and DIA employees or about 1/4 of the CIA's total staff. McDonald has been secretly meeting Lundahl at these outdoor venues at Lundahl's suggestion since at least 1968 whenever McDonald came to Washington. Lundahl has told McDonald previously that he wants their meetings away from his CIA office (at the classified NPIC headquarters at Bld. 213, Washington Navy Yard). (Brad Sparks; McDonald files)

May 4, 1970 CIA Official Lundahl Meets McDonald and Richard Hall
CIA NPIC Director Arthur Lundahl (see Apr. 28 entry) meets with Dr. James McDonald and former NICAP Asst Director Richard Hall at Hall's home. (Brad Sparks; McDonald files)

May 18-25, 1970 McDonald's Third Trip to BB/Files
McDonald makes his third trip to visit Project Blue Book or the BB files since June 1966, this time a weeklong excursion to the newly moved BB files at Maxwell AFB, Montgomery, Alabama. The AF had deliberately moved the BB files to Alabama after BB's closure on Jan. 30, 1970, to make them as inconveniently remote and/or inaccessible to UFO researchers as possible (see Bolender Memo entry, Oct. 20, 1969, where this AF intention to hinder UFO research access was discussed). McDonald copies and reviews about 400 mostly radar cases from his prepared list of 536 BB cases of interest. (Brad Sparks; McDonald files)

Occurred: June 9, 1970 Approx. 1500hrs.
Location: Eureka, CA
Witness: 1
Shape: Oval
Duration: 1 minute
Silver globe spotted over Humboldt Bay
The Coast Guard had their helicopter up and hovering, it brought me out back. I live just off the 101 and have nothing in between my home and the ocean. The chopper noise brought me outside. I saw a silver "globe" fly through the air, it moved like something out of a video game before leaving my view. I'm not really sure what brought my eye to it but like I said the way it moved through the air was different then anything I've ever seen.
((NUFORC Note: Witness indicates that the date of the sighting is approximate. PD))

Occurred: June 15, 1970; 2000hrs.
Location: River Pines, CA
Witness: 2
Shape: Disk
Duration: 10 sec
Grey to silver saucer shaped craft flew directly at and over vehicle at altitude of few feet.
Wife and me were driving from Plymouth Ca to River Pines Ca on E16. About half way there, a saucer shaped craft flew down and straight at us as if it was going to hit us directly in the windshield. Then it veered up and flew directly over us at an altitude of a few feet. There was no sound or any other affects. It scared the bijous out of both of us. I turned to my wife and asked her if she saw what I had seen, she replied by asking if I had seen what she had seen. It was a totally frightening experience for both of us.
((NUFORC Note: Witness identifies himself as a retiree. PD))

Occurred: July 15, 1970; Approx. 2200hrs.
Location: Roseville, CA
Witness: 2+
Shape: Disk
Duration: 2 to 3 minutes
Silent, disc shaped object with rotating vertical bands of light, accelerated away with amazing velocity.
It was a disc shaped craft, approximately 150 feet in diameter and about 50 feet thick. There were vertical bands of light, which would rotate around the perimeter then go out, only to repeat in a few seconds.
The object was hovering above the trees on the other side of Atlantic Street near Roseville High School, about 100 feet horizontally from us and 200 feet vertically.
We watched it for 2 to three minutes before it accelerated away from us and up. It moved so quickly that it looked like it shrank to the apparent size of a star, but left no trail.

You could see a discrete point of light moving vertically up until it disappeared. It made no noise that any of us could detect.

One of my friends called the police department who informed him that they had had other reports as well from surrounding areas.

((NUFORC Note: Witness indicates that date of event is approximate. PD))

National UFO Reporting Center

1971

A LITS Reader's 1971
Occurred: within 1971; Approx. 0300-0400hrs.

UFO Sighting in Red Bluff, CA

Witness: 3
Shape: Lit orb
Duration 14-15 minutes

Thought I would send along a little of what I seen in 1971.

I was working with 2 other friends in a restaurant. It closed at 11 but we were tasked to do some cleaning and painting until the bar closed a 2.

East of Red Bluff was a residential area know as Antelope. Going through this on 99E a turn left would take you through houses, small farms then wind up hill to a spot where the road dead-ended at a fence. This was called Hill 9. I believe this was maybe the road that would be moved to the present day Highway 36. There was a private road that went down to a house on the river at the top but the perfect place for kids to drink a beer or make out. You would be high up on a hill and be able to see anyone coming from a mile off.

We were sitting on the road at around 3 or 4 in the morning. A couple miles off and below us was a grammar school with a Spanish name. On the edge of where houses got thick again. It had a streetlight or two.

For may 10 minutes I had noticed a light over that school that seemed brighter than it should be but did not pay much attention. The light was white and at some point there became a white cloud around it. When I noticed the cloud there was a smaller cloud off to the left of it that was dissipating downward sort of like dry ice in a bowl. That got my attention and we all started watching the light in the middle of the cloud. The Light did not seem much bigger than a streetlight but brighter. Hard to say but the cloud looked about the size of a movie screen and maybe 1 or 2 times the height of a telephone pole above the ground.

The light started slowly moving to the left or west out of the cloud and the cloud started to dissipate toward the ground like the first one. We were really watching now.

It moved maybe 100 yards or more and shot northwest and upward to be maybe above Id Adobe Park across the river. I do not think it was too much farther out as it did not get vary much dimmer. The speed was amazing and it stopped instantly. Stayed there half a minute and shot southeast and up and stopped again. Now it was very far away and very high. The brightness was down to a medium star. Stayed there half a min. or so when the light turned red. The red light shot across and stopped for a moment, began to blink. It was now so far out it was getting hard to see. The light then went in circles going up and south until it was out of sight. A long time ago but I think this is pretty accurate. The speed was amazing. Like watching a falling star but having it not fade out, have a tail and stop instantly. The red light was very red. The time frame is hard to say. From the time it started to move maybe 3 minutes before it was out of site.

UFO IN BERMUDA TRIANGLE

This encounter occurred in 1971, while aboard the aircraft carrier, USS John F. Kennedy CVA-67 (now CV-67) in the Bermuda Triangle. I was assigned to the communications department of the Kennedy and had been in this section about a year.

The ship was returning to Norfolk, VA after completing a two-week operational readiness exercise (ORE) in the Caribbean. We were to stand down for 30 days, after arriving in Norfolk, Virginia, to allow the crew to take leave and visit family before deploying to the Mediterranean for six months.

I was on duty in the communications center. My task was to monitor eight teletypes printing the "Fleet Broadcasts". On the top row were four teletypes each printing messages from four different channels. On the bottom row were four more doing the exact same thing except the signal was carried on different frequencies.

If one of the primary receivers started taking "hits" I would be able to retrieve the message from the bottom one.

I also notified Facilities Control of any hits so they could tune the receivers. On the other side of the compartment (room) was the NAVCOMMOPNET (Naval Communications Operations Network).

This was the Ship to Shore circuit with the top Teletype being the receiver and the bottom as the send (known as a duplex circuit). Next to this was the Task Group Circuit for ship-to-ship communications (task group operations or TGO).

It was in the evening, about 20:30 (8:30 PM) and the ship had just completed an eighteen-hour "Flight Ops". I had just taken a message off one of the broadcasts and turned around to file it on a clipboard. When I turned back to the teletypes the primaries were typing garbage. I looked down to the alternates, which were doing the same.

I walked a few feet to the intercom between the Facilities Control and us. I called them and informed them of the broadcasts being out. A voice replied that all communications were out.

I then turned and looked in the direction of the NAVCOMMOPNET and saw that the operator was having a problem. I then heard the Task Group operator tell the watch officer that his circuit was out also.

In the far corner of the compartment was the pneumatic tubes going to the signal bridge (where the flashing light and signal flag messages are sent/receive).

There is an intercom there to communicate with the Signal Bridge and over this intercom we heard someone yelling "There is something hovering over the ship!"

A moment later we heard another voice yelling. "IT IS GOD! IT'S THE END OF THE WORLD!"

We all looked at each other, there were six of us in the Comm Center, and someone said, "Lets go have a look!" The Comm Center is amidships, just under the flight deck, almost in the center of the ship. We went out the door, through Facilities Control and out that door, down the passageway (corridor) about 55 feet to the hatch that goes out to the catwalk on the edge of the flight deck (opposite from the "Island" or that part of the ship where the bridge is).

If you have ever been to sea, there is a time called the time of no horizon. This happens in the morning and evening just as the sun comes up or goes down over the horizon.

During this time you cannot tell where the sea and sky meet. This is the time of evening it was. As we looked up, we saw a large, glowing sphere. Well it seemed large, however, there was no point of reference. That is to say, if the sphere were low; say 100 feet above the ship, then it would have been about two to three hundred feet in diameter.

If it were say 500 feet about the ship then it would have been larger.

It made no sound that I could hear. The light coming from it wasn't too bright, about half of what the sun would be. It sort of pulsated a little and was yellow to orange.

We didn't get to look at it for more than about 20 seconds because General Quarters (Battle stations) was sounding and the Communication Officer was in the passageway telling us to get back into the Comm Center. We returned and stayed there (that was out battle station). We didn't have much to do because all the communication was still out. After about 20 minutes, the teletypes started printing correctly again.

We stayed at General Quarters for about another hour, then secured. I didn't see or hear of any messages going out about the incident.

Over the next few hours, I talked to a good friend that was in CIC (combat information center) who was a radar operator. He told me that all the radar screens were just glowing during the time of the incident. I also talked to a guy I knew that worked on the Navigational Bridge.

He told me that none of the compasses were working and that the medics had to sedate a boatswain's mate that was a lookout on the signal bridge.

I figured this was the one yelling it was God. It was ironic that of the 5,000 men on a carrier, that only a handful actually saw this phenomenon. This was due to the fact that flight Ops had just be completed a short time before this all started and all the flight deck personnel were below resting. It should be noted that there are very few places where you can go to be out in the open air aboard a carrier.

From what I could learn, virtually all electronic components stopped functioning during the 20 minutes or so that what ever it was hovered over the ship. The two Ready CAPs (Combat Air Patrol), which were two F-4 Phantoms that are always ready to be launched, would not start.

I heard from the scuttlebutt (slang - rumor mill) that three or four "men in trench coats" had landed, and were interviewing the personnel that had seen these phenomena. I was never interviewed, maybe because no one knew that I had seen it.

A few days latter, as we were approaching Norfolk, the Commanding and Executive Officers came on the closed circuit TV system that we had. They did this regularly to address the crew and pass on information. During this particular session the Captain told us how well we did on the ORE and about our upcoming deployment to the Mediterranean.

At the very end of his spiel, he said, "I would like to remind the crew, that certain events that take place aboard a Naval Combatant Ship, are classified and are not to be discussed with anyone without a need to know". This was all the official word I ever received or heard of the incident.

Being young and excited about my visit home and going to the Med, I completely forgot about it until years later when my wife and I went to see "Close Encounters of the Third Kind" at the movies when it first came out. In fact the friend that had been the radar operator was with his wife and went with us.

As we walked across the parking lot to my car, I ask him if he remembered what we had experienced years earlier on the ship.

He looked at me and said he never wanted to talk about it again. As he said it he turned a little pale. I never talked about the incident again. When I discovered "Aliens and Strange Phenomenon" on MSN and started reading the posts I started thinking about it again. Now I seem obsessed in finding out all I can about these phenomena.

Source:
Jim Kopf Mt. Airy, Maryland
Submitted to UFO Casebook

June 13, 1971, Dr. James E. McDonald Dies.

Dr. James E. McDonald took his own life at Tucson, Arizona. His studies of the UFO problem had earned him international recognition as an authority on the subject. (UFO Investigator/ June 1971, page 4)

WAS IT A SUICIDE?

Occurred: September 1, 1971; Approx. 0400hrs.

Location: FREEWAY BY AIR FORCE BASE, CA

Witness: 1
Shape: Cigar
Duration: 30 MINUTES

FLARE MOVED ALONG HILLTOP AND TURNED INTO A CIGAR 5 STORY SHIP WE WERE TRAVELING NORTH ON FREEWAY 4 AM AND SAW ON THE HILLTOP COMING TOWARD THE FREEWAY A BRIGHT RED FLARE. NOT SURE WHY IT WOULD BE THERE OR WHAT WOULD HAPPEN WHEN IT REACHED THE FREEWAY BECAUSE THE HILL ENDED AT THE FREEWAY WE WATCHED IT AS WE DROVE THEN IT REACHED THE EDGE OF THE HILL AND THE FLARE DISAPPEARED AND A 4 OR 5 STORY CIGAR SHAPED SHIP STOOD STRAIGHT UP AT THE TOP OF THE HILL. IT GLOWED REDDISH GOLD WITH NOT APPREANT LIGHT SOURCE. IT STOOD THERE WITH NO MOTION AND NO SOUND FOR SEVERAL SECONDS THEN SHOT STRAIGHT UP AND OUT OF SIGHT WITHIN SECONDS. I WANTED MY HUSBAND TO PULL OVER AND SEE IF ANYONE WOULD STOP SO WE COULD SEE IF ANYONE ELSE SAW WHAT WE SAW. HE WOULD NOT. HE DID NOT ACT SCARED BUT REFUSED TO STOP.

((NUFORC Note: Witness indicates that date of incident is approximate. We spoke briefly via telephone with the party who submitted the report, and she sounds to us to be a quite serious-minded person. She is going to attempt to determine the U. S. Air Force base she and her husband were near, at the time of the sighting. PD))

1972

Occurred: June 1, 1972; Approx. 1920hrs.
Location: Shelter Cove, CA
Witness:
Shape: Disk
Duration: 30 min.
Light over ocean traveled at low speed until directly overhead at 200 to 400 ft. classic saucer with dome on top appeared cast of gold and lit from within, no lights or windows. Resolution to where a 6-inch fitting would be obvious. No sound, lit beach for a radius of 75 yards. We panicked and ran for car a mile away. Followed us to car. Ignition functioned and I persuaded a terrified 4 people to continue to observe, After 6 minutes it rose at an incredible speed straight up, becoming a star within 6 to 10 seconds. No sound, sonic boom or trail. The diameter was 150 feet dome-to-dome height perhaps 40 to 50 feet. Stress not a blurry light. This was a vehicle that saw us on the shore and came in to examine us. One anomaly. We all sensed a force intruding on our minds once we reached car. I was a science student and amateur pilot at time, a good observer, rational as hell. This sighting remains crystal clear after some 34 years and I have zero doubt it was a craft from a distant neighborhood. That's a joke. Area is known as the lost coast of Ca. for its isolation
((NUFORC Note: Date may be approximate, although the witness does not indicate that fact. PD))

Occurred: June 9, 1972; Approx. 1300hrs.
Location: Anderson, CA
Witness: 1
Shape: Circle
Duration: 7-10
Seconds
Two round reddish orange metallic disks few overhead and were seen for about seven seconds.

I was in the 7th grade at the time. I was out in a large open playing field at my school during lunchtime. It was a clear sunny day. The bell had just rung (ending lunch break) and I noticed that I was the last person out on the playing field, so I started walking back to class. For some reason I then looked up into the air and saw two round reddish orange metallic disks flying overhead. At first I thought they were balloons, but I quickly dismissed that idea as balloons do not fly in formation (side by side) and they do not fly into the slight wind that was blowing. I then thought they were two fighter aircraft, because of the small size, but the objects were round, with no wings, no tails, no projections of any kind and they were completely silent. I stopped walking and started tracking them with my eyes and turning my head to keep them in view. They were moving from my left to my right almost directly overhead, ~ 850. After watching them for about seven seconds or so, t! he disks started to approach the area of the sun in the sky. I turned away when the glare became too great. I waited for about three seconds, thinking that I could see the disks again as they came out on the other side of the sun, but I could not find them. I looked all over the sky for several minutes but they were gone. I went back to class (no one missed me) and told my teacher I had just seen two UFOs (no reaction). Back at my desk I wrote out a short description of my experience.   ((NUFORC Note: Witness indicates that date is approximate. PD))

Occurred: July 15, 1972; Approx. 2200hrs.

Location: Medford, OR

Witness: 1
Shape: Light
Duration: ten minutes
Projecting-retracting-light beam from no visible source
Traveling south towards the foothills to see my Girlfriend when I noticed a light against the silhouetted mountain. At first I thought it was someone's spotlight, but suddenly realized that it was much larger as it broke into a beam of light projecting down into the city. I pulled over and got out to see what the heck this was all about and it was absolutely silent. No sound from the City, no barking dogs, not any noise.
It was weird, so quiet. I thought what could that be? Is it a large light that the carnivals used to use to shine a light beam up into the sky to attract people? As I studied this light, it rose above the horizon of the mountaintop and I could see stars below it, so I know it was not a land-based object. After a few minutes, it receded back to itself and vanished. All that remained in that location of the night sky was a bit of fog looking stuff, which slowly dissipated. At that moment, all sound returned. The sounds of the city and freeway traffic far off, and closer was dog barking, lots of dogs barking all around at all the home sites in the foothills. The hair on my arms stood up and I chilled all over. After thinking more about what I saw, I realized that the size of that object was huge and the beam of light did not simply turn on and off, but projected (visually extended and retracted like a telescope). It certainly did not travel the speed of light. I am reporting this because I just ran across your site for the first time and never knew there was a place to report the incidence. I am having chill bumps just reliving the experience. What ever that thing was, it controlled sound and was silent itself. The light beam that it projected was a different light than I have ever seen, it was thick and intensely luminous even though it source had to have been at least two or more miles away. I observed it from a side view, it never shone on me, but I could see what it was illuminating at a great distance. Thanks for a place to finally tell of my encounter after 30 years.   ((NUFORC Note: Date is approximate. PD))

1973

September 20, 1973; ORBIT
On the 59th day of flight Skylab III, the three-man crew saw and photographed a strange red object (see photos). Not more than 30-50 nautical miles from them, Alan Bean, Owen Garriott and Jack Lousman reported the object was brighter than any of the planets. Unexplained.

Occurred: October 1, 1973; Approx. 2100hrs.
Location: Roseburg, OR
Witness: 2+
Shape: Cylinder
Duration: not sure, maybe 2 hours
We were on the outskirts of town, just passing the sewage treatment plant, when this object with blinding white light enveloped the car
My ex-husband and I were on the outskirts of town right by the sewage treatment plant. It was pitch black and all of a sudden there was this blinding white light that enveloped the car and I remember being terrified.
The next thing I know we are down the road a good 4 or 5 miles from the plant, I don't remember how we got there, but now we are at the old Oak St Bridge. Anyone from that area should remember that old metal bridge, we were going so fast across it we were fishtailing, it had been raining. I know we were both terrified and I looked up at the sky and there were 3 or 4 of them following us all the way thru town.
We finally got to my Aunt & Uncle's house where we were staying the night, I remember being so afraid to look up in the sky, but as we ran into the house I saw them go zip, zip, zip and they were gone. I don't know how else to describe it.
Once inside the house, we immediately started telling the other adults there what we had seen. Of course we were met with jokes and sarcasm. My ex couldn't take the ridicule and so he did an about face and said he really hadn't seen anything. As you can imagine, I was furious with him for not standing up and telling them the truth.
Years later, after we divorced, we met for drinks and he admitted to me that he had seen what I had seen, but could not take the ribbing from the other men in the room.
((NUFORC Note: Witness indicates that date of incident is approximate. PD))

October 25, 1973; FBI Letter
A Letter from the FBI Director denying the Bureau's involvement in the investigation of UFOs. In 1976 the FBI released 1,000 pages of UFO-related documentation.

November 28, 1973; Gallup Poll Showed that 51 percent of Americans believe UFOs are "real," 11 percent claim personal sightings, and 93 percent are aware of the subject.

1974

During the next three decades there are many UFO reports that are satellites, the international space station, iridium flares and the chevron shaped formations of the current STEALTH spy planes that made its secret debut in US air space. Between the U2 the SR-71 and other craft designed by the well-paid government contractors who operate in conjunction with the pentagon, are top-secret projects so any speculation is good from their viewpoint of secrecy. Although many sightings are of genuine UFOs we have left sightings of these planes in the 70s-90s to illustrate the varieties of phenomena out there, which is mistaken.

The military also used echo aluminum balloons to help skip radio waves around the planet that show up as orbs, but they do not speed around. The idea here is to become more familiar with the space junk we are contributing, to be more concise in sightings by applying logic and visual identifiers to ones sighting.
NOW THAT PROJECT BLUE BOOK IS GONE AND DR. McDONALD IS NO LONGER FILTERING THE REPORTS, THE GUIDELINES AS YOU WILL SEE HAVE CHANGED, ALLOWING MORE RANDOM REPORTING TO OCCUR. Part of this is due to the reorganization. It is also because the interest in UFOlogy has gone underground as much as it is trendy, and is being taken care of quietly without the public sector involvement, deliberately. It has taken a level of credibility out of the reports, since now the public is out of the serious investigations. The government is humoring the public, as they laugh at all the incredible misinformation being believed. They arrogantly cut us all out of the data flow which means people are apt to arrive at ignorant conclusions from lack of real information. The convenience of this is to shroud secret technology in confusion and mystery. Where once ridiculing the public for believing in UFOs, it is now encouraged, to keep everyone's focus off the technologies our own military has created and are testing. While legitimate organizations like NUFORC are filtering reports, they too, desire to allow the public to learn what is being seen, and comment to keep the public informed on what some of the phenomena is. Thank You.

AUTHOR'S NOTE: I adamantly must stress if any reports are submitted to NICAP, MUFON, NUFORC, CUFON, Please SPELL CHECK your work. Do NOT use abbreviations, small (i) for your identity and use proper grammar. A good sighting will be tossed for negligence to these details. The ones on the receiving end must translate and decipher what is being said and do not always have the time. It took me 1000+ hours to CORRECT simple punctuation, grammar, spelling errors alone in the sightings we used in this book from some of our sources. The Author

Occurred: June 1, 1974; Approx. 0600hrs.
Location: Lakeside to North Bend, OR
Shape: Light
Duration: 30 min.
Cone shaped light
I was on my way to work. It was dark. I was heading south on US 101 from Lakeside to Coos Bay. I spotted the light shortly after I got on the highway. It was cone shaped with the wide part toward the ground. It appeared very large but as I drove towards town it never changed in size and I didn't seem to get any closer to it. It covered a large piece of the sky bigger than several football fields. There were several other cars on the road but I never heard anyone, including myself mention it. It wasn't in the newspapers or on TV. I didn't say anything about it for more than 20 years later. I did a lot of thinking about it and came to the conclusion that it may have been the reflection of a Satellite launch in California or possibly Cape Kennedy Florida.

Occurred: June 1, 1974; 2200hrs.
Location: Placerville, CA
Witness: 1+
Shape: Chevron
Duration: 3 seconds
Chevron shaped objects in formation
Similar or the same as Lubbock Texas lights, at very high altitude, extremely bright, possibly in sunlight at 10 pm no noise. It could be seen keeping formation, so un-earthly as to cause considerable fright among observers.
((NUFORC Note: Date is approximate. We will encourage the witness to have the other observers submit reports, as well. PD))

Occurred: August 1, 1974; Approx. 2100hrs.
Location: Chester, CA
Witness: 3
Shape: Triangle
Duration: 2 minutes

Three white lights in equilateral triangle move across night sky, multiple witnesses, and summer camping trip 1974.

On our annual family camping trip near Chester, California, my sister, my cousin and I would occasionally see what we thought were satellites floating across the night sky. It has always bothered me that I have never seen anything similar to these "satellites" since then. After stumbling upon an online account of "black triangles" I came to the realization that what we saw was probably NOT a satellite at all. In fact, I think is scientifically impossible for what we saw to be a satellite. So now, 38 years later, here is my belated report on the "three-star satellites" witnessed over the Potato Patch Campground near Chester, California in July/August of 1974. I can remember it as if it were yesterday--one of those clear childhood memories that stick with you in a very visual way. I remember talking about it with my sister as we lay there on our sleeping bags, looking up at the stars and wondering what those slowly moving triangle of stars were. We have talked about this as adults, though not in recent years.

The first time we saw them, it must have been between 9-10 PM, as it was fully dark and the stars were out. We were just lying there, looking up at the night sky when we saw what looked like stars formed in the shape of an equilateral triangle drifting overhead. There were three white lights, one in each corner of a triangle. I could not see any body or an edge to a triangle, just the three lights. What made them unusual is that they moved, and that they moved together in such a way that you knew they were physically connected. It must have been really high up, because the lights looked the same size as the stars, slightly bigger, but also slightly brighter than the stars. Anyways, it moved slowly against the other stars in what I think was a northern direction. The three unified lights would speed up then slow down along its path and slowly turn or rotate around itself. The movement was very graceful, not at all like an airplane. It just looked like it was floating and spinning past us up there in the night sky. It seems like it took about 2 minutes before it went behind the trees.

I can remember seeing this at least twice for certain, and probably more like 3-5 times though I can't recall the exact instances of each. I do remember that it became commonplace for us to see these that summer. I remember that we would lie out and look for satellites, and we were successful several times. We were not all excited about the satellites. I am sure a shooting star would have been more exciting. We certainly didn't think we had seen a UFO. We just figured they were "three-star satellites" (that's what I called them) orbiting the earth. But, satellites don't speed up and slow down, and as far as I know satellites don't come in sets of three that move synchronously about at various speeds across the night sky. In retrospect, If it were indeed a satellite, then it would have had to be miles wide.

I have not mentioned this Black Triangle correlation to my sister. Though we have discussed the satellite sightings as adults, I have not discussed this with her in any detail within the last 10 or 15 years. It would be interesting if a third party could interview her to correlate the details of this story as I remember them.

((NUFORC Note: Witness indicates that the date of the sighting is approximate. PD))

Occurred: August 15, 1974; Approx. 2000hrs.

Location: Gold Beach, OR

Witness: 3
Shape: Circle
Duration: 30 seconds
Ball of light shining, like a cone of light on the beach.

My two friends and I were returning to college in Louisiana after working the summer in Washington State. I was driving along the coast highway at dusk. Suddenly we all noticed a motionless hovering white ball of light, shining a cone of light on the beach. After approximately 10 seconds the ball of light shot off toward the horizon - west over the ocean - at an incredible speed. The cone of light remained, slowly becoming dimmer and assuming a "smoky" appearance while maintaining the cone shape. It gradually faded away. I do not know the time of the event, but it was dusk. It was a clear day and little if any clouds. The first impression (dictated by logic) was that it must be a helicopter with a searchlight, but try as we might, there just was no helicopter (or anything else) there - just a ball of light. The distance from us to the light is difficult to estimate, but in order to provide some context; I would guess it was between 70 and 100 yards away. The size of the ball of light is also difficult to gauge, again I would estimate approximately 3 feet in diameter. The length of the cone of light was approximately 30 feet. We stopped at the next town (heading south), Gold Beach, Oregon, and reported to the police what we had seen. They told us that a lot of people had been reporting UFOs recently but what we saw was a meteor (!). With that, we returned to the car and continued on our journey. ((NUFORC Note: Date of incident is approximate. If, in fact, the 1) object was hovering for a period of time, and 2) it was visible for 30 seconds, or so, we doubt that it was a meteor. PD))

Occurred: September 15, 1974; Approx. 0300hrs.

Location: Chico, CA

Witness: 1
Shape: Chevron
Duration: 5 min
In 1974, I saw a white triangular object moving low and slow over orchards and fields north of Chico CA
At approx. 3:00AM walking home through a newly planted almond orchard. Looking north toward the airport I saw what I took to be the red and blue wing lights of an airplane a few miles away, after a min. I saw it was much closer, about 300 yards realizing the lights were on either side of a cockpit. The object was coming straight toward me, about 30ft.above the ground, moving about 15 m.p.h., with a very quiet wiring sound which only became audible when nearly overhead. Its shape was triangular like the letter A if it were wider at the bottom, edges rounded, only taking the circumference, lowering the cross stroke, widen the legs, It was white and smooth, it crossed Nord hwy going south turned west went a few miles turned north diagonally up until it was too dim to see. Weather was clear, crisp with little light pollution.
((NUFORC Note: Date is approximate. PD))

REPORT #2-YAKIMA, WASHINGTON at 19:50 hrs. (local) on October 07, 1974. Huge search light beam from 4 miles high in the sky illuminating a one mile diameter circle on the ground. Passenger asked pilot what was a light in the sky overhead. Flight was direct from Pendleton, Oregon, to Sea-Tac (Airport) at 10,500' msl. The light source seemed to be sourced at about 20,000' msl like a huge landing light or search light that illuminated the ground below about 1 mile in diameter. The source was not visible as though the fuselage blocked a wing-light, but the cone of light got brighter and ATC denied any traffic in the area. At this point I elected to change course and fly around, rather than through, the light beam area. Air to air queries to other pilots affirmed that the anomaly was visible to others that night. At this point "Deep Throat" advised to "can the chatter if you want to keep flying". The beam was visible until descending into Sea-tac. The next day a Seattle radio station reported that some hunters in Winthrop, WA, called in to report seeing the same light to the south and wondering what it was.(NUFORC Note: We have contacted the source of the report, requesting more information about the incident. PD)

October 14, 1974; Robert Spencer Carr Announces Crashed Saucer (Alleged Roswell parts, pieces and alien remains)
Robert Spencer Carr, a NICAP investigator and community college professor, announces that the U.S. government has recovered a crashed saucer and dead alien bodies from a site in New Mexico. After worldwide publicity is kindled Carr releases more details over the next few days and weeks, that the saucer and the bodies are kept in "Building 18" or Hangar 18 at Wright-Patterson AFB, Ohio, that he first heard the story about 1961, etc.

1975

Occurred: 1975

Location: HAPPY CAMP, CA

Witness: 5
Object: several aliens and a single orange glowing disc
Duration: 5 minutes or so

1975 - In Happy Camp, Siskiyou County, **California** just after midnight Steve Harris, San Gayer, and Carl Jackson returned to the spot of their "creature" sighting earlier that evening, accompanied by Helen White and Rick Pool. They shone spotlights around, but saw nothing. Then Steve fired his rifle several times. Immediately they heard a loud "wow-wow-wow" sound, and turned around to see in their spotlights the dark, non-reflecting forms about 5 feet tall. They were vaguely human shaped, with a glow around the edges, and about 30-35 ft away. These forms were moving about slowly, but staying at the same distance. They seemed to absorb the light from the flashlights without reflecting it, but occasionally an entire shape would start glowing slightly. No details could be made out. After several seconds, Helen failed to use her Polaroid camera, Helen and then the others began to feel a choking sensation, "as if the oxygen was being depleted from the air." After another three or four minutes of this, all five of them piled into their Ford Bronco and drove away. Looking back, they saw a glowing orange disc-shaped object rise up from near where they had been. It was about 40 feet in diameter, and it followed them at a distance of 150 feet and right above the treetops, until they reached the highway. (Source: David F. Webb & Ted Bloecher, HUMCAT: Catalogue of Humanoid Reports, case # 1975-54, citing Paul Cerny for MUFON).http://www.ufoinfo.com/onthisday/October26.html

The Walton abduction, encounters with aircraft, SAC overflights and NORAD on top alert...all in 1975Over a period of about three weeks in October and November of 1975, several Strategic Air Command (SAC) bases in the northern tier states were placed on a high priority (Security Option 3) alert because of repeated intrusions of unidentified aircraft flying at low altitude over atomic weapons storage areas. The Commander-in-Chief of North American Air Defense Command (NORAD) sent a four-part message to NORAD units on November 11, 1975 summarizing the events: "Since 28 Oct 75 numerous reports of suspicious objects have been received at the NORAD CU; reliable military personnel at Loring AFB, Maine, Wurtsmith AFB, Michigan, Malmstrom AFB, Mt, Minot AFB, ND, and Canadian Forces Station, Falconbridge, Ontario, Canada have visually sighted suspicious objects." A Teletype message to the National Military Command Center in Washington, D.C., said: "The A/C [aircraft] definitely penetrated the LAFB [Loring Air Force Base] northern perimeter and on one occasion was within 300 yards of the munitions storage area perimeter."

Note official documents cited below with appropriate incident reports beginning in late October, plus the NORAD Command Director's Log (1975) listing the incidents in zulu time which we later converted in our chrono below to local time to show the sequence of events. Note also that the Air Force admitted destroying the documents regarding the Malmstrom incidents! Click here to access only the The 1975 SAC Base Northern Tier Overflights..Or continue below and watch this blockbuster year unfold.

Francis Ridge
NICAP Site Coordinator/Nuclear Connection Project

January 1, 1975; Map of U.S. Showing CUFOS/MUFON UFOI
Areas equipped for investigation as of this date. Map prepared by Center for UFO Studies shows MUFON and CUFOS investigators.

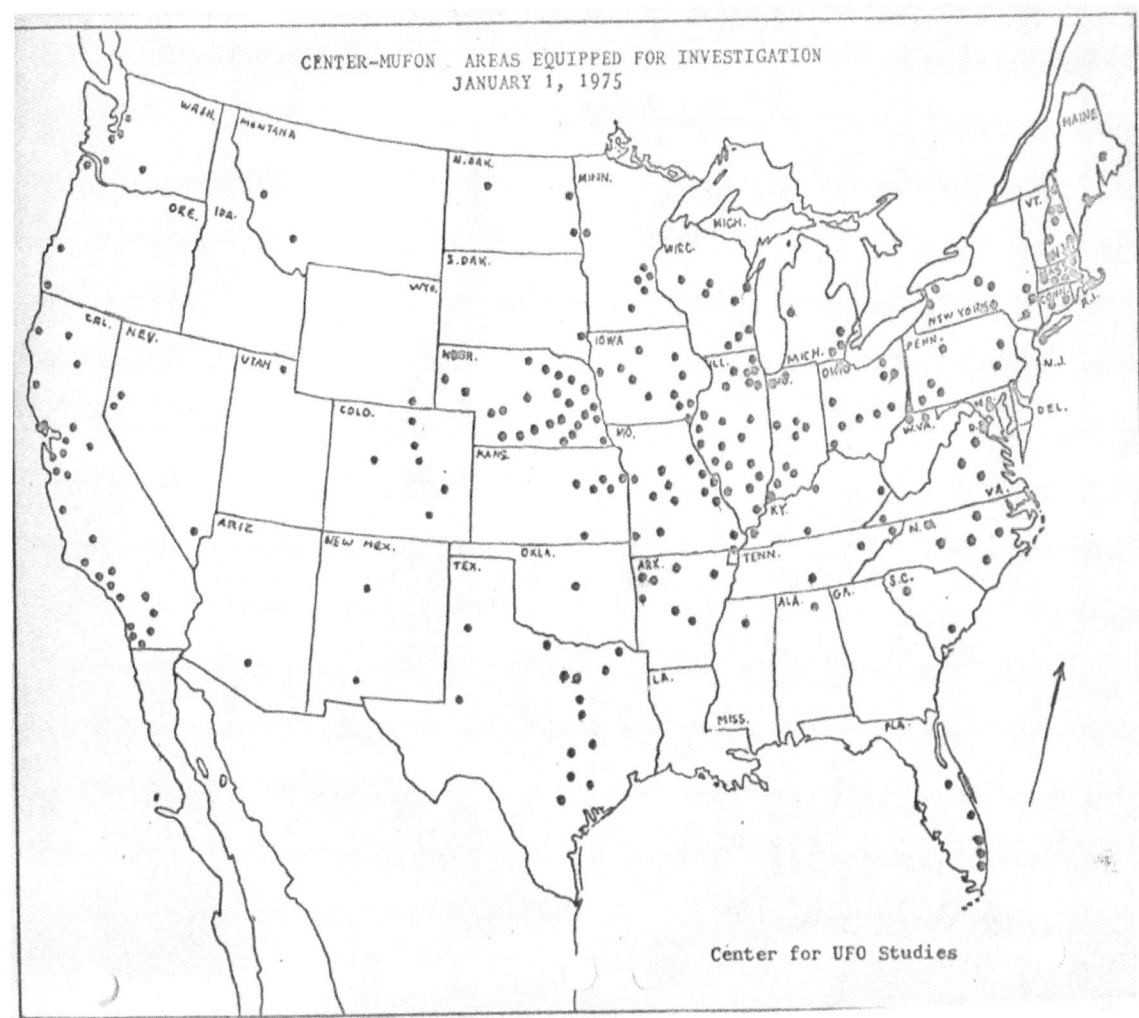

January - March, 1975; Message to Henry Kissinger www.nicap.org
From the American embassy in Algiers, Algeria. The UFOs in this report were seen by multiple witnesses. The objects landed and took off. They were also spotted on radar and seen visually simultaneously.

March 1975; Project Blue Book Files Transferred

When Project Blue Book was closed down in January 1970, the original files were transferred to Maxwell AFB in 1975 but were made available on request for public viewing sometime in 1976. In 1975 these documents were microfilmed by the Air Force for internal use and then transferred to the National Archives for public release. Before microfilming these documents for public release, however, the Air Force blacked out witness names and other personal information in accord with its policy of protecting the privacy of witnesses despite the fact that the files had been available for copying and inspection for years without these deletions. In 1998 a set of the original un-redacted Air Force microfilm was discovered at the National Archives. In addition to witness names and information, it has been confirmed that these rolls contain some pages that are not on the NARA rolls.

Occurred: March 1, 1975; Approx. 2100hrs.
Location: Roseville, CA
Witness: 3
Shape: Disk
Duration: 1 minute
I saw a disk shaped saucer with dome on top with windows. Lights were on. In Roseville, trees on either side.
It was a disk with a dome on top. It had windows all around the dome and the lights were on. We were driving up to Roseville. There were tall trees on each side of us. As we were driving it just appeared, hovered about 1 minute and then disappeared.
We didn't see it coming or going. It was about a third of a mile ahead of us, and about 2 light posts high.
2 kids and myself saw it from my high school, (La Sierra High School) and my 18-year-old boyfriend.
((NUFORC Note: Witness indicates that the date of the sighting is approximate. PD))

Occurred: June 1, 1975; 2000hrs.
Location: Williams, CA
Witness: 2+
Shape: Oval
Duration: 15 min
2 gold /yellow oval /egg shaped lights
We lived towards the end of town and use to play football in the schools football field at night. It was getting to be dark, and night fell quick around the rice fields. We had sat down on the concession stand trailer and noticed 2 oval egg shaped Bright yellow lights coming towards us at about 100 feet or so up traveling from SE to NW these lights where not flashing they where just as someone had put 2 eggs but yellowish gold on grey paper. They flew in a formation like fighter pilots would there was absolutely no sound, none. They flew over us and towards the coastal mountains then straight up after about 5 miles away from us. We ran.
I myself was raised around USMC bases and know what all kinds of aircraft look like at night. These where not like any thing I have seen. It truly changed me, and I will think about this till the day I die.

September 30, 1975; Orland (Corning), CA www.nicap.org
Domed disc hovered, illuminated area "like daylight," cows fled; emitted bright red light, humming sound, took off at high speed (NICAP UFO Evidence II, Sections IX, X).
Group /Category 4 Animal Effect Cases Case Directory

Occurred: September 30, 1975

Location: Orland, California

Animal Reaction Feature: A domed disc emitting a moderate humming sound descended to near the ground in a pasture. A red light on top illuminated the ground around the craft. Holstein dairy cattle in the pasture bolted, scattering to the far ends of the field. About 200 yards away, cows in a holding pen stampeded, raising clouds of dust. Just beyond the holding pen, about 120 cows in an open-sided milking barn backed up in the barn and packed themselves tightly together, pushing each other against interior barriers. Motionless, each cow's head was riveted in the direction of the disc, ears pointed. The cows remained packed together for some time after the object departed.

Milk yield was less than average the next day. The animals showed signs of some nervousness for the next 2 days.

Joan Woodward, Animal Reaction Specialist: On a dairy ranch, at 4 am on a clear night with a quarter moon in the sky, Hubert Brown, 22, and Tyron Philips, 37, were busy moving cows and milking. Brown was in a field to bring in the last string of cows when he saw a bright red glowing object descending rapidly toward him at about a 35-degree angle from the northeast. The object came within a few feet of the ground near Brown. The cattle bolted as did Brown, who ran to where Philips was working in the milking shed about 200 yards away.

As they watched, cattle in the holding pen stampeded, and the cattle in the milking barn backed up and crowded together as described above.

The clearly outlined domed disc was estimated to be 60-70 feet in diameter and had a bright red light on top that illuminated the craft and the ground around it. There seemed to be windows, but the red glow was so bright, it was difficult to see. There was a moderate humming sound. The object hovered and wobbled just above the ground for 30-40 seconds before it suddenly took off at about a 45-degree angle into the sky, disappearing to the west. No burned or discolored areas were noted on the grass.

No EM effects were reported.

An unusual CE-1 UFO encounter on a California dairy ranch involved two ranch employees and dairy cattle. Definite animal reactions were noted from a large, domed craft that hovered close to the ground in the pre-dawn hours showing a bright red light

View through dairy barn toward UFO

Occurred: October - December 1975
Location: Out in space over the Indian Ocean
U.S. satellites were illuminated or "blinded" over the Indian Ocean. Source 10-1,000 times as strong as natural sources. 5 incidents. One sighting was 4 hrs long.

Metropolitan D.C. Besieged With UFO Activity 1975
Strange lights, unexplainable maneuvers, and craft-like shapes are continuing to puzzle residents of the Washington, D.C. area. Reports began in October and have continued through December at the NICAP office, many of the reports have similarities while others are an entity onto themselves. (The National Military Command Center [NMCC} is located in Washington, D.C.)

Occurred: October 16, 1975; Approx. 2020hrs.
Location: Davis, CA
8:20 p.m. seven witnesses including the main witness named Landeros sighted two UFOs maneuvering in the sky. The first object was triangular in shape with white and yellow lights. It went down and up rapidly, then back-and-forth. The second UFO was a domed disc; it came from same direction as the first and had red, green, and blue lights spinning on its rim. It hovered, tilted, and shot off toward the west. (Source: CUFOS News Bulletin, February 1976, p. 9).
When Project Blue Book's closure was announced in Dec. of 1969* (with full knowledge of the March 1967 missile incident), the media and the public was told that "No UFO reported, investigated, and evaluated by the Air Force has ever given any indication of threat to our national security. There has been no evidence submitted to or discovered by the Air Force that sightings categorized as 'unidentified' represent technological developments or principles beyond the range of present-day scientific knowledge.
There has been no evidence indicating that sightings categorized as 'unidentified' are extraterrestrial vehicles." The incidents listed in the latter part of October and the first few weeks of November dramatically illustrate that at least two of these claims are false.

Target: Loring AFB, ME Although it is no longer an active Air Force Base today, in 1975 Loring AFB was a Strategic Air Command Base and a storage site for nuclear weapons. The nukes were stored in fenced weapons dump consisting of small huts covered with dirt for camouflage from the air. The 42nd Security Police Squadron patrolled it day and night.
Intrusions at Loring (AFB) - (CLEAR INTENT, 16-26; Barry Greenwood and Larry Fawcett) www.nicap.org

Mount Shasta Weed, California Huge Orange Ball & Tail Making Turn

Occurred: December 23, 1975; Approx. 2345hrs.

Mt Shasta/Weed, California (Huge Orange Ball & Tail Making Turn)

Witness: 15
Objects: 1 orb
Duration:

Full Description of event/sighting: The night this took place I was just leaving work with 6 friends, they came up to the golf club bar/rest. We were going to a Christmas party at a friends place on the 17th hole. From the bar you can see his house. My girl friend was there and the rest of the people that saw the Orb too! (By the way I don't drink/drugs) The week before we saw the Orb, it had been snowing; there was 3.5 feet of snow on the ground. The golf club is down in a valley on a lake (Lake Shastina off hwy.97 10 miles N from Weed), I-5 is to the west about 8-9 miles. We walked outside, I looked up to see clear skies and a full moon, and the moon was in the south sky to the right of Mount Shasta. Mount Shasta has two peeks one at 14,950 + and Mount Shastina at 10,000 +. WE were standing looking down to my friend's house to the west, there we could see and hear our friends every word that was said. It was so quite and still and very bright out.

I turned to look at the mountain when someone said look a shooting star. I turned to see where he was looking, and it was unreal. It was a huge ball of orange with tail miles long of every color of the rainbow. It was coming out of the east by north east about 22 degrees above the horizon. It was moving fast and it looked to be 150-200 miles away. (By the way I have been a pilot since I was 19, and at the time of the sighting I was 25). I thought it was a meteor but then it slowed down and stopped! It went from a deep orange to a white, then turned to the south and started to move faster. We were standing there asking each other what the hell is it? As it was coming towards us we could hear all our friends yelling what the heck is that! It was not easy to tell the altitude but the speed was a little easier to tell. I thought it was 400-500 mph.

As it went by us, still going south, it then started a left two minute turn some where over Mount Shasta City. At that time I could see the altitude as it was very close to Mount Shasta. I could see the peek of Shasta above it, and it was even with the peek of Shastina, 10,000 altitude. While I was trying to see the altitude the Orb was changing it brightness to a soft swirling marble of lights, and darks, (mother of pearl) the size of the Orb was like if I held a soccer ball at arms distance. It was big! At that point it was now headed north by northwest up I-5. We stood there with our mouths open until it was gone. That night it was on the radio and the next day T.V. and the newspaper. Thank you. Thank you to the witness for the report. Brian Vike, Director HBCC UFO Research http://www.ufoinfo.com/sightings/usa/751223.shtml

Occurred: December 24, 1975; Approx. 2100hrs.
Location: Winters Freeway, CA
Witness: 1
Shape: Rectangle
Duration: 2 to 3 hours

I lost about 2 to 3 hours when I drove off I felt sad almost in tears like I lost a dear friend, I was driving down winters freeway at 9:00pm at night, on Christmas eve in 1995, going to Burney Ca. to a Christmas party with my ex wife, when I noticed off to my left in a field a large dark object, floating up and down, shining three round lights all around the area, the light coming out was strange because they were flat like a laser show which weren't invented till years later. I flashed my head lights high and low several times, then suddenly it shot right over the top of my car and stayed there shinning those weird lights inside my car scaring the hell out of me, I went as fast as I could go but it stayed there over my car, suddenly I felt happy, pulled over and got out and looked at this ship 4 feet above me it was dark grey with small fins in back arcing electricity, then I was leaning on the car watching in the field again. ((NUFORC Note: Witness indicates that date is approximate. PD)) Note *abduction *lost time

1976

Occurred: February 18, 1976; 1845hrs.
Location: Rancho Cordova, CA
Witness: 2+
Object: 8 to 10 discs
Duration:
6:45 p.m. PST 8-10 discs were sighted flying in a V formation over Rancho Cordova, California. The objects were 3 to 6 meters in diameter, gray in color, and made no sound. Afterwards the witness saw ten aircraft circling that portion of the sky. (Source: CUFOS investigation file, report dated March 1, 1976).

Occurred: February 19, 1976; 1900hrs.
Location: Marysville, CA
Witness: 2
Object: 4
Duration: 5 minutes
7:00 p.m. Sutter County sheriff's office received a call reporting a UFO being observed by two witnesses. Deputy Gerald Teplansky was dispatched to the scene and was able to observe the objects for about five minutes. The deputy stated he saw three red-orange glowing objects hovering in a triangular formation. He soon saw another object streak northward across the sky before all four objects disappeared.
(Reference: UFO INVESTIGATOR, July 1976, page 3)

Occurred: March 5, 1976; 2125hrs.
Location: Redmond, OR
Witness: 1
Object: 1
Duration:
9:25 p.m. A 23-year-old woman saw an object high in the sky, which then dropped down to an altitude of only 75 feet. The object was disc-shaped, about 26 feet in diameter, and had red and white lights revolving around its exterior. Two light beams were directed from the object to the ground. The radio in her car went dead when the object came near, and although the woman kept driving, the radio did not work again until two miles down the road. The object did not follow the car. (Source: Mark Rodeghier, UFO Reports Involving Vehicle Interference, case 395, citing CUFOS)

April 14, 1976; CIA Memo
The CIA has its own UFO experts, is proven by this 1976 memo.

OFFICIAL UFO Article, May 1976 (Original version in PDF)
"Are UFOs Here to Survey Our Strategic Areas?" by Richard Hall. Our military facilities, electrical generating stations and UFOs, which keep careful watch on the important activities, haunt fuel reserves. A special study of close-range UFO sightings suggests that the supposed extraterrestrial pilots may be interested in manufacturing areas, military installations and electric power stations in the United States. These conclusions are highly tentative since the mapping project that led to them is incomplete and continuing. However, they may indicate some important lines of investigation if the patterns hold up and are borne out by parallel studies in other countries. HTML version www.nicap.org

Occurred: May 10, 1976; Approx. 1100hrs.
Location: Covelo, CA
Witness: Multiple
Shape: Disk
Duration: 6 seconds
BRIGHT OBJECT ENTERS VALLEY WHILE WE ARE ON THE PLAYGROUND. I WAS IN 5TH GRADE AT THE TIME. WWE WERE OUT ON THE PLAYGROUND PLAYING WHEN SUDDENLY A OBJECT AS BRIGHT AS THE SUN CAME INTO THE VALLEY AT A SPEED WHICH WAS PROBABLY THOUSANDS OF MILES PER HOUR.IT STREAKED ACROSS THE SKY SILENTLY AT 3 RIGHT ANGLES, REVERSED ITS SELF BACK OVER THE SAME ANGLES AND DISAPEARED. THE EVENT WAS SHORT BUT I WILL NEVER FORGET IT. IT WAS A CLEAR WEATHER DAY.COVELO IS ALSO CALLED ROUND VALLEY. AFTER THAT THERE WERE ALOT OF MILITARY PLANES [FIGHTER JETS] PRESENT.THE TEACHERS RUSHED US IN TO THE CLASSROOM AND PASSED OUT PAPER.SHE THAN INSTRUCTED US TO DRAW WHAT WE HAD SEEN.WE DID SO.TO THIS DAY I DONT KNOW WHAT HAPPEND TO THE PICTURES THAT WE ALL HAD DRAWN.IT STILL MAKES ME NERVOUS WHEN I THINK ABOUT IT, MY NAME WAS ON THE PAPER AS WELL AS MY FELLOW STUDENTS.
((NUFORC Note: Witness indicates that date of sighting is approximate. PD))

Occurred: July 15, 1976
Location: Cave Junction, OR
Witness: 5
Shape: Triangle
Duration: 12 hours

Year 1976---The Real Man In Black Time for a real vacation! I had been working steady with very long hours, six day a week for over six years-, in a large lumber mill as a computer tech and electrician., previous employment as a Electronic Designer and Engineer.

Now I'm a rare divorced dad, with full custody of three children. So I asked them where they would like to go, "lets get Grandma and go to Salt Lake City as she always wanted to see it" was the unanimous decision, my Olds Cutlass Supreme was a great road car and in perfect condition so it was loaded with camping gear and ready to go! Sunday morning after Church off we went, up to Hat Creek for the night and in the morning down to Reno, after seeing the town then drove out to Winnemucca, Nevada for a late lunch, continued driving to Salt Lake City where we stayed for the night. We saw the Mormon Temple and Temple Square and all had a good time. After sightseeing I remembered a lake campground east of Salt Lake City so we went there, a lot of fun! That evening there was a bad storm warning for late the next day, on the radio. We left early.

We decided to drive to Pendleton Or. and while there stocked up on some really neat clothing! How about Portland, down the Columbia river, we went to Meryhill, WA., where the Stonehenge replica stands--We were having a great time, and all the above detail is to show our frame of mind.

While driving from Maryhill on the freeway {84), I was thinking about the Hanford Atomic facility somewhere to the north, as we were headed west. Grandma said "Look at those three funny airplanes" , looking north,

I saw three, black, triangular objects, above the other side of the river gorge, headed our way, and grandma and kids were talking with great excitement! Suddenly, as my eyes were on the road, everything quit working, engine, dash lights, radio, CB, so I coasted over to the side of the road--wondering what's going on?

Next thing I knew, I was standing in front of the car, wondering what I'm doing out here, Gradma and the kids, in the car "were looking at me like what are you doing out there"?

I got in the car, started up fine, everything working, except the dashboard clock--we all acted dazed and dis oriented--we decided to stop in the next town and get something to eat--and then continue on and see the dinosaurs at Fossil Or, that excited everyone, so we stopped at a drive in at The Dalles,

The Man In Black! While munching our burgers and fries, the hair on the back of my neck stood up as if some one was behind me, I turned around and this Man was sitting two tables away, not eating but just staring at us, nothing was on his table and I sensed danger! I stood up and walked over to him, he was thin (like a person that had rickets), olive-brown skin, coal black hair and beady eyes framed by bushy-black eyebrows and pointed chin! Black shoes and suit, white shirt--feeling this was unreal asked "May I help you with something?" He shot back crisply, "Where are You Going"? Feeling very uneasy I stated, "going to take the family down to Fossil, and see the sights"! He replied, in a low but distinct voice, 'DO NOT GO DOWN THERE, THINGS BIGGER THAN HOUSES, COMES IN OVER TOPS OF THE TREES!" With that he stood up and went out the door, stunned, I went back to our table and discussed the events of the day and it dawned on us that some how we were unable to account for three hours of time, and then this weird guy, Then we all saw him drive very slowly by the side glass door of the diner, in a very black- highly polished utility truck. (Painters vans we called them then}, just staring at us as he passed by.

Quick vote, panic setting in, lets go home quick! Down highway 97 we raced, past Bend Or, we started feeling a little safer, Forest road, star lit sky, was beautiful, Then a Big Black Car passes us like we were standing still, with that same face looking at us as he passed, (I was doing 65 mph). And was quickly out of sight. Everyone was quite upset again! I mashed the gas pedal up to 80 and let it roll quicker out of there the better! Again that black car, with that stern face passed us again, Panic now, find a place with people, Grandma found Dechutes State Campground on the map just ahead., quick right turn, and there standing by the side of the road, was that strange man.! This Is Nuts!! Shaking, I quickly found a campsite near the bathroom, removed my 30-30 from the trunk, and stood guard the rest of the night!!! Next morning we traveled on home, wondering what had happened! Still don't know! The Oldsmobile never ran after that trip, too many problems with it! The storm? That when the Estes Dam broke. Was the man in black bad or good, good I think, but I don't want to meet him again, Thank you! The missing three hours, just gone, not sure I want to know!

Other strange experiences, yep, since I can remember, UFO perhaps, don't know! Sincerely.

Occurred: September 10, 1976

Location: Colusa, CA

Low-hovering disc with dome, rotating rim and bright lights, power failure. Object moved away, shone beams of light down (Ref. 1, Section I).

1977

1977 brings with it the debut appearance of, The STAR WARS EPIC and with it, the plethora of Space Alien movies and sequels that dominate our culture well into the 21st century. The discomfort of aliens is now replaced with the cultural mythology between good and evil dominating our cosmos.

Carl Sagan "COSMOS" with his books and TV shows expand the minds of earthlings with new concepts of science and physics. Eric Von Daniken with his medley of books about the Ancient Civilizations, "Chariots of the Gods" and their connection to UFOLOGY had planted seeds into the culture for the budding future researchers that now have found further evidence of our earth/space connection.

This is the year that the MADAR Project got its real test. Go to July 10th and listen to the data tape recording! There were seven detections of magnetic anomalies by the Multiple Anomaly Detection and Automated Recording system at my Mt. Vernon, Indiana facility, within a six-week period! There were UFO sightings within 60 miles NW and 85 miles SW at the same time MADAR was triggered. **Francis Ridge NICAP Site Coordinator**

April 20, 1977 - CIA Memo Declassified
Memo from CIA Director Walter Bedell Smith to Director, Psychological Strategy Board, declassified 20 April 1977 (Good, Above Top Secret, 511)

November 16, 1977 - Close Encounters of the Third Kind was released.

1978

This year "Mork and Mindy" debuts on prime time. Now an Alien stars an award winning prime time slot that lasts for 5 years.

Occurred: January 5, 1978
Location: Klamath Falls, OR
V-formation of 10 lights passed plane

January 28, 1978; AIAA Symposium
American Institute of Aeronautics and Astronautics symposium on Space and UFOs in Los Angeles, CA.

February 1, 1978 NASA
NASA Information Sheet perpetuates the myth that no government agency is engaged in UFO research.
Since others deny UFOs existence then you and your sightings don't exist, and you're crazy, as they are the authority right??? NOT

Occurred: March 4, 1978; Approx. 2200hrs.
Location: Yreka, CA
Witness: 9
Shape: Triangle
Duration: 10 minutes
Low flying large darkened triangle with no sound seen by nine policemen
I was a police officer in my small town. A group of us from the same Department were on a recreational basketball team that played in the evenings. As we stood in the parking lot on a very cold night, talking, all 9 of us saw: Approaching from the South, an area of sky that was dark, meaning the stars were covered by an object. It was low in the sky, moving very slowly. As it came closer we could all see it was a huge triangular shaped object, wider than it was long, similar to a B-2 bomber, but much larger. I estimate it was at least 100 yards from "wing" tip to "wing" tip, and about 25 yards long. There was no light of any type, just the obvious blotting out of the night sky, which was full of stars in our small town.

As it came directly overhead, and slightly East of our location, the most blatant thing that stood out was the lack of any sound from the object. I estimate its altitude at 500-1000 ft. Even as it slowly passed, and continued north, there was no sound at all, and no lights. While we had barely said a word while it was in sight, we all started talking about it, comparing details. We all agreed that, due to the very slow speed, and lack of sound at its low height, that it was not an aircraft. (Our small town was on Victor 32, an airliner "freeway" from California to Seattle, and we had all seen and heard a lot of aircraft before.) As we were all cops, we more or less agreed to not talk about, for fear of being ridiculed.

((NUFORC Note: Witness indicates that the date of the sighting is approximate. PD))

Occurred: March 21, 1978; 0400hrs.

Location: Medford, OR

Witness: 1
Shape: Disk
Duration: 7 or 8 minutes
In flight formation with two disks

I'm currently a 25,000-hour pilot (2009), but on March 21, 1978 I only had around 5,000 hours. I have reported this incident to Dr Haines and Dr. Friedman, no one else.

I was flying an Aero Commander 680, on a the last leg of the day on a freight run from Klamath Falls, OR to Medford, OR at about 4am. I had climbed to an altitude of 15,000 feet on this night in order to top icing conditions resulting from a fast moving cold front that was at that time about half way between the two cities. I had a passenger with me sitting in the copilots seat. About 15 miles from the Medford VOR, I broke out the side of clouds into a perfectly clear rain washed sky, with full moon setting on the horizon ahead. This of course, left me very high for landing at Medford...so I canceled my IFR flight plan and headed directly to the airport, arriving there at an altitude above 8,000 feet where I began circling to descend for the traffic pattern.

As I was passing about 7,000 feet, I was circling to the left over the airport on a southeasterly heading when I noticed what looked to me like the landing light of a relatively slow moving aircraft flying in the opposite direction. I called the tower and asked if they had other traffic in the area, they said they did not. I was not too surprised to hear that in as much as we were both above the towers airspace so contact was not required. As we passed abeam each other at about the same altitude and within about one mile, the craft silhouetted against the full moon on the western horizon, when I could see that the craft had no tail and a light on the back as bright as the one on the front that I originally thought was a landing light. Thinking this was unusual; I turned toward the craft, to the right. As I turned the craft, also turned to the right and accelerated. When I started the turn, I was southeast bound, as I turned steeper and steeper to keep the craft in sight, it continued to accelerate in a turning arc about 2 miles out side my turn until I lost sight of it over the top of the aircraft, then in about a 60 degree banked turn then headed to the northeast. I continued the turn all the way around to a northwesterly heading when I rolled out, not having caught sight of the craft again in the turn.

Shortly after rolling out of the turn heading to the northwest.... two flying disc pulled up on my left wing, no more than ten feet off my wing tip and perhaps 5 feet higher, where we flew together in formation for perhaps 15 to 20 seconds. The furthest one was lit slightly ahead of the nearest one, which was not lit. The lights of the city were shining up from the ground in the rain washed air. I could not see much detail on the lighted craft due to the bright lights on the front and back, but I could see good detail on the nearer one, both craft appeared to be the same. They were disc shaped, white or light gray in color and I could see what looked like scotch marks on the bottom. They were about the same diameter as the wingspan of the Aero Commander or about 55 to 60 feet across. I could not see any openings of any kind in either of the craft and could not hear any sounds from the craft over the sound of the planes engines. At the time I was flying at about 160 knots, after the 15 or 20 seconds of formation flying the two craft accelerated to about 190 knots and pulled out in front of us about 2 miles when they instantaneously accelerated to a streak of light heading into space. I looked over to my passenger, who was a little white in color. I'm not sure if he was sick from the high G turn or stunned at what we has just witnessed. After we landed it took a couple of minutes to get him out of the plane. I saw him about 5 years later. he still did not wish to talk about it. As this was occurring I was telling the tower these events as they happened, who never did see the craft. I cannot understand why.

Occurred: June 1, 1978; 0000hrs.
Location: Grants Pass, OR
Witness: 3
Shape: Disk
Duration: 2 hrs.
Object landed on corner of street. It stayed there approx. 2 hours. Had red, blue, yellow, green lights along middle that spun approx. 10-15 yards from where we were standing.
Two of my friends and I were sleeping outside in the front yard in the summer. We were 7, 10 and 8 years of age at the time. One of my friends sat up and start crying and pointing to the end of the driveway. My other friend and I looked up and saw an object that looked like two bowls on top of each other with lights and gray squares (windows???) in the middle. The lights spun around the object. There was a very low hum coming from it. We went into a panic and started trying to get into the house but it was locked because my friend's parents were gone for the night. So, we tried to think of a plan to get around the craft so that we could run down the street to my house, which was two houses down. But, we were too scared so we just climbed back into our sleeping bags and pulled it over our heads. The dogs were going crazy in the house. One kept on trying to jump through the window. We stared at it for quite awhile. It finally started hovering about 5 feet from the ground and slowly took off and hovered above the house across the street for about 10 minutes then it just took off so fast that we had a hard time following it with our eyes. It went over the horizon and we booked it over to my house and never did sleep outside again that summer. Too bad I didn't have a camera it would have been the best view and the closest of a UFO ever taken.

Occurred: June 15, 1978; 1200hrs.
Location: Hoopa, CA
Witness: 1
Shape: Oval
Duration: 45 min

Twinkling lights, oval shaped craft, no sound, hovered for 45 min. around midnight, 1978
I was at a dinner party talking about a recent trip through Nevada and the infamous area 51 when one of the guests mentioned an experience 20 years ago that changed her life. She lived in Hoopa, a sort of Indian reservation in northern California right near the Trinity River. At around midnight she looked out her bedroom window and saw tinkling lights in the shape of an oval just hovering near the western ridge of the valley. At first she thought it was some type of helicopter but she didn't hear the pounding sound that usually accompanies one. All her other family was asleep so one else witnessed the event. She lay in her bed and watched the craft just hover for 45 minutes but then she fell asleep.

Occurred: July 20, 1978; 2000hrs.
Location: Pine Grove, CA
Witness: 3+
Shape: Formation
Duration: 5 min.

3 unknown objects overflew our house just after dark. They were in formation.
As me and my extended family were sitting on our front porch after dinner, just chatting away, we noticed 3 large white lights come from the western horizon and stop instantaneously overhead. The rate at which they traveled was estimated by us about 400 mph. When they stopped, they hovered for about 3 min. and pulsated, and then they shot off towards the east at 2 to 3 times the speed by which they came from the west.

Occurred: November 27, 1978; Approx. 2000hrs.

Location: Ukiah, CA

Witness: 2+
Shape: Sphere
Duration:10 minutes

UFO over Ukiah, sphere with beams coming out from it...then I forgot the entire sighting until I saw a report in the paper. I was driving home after dark listening to the radio. It was either the night of Nov. 18, 1978 or Nov. 27, 1978...the 18th was the day of the Jim Jones People's Temple killings and the 27th the day of the Mayor Moscone/Harvey Milk killings in SF. I had been listing to news of Jim Jones on the radio all week, then of the Moscone/Milk murders, so I am a little mixed up as to which night it was, but am pretty sure it was the 27th, because it was the 2nd huge event in a week. (The sighting was reported in the Ukiah Daily Journal on the front page the next day, after being reported by others). I was driving south on Hwy 101 just south of town.1/2 mile, after dark listening to the radio. Then I noticed that the top of my windshield was lit up a little. I looked up and there was a cloud, which seemed to glow. I pulled over to get a better look at it and noticed a hitchhiker up the road from me about 300'. He was looking up too. I got out of the car and this huge cloud looked as if the moon was behind it, but the moon was behind the mountains to the southeast and hadn't risen over them yet. As I watched, a sphere began to show inside the cloud. There was no sound at all. Then beams of light came out from it like a Star of David over Bethlehem. At first it seemed more like a religious experience. I felt this overwhelming feeling of awe. Then I looked into my car at my camera on the seat thinking that I should take pictures. But, amazingly i didn't. I have no idea why i didn't and after I thought to get the camera, I have no recollection of thinking about the camera again. The cloud seemed to clear and I could see this ball of light. Then the beams went back inside the sphere and the clouds went around it again and it moved west and back around the mountains. I was so amazed and jumped in the car to go home. I was thinking, "Wait till I tell ((name deleted)) (my sister) about this. I also thought I would pick up the hitchhiker and ask him about what he saw. I don't remember driving past him or seeing him, or don't remember anything about the 5 minute drive home until I pulled up to my mailbox. I went ino the house with NO memory of the saucer!!! I never called my sister about it. I didn't think of it or remember anything about it. I just turned on TV and that was that! But, the next day when I got the paper, here was the story. Other people had seen the same thing. Then, when I saw that story I though, "What? I saw something too, why didn't I remember that?" I was dumbfounded! It took about an hour for me to get the details to come back to me. At first I just knew I saw the cloud with some shape and it took the hour for it all to come back. Then I couldn't believe I had forgot it and didn't pick up the hitchhiker, didn't take a photo and didn't call my sister and didn't even remember this...one of the most amazing experiences of my life! So, they must have done something to block my memory. I have no idea if I lost any time either. I also saw a UFO when I was a little boy north of Buffalo in Niagara County, at night. Then my sister and neighbor saw it again over the woods the next day. But, the article may still be on file on microfilm at the Ukiah library or at the daily journal. I have seen photos of the same thing from Mexico City and Montreal.

Occurred: December 15, 1978; 2000hrs.

Location: Sisters (near), OR

Witness: 2
Shape: Other
Duration: possibly 4- 5 seconds

Neon multi-colored jellyfish type hovering object in Oregon late 70's -first report of this incident Very clear, close up colorful moving object shaped like a giant jellyfish hovering in forest. This is very old that was never reported; I don't think I ever made a formal report of this before -- I really don't know why not-- maybe I thought people would judge me in a bad way. Anyhow, my brother and I saw the object from the back seat of our car as we were on a trip with our parents from Nevada to Oregon.... it was really late and the other kids in the car were asleep, but not us 2. It was between about 1977 and 1979 and I think it was on Christmas break or possibly summer break... exact date and year or time of night I am not sure of, but the sight was definitely clear. It was possibly 30 meters or so from us on the right hand side of the road down in the forest... we were situated at a level slightly higher than the object and the road was pretty curvy. The thing was about 10 feet across maybe... dome shaped with hanging down undulating legs. The whole thing was like neon glowing in the dark, even its legs. The top dome had different colored (I think oval) lights going all the way around the dome. It was so long ago I wish I had drawn an exact likeness, but I definitely remember these things. It was such a beautiful object hovering just above the forest floor. We only saw it for a moment but it was clear... no distant nebulous lights or anything. Still, my brother and I really had not discussed it much over our lifetime. I think we might have tried to tell our parents later on, but they must have laughed it off and no report was made. My brother and I are both highly educated. We are not redneck storytellers or something. I have no wish of any notoriety at all. I am just curious. I even called my brother this evening to ask about it and he laughed and said yeah that octopus thing ...that was definitely strange. (He was 2 years younger at the time we saw it and maybe had a less clear memory of it, but he definitely confirmed that he had never seen anything like it and he still thinks it very odd). I would like to know if anyone has seen this specific type of object besides us.

1979

This year was the second and final season for Project UFO, which was an NBC television series. Based loosely on the real-life Project Blue Book, the show was created by Dragnet veteran Jack Webb, who pored through Air Force files looking for episode ideas. Col. William Coleman, who had had his own spectacular UFO sighting while in the Air Force, produced this show. Francis Ridge NICAP Site Coordinator

Occurred: March 1, 1979; 0400hrs.
Location: Medford, OR
Witness: 1
Shape: Unknown
Duration: 5 min
Summary: While in flight landing at Medford from Klamath Falls, I spotted what appeared to be another aircraft flying the opposite direction. Since the moon was setting, when this craft silhouetted against the moon, I could see it had no tail. I turned toward the craft, and it turned toward me. The encounter ended with the craft flying formation with me for about 30 seconds, where I could see much detail of two craft.

This encounter may be too old to be of interest to you. I did report it to a University the day after, but I do not remember which University. I got the phone number from the Medford Police for the University. About 4 years ago I reported the encounter to Dr. Haynes. I only mention it to you now, because I still have relatives in Medford, who sent me a newspaper clipping of a recent sighting there, that mentions the name of your organization. If you have an interest in more detail, let me know, I will email you a full description. Russ Smith.

Occurred: May 25, 1979; 0110hrs.
Near the Sacramento River in Colusa, CA
Witness: 2
Shape: Bullet like
Duration:
1:10 AM. A silver colored, heel-shaped or bullet-shaped object hovered between fifty and five hundred feet above some trees near the Sacramento River for three minutes. It had a row of red and blue-green lights along its leading edge between two headlights. It also had a smooth metallic surface. It made a sound like an electric motor as it moved very slowly toward the two witnesses, Ruben and Carlos Genera. Many animals reacted to its presence. The close encounter lasted seven minutes. (Sources: Paul C. Cerny, MUFON UFO Journal, May 1984, p. 3; UNICAT database, case 235, citing Paul Cerny).

May 26, 1979; Colusa, CA

12:15 AM Two fishermen observed a hemispherical object with blinding white headlights and a row of body lights. It hovered in one spot for 2-3 minutes and then flew overhead emitting a humming sound like an electric generator. (Colusa Sun-Herald, May 29, 1979).

Occurred: July 1, 1979; 0100hrs.
Location: Chico, CA (Hwy 70 at 179 junction)
Witness: 2
Shape: Other
Duration: 3-5 minutes

A large slow moving craft was observed, first at a distance as a bright light, then passing directly overhead.

Early in the summer of 1979, a friend and myself had just entered Hwy 70 from a back road and at the Chico cut-off. We noticed a bright light in the distance - I don't remember the color - moving oddly in some fashion. I am not sure what actually caught our attention. Perhaps the object was moving too slowly or in an unusual flight path. In any case we wondered what it was and pulled over to watch for a minute. The second we got out of the truck, the thing changed direction and came directly for us. It moved slowly, silently and steadily until it passed directly over our heads. Since we had no perspective for distance against the night sky, I cannot estimate the size or altitude. It appeared to be as large as 747 and flying at say 1000 feet. I clearly remember that it looked to be the shape of an eye with a hole in the middle where the iris would be. I can remember seeing stars right through the middle of the object. I clearly remember the thing looking to be diamond or eye shaped, however, with so many reports of wishbone or chevron shaped objects, I must admit that my own eye/brain may have filled in detail where no structure existed. No sound could be heard and it continued to move slowly into the SW until disappearing. It appeared to be black and I think some lights could be seen...but not ordinary aircraft lights. I guess that it had a very bright light shining forward but I am not sure which lights had actually caught our attention. I am quite convinced that I saw an experimental [stealth] blimp or otherwise operating out of nearby Beale AFB. Maybe it was a stealth blimp piloted by aliens. (Lockheed and others DO make chevron shaped, black, secret blimps)

Occurred: July 1, 1979; 0300hrs.
Location: Chico, CA
Witness: 2+
Shape: Rectangle
Duration: 4 - 7 minutes

Summary: A huge solid black mass, silently glided directly over us from the NNE going SSW then did a perfect 90 degree turn directly over us and went due E. No lights, no sound, just a monstrous huge black vehicle. My impression was it was like a rectangle. We only knew it was there because it was blocking out the stars (there were no clouds that night), that is what drew our attention to it, blocking out a large section of the sky and it's movement. I have no idea how high it might have been although it was at least above 3500 feet. My associates, and myself were all standing together talking when one of the guys said what is that coming from the "North?" We all looked and then could make it out, and it was awesome. We were all "on duty police officers" taking a break in our favorite spot where the ambient lights of the city didn't block the stars. The north valley is known for it's clear nights and fantastic views of the night sky. If we hadn't been taking that break and watching the sky while we talked we would never have seen this object. And no, I have not said anything before this time. We all looked at each other and basically said, "What the hell was that thing!"

We were all Chico or Butte County Deputies at the time that often worked the same areas of the county/city together. This location we were standing in was on a sight rise or hill on the SE side of town, leading up to the foothills of the Sierras. It was away from public view where we could park four or five police cars, drink some coffee and relax for a few minutes without scrutiny or criticism from passing motorists; and the sky gazing was superb from that location. We were all veteran cops as well as military veterans. I was a combat air-defense control officer in the military and have somewhat of a good idea of what anything man made can accomplish in the air. This was not anything that we could possibly have built. I don't know why I thought it was a rectangle, but I just had that feeling about it. Whatever shape it was, it was huge!

Occurred: August 9, 1979

Hayfork, CA

Pilot observed two disc-shaped objects below his plane, contour flying "on the deck" through hills and valleys (sections III, X).

October 26, 1979, Colusa, CA

Roosters crowed, ducks quacked, geese honked, and bullfrogs started croaking loudly in response to silver bullet-shaped craft that hovered briefly Form 97-AR

Occurred: October 26, 1979; 0015hrs.

Location: Colusa, California, Cat: 4

Witness: 2
Object: 1 bullet shaped disc, silver humming, glowing,
Duration: 2 to 3 minutes
To: NICAP
Animal Reaction Feature: Joan Woodward, Animal Reaction Specialist

Just after midnight, a silver bullet-shaped craft with two beams of light hovered and then flew overhead. It emitted a steady hum. Animals along the river and at a nearby farm vocalized. Roosters crowed, ducks quacked, geese honked, and bullfrogs started croaking loudly. [The bullfrog reference is bothersome; October is past the time of the mating chorus of bullfrogs in California— and "croaking loudly" does not sound like an alarm call?

The sighting: At 12:15 a.m., two brothers saw a silver colored bullet-shaped craft with two intensely brilliant beams of light hovering over the Sacramento River north of Colusa. It appeared to arrive out of nowhere and hovered for 2 to 3 minutes before flying over the witnesses' heads. It made a humming sound that the witnesses compared to a big electric generator.

No distance estimates were given. No EM effects or physiological effects were reported.

Source: Gribble, Bob, 1989, Looking Back, MUFON UFO Journal no. 258, October 1989, page 24 [Copied below]: Silver, bullet-shaped craft with two intensely brilliant beams of light was observed hovering over the Sacramento River north of Colusa, California, by Carlos Genera and his brother Ruben on the 26th [Oct. 26, 1979—jw]. "At about 12:15 a.m., it came out of nowhere and stayed at one point for about two or three minutes, and then came over our heads," Carlos said. "It emitted a nice steady hum like a big electric generator." Ruben said the vehicle created a considerable stir among the animals along the river and nearby farm. "Everything started getting excited all at once. There was a big commotion right away

Occurred: November 14, 1979; 2200hrs.

Location: Hat Creek Campground (Near), CA

Witness: 2
Shape: Rectangle
Duration: 20 to 30 minutes

Watched the object cross diagonally across the sky over our car, drove up a logging road to see where it went. It was HUGE, rectangular and it had lights all around the edge only...It was quiet, very quiet and moved quite slowly from the direction of Mt. Shasta, south towards Redding On or about the 14th of November 1979, my friend Vic and I were headed towards Susanville from Medford Oregon, on our way to Reno (Vic had a job interview), it was around 10pm. Vic had a convertible and it was an unusually warm evening so we had the top down... In the area of Hat Creek Campground, on Highway 89, I was leaning back with my head on the back of the front seat when I saw a HUGE object moving diagonally over the highway above us. I was saying Oh my God, and Vic was trying to see what I was looking at while trying to keep the car on the road. When he realized what I was looking, he immediately pulled over to the side of the road. For about a half a minute we watched it fly over the woods to our right, and almost out of sight, so Vic said, "Let's follow it". We found a logging road and drove up until we reached the edge of a canyon. We sat on the top of the back of the front seats and just watched the huge, silent, rectangular object with lights that went all around the edge of it, (I can't remember if they were in color). When we realized that if we could see them, then they could see us, we decided to leave the area...It was the most amazing thing I have ever seen in my life. I never reported this to anyone of authority for obvious reasons. ((Name deleted)) Medford Oregon
References:
1. Volume II, The UFO Evidence, A Thirty-Year Report - Richard Hall, 2000
    2. CUFOS report, Rodeghier,

July-December 1979

Los Angeles and San Fernando Valley CA. Flurries of sightings (section VIII).

SECTION 8: 1980-1989

Occurred: January 22, 1980; Approx. 0200hrs.
Location: Murphy/Arnold, CA
Witness: 1
Shape: Sphere
Duration: 45 Minutes
Tracked by a UFO
Around 1980 in early winter, I was driving home from Angels Camp, California, on Highway 4 eastbound. It was around 2:00 in the morning and snowing. I was just outside of Murphy's, when I pulled over on to a large snow covered dirt shoulder to tighten my tire chains that were hitting against the wheel well of my car. As I kneeled down to tighten the rear passenger side tire chains, a bright white light flashed on me, lighting up everything around me. At first I thought it was the headlights of a vehicle coming, but noticed there wasn't one. I didn't think anything of it, but a few seconds later it happened again. This time the light stayed on, and I immediately stopped what I was doing and looked to see where it was coming from. I noticed it was coming from directly above me shining through the snow-covered clouds. There was no noise and I couldn't see anything other then the beam of light shining directly on me like a spotlight.
I wasn't sure what was happening. Suddenly I felt a sense of panic. I jumped back in my car and tried to get out of there as fast as I could. I was about 15 miles from home, but because it was snowing, and the roads weren't plowed, I couldn't drive any faster then 25 miles an hour. During the trip back I didn't see a single vehicle in either direction. I felt scared and alone.
When I finally got off the highway and into the subdivision where I lived, I felt a sense of relief. But when I turned the corner to go up the driveway to my house, I saw a bright light at the end of the street shining through the clouds. It looked like a really large star. But I knew this wasn't possible, because of the low cloud cover from the snowstorm. I was really frightened by the sight of this. I felt like what ever it was that shinned its light on me earlier, followed me home. I ran into the house and immediately woke my parents to tell them what I saw. They didn't think much of it, and just wanted to go back to sleep, reassuring me that everything was okay. That night when I went up to my room I was really excited and nervous. I was so wired I couldn't sleep, and stayed up all night into the next morning.
As a side note; I lived in Arnold, California for 7 years. It's about 4000 feet above sea level on the western Sierra Nevada mountain range. Living there I experienced many snowstorms. When it snows there, a thick cloud cover blankets the sky. The clouds are very low to the ground, and you can't see through them. Whatever I saw that night somehow penetrated this blanket.
((NUFORC Note: Witness indicates that the date of the sighting is approximate. PD))

Occurred: May 15, 1980; Approx. 0100hrs.
Location: Hat Creek, CA
Witness: 2
Shape: Rectangle

Duration: 2 hours
Do not have a clue what took place with my friend and I
This event has been very hard to talk about. My friend J and I purchased property in a remote area outside the town of Hat Creek Ca. We were in the process of putting in a road to it. We were camping out rather that traveling back and forth to town each night. We had returned to our camp sight just before dark started a small campfire had dinner and turned in for the night. I'm not real sure about the time anyway during the night I woke up and had to go (((to the toilet)), I stepped out the tent the campfire was cold. We were camping at the foot of Hog Back Ridge the moon was coming up from behind the ridge. The moon lit up the night I saw something coming very large coming from the east it was huge I called my friend J you have got to see this he thought I was nuts until he got out of the tent. The thing came right over the top of us it was huge it made no sound. I to this day do not know how but J and I were on board this craft we were greeted by four humans at lest I think they were they looked just like us. They spoke to us for quite some time I do not remember what they talked about but it seemed to be important. My memory from there on is pretty much a blank. We woke the next morning back in our tent. This is the first time I have told this story out side of my family. I have no explication. ((Name deleted)) ((NUFORC Note: Date is approximate. PD))

Occurred: May 15, 1980; Approx. 1600hrs.
Location: Eureka, CA
Witness: 2
Shape: Cigar
Duration: 2 minutes?
Rode the school bus with friend Joel to his house after school. When we stepped off the bus, we saw a shining craft in the sky, an elongated cigar-shaped craft with no wings or contrail or anything. It was flying in strange way, going up and down like a sine wave or something, straight ahead the whole time but in an up-and-down way, like it was riding on waves or like when you skip a rock across a river or something. We could tell that it wasn't a regular airplane. Joel ran into his house to get his mom and her camera, but by the time they got back out there the craft had flown off past a bunch of trees and houses and we just weren't able to see it any more. We went in their house and called the local airport to see if they had any unusual airplanes in the area or something but the guy who answered said no, asked us where we seeing it from and acted nervous and hung up pretty quickly. Which seemed odd.
((NUFORC Note: Date is approximate. One of four reports that came from the same source.PD))

Occurred: June 1, 1980; Approx. 2000hrs.
Location: Oroville, CA
Witness: 2
Shape: Sphere
Duration: 1-2 minutes
Saw 2 glowing spheres going from West to East for about two minutes.
I was outside my house at XXXX Ashley Ave. after dark when I saw two spheres approaching from the West going east.

I called my wife and she came out and saw them too. They were about 3 feet in diameter and less than 1000 feet in altitude. One was above the other and the lower one was in front. Their color was a very pale yellow and they glowed. Their speed was about three times as fast as a small single engine airplane. There was no sound.
((NUFORC Note: Witness indicates that the date of the incident is approximate. PD))

Occurred: June 25, 1980; 1000hrs.
Location: Cave Junction, OR (just outside town)
Witness: 1
Shape: Unknown
Duration: 5 minutes
Men In Black Encounter in Oregon
In approx. 1980, on a late spring day, I was driving up to Oregon when I somehow got off on a logging road, which was winding along a steep mountain. After a while, I noticed that the road was narrowing, and I was afraid to follow it any further. I decided to try and turn my car around, and drive back to Cave Junction, Oregon that was about 30 minutes away. As I was backing up the car, it went off of the road, and the rear end of the car was off the road and the car was tottering. I slowly got out, and was trying to figure out how to get my dogs out, which were in their kennel in the back of the car. If I tried to reach them, the car could tumble down the cliff. While I was trying to decide what to do, a big car pulled up, and 4 tall oriental men, with a strange skin tone, (too orange) got out of the car. In spite of the 100+ degree heat of the day, they were all wearing black trench coats. I said Hello, and they did not reply. Without a word, they went to the front of my car, and picked it up, and sat it back in the road, facing the direction back to home. I do not think that even twice that number of humans could have lifted my car from the front end only and place it on the road. I do not remember how I got past their car, or they passed mine, as they were headed up the firebreak road as I was originally as well. The after thought of the encounter left me really scared, and as soon as I was back in town, I rented a hotel room and slept for a couple of days. This experience is also written about in Dongo's The Alien Tide (pg 16) I apologize for reporting this some 40 years later. Those weird men on the mountain saved our lives, but they still scared me.((NUFORC Note: Witness indicates that the date of the incident is approximate. PD))

Occurred: August 15, 1980; Approx. 2000hrs.
Location: Bend, OR
Witness: 2
Shape: Unknown
Duration: 10 minutes
Bright round object seen at night, shot upward at a 45% angle, then got dimmer and dimmer until it faded away.

My husband and I were travelling north at about 8 p.m. in our car at night and were starting to descend into Bend, OR. I saw a round, bright light in the sky. All of a sudden, the object shot upward, but not in a straight line. It shot at a 45% angle to the right. My husband was asleep, but I woke him up. He saw it too. The object started to get dimmer and dimmer, then a cloud came over it and we could no longer see it. We don't know if it faded out completely, or if the cloud was just not allowing us to see it anymore. It really shook me up. My husband, also commented that what we saw was very strange, not like what an airplane would look like in the sky at night. I remember stopping to get a cup of coffee, in Bend, before continuing on our journey to the Oregon Crater, located to the east of Bend.
((NUFORC Note: Witness indicates that date of incident is approximate. PD))

1981

January 9-March 7, 1981; N. California Local concentration of sightings (Volume II, The UFO Evidence, Section VIII).

Occurred: January 17, 1981; Approx. 2100hrs.
Location: Oroville, CA
Shape: Diamond
Duration: 30 sec
The craft was 250ft of the ground, moving north directly overhead slowly. There was no sound! Headed over town, made a left bank, and went south at a very high rate of speed. Very soon after that all the neighborhood dogs started to all bark at the same time. I know what aircraft look like that was no plane! or chopper!

Occurred: June 15, 1981; Approx. 2100hrs.
Location: Red Bluff, CA
Witness: 2
Shape: Unknown
Duration: 10 minutes+/-
Unknown lights flying across sky

I'm not sure exactly what time it was. It was nighttime and I was going out to feed the animals when I saw a pair of lights in the southeast heading towards the northwest. At first I thought it was a pair of lights on the wingtips of a low flying airplane so I didn't pay much attention to them. But then it occurred to me that I couldn't hear any sound of an engine so I looked at them again and noticed that there was nothing in between the two lights. I then thought that it might be a pair of satellites moving in the same direction. Since I lived in the country, I had an excellent view of the night sky without any street lights or lights from buildings to obstruct my view. As I took a harder look at the lights, I noticed that they left a trail, like a jet plane would leave, trailing behind them, but there were only two lights and I could plainly see three trails. I called out to my brother to come and have a look, but he wasn't able to come outside, but my mother did and she saw them too. I don't know what they were, but I don't understand how two lights can leave three trails. As I was looking southeast, one light was slightly on my right, the other on my left, both were moving at the same speed and in the same direction. The two lights left thin trails behind them as they moved across the sky to the left of the light on my left, I could plainly see a third trail, a wider trail, but there was nothing there to cause it. Another time I saw something was when I was out driving truck cross-country. My trainer was in the bunk sleeping and I was driving the night shift. Just as I crossed the state line from California into Arizona, I saw a ball of light streak across the sky in front of me and appear to crash into the hills. The ball of light was green. The first thing I did was turn on my CB radio and listened to see if anyone else had seen it. No one mentioned anything about it on the radio. My first thought was that maybe I had seen it and no one else did or maybe it was my imagination. Then I figured it might have been a meteorite falling to earth. The green glow could have been cause by copper deposits in the meteor burning as it went through our atmosphere.
((NUFORC Note: Date is approximate. PD))

Occurred: October 31, 1981; 1800hrs.
Location: Weed, CA
Witness: 5
Shape: Disk
Duration: 10 minutes
10/31/1981 Saucer Shaped UFO hovering between the mountain Mt. Shasta and the city of Mt. Shasta California
I grew up in Weed, California, Siskiyou County, and heard numerous stories of UFO sightings and in particular, stories from my mother regarding "St. Elmo's Fire." The sightings of St Elmo's fire were always located several miles south of Weed in an area that bordered the then Kellogg Ranch. Located west of I-5. My mother described St. Elmo's fire to be as large as a car, hovering approximately 5 feet off the ground. The color was always a brilliant emerald green mixed with red/yellow flames. She stated that the only way you knew St. Elmo's Fire was coming was because the air surrounding you would become charged with electricity causing the hair on your body and your head to stand up.

My story begins early evening October 31, 1981. Fall was settling in and when I was growing up, November 1st always brought snow; thus, there was coolness to the air with the smell of wood smoke. I was newly dating someone who lived in Mt. Shasta, California and I was headed to a dinner party that friends of his were having. Just as I approached the Kellogg Ranch area headed south on Interstate 5, (which at this date of 1981 had a McDonald's and a hotel located on the east side of the highway) I felt the crackle of electricity in the air with it being stronger to my right, which would be due west. I was slightly confused due to this new sensation and as I scanned left to right I saw a huge ball of emerald green fire shooting from the west headed due east about 2/10's of a mile in front of me. The ball of fire was as large as a sedan and approximately 5 feet off the ground; just as my mother use to describe it in her stories. As I visually followed St. Elmo's Fire from right to left, the mountain, Mt. Shasta was in my line of vision to the east. To my utter amazement, there was a saucer shaped UFO hovering on the side of Mt. Shasta closest to me. It was huge, sliver in color and slightly dipped downward on the left side of the UFO. The UFO was located toward the base in the area we call the "timberline." As I traveled south toward Mt. Shasta, I was able to keep it in my line of vision for approximately 15 minutes and then it shot straight up in the air and disappeared. The light during this entire sighting was bright enough for me to not misinterpret what I was seeing. I had hoped that the UFO would remain where it was as the house I was headed for would have had a front and center view of a lifetime.

The house where the dinner party was being given was located almost at the base of Mt. Shasta on the Everett Memorial Highway. The house is located on the west side of the highway and Mt. Shasta High School is still located across the road from this house. Upon entering the house everyone was excited and asked me if I had seen anything strange on my way to their home. Because I didn't want them to think that I was strange, I told them to tell me what they saw. They stated that they felt a charge of electricity and when they looked out the picture window that faces Mt. Shasta, they saw a UFO and described it exactly the way I saw it. None of us saw lights that were lit up on the UFO-could be due to the fact that it was still light outside that none were needed. I then told them about seeing St. Elmo's fire and then the UFO. My friends apprised me that, though they had never seen St. Elmo's fire, they had heard stories of sightings in the same area where I saw it.

The four individuals that saw the UFO besides me were approximately 8 to 15 years older than me as I had just turned 21 that April. Two of the individuals were schoolteachers and the other two were business owners.

This was not the first time I had seen a UFO while living in Weed, California, and the other sightings were also witnessed by my siblings and friends and on one occasion, we even called law enforcement, who did respond, but found ways to discount what we were seeing and what they were seeing.

References:
Volume II, The UFO Evidence, A Thirty-Year Report - Richard Hall, 2000
2. CUFOS report, Rodeghier

1982-1989

Occurred: March 24, 1983; 1645hrs.
Location: Happy Camp, CA
Witness: 1
Shape: Circle
Duration:
4:45 PM. California Highway Patrol officer Harold R. Chandler was on routine patrol near the Northern California border with Oregon when he spotted a dark, circular object, flat with no protuberances, flying overhead across his path. "I was looking up at the bottom of the object, it was about 800+ feet above me. It' speed was slow, maybe 80-100 m.p.h. The craft stood out clearly against the sky and appeared dark gray." (Investigation by Paul Cerny; MUFON report form.)

Occurred: May 15, 1983; Approx. 1930hrs.
Location: Willits, CA
Witness: 2
Shape: Diamond
Duration: 1.50 minutes
Saw dark diamond shaped UFO for 1 and half minutes at tree top level.
While driving up Sherwood Road going outside of Willits California, the driver (Jeff) saw something weird as we crossed the top of Sherwood by the cross.
As we traveled about 50 yards, the driver decided to pull over and I do not know if the car stopped running or if it was on his own accord, the passenger (Martha) saw the object too. We exited the vehicle (a 1979 Datsun 210), which is when I saw the object moving across the sky. I previously was in the back seat and could not see what the others could see.
The object we saw moved very slow, about 15 miles an hour, and did not have much sound to it. It was at treetop level (about 100 to no more than 150 feet above) Lights shot from the pointed end to the back of the object and then again to the forward point in different colors. It made a low humming sound.
As it passed over us, we maneuvered around the tree's, backtracking from where we had come from to get a better view of the object. We then ran to see it from around the tree's and the object had now disappeared.
Cars were now coming and we ran back to our car, jumped in, and left the sight area. None of us ever forgot that experience.
((NUFORC Note: Witness indicates that the date of the sighting is approximate. PD))
((NUFORC Note: Witness elects to remain totally anonymous; provides no contact information. PD))

Occurred: June 1, 1984; Approx. 2100hrs.

Location: Bandon, OR

Witness: 2
Shape: Rectangle
Duration: 5 minutes

Oregon coast rectangle football field sized white square lights

A friend and I were looking out our parked car window at about 9 PM. We both noticed what seemed to be a star moving from north to south. We got out to look it approached very quickly and then slowed to what seemed like a few mph. We saw 2 vertical square white lights approx. 100 feet above us. As the object passed over us the vertical lights went out of view and we could see a set of horizontal white lights. Then a set of blue square lights swept back at not quite a 45-degree angle. Then another set of white lights. All of the lights on the craft, were non radiant but bright and solid. As far as I could tell the object was about the size of a football field. It moved from then north to south-southeast. As soon as it had passed over us it appeared as a star and quickly moved over the horizon. It made a slight humming noise. The night was very clear and dark stars where extremely visible. I recently talked to someone who gave the same description of an object during the same time period but in broad daylight within a mile of my sighting. The only difference was that he and the person with him heard the object had never told anyone of my sighting so I was shocked to hear another account.
((NUFORC Note: Witness elects to remain totally anonymous; provides no contact information. PD))

Occurred: June 19, 1985

Location: Redwood Valley, California

Witness: Multiple
Object: Round disc
Special Features/Characteristics: Children, Sound, and Witness Photo

Three children were camping in their Redwood Valley backyard when they heard a "whoosh" and saw a bright light. The flying "ball of fire" made three circles above their yard. Debbie, 11, said the object was bigger than a beach ball. The object began to get smaller and increase its speed until it disappeared in the southwest.

View full report

Source: Ukiah Daily Journal (Ukiah, CA), June 20, 1985

Flying 'ball of fire' seen by 3 children camping in backyard in Redwood Valley, CA

Location: Redwood Valley, California. Three children were camping in their Redwood Valley backyard when they heard a "whoosh" and saw a bright light. The flying "ball of fire" made three circles above their yard. Debbie, 11, said the object was bigger than a beach ball. The object began to get smaller and increase its speed until it disappeared in the southwest.

Brother and sister Jessica and Shawn White, listen as their friend Debbie Crumrine describes a flying "ball of fire" that circled the White's yard while the three kids were camping out with sleeping bags and lawn chairs. (Source: Bev Reeves, Ukiah Daily Journal)

Source: Ukiah Daily Journal (Ukiah, CA), June 20, 1985

"UFO startles camping kids"

By MISSY CHESSHER Journal staff writer

Ukiah Daily Journal

When three youngsters camping in their Redwood Valley backyard Tuesday night heard a "whoosh" and saw a bright light, they at first thought it was a jet and then a fireball. Now, they don't know what it was.

"So when do jets grow larger and larger?" asked ll-year-old Debbie Crumrine to her camping companion, 8-year-old Jessica White, who suggested the noise and light about 1 a.m. could be a jet.

Shawn White, Jessica's 11 year-old brother, thought the noise was traffic until the bright light made him curious. "I thought they were shining the flashlight in my face," Shawn said.

"It just appeared from thin air," Debbie said. She said the flying object came from the south and was traveling north. The object then paused below the treetops in their neighbor's yard "just like it had a mind of its own," she said. Debbie said the object was bigger than a beach ball. She said the object was red on the outside and bluish-green on the inside - "Like the color of fire." When the object paused, she said the middle stopped while the outside continued turning.

"The thing that scared me the most was when it paused," Debbie said. "She was crying, and I thought I was going to faint," Jessica said. After the object paused, Debbie said it continued traveling on its path and then returned and made three circles above their yard. The girls had seen enough. They covered their heads with their sleeping bags, but Shawn continued to watch. Shawn said after the object had circled their yard, it began to get smaller and increase its speed until it disappeared in the southwest. "It went from a big rock, to a Softball, to a star faraway," he said. He said the object had a muffled sound like a jet and a streak of fire behind it. "It was so big; it was so bright," Shawn said. "I thought I was going to get burned up."

Debbie said it was a cool night but the air seemed warmer while the object circled in the sky. She said the incident lasted about two or three minutes. After the object disappeared, all three children spent the rest of the night on the floor in the house. Debbie said she never wanted to spend the night outside again. But Jessica and Shawn both said they would like to see the object again. "I wished they had beamed me up," Shawn said.

Debbie was anxious to see their story in the paper because she assumed other people would report the incident too. "I know you don't believe me," Debbie told her mother, "But just wait till you see the paper tomorrow."

Spokespersons at the Ukiah Airport, Federal Aviation Administration, and Mendocino County Sheriff Department each said nothing similar to the incident had been reported to their agencies. Robert Gribble, a staff member of the UFO Reporting Center in Seattle, said the center had not received any reports from the Redwood Valley area. "It's not a real common thing, but we have gotten reports similar to this in the past," he said. Debbie and Shawn attend Redwood Valley School, and Jessica will attend the school this fall. Virgina Crumrine, Debbie's mother, said, "My daughter's not prone to making' things up."

((NUFORC Note: Date is approximate. PD))

Withdrawal Notice - Dated 28 June 1985 www.nicap.org
Top Secret UFO reports are being withheld. This is an example Intelligence withdrawal notice, referencing an October 15, 1955 document. (Source: Above Top Secret, Good, 529)

Occurred: August 20, 1985; Approx. 2100hrs.

Location: Kelseyville, CA

Witness: 2
Shape: Light
Duration: 30 minutes

All craft were spotted in lake county, California in a span of about 20 years

I do not remember the exact date as it was quite a while ago. The craft was a large bright orange ball hovering above the lake "Clear Lake". My friend and I watched this object for a good half hour just sitting above the lake. There were no noises of any kind at all. When the craft it it was gone in 5 seconds. The unusual thing too was that I was watching the news two weeks later and the same object was sighted over New Zealand. This was not the first time I have seen crafts in the area as I have seen at least two others in same area.

((NUFORC Note: Witness indicates that date of incident is approximate. PD))

Occurred: April 21,1986; Approx.1830hrs.

Location: Marysville, CA

Witness: 2
Shape: Glowing Sphere
Duration: 10+ minutes

We saw a 3-foot diameter glowing sphere hover over water, land and then straight up to the sky. I've waited long enough, so here is my sighting and I hope someone else saw it or can help me figure out what it was that I saw.

It's been awhile, springtime in 1986 and I think it was past mid April, and I must admit, I don't recall the exact date, but it didn't seem to matter at the time, the weather was nice. The place was a backwater slough at the south end of Plumas Arboga Rd just past the over pass of highway 70 in what is considered part of the town of Marysville California.

My younger brother, his friend and my cousin (all about 16 years of age) went cat fishing at a jack slough that evening around 17:30. We were on the east side of the slough which ran north and south, and would be facing east while we fished, (towards Beale AFB at that) and were gathering firewood before the sun went down.

My cousin was next to me and suddenly said, "Hey you guys, what's that?" I was hunched down toward the east getting ready to start the fire, and turned around and looked back toward the west. The sun just went down past the levy and the sky was clear, there was a little glare, and when I first looked back I didn't see anything. My brother and his friend started yelling, "what is it, what in the hell is it?" I looked at them and realized they were looking almost right over our heads, I looked up and I saw a glowing sphere, approximately 3 feet in diameter, about 10 to 12 feet over our heads! I was in awe as I quickly began to rationalize what I was looking at because I am skeptical when it comes to this kind of thing, but this was not ordinary and it was right there in plain, clear site!

It was a round bubble looking sphere, transparent, with a small bright glowing light in the center. That outer perimeter of it was slightly glowing because of the small light from within it. It was moving dead east about 2 to 3 miles per hour on a straight course for the AFB. It did not make any noise, did not move up or down or side-to-side, just straight and very steady. The younger boys were getting excited and kept asking me what it could be. I kept staring, watching it almost every single second, and paid attention to everything that was happening to the best of my ability.

It moved steady over the water, I believe the cooler air would cause some sort of fluctuation or rift with any man made floating craft especially something so small, but nothing, straight it went. There was a reflection in the water, and a shadow as it went over the opposite side and towards the opposite levy from us. Over that levy was nothing flat, plowed up farm land as far as the eye could see, and I knew it, so we ran south toward the bridge, went eastward over the bridge and watch it float over the field. We didn't know what to think.

We watched it travel slowly for about 10 more minutes and then, without a sound, about ¾ of a mile away, it curved up at 90 degrees and shot away! We talked about calling someone, but couldn't come up with anything that was going to sound serious or rational. I should have called anyway, and I regret it to this day.

But it didn't stop there. A few days later, my cousin, who lived across the street, and about 75 yards from the first sighting, called on the phone screaming for me and my brother to get to his house right away, the light was back.

We ran as fast as we could, and when we got there we could hear him and his mother in the back yard, so we ran to the back yard and there it was, the same sphere, floating just like before, over his fence, over the small field, over the levy, over the exact spot where we first saw it, over the slough, the east levy, over the field toward the AFB and then disappeared just like before. To my knowledge, it was never seen again. I will send some images via email. Thank you for your time. I was 23 years old.

Occurred: October 1, 1986; Approx. 0100hrs.

Location: Klamath Falls, OR

Witness: 3
Shape: Circle
Duration: 45 minutes

An object was dropping firecracker like pieces from it. It seemed to know I was watching it and it vanished in a type of explosion.

A light in the southern sky over an air force base was hovering and seemed to be dropping firecracker-exploding objects from it My friend ran and called to place a police report, came back out to watch with me. We felt that it knew we were watching and I spoke out loud, " I know you can see us." The object seemed to acknowledge this and darted across the sky and hovered over our house. A few minutes later the craft seemed to explode and disappear and where it was left a perfect ring, or circle. (Like a portal trace) It took several hours later for the Klamath Falls Police to call us back and informed us that one of their officers also saw the same object. I was watching the TV series "Sightings" several years later and a group of people were watching a ufo that behaved identically to the one we saw and was also talking to it.

At he time the police suggested we call the Air Force base and talk to the tower commander who denied seeing anything. (Authors note "Typical deniability") The tower commander gave us another number to talk to " someone who is more versed in the subject "and after I tried to find out if anything he said that they get lots of calls from this area all the time. I asked him if the explosion was the craft leaving at a great rate of speed or what? (Authors Note: That cracking is the opening and closing of a wormhole, portal, or inter-dimensional door known as a "Tesla based teleportation protocols, utilizing high energy discharges, which crack when initiated).

((NUFORC Note: Date is approximate. PD))

November 17, 1986; NE Alaska
HUGE CITY SIZED UFO SEEN BY ALL on JET

Japan Airlines (JAL Flight 1628) freighter aircraft encountered lighted maneuvering objects, Saturn-shaped; bright illumination, heat, radar-visual, E-M effects, satellite objects (Ref. 1, Section III).
RADCAT Case Directory Category 9, RADAR Preliminary Rating: 5
RADCAT is a revitalized special project now being conducted jointly by NICAP & Project 1947 with the help and cooperation of the original compiler of RADCAT, Martin Shough, to create a comprehensive listing of radar cases with detailed documentation from all previous catalogues, including UFOCAT and original RADCAT.
Fantastic Flight of JAL 1628 Nov. 17, 1986 NE Alaska
Richard Hall: Japan Airlines B-747 encounter with several UFOs, tracked on FAA radar. Crew saw giant Saturn-shaped object.
Fran Ridge:

In the late afternoon of November 17, 1986, Japan Air Lines flight 1628, a Boeing 747 with a crew of three, was nearing the end of a trip from Iceland to Anchorage, Alaska. The jet, carrying a cargo of French wine, was flying at 35,000 feet through darkening skies, a red glow from the setting sun lighting one horizon and a full moon rising above the other. This sighting gained international attention when the Federal Aviation Administration (FAA) announced that it was going to officially investigate this sighting because the Air Route Traffic Control Center in Anchorage, Alaska, had reported that the UFO had been detected on radar. Captain Terauchi was featured on numerous radio and TV programs and in People Magazine. Within a few months of these events he was grounded, apparently for his indiscretion of reporting a UFO, even though he was a senior captain with an excellent flying record. Several years later he was reinstated. The UFOs in this case were tracked on both ground and airborne radar, witnessed by experienced airline pilots, and confirmed by a FAA Division Chief. Audiotapes of the transmissions and AF radar confirmation were presented on the History Channel's "Black Box UFOs" in August of 2006.J
Detailed reports and documents
Gigantic Saturn-Shaped Object Dwarfs Boeing 747 Over Alaska - UFOE II – Hall

January 30, 1987; Alaska www.nicap.org

A U.S. Air Force KC-135 airplane flew inbound from Elmendorf Air Force Base in Anchorage, Alaska to Eielson Air Force Base in Fairbanks, Alaska. At 20,000 feet the airplane crew sighted a very large disk-shaped UFO that was similar to the one encountered by the JAL Flight 1628. The UFO moved to within 13 meters' (40 feet) of the KC-135. And as can be determined by the U.S. Air Force Pilot's question 5:29 minutes into the video, it's apparent that he has been briefed about the JAL Flight 1628's UFO encounter. The answer by the FAA official was also very interesting. He said, "It's very rare seeing the lights (UFOs) up there". He thereby admitted that UFOs were seen from time to time, albeit rarely! And the silly notion that the U.S. Government (here the FAA) is not investigating UFO sightings or is not interested in UFOs, can once and for all be ruled out by listening to the message the U.S. Air Force Captain received from the FAA (5:42 minutes into the video). The FAA message to the U.S. Air Force Captain said, "Call the local FAA office in Anchorage after you have landed at Eielson Air Force Base". Then the Captain asked, "That's concerning the object we were looking at?" "Affirmative", answered the Eielson Air Force Base official.

January 31, 1987; Alaska www.nicap.org

Alaska Airlines Flight 53 encountered several very large disk-shaped UFOs that trailed them. The UFOs eventually flew away from the airplane, moving about a mile a second!" Flight 53 was told the object was not being tracked on ground radar, but the crew had picked it up on their onboard radar. The Alaska Airlines segment can be seen 6:00 minutes into the video.

March 12, 1987; Army Interplanetary Unit www.nicap.org

US Army Intelligence letter confirming the existence of the Interplanetary Phenomenon Unit, alleged to have been involved in UFO recovery operations. (Timothy Good, Above Top Secret, page 484)

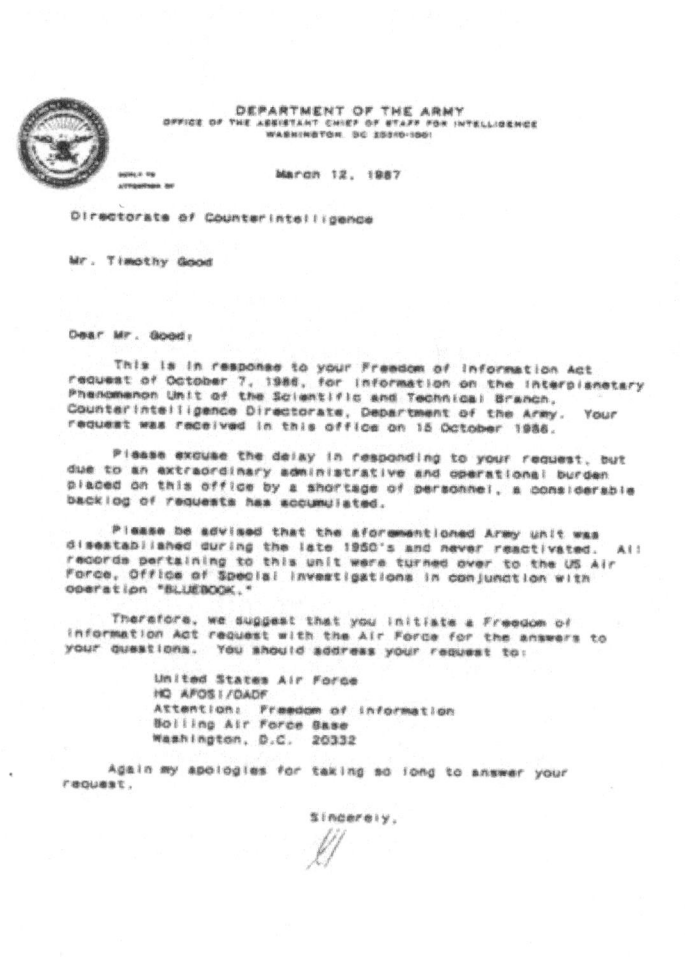

Occurred: June 1, 1987; 0100hrs.

Location: Coos Bay, OR

Witness: 3+
Shape: Disk
Duration: 1 hour

Dome type object went straight up with a little round object along side and below. Object kept moving up and down side to side and changing colors with little dot moving proportionately. After approx 30 minutes a long brown object came straight up from same location, stayed with other for about 5 minutes then shot straight to the east a full 90% in about 10 seconds, stopped on a dime and went straight down. We all had either riflescopes or binoculars and compared notes, all matched.

Occurred: June 1, 1987; Approx. 1600hrs.
Location: Redding, CA
Witness: 2
Shape:
Duration: 1 1/4 hr
Rectangular craft trails auto for over 100 miles.
WHILE RETURNING HOME FROM MT. SHASTA, CA TO BAY AREA, CA I, PASSENGER IN CAR WATCHED A RECTANGULAR SHAPE WITH 7 LIGHTS ACROSS FRONT FOLLOW SILENTLY ALONG SIDE OF HIGHWAY AND MAINTAINING A DISTANCE OF ABOUT 200'.
IT TRAILED CAR FOR OVER 100 MILES AND THIS TRANSPARENT WHATEVER MOVED CLOSE TO THE GROUND WITHOUT MANUVERING AROUND TREES, ROCKS, ETC.
IT MOVED TROUGH THEM. WE PULLED INTO A TRUCK STOP TO EAT. AS WE LEFT THE HIGHWAY I NOTICED THE VEHICLE COME TO REST IN A COW PASTURE ACROSS THE HIGHWAY. I SHOULD ADD THAT THE CRAFT TRAILED BEHIND TO THE LEFT OF THE CAR UP TO THE TIME WE PULLED OFF HIGHWAY. IT WAS GONE WHEN WE RETURNED TO CAR.
THE NEXT YEAR THAT RESTAURANT HAD SPACEBERGERS AS A SPECIALTY.

Occurred: October 1, 1987; Approx. 2000hrs.
Location: Roseburg, OR
Witness: 2
Shape: Oval
Duration: 30 minutes
Two small glowing oval orange objects and one larger triangle craft sighted in rural Oregon.
It was Friday before hunting season- so it was either September 25 or October 2, 1987- since Oregon's hunting season is always the last weekend of September or first weekend of October (I checked online to see if I could figure out what date is correct using Oregon Dept. of Fish and Wildlife with no success.)
My friend and I were seven miles northwest of Roseburg, Oregon, in a sparsely populated area (no houses for at least a mile away) checking out a hunting location that I was very familiar with and it was just before twilight. We sat down to rest next to a creek-we were in a fairly deep, narrow valley between two hills- when my friend noticed two yellowish lights at a distance about two thousand yards up the valley.

At first, I did not think much about it since I knew of a gravel road at the end of the valley and thought it was probably just headlights of a car, and told him of such. We were about to get up since it was now almost dark-and had about a quarter mile to hike back to the truck-when I again notice the two yellow lights were now closer and heading directly down the canyon towards us.

I still did not yet think they were UFO's or anything out of the ordinary when all of sudden we could hear animal(s) moving in our direction away from the direction of the yellow lights- I couldn't see the animals since it was almost dark now and the area had enough brush and trees to hide them as well.

At the same time we were listening to the animals, I also started to see small (about 1 inch) red glowing "sparkling" spheres appearing at the base of a tree twenty feet away from us, towards the direction of the yellow "lights." These red spheres, suddenly appeared at the base, then started to swirled around the tree, all the time rotating upward, and then just as suddenly disappear once they reach the top of the tree. More and more red spheres appeared around the trees in front of us when I started to asked my friend if he was seeing what I was seeing (I was worried that I was losing my mind) when he shakily replied, "Yes."

I once again looked at the yellow "lights," which were now close enough to see that they were not yellow lights at all, but two glowing, yellowish/orange, oblong spheres or "crafts,"- about six feet long and 3 feet high. The crafts were about ten or fifteen feet above the oak trees and twenty feet apart- still fixed on the same course, now about eighty yards away, and still heading down the valley towards us.

These oblong crafts did not make any sound and did not rotate that I could see whatsoever but maintained a slow, methodical speed and spatial distance between themselves and the trees, all the while moving in our direction. There was no propeller, wing, rudder or prop whatsoever on these "crafts" that I could see or detect. They did not appear to be flimsy like a balloon nor rigid metallic either, but almost smooth, round-cornered plastic looking objects. In addition, they did not glow from within nor from without, but within the skin of the craft itself. Also, underneath the two spheres the ground was brightly lit up. The light that they emitted was strange in that you could not see the source of the light emitting from the spheres itself, like you would from a headlamp, instead the it was "magically" lit up underneath the spheres and the edges of the light did not slowly diffuse but stopped very abrupt… almost like it the light was contained!

This was becoming too much for my senses… (I will admit I was starting to get freaked out, feeling quite alone even with my friend there in the woods and even though I hunted in this area alone all my life) we both stood up and I said let's get the heck out of here. As soon as we moved, the spheres moved in an instance directly overhead and both of us were now bathed in the light. I stood, in awe, frozen, looking at these THINGS, when one of the spheres shot out a red glowing disk (the same color of the smaller red spheres) straight upwards and then the red disk exploded in a brilliant flash of red light.

I felt frozen as if I was knocked unconscious and the next thing I know I am running. My friend is also running beside me and we are both running towards my truck not saying a word when I reach the top of a hill I see another object or craft that is triangular in shape that appears high and quite far in the distance further past where my truck is located.

This triangular craft is hard to describe how big it is because I cannot reference how far away it is, but it is definitely triangular and of decent size, as well as, far away. The craft appears to be more metallic looking, in that it is straight lines and appears to be rotating or spinning and has three sides like a pyramid. It either has three lights or sides or corners of red, silver, and blue that either interchanges or spins giving it an appearance of blinking or spinning motion. It also appears that the bottom side arches or wobbles back and forth, giving it the "blinking" or jittering effect? I cannot describe this craft any further as to the fact that I was moving rather fast. ((NUFORC Note: Witness indicates that date of incident is approximate. PD))

Occurred: June 1, 1988; Approx. 0400hrs.
Location: Trinity Village, CA
Witness: 1
Shape: Unknown
Duration: 2 minutes
Satellite in orbit takes sudden right angle turn and disappears.
I was sleeping under the stars in July, or early August after a long night of amateur astronomy. My eyes opened about 4:00 AM, judging from the constellations overhead. I looked directly above me, and was caught off guard by a very slow moving satellite in orbit exactly overhead. It was pretty faint. I was barely able to detect its movement except that I watched it for over a minute, and I could see that it had slowly moved against the backdrop of stars, making its way southward. Then, without warning, the satellite took a sharp 90-degree turn and accelerated to an incredible amount of speed in a westward direction. Within a fraction of a second it was gone. From what I remember I only saw it for a few moon lengths after it had taken the right turn before it was gone from view. I was in my early teens at the time. I refused to believe I had seen something extra terrestrial since I was a big science buff, and had always observed the scientific consensus refute UFO's as having been extra terrestrial spacecraft. For over 20 years I didn't even think about what I saw until I began hearing the myriad of other UFO recounts that described lights taking sudden right degree turns like the one I saw. It's still hard for me to accept that what I saw was extra terrestrial, but whatever I saw it seemed to violate the laws of physics, at least in terms of propulsion.
((NUFORC Note: Witness indicates that the date of the sighting is approximate. PD))

Occurred: October 18, 1989; 2130hrs.
Location: Shasta Lake, CA
Witness: 2
Shape: Diamond
Duration: 15 sec.
A close encounter sighting with missing time happened right before the '89 earthquake in San Francisco.

A friend and myself were looking for something on the TV. About 17:00. I noticed that the all California World Series was just starting. However being that we didn't like baseball at the time we decided to turn the TV off. As soon as I turned the TV off, everything went dark and we both heard a noise loud enough to shake the whole house. It sounded like a helicopter that was trying to turn blades that were way too big for it. A whoosh, whoosh, whoosh, kind of sound slowly emanated loud enough and scared us so badly that we ran outside. About a hundred yards above us was a "craft" shaped like a big black kite with red "chaser" lights like on a Christmas tree in a cross pattern. This object was about 2 acres big and ascending rapidly when it was about the size of a quarter it took off west very quickly and disappeared. Both of us were scared and we ran back inside and realized it was 21:30 and neither of us knew what had happened to the last 4 1/2 hours. I was late, I was supposed to be home at 21:00 and since I was at his house and had to walk home. I asked him to walk me halfway and he said no. I walked home through the woods and arrived at about 22:05 that's when I learned that there had been an earthquake in San Francisco. For some reason I never discussed it at home or with him at all. Five years later I remembered the incident while having a conversation with my mother. But I still don't know what happened to the 4 1/2 hours.

SECTION 9: 1990-1999

Occurred: July 15, 1990; Approx. 2000hrs.
Location: Chico, CA
Witness: 3
Shape: Disk
Duration: 1-2 minutes
Fully visible flying saucer
My two cousins and myself were out in front of our Grandmothers playing with remote controlled cars. We for some reason looked up to the night sky to see a disc shaped craft approximately 100 to 200 feet in length and maybe 50 feet in height. The craft was very low in the sky maybe fifty feet taller then your average telephone pole and It had several different colored lights circling the craft and several black "windows" around the saucer. The craft was a light grayish blue color that to this day I have never seen on any aircraft. It passed over our grandmother's house making no sound at all. We witnessed it in the front for about one minute moving very, very slowly as if it was studying us. We then ran through the house to the back yard and the saucer was gone. I spent four years in the United States Marine Corps working in and around aircraft from several branches of the military. I have never seen an aircraft that matches the color or metal construction to this aircraft. The saucer appeared to be one smoothly constructed craft something that I have never seen since. I have seen other things in the sky since then and have heard numerous stories from around this area, even a couple of abduction stories, and was wondering what information if any you guys had on this area.
((NUFORC Note: Witness indicates that date is approximate. PD))

Occurred: August 23, 1990; 1845hrs.
Location: Redding, CA
Witness: 1
Shape: Changing
Duration: 3 to 5 minutes
A symmetrical group of orange dots in night sky, no sound at all, no break in flowing formation at all
I was on my balcony when I looked up and saw what seemed to be many little orange dots moving upward in a flowing motion. I could see the shape shifting smoothly like the tail of a jellyfish or a flock of birds. But the distance between the dots never shifted or broke pattern. There was total silence and steady movement. I could see the spaces between the dots clearly and as it got further away the dots got smaller. But the whole time it was totally silent and it was dark out but I could see each of the dots clearly. What caught my attention was the amount of dots and the flowing movement the group of dots made. It was not sharp movements but smooth flowing, rotating movements.

October; 1990; Ochoco National Forest, OR

Coyotes spooked by descending blue white light. (Woodward)

Group /Category 4

Animal Effect Cases

Case Directory

Ochoco National Forest, Oregon

October 1990

Animal Reaction Feature:

A hunting party composed of three men had made camp, and two of the men had gone inside the tent to turn in. The third man remained by the campfire and saw a silent brilliant blue-white light flying to the north on a downward angle and then abruptly turning east. After the turn east, the light lit up the entire canyon so details could be seen as clearly as in daytime. The blue-white light was estimated at no more than 4 miles away when first seen and ½ mile away when it turned east.

The witness watched for a few more minutes and seeing nothing further, he entered the tent. He told one of his friends what had happened, and then lay there thinking about what he had seen. At this point, a pack of coyotes started "yipping and yelping" and making typical coyote noises. The noises were coming from the east (the direction the blue light had been traveling when it disappeared). The coyote vocalizations became so loud that all 3 men got up. The animals, still vocalizing, ran by the camp going west. The witnesses did not know if the coyote behavior was associated with the blue-white light he had observed, but he noted that he had never know coyotes to come into a camp with a fire still smoldering and that in general they avoided human contact. He thought they were acting as if they had been spooked.

Joan Woodward, Animal Reaction Specialist:

This incident took place about 30 miles ESE of Prineville, south of route 380, which parallels the Crooked River. When the bluish-white light was initially seen, it was going north and was traveling downward at an angle of 25-30 degrees. During at least part of this time, the light was between a mountain range to the east and the witness, making it no more than 4 miles away. It was traveling so fast it seemed to be leaving a trail of light behind it, but the witness felt this was an optical illusion caused by its speed. When it got to the highway area (380), it made a hard right turn without slowing and flew east lighting up the canyon (at this point the distance was estimated as ½ mile from the witness). The bluish hue of the light reminded the witness of an arc welder. The time from first sight to arrival at the highway was estimated as 4 seconds, and then the light was in sight for an additional 2-3 seconds as it traveled east.

No sound, no EM effects, and no physiological effects were reported.

[There is a question if the coyote behavior was related to the blue-white light. Wild animals moving away from an unknown have been reported—jw]

April 1991; Space Shuttle Atlantis, mission STS-37

During the mission of the Space Shuttle Atlantis "STS-37", a metallic object arrives from the top and then swings in front of the video camera (it moves to the right and to the left, up and down). In the background we can see the earth.

Occurred: April 15, 1991; Approx. 2300hrs.
Location: Placerville, CA
Witness: 1
Shape: Light
Duration: moments
Spring 1991 Placerville, Ca. Light 5-10 seconds view from patio clear sky Light outruns 2 jets!! Single light swerving back & forth with forward momentum, followed by 2 military jets that sounded like they were near full throttle. "Very Loud" As the jets followed the light across the horizon "2-3 seconds" the light silently went up and away from the jets at an incalculable speed. As though the jets were moving 1 m.p.h. and the light 100 m.p.h. I know these are not near actual speeds that are just my way of describing the situation.
((NUFORC Note: Witness indicates that the date is approximate. PD))

Occurred: June 1, 1991; 1200hrs.
Location: Roseville, CA
Witness: 2
Shape: Cigar
Duration: 10 minutes
The cigar-shaped silver craft was stationary in the sky, and then disappeared.

One cigar-shaped, silver craft, stationary, high above a popular weekend open market (Denio's) held near downtown Roseville. My husband and I were driving eastbound on I-80 and pulled off into a shopping center to better observe the craft. After another minute or so, it just disappeared from the sky. This was noon on a clear day. The object seemed to be very high up and was clearly visible against the bright blue sky. My husband and I were both employed at the time (now retired). He was a heavy equipment operator for the City of Sacramento and I was an analyst for the State of Calif.

Occurred: September 1, 1991; 2200hrs.

Location: Grants Pass, OR

Witness: 1
Shape: Light
Duration: 8 hrs
Light appeared over mountain
I was outside watching meteor shower when a very bright light appeared over the mountain in front of me. It hovered there for a while and then went down behind the mountain. Then it would reappear and go back down. It repeated this for quite awhile. It finally came back over the mountain and just stayed in one place for a long time. It would get very bright and then so dim that you could barely see it. It slowly went across the sky and stopped. There were clouds coming in and it would get covered in the clouds but didn't move other than getting brighter and dimmer it looked exactly like a star. I watched it until the sun started coming up and I couldn't see it anymore.

Occurred: February 1, 1992; 2000hrs.

Location: Grants Pass, OR

Witness: 2
Shape: Light
Duration: 10 min
Going across sky. Thought it was airplane at first...until it started melting.
Had just gotten home and was outside smoking a cigarette when I saw a light going across the sky. I thought it was a plane. It stopped all of a sudden and just stayed there. All of a sudden it turned bright red and appeared to be on fire. It started dripping like plastic would if heated. I ran in and called 911 because I still thought it was an airplane. The operator thought I was a crank call. She put me on hold for about 2 minutes and got back online and asked me to tell her exactly what I was seeing. I explained it to her and she told me that she had about 50 other calls coming in on the same thing. My husband at the time was out videotaping it. She asked me if she could have the fire department come over to look at the tape. The fire chief arrived and watched the tape and shook his head and asked what we thought of UFOs.

Occurred: February 25, 1993; Approx. 2330hrs.

Location: Chico, CA
Witness: 1
Shape: Triangle
Duration: 20 seconds

Three silver/grey triangle crafts, at very high altitude, no sound, no lights. Chico Ca. 1992/93 I was attending Chico State University in 1992/93. I was a junior living on campus in Shasta Hall. I had just arrived at a party one cold night and did not really know many people there. I went outside to the backyard where I stood on a wooden deck. As I was looking up to the very clear, cold, and starlit sky; I saw a faint object out of the corner of my vision. As I looked up and towards my left I saw 3 grey/silver triangle objects exact in size flying in a line, one in front of the other. These objects had no sound, and no lights. They seemed to be in exact distance from each other. They also appeared to be very high in altitude. My sighting lasted about 20 seconds before the crafts moved across the sky and then out of sight, behind a large oak tree. Something very strange happened as I was viewing these objects, everything seemed to go silent as I watched them fly over me. As I remember everything went silent as I viewed them. After they moved out of sight all the party noise (music, laughing, people talking) returned in an instant. I was astonished! No one else had witnessed what I just saw. I felt amazed, confused and a little scared. I had never seen anything like this in my entire life. Later that night as I walked home to my dorm I could not stop thinking about what I saw that night. When I got to my dorm, I went to my friend's room, woke her, and told her what I had witnessed. Since that night I have only told a few people about my experience until now. I will remember that night for the rest of my life. I often still think about what I saw that night, 10 years ago. I have many questions, about this unique moment; I just don't know who has the real answers.
((NUFORC Note: Date is approximate. PD))

Occurred: June 15, 1993; 0130hrs.
Location: Anderson, CA
Witness: 1
Shape: Other
Duration: 5 seconds

All I saw was a blue-green light flash from one part of the sky to the other in a north/south direction. This flash started in the atmosphere and ended in the atmosphere leaving a trail of light that slowly faded over the period of a few seconds. This did light up the sky too, there seemed to be a trail of exhaust like sparks emitted from this "thing".

Occurred: August 12, 1993; 2230hrs.
Location: Ashland, OR
Witness: 3
Shape: Light
Duration: 30-45 min
SUMMARY: 3 luminous objects moving extremely fast, instant directional changes, changed color

My parents and I were observing the Perseid meteor shower in a winter recreation site. We decided to go there because it was very dark and it was basically a large field in which we could park our pickup. We had binoculars and a mid-size telescope with us. As we were watching the meteor shower, we all noticed moving, star like points, which changed color from white to red. There were 3 such objects, and they appeared to be at an extremely high altitude (at least where commercial aircraft would fly). They were mostly directly above us or at least 3/4 up from the horizon. We were not certain of their altitude since there were no reference points. Through binoculars the objects appeared to be executing extremely high-speed maneuvers (looping and passing close to one another) and they could accelerate or turn VERY quickly. The telescope was impossible to use while viewing the objects because of their speed. There are no way these were planes, helicopters, meteors, or flying insects like fireflies (there are no fireflies that change color or live west of the Rockies). I have no clue about what they were. I would like it if someone e-mailed me regarding this, if it's possible

Occurred: August 15, 1993; Approx. 0300hrs.

Location: Cloverdale, CA

Witness: 3
Shape: Flash
Duration: 3-4 minutes

Bright stroboscopic light slowly passes by within 100 yards, and leaves with an otherworldly piercing whistle.

I grew up in a small town in northern California. Cloverdale is situated 90 miles north of San Francisco, at the north end of Sonoma County it is surrounded by rolling hills and vineyards. The geysers (at one time the world's largest geothermal power plant) sits about sixteen miles to the east. The Indians of the area considered this area sacred ground and they made trips there for healing and ceremonial rites.

My father worked at the geysers for about twenty years. During his many night shifts he and others had seen several un explainable lights in the sky. During the early nineties the area known as the geysers, along the borders of Lake, Sonoma, and Napa counties was known as a UFO hotspot. I had many sightings alone and with other people beginning around 1989.

This particular event happened in the summer of 1992, along the road to the geysers at roughly 3 am. I was with two friends C and G who were parked along the side of the road. We were on a slight bluff overlooking Geyser's creek, facing up the river valley towards the east and the geysers. Suddenly C. or G. said, "What's that?" There was a super bright white/blue stroboscopic light, blinking two or three times a second heading just towards us from up stream. The light appeared suddenly from a hill around a bend in the river. It was slowly following the path of the creek, at a very low altitude and slower than a plane. It was lower than the tops of the hillsides on either side of the river valley, maybe 100 feet high… It kept getting closer and closer and all I could see was the light blinking almost hypnotically… The light was so bright that one couldn't make out a shape. It was as if there were no craft, but only a light. Eventually the light reached where we were and passed us just above the treetops and disappeared behind the hill to the west. It was heading in the direction of town… It floated by extremely close, within a hundred yards or so and we all noticed that it didn't make any sound… none… zero… if it had been a plane or a helicopter or whatever we would have heard it the entire time we saw it, it was that close.

This is where it gets weirder. My friends (who themselves are avid ufo observers) got in their car and turned around to try and follow it. I was left alone with my truck. I went to relieve myself behind my truck, feeling quite dismayed by what we had just seen. All of a sudden I heard a sound that seemed to be the birds in the trees chirping in unison! Now this is at three in the morning when birds do not chirp. Suddenly, the chirping became louder and so intense that I could feel the vibrations turning my stomach, not unlike when at a concert, the bass rattles your pant leg. Although this was an otherworldly sensation, the chirping was at such a vibratory wavelength that it made me feel as if my head were going to implode… I almost fell to my knees from its force… slowly it faded, sounding like a cross between a flute and a whistle. That is the closest description I have of the sound….
((NUFORC Note: Witness indicates that the date of the sighting is approximate. PD))

Occurred: September 3, 1993; 1500hrs.
Location: Burnt Ranch, CA
Witness: 2
 Shape: Disk
Duration: 30 min.
Huge disk with hour glass directly above (center) silver type ball, moving erratically, but in some sort of sync with each pattern different colored lights would react accordingly, and above the ball a gigantic looking skyscraper. People kept passing us on the highway and stopping but these people were giving off a feeling to me as though they missed the bus. Will never forget that smell (Corona) from the intense energy
it was at tree top level gliding alongside the highway as my wife and I drove.

Occurred: January 1, 1994; Approx. 2000hrs.
Location: Redding, CA
Witness 1+?
Shape: Triangle
Duration: all night
Object came from mountain to our car.
Base inside mountain close to Shasta dam
((NUFORC Note: Witness indicates that the date of the event is approximate; provides very little information about the incident. PD))

Occurred: June 15, 1994; Approx. 2300hrs.
Location: Drain, OR
Witness 2
Shape: Light
Duration: 3-5 min.
Large circle of light in the sky flashes brightly, then shrinks to star size and zigzags away.

My friend and I were trying out her new telescope. Couldn't get it to focus clearly. The pet cats were climbing all over us, and they usually don't want to cuddle with me, so I thought that was odd. I got up from my chair in the driveway and looked up above the tall, surrounding trees to see if I could find the moon, easier to focus on. I turned around to look behind us, and above the Douglas firs and the house was a bright white light.

Complete circle, briefly thought it was the moon, but it had an electrical glow. I asked my friend "what's that?" She said, "She didn't know- not a lamp."

She tried to focus on it with the telescope, and when she put her eye to the eyepiece the light flashed brightly. My friend jerked her head away and exclaimed that the light had hurt her eye. Then the light shrunk down to star size, zipped diagonally to the left. Then took off up further into the sky and zigzagged to the left and disappeared. ((NUFORC Note: Date is approximate. PD)

Occurred: August 10, 1994; Approx. 0130hrs.

Location: Grizzly Mountain, CA

Witness: 2
Shape: Light
Duration: 2hrs.

An eerie and unexplainable experience on Grizzly Mountain One day my wife and myself decided to go camping above Oakhurst and Bass Lake up in the mountains, she was pregnant with or son at the time. We headed up to Grizzly Mountain where I've previously hunted since a teenager. It was late when we got there and I set up our tent and started a fire. While we sat there enjoying the great outdoors I began to feel very uncomfortable. That feeling you get as if someone's watching you or your not alone. I told my wife how I felt and she shrugged it off saying I'm probably just tired. I agreed and went and sat in the front seat of our car to look for a something I could use as a weapon so I could make myself feel as comfortable as she was feeling. While sitting in the front seat I noticed a light to the left of a treetop. I thought it was just a star but then the light moved to the right of the tree top and I began to rationalize it as the rotation of the earth or something or wind blowing the tree tops, yet I felt no breeze and none of the tree tops seemed to be swaying. I watched the light for a while as it kept moving from one side of the treetop to the other, until I finally heard my wife call my name and ask what I was doing. I told her to come sit in the car next to me and watch what I thought I was seeing. She was quite annoyed by it and felt like I was trying to play a prank or something which she didn't find funny at all because she was pregnant and wasn't in the mood for it. When she sat in the car I asked her to watch the star/ light and tell me what she thought of it. After a few minutes she got real close to me and said she wasn't feeling very comfortable up here in the mountains anymore. She told me she things the light keeps moving from left to right of the treetop and then back again. That settled it for me. I knew I wasn't just seeing things then. Then I felt boldness and got out of the car and started yelling up at the light saying come on down we're friendly and we mean you no harm. I want to see what you look like and stuff like that. Truly I was completely scared but by doing this it kept my nerves steady. Then the wind began to blow real badly and I thought I sensed a movement on the ground it the surrounding trees but I couldn't tell.

The light of the fire was making it too hard to see the surroundings. I put the fire out real quick and waited a bit so my eyes could adjust to the dark so I could try and see if I noticed any movement. A few moments later my wife freaked out saying she felt weird and was scared for her unborn baby and ranted about other things I don't remember now. Then I heard a lot of movement in the trees about 30 yards away and it freaked me out. I jumped in the car with her and told her screw this we're getting out of dodge. I tried to start the car and the freaking car wouldn't start. We were both really stressing out then. We were parked facing up hill where are camp was and my car was a stick shift, so I put it in neutral and let it roll back to pop the clutch. It worked and the car fired up and we took off down the way we came, sky ranch road. On our way up the road was really clear, but as we sped off back down the mountain trees started falling over next to us and branches were falling off of trees and I couldn't make any sense of where I was. I knew those roads like the back of my hand and due to panic or who knows what else, I felt a bit lost. Then my wife said stop the car, we just passed that same spot a while ago and I think we just went in a big circle. I stopped the car and we got out of the car and we started calming each other down trying to reason all this out. Then I became stressed out again because I knew that there was no way we could have went in a circle. There was only one way back down that mountain, sure there were plenty of off roads to go elsewhere, but none of those roads would have took us in a circle based on where we stopped. I couldn't make sense of any of it. For some reason we both looked up about the same time and we both saw a blue glowing light with an oval/cocoon shaped object spinning within it. I was in awe. It was so pretty yet so weird looking. My wife jumped back in the car and said drives damn it. I jumped back in the car and told her we aren't going any further. My instincts told me that when you're panicking its best to just stay put and calm down regardless of the situation. Think things out. I checked by watch and I knew it was almost morning time and we could just wait it out.

Whatever it was it was sure to leave when the sun came up. While waiting we both faded in and out of sleep until I noticed things were getting a lot lighter. Thank God the sun was coming up. When the sun came up we just sat there in the car holding each other. I asked her if we just both had the craziest dream or what? She said there's no way we could have had the same crazy dream and she wanted to leave. My stubbornness wouldn't let me just leave though. I wanted to go back to camp and get my tent and belongings. She said forget them, we can get more, but I wanted my stuff and my curiosity was killing me. I wanted to troll around the camp in the woods and see if I found anything strange. We got back to camp and packed things up and walked around a bit and noticed a large burnt area up away from was we camped. We felt real creepy again and both agreed to leave. We told my Mom about it and my Step Dad when we got back. My Mom was way curious because she knew I never was a storyteller and never made things up. My step Dad was probably thinking just the opposite after hearing what we just them. We never told anyone else though. Thought people would laugh or make fun of us. A few months later my Mom talked her husband into taking them up to the same spot to camp. He knew exactly where it was. He's been hunting up there for years. The next day we went by there home and they told us they went up there to camp, but they didn't stay. They both told us that they felt really uncomfortable at that camp spot and felt like they were being watched. They came off the mountain the same night and came home. I haven't been back to that sight yet and don't plan on ever going there again.
When I found this web site right now I couldn't resist telling our story. It feels real good to be able to talk about it again. Thanks
((NUFORC Note: Date may be approximate, although the witness does not indicate that fact. We will attempt to confirm the date. PD))

Occurred: September 9, 1994; 0630hrs.
Location: Chico, CA
Witness: 1
Shape: Fireball
Duration: 5-7seconds

A large solid spherical, bright emerald green with yellow fiery tail, coming across sky at a slow downward angle descended and no impact was felt or heard. It seemed like it was expected. Husband and I were by window when large fiery emerald green type globe with fiery yellow tail started to descend gradually across sky. It appeared to be close and though it went behind trees, we expected to hear or feel impact or have a fire due to the dryness of our area and the fieriness of the object. Neither hearing nor feeling an impact occurred.

When I phoned the local sheriffs dept. to tell them there may be a fire in the vicinity due to the object colliding in the outskirts, they referred me to call NUFORC.

I then proceeded to call that #. Later I sent report to both NUFORC and Mufon and contacted the local paper. All items mentioned were saved in my files and when I spoke to NUFORC recently it was suggested if I still had these items to resubmit my sighting on this net site and send in articles, which I have done on 9/27/99.

((NUFORC Note: We did speak via telephone with this witness, on the date indicated, and we consider her to be exceptionally reliable. NUFORC received many other reports from California on this date, including a report from an FAA ARTCC supervisor. PD))

Occurred: September 20, 1994; 2000hrs.
Location: Rancho Cordova, CA
Witness: 3
Shape: Fireball
Duration: 1second

While driving East on highway 50, a fireball dropped out of sky.

I was driving, my wife was in the passenger seat, and my friend John was seated in back seat. We were taking John home. We were approaching Sunrise Ave. When this fireball dropped out of the sky on the horizon, my wife and I both saw it, as it was right in view. I think it was partly cloudy. We were both silent for about what seemed like a minute, then I asked if anybody else in the car had seen what we saw. John did not as his vision was cut off from the roof of the car. I was really relieved when my wife asked me what was that. It looked like the sun, it was hard to tell how large it was, it was large enough to get our attention. My wife was a student, and I worked in maintenance at a local store.

Occurred: January 15, 1995; 0230hrs.
Location: McKinleyville, CA
Witness: 1
Shape: 50 bright blue objects
Duration: 45 min.

A man is awakened from sleep by barking dogs and goes outside to investigate. As he looked at the storm clouds overhead, he noticed approximately 50 bright blue objects in the night sky. At first he thought they were stars, but then realized they were moving and occasionally forming into "star like" clusters. Some moved west toward the ocean, and others appeared to rise up higher in the atmosphere, at which point they would be joined by smaller, green objects, that appeared to fly in from out in space. The larger objects appeared to be emitting a pulsing light. They also seemed to be flying into, and between, the tops of the storm clouds in the area. The observer tired of the events after approximately 45 minutes and went back to bed.

National UFO Reporting Center Case Brief
McKinleyville, CA
January 15, 1995

As if the torrential rains over California during early 1995 were not enough, the residents there experienced a number of intriguing UFO sightings, as well. Perhaps the most dramatic was a sighting report we took from a man in McKinleyville, CA, located on the Pacific Coast just south of the Oregon border. He reported that at about 0230 hrs. On Sunday morning, January 15, 1995, barking dogs awakened him in his neighbourhood, and he went outdoors to smoke a cigarette. As he was smoking, he peered up at the storm clouds overhead, and his attention was immediately drawn to a cluster of perhaps twelve "cobalt blue" objects in a loose formation, which appeared to be flitting between the tops of two adjacent storm cells. After several minutes had elapsed, one of the objects appeared to climb to a higher altitude. After having risen (seemingly) to a considerably higher altitude than the other blue objects, it appeared to hover briefly, at which point it was joined by a cluster of considerably smaller, distinctly green objects, which appeared to streak in from all directions at high altitude. The smaller, green objects suddenly "locked" in formation with the blue object, at which point the formation proceeded to drift west out over the Pacific Ocean. The observer reported that he watched this process of assembly take place perhaps a dozen times over the course of approximately 45 minutes, at which point he became tired of the spectacle and went back to bed! (Note--Over the last nine months, the Center has received a number of calls regarding alleged sightings of formations of large numbers of bizarre objects or lights in the sky. Among the most dramatic was an hour-long telephone report received on October 16, 1994, from southern Michigan, during which up to 50 (!) strange, round, brightly lighted objects were visible in the sky. The telephoned report was tape-recorded

Occurred: February 2, 1995; 2300hrs.
Location: Shady Grove, OR
Witness: 2
Shape:
Duration: 15 min
Man and wife witness very bright, moving light over ridge to southwest. Flashing green & red lights. Good rept.

Occurred: February 20, 1995; 1900hrs.
Location: Redding, CA
Witness: 3
Shape: Triangle

Duration: 30 minutes

Craft hovered at tree line above our private road. Then circled in a LARGE oval pattern about 8 times. Came within a rocks throw of our 2nd story deck. WAS SILENT.NO RUNNING LIGHTS.VERY BRIGHT LIGHTS UNDERNEATH.REMINDED ME OF GLASS BRICKS.TOO BRIGHT TO LOOK AT FOR LONG? Called Ridge Apartment Tower. The A.T.C. said it was a type of U.S. Military craft I can't remember. C.A.P. confirmed. Next summer pix of same craft in paper didn't look AT ALL LIKE CRAFT I DESCRIBED TO THEM!! The A.T.C. said we live in a blind radar area, by the way. I feel this is the new stealth, but who knows? COMPLETELY SILENT AND COULD HOVER SILENTLY?

I worked at Stanford University Hospital for 22 years. Ten of those years were in Administration and the rest were as a Ward Clerk on various nursing units. I am now retired. I am 52 years old and am female. My son, now 15 and my daughter, now 17 both saw it. My daughter came running in on a clear, cold February night yelling to me that there was a UFO hovering very low over our private road. She was pretty shook up. By the time I got there, it had gone on. However, I went outside and checked the sky. Lo and behold, I saw this very large craft shaped like a stingray fish without a tail flying very low in an oval pattern. The craft appeared to be in a training mode, because the pattern never varied. The craft slowly flew the same pattern for about 30 minutes. Our house was located just before the craft turned to the left (east), and was flying so close to our 2nd story deck I could have hit it with a rock! There were no running lights. The entire underneath had what looked like very large glass bricks. They were so bright; it was difficult to look at them for any length of time. There was a HUGE bubble where a cockpit would be. It was VERY tall and narrow (that's the only way I can describe it). Color was either grey or green. There were no windows apparent. I think I saw a tail fin, but I might be mistaken. It was dark. My daughter however says she didn't see a tail fin. Unfortunately, I didn't own a video camera or any kind of camera at the time to record this incident.

Occurred: March 31, 1995; 2145hrs.
Location: Placerville, CA
Witness: 1
Shape:
Duration: 5 sec.
Ex-USAF/aerospace employee (ret.) sees very bright, pure white light descend vertically. Extraordinary sighting. Good rept.

Occurred: June 12, 1995; Approx. 2300hrs.
Location: Drain, OR
Witness: 2
Shape: Circle
Duration: 5 min.
Bright light flashes and shrinks before zipping off.

My friend and I were sitting in her driveway in Drain, testing her new telescope. Clear night. My friend's cats started climbing all over us and we thought that was odd, because they usually hide when company is over. We couldn't get the telescope focused on anything very well, so I got up to see if I could find the moon, easier to focus on. I walked down the driveway and turn around and looked above the house. She had tall Douglas firs between her house and the road. Above the trees was a round white light. I asked her when they got a new streetlight and she said they didn't. So I asked her to look and tell me what the light was. It was NOT the moon. She said she had no clue what it was and grabbed the telescope. When she bent down to look at it through the telescope, the 'light' (which was a bit smaller than the moon) flashed very brightly. My friend jerked her head back and held her hands over her eyes, saying the light had hurt them. Then the light shrunk to start size, zipped diagonally down and to the left and then took off way up into the sky, where it continued to zigzag across the sky and then disappeared.((NUFORC Note: Date is approximate. PD))

Occurred: July 28, 1995
Location: Cottonwood, CA
Witness: 2
Shape:
Duration: 90 sec.
Two boys report witnessing 2 bizarre yellowish-white objects fly overhead. Reported to local sheriff's office.

July 28, 1995; The GAO Report on the Air Force Investigation of Roswell
The two Air Force reports on the Roswell incident, published in 1995 and 1997, form the basis for much of the skeptical explanation for the 1947 incident, the purported recovery of aliens and their craft from the vicinity of Roswell, New Mexico. The first report, The GAO Report on Roswell (Actual report in pdf file) "The Roswell Report: Fact versus Fiction in the New Mexico Desert," identified a secret military research program called Project Mogul. As the source of the debris reported in 1947, More background information can be found in two IUR articles: The GAO Report, July/Aug. 1995 IUR Vol. 20, No. 4 & What The GAO Found; IUR 1995, Vol. 20, No. 4 www.nicap/roswell.org

Occurred: June 29, 1995
Location: Klamath Falls, OR
Shape:
Duration:
Man reports, bright object overhead, "brighter than Jupiter." Moves slowly to SE, enters haze at 15' K, disappears from view.

Occurred: June 30, 1995; Approx. 1500hrs.
Location: Medford, OR
Witness: 5+
Shape: Disk
Duration: 3-4 seconds

A metallic disc with a brilliant amber colored light passed directly overhead at incredible speed at low altitude.

It was mid-afternoon on a clear, sunny day in the mid 1990s. My kids were in the front yard talking to my in-laws who had stopped by to visit. I walked to the curb to bring the trashcans back up to the house and noticed a brilliant amber colored light low on the horizon over Mt. Ashland coming directly toward me at an incredible rate of speed. It looked like a huge fireball in the sky and I was scared that it was a meteor about to hit.

It only took a couple of seconds to reach us and I was shocked when I looked up to see a round metallic object streaking by directly overhead at a low altitude. (It was about the size of a tennis ball held at arms length from my perspective) There was no contrail or fiery tail and it made absolutely no sound at all. The light source must have come from the top of the craft because there was no light coming from underneath. I watched as it streaked by and disappeared from view within a second or two. It went from one horizon to the other in the span of about 3 to 4 seconds. I asked my in-laws if they had seen it and they said they hadn't been looking at the sky and hadn't noticed anything.

I had never believed in UFOs and was a bit shaken by the encounter. I listened to the news that night hoping for a logical explanation to what I had seen, but there was no mention of anything at all. Out of fear of sounding like a kook, I've kept the incident pretty much to myself all these years, but I've never been able to get it out of my mind.

((NUFORC Note: Witness indicates that date of incident is approximate. PD))

Occurred: July 20, 1995; 2200hrs.
Location: Redding, CA
Witness: 16+
Shape: 4-cigar shape
Duration: 20 minutes
16 people observe 4 bizarre cigar-shaped, metallic-appearing objects fly slowly S to N over home. No wings, tail, or engines!

Occurred: November 8, 1995; Approx. 1900hrs.
Location: Pilot Rock, OR
Witness: 2
Shape: Light
Duration: 1 hour
Two brothers were returning to Pendleton, OR, and Dixie, WA, after hunting in Oregon. Upon approaching Pilot Rock, OR, the brother driving the lead vehicle, a Suzuki (Samurai?) noticed a peculiar cluster of blue-white colored round lights, positioned in a square or rectangular pattern, ahead. He called his brother on CB radio (Ch. 4) to ask is he saw it, and the brother answered that he certainly did. The lights disappeared from sight, then reappeared on the other side of the road. They passed through Pilot Rock, and the object became visible again as they were leaving that town. The object had the same appearance as it had had before they arrived at Pilot Rock, and it stayed with them (according to their conscious recollection, for another 30 minutes.
The brothers made several interesting observations at the time of the incident. Most interesting was the fact that even though they were talking over a CB channel that traditionally is used very heavily by loggers, and other users in that area of Oregon (Ch. 4), they heard no other users during the entire experience! Also, they saw no other traffic on their side of the highway, even though they remember seeing oncoming traffic in the other lane.
When they got home, they topped off the fuel tanks in their trucks. Craig's vehicle indicated that he had driven 98.7 miles from deer camp, where both he and his brother had filled their vehicles' fuel tanks from fuel cans. When they got to Pendleton, his vehicle, which was getting about 25 miles per gallon, took less than 2 gallons of fuel!
Craig had spent 9 years in the military, during which time, he said, he had seen many different types of weapons used. This aerial event, or object, was like nothing he had ever seen before, he said.

Occurred: December 1, 1995; 2100hrs.
Location: Mount Shasta, CA
Witness: 1+
Shape:
Duration:
Commercial pilot reports witnessing multiple peculiar flashes of light from N of Mt. Shasta to Eugene, OR. Bursts moved northwest from aircraft.

Occurred: January 22, 1996; 2100hrs.
Location: Redding, CA
Shape:
Duration:
Telephoned Report:
He called to report having seen a huge, triangular ship go by. It was moving slowly, he estimated only 20-30 mph.
The object had rows of very bright lights (on the side?), and it had "rainbow light" in the center.

The caller got a very good look at the back end of the craft. It seemed to have a very small "afterburner" on the trailing edge, which was round.

Occurred: January 25, 1996; 2115hrs.
Location: Redding, CA
Witness: 1
Shape:
Duration: 12 min.
Young boy called to report that he had been witness to the object that had been reported a week earlier over Redding.
The caller saw a circular ship, principally the bottom part of the object, and it had no "belly lights." It was not an aircraft. He was emphatic on this point.
It was apparently moving very slowly, since he reported that it approached his location for up to 5 minutes, or so, he thought.
He saw the craft again on Tuesday, 29JA96.

Occurred: February 8, 1996; 0015hrs.
Location: Redding, CA
Witness: 1
Shape:
Duration: 15 min.
Man reported witnessing "upside-down snow cone" pass overhead slowly. He heard a "popping sound" emanate from it, and he saw flames shooting down from it. It wobbled somewhat in flight, and it appeared to be descending.
The object was a yellow color, and it appeared to have 3 lights on the bottom. It may have crashed, he thought.

Occurred: February 10, 1996; 1700hrs.
Location: Mount Shasta, CA
Witness: 1
Shape: Fireball
Duration: 10 seconds
Driving on the road, which ascends Mt. Shasta, first mile in. Object seen above and to front of car.
A green 'fireball' seen at about tree or power-line level moves south to north above road in front of vehicle. Object appeared and disappeared in approximate 10-second time frame, moving swiftly. Object left slight trail.

Occurred: April 5, 1996; 1200hrs.
Location: Redding, CA

Witness: 1
Shape: Disk
Duration: 1 min.
Summary: I saw a disc hover for a 1minute and then it take off at a high speed. It was a saucer shape object with two lights on the sides and a clear canopy.

Occurred: May 25, 1996; 2300hrs.

Location: Chico, CA

Witness: 1
Shape: Triangle
Duration: 6 Seconds
Silent, dark triangle shaped UFO sighting
I was standing on my apartments balcony, drinking some cold tea, as it was a pretty hot night that night. My apartment was situated very near the central part of the Chico, a well lit, bustling with nightlife, college town, and suburban atmosphere city in Northern California. Suddenly from behind my apartment or coming from the opposite direction from where I was looking out, a massive, triangular shaped object flew over my apartment, moving towards the central part of the city. What scared me almost as much as it's unnatural appearance was the fact that it didn't make even the slightest amount of noise. There were no lights on it that I could see in the short amount of time it was visible to me, and I believe it was black in color, although I could be wrong, as it was quite a dark night. The funny part of this story would be that I'm pretty much a non-believer in everything, but my boyfriend, who was in the living room when this occurred, believes in UFOs, aliens, and all of that kind of supernatural stuff. It was with that, that I entered the living room, told him what I had seen, only to find that he didn't believe me! Am I the only one whom has seen such a craft?

Occurred: June 1, 1996; Approx. 2000hrs.

Location: Eugene, OR

Witness: 1
Shape: Triangle
Duration: 2 minutes
Diamond broach shape moves over Eugene, OR.
I was returning from taking my kids to ballet class. It must have been winter, because it seems to me it was dark early. As I pulled into my driveway in the north of town, I saw a huge "diamond broach" shape in the sky, south-southwest. It was humungous. If I had put my hand up, my palm would have just covered it. It was a triangle, point up, with lights in rows inside the triangle, like diamonds. I don't recall how many. It could have been 6 (1 at the point, 2 in the next row, then 3 on the bottom row), or there was another row, making 10. It wasn't moving and didn't make a sound. I could see lights from helicopters and small planes heading towards it from every direction. I jumped out of the car to see it better, but my kids freaked. They ran to the house, begging me to open the door and come in. They didn't want it to come get us. My kids were very young at the time, the oldest about 7, and the youngest at the time 4. I don't know why they thought this thing would take us away, and I wasn't showing fear, but rather fascination, so I'm not sure whey they reacted with such terror. For them I went inside.

Later that night, or possibly the next night, I chatted with someone on Prodigy (remember that!) who claimed to work out at the airport in Eugene (the airport was north of the object, behind me while I was watching the triangle). She said there were some sort of sighting, a report made even, and no official announcement of what it was, but supposedly it was noticed by the airport, or perhaps reported by them.

Again, I can't be sure of the date, but it must have been fall or winter of 1996, or perhaps late winter of 1997, although I don't remember it being cold, just dark early. It could even have been the year before.

((NUFORC Note: Witness indicates that the date of the incident is approximate. PD))

Occurred: June 1, 1996; Approx. 2000hrs.
Location: Klamath Falls to Medford, OR
Witness: 4
Shape: Light
Duration: 1 hour
While leaving Klamath Falls Oregon by car I noticed a star to my right. Or which I thought was a star, maybe Venus. Anyways i was talking to my friends 2 adults and a their daughter. We were going to K mart in Medford to get their daughters Brownie outfits. We were about 4 miles from downtown Medford and I noticed the light was extremely close and much larger. Like a headlight (very bright) and it was moving as fast as we were it was maybe a mile away from us. Parallel to the highway. When we reached the part of the highway parallel to the Airport it stopped above the Airport maybe 2 miles up and extremely bright. I was changing the radio stations to see if anyone had reported it. When we pulled into the Parking lot to K mart we watched it hover for 45 min. No noise. We didn't notice but across the street from Kmart many transformers had blown before we got to Kmart. The power was on at Kmart so we didn't notice till we saw many Police looking around. I crossed the street and approached an officer and said I know what caused this power outage. I pointed at the bright light hovering and he looked up, but at his angle he couldn't see it because of a rooftop, he told me to get lost. The light slowly moved across town and fog set in. we watched it for 20 more min then we decided to leave. I heard of no reports on this incident, I cant remember the date or month but I'm sure it would be easy to find out the day the transformers blew on the summer evening in Medford 1996. Thank you
((NUFORC Note: Approximate date. PD))

ANG Pilot: TWA Jet Hit by Object

DAVID A. FULGHUM/WASHINGTON

Witnesses believe an explosion ripped apart Flight 800 over Long Island, but deny involvement in a cover-up

Two New York Air National Guard pilots, with the best view of the crash of TWA Flight 800 last July, are disagreeing about what they saw immediately before destruction of the Boeing 747-131 jetliner. One believes the airliner was struck by a fast-moving object coming from the east, while the other saw only a fiery trail from the west.

However, both believe a violent explosion ripped the aircraft apart, propelling some of its passengers high enough that they did not hit the water's surface until 3-4 min. after the initial explosion.

Maj. Frederick C. Meyer, pilot of an HH-60 helicopter from the ANG's 106th Rescue Wing, has just been freed from an FBI gag order preventing him from giving interviews about the 1996 disaster off Long Island, N.Y. The copilot, Capt. Christian Baur, remains under FBI restrictions not to speak about the accident. But two officials familiar with his testimony told *Aviation Week & Space Technology* in detail what he told investigators.

In the days immediately after the accident, before being ordered not to speak, Meyer discussed his initial impressions with news media (*AW&ST* July 29, 1996, p. 32). Last week, he chose Aviation Week as the first news organization to hear a detailed account of his recollections and his testimony to federal investigators.

Meyer and Baur were in one of the wing's two aircraft operating north of the crash site. The helicopter was operating over Long Island about 12 mi. north of the TWA crash site. Baur, the copilot, was at the controls practicing instrument approaches. The crew was awaiting darkness so they could begin training with night vision goggles.

The key point on which the two pilots disagree is whether a streak of light appeared from the opposite direction of the flight of TWA 800 (which was flying from west to east after takeoff from Kennedy Airport), a possible indication of an intercepting missile or some other object.

MEYER'S ATTENTION WAS first called to the area of the sky where the accident occurred "by a streak of light moving from my right (west) to my left (east)," the same direction as the TWA flight, he said.

Baur's account differs on this point. According to the two officials who have heard both pilots' accounts, Baur, on the left side of the cockpit, saw a streak moving from left to right toward the approaching TWA aircraft before the initial explosion.

"Almost due south [of the helicopter], there was a hard white light, like burning pyrotechnics, in level flight," Baur told investigators from the National Transportation Safety Board, FBI and a Federal anti-terrorist task force. "I was trying to figure out what it was. It was the wrong color for flares. It struck an object coming from the right and made it explode."

Baur's first impression was that there had been a midair collision, possibly between two light aircraft that tow banners along the beach.

"They had witnessed these aircraft come very close to each other at that time of day, and that's what they assumed," the second official said.

NTSB investigators have suggested unofficially that the streaks the pilots saw could have been light reflections from the skin of the aircraft, tongues of flame from the airliner or the forward door of the aircraft popping open, a possibility that still intrigues investigators, the second official said.

MEYER COULD NOT actually see the aircraft, but only the streak, and he admits that Baur, a younger man, has better eyesight. Moreover, Meyer adds, "Whatever Chris saw on the left side I didn't see because he blocked my view." Baur disputes this, saying that the explosions and crash were virtually dead ahead of the aircraft.

The helicopter was executing a missed approach and was about halfway down Runway 24 at the Francis S. Gabreski International Airport at Westhampton Beach, N.Y. It had started a climbing left turn to the south when the accident occurred. The Sun had not yet set and the sky was still bright.

According to Meyer, the streak was about 15-20 deg. above his line of sight and perhaps 15 deg. left of the aircraft's centerline.

"I don't know if it was a missile that struck the airliner," Meyer said. "Nothing at that moment said 'missile' to me. I spent a number of years in Vietnam and had seen missiles fired, some of them at me. But, that was 25-year-old missile technology, which

Occurred: August 15, 1996; Approx. 1200hrs.
Location: Gold Beach, OR
Witness: 2+
Shape: Other
Duration: 2 - 5 minutes

My family and me were on our way to the county fair and we saw the UFO
The object was metallic silver and the sun shined brightly off of its surface, it was shaped like a rounded rectangle more than a sphere. As long as it was in site it was motionless. It was quite a ways up in the sky, and the sky was clear. Me and my family were going to town to the fair to help some friends and we came around the corner and I looked up and I saw the silver spheroid and pointed it out to my family, at first I thought it was a blimp or balloon but it didn't move so I asked if anyone else could tell what it was, but no one knew what it was. We went behind a small hill and when we emerged it was gone and the sky was empty, we hadn't been behind the hill for more than a few seconds so I looked around the sky to see if it had moved but I didn't see anything in the sky. The color of the object was like shiny chrome but the edge of the object was reflecting the sun like chrome does but only the edge, as if it were a plate stood on edge with the center painted silver and then out side a glaring white.

Occurred: September 6, 1996; Approx. 2000hrs.
Location: Lake Shasta, CA
Witness:
Shape: Light
Duration: 1-2 minutes
Summary: Object speeds through night sky.

I'm just a 45-year-old stargazer, observing the heavens. We were on a houseboat on lake Shasta viewing the north westerly star that flickers a lot and is one of the brighter stars when shooting up off the horizon was a star like object moving at an incredible speed. We both had binoculars and stated to each other wow what is that as we followed it. Once it became almost straight up from N/W heading S/E another object just as fast came from the east heading due west. My friend followed it and I removed my binoculars to see if the object heading S/E could be viewed with the naked eye, which it could not, so I captured it again with the binoculars until it went out of sight. According to the docudrama asteroids there are quite a few asteroids orbiting the earth so I wasn't too sure if I should even list this one. However I've never seen anything like this before moving at the speed that it was so here it is. Thx for existing.

Occurred: September 21, 1996; Approx. 2130hrs.
Location: Maxwell, CA
Witness: 1
Shape: Other
Duration: 10 minutes

Large, motionless, thimble-shaped object radiating bright light from big windows, 1000 ft. over I-5.

My UFO sighting began when I noticed an extremely bright light in the night sky over I-5 in northern California on or about 9/21/96, between the towns of Willows & Maxwell. I was driving to work about 9:30 p.m. and saw what I took to be a CHP helicopter illuminating the freeway with a searchlight about 5 miles ahead. (This area is flat and the freeway lies in an almost straight line for miles.) As I drove closer, I realized the light was not circling the freeway but remained absolutely motionless, and that instead of a searchlight, what I was seeing was radiated light coming from inside some type of large craft or object. It was hovering about 1000 ft. above the freeway and as I drove almost directly under it, I could see it had a "thimble" shape (larger end up). I could view it clearly because of all the light that was coming from the large windows that surrounded it. I thought about stopping for a better look but didn't have a camera (naturally), and also thought to myself, "Sure, get a real good look, like Travis Walton did," so I kept going. I tried to watch it with my head out the window as I passed under it, but almost lost control of the car so I focused on driving and lost sight of it--nothing in the rear view mirror to indicate whether it was still there after I'd passed. There was normal traffic on the freeway that night and I regretted not having a CB radio, because the truckers sure must have been hollering about it...

Occurred: December 18, 1996; Approx. 0430hrs.

Location: Eugene, OR

Witness: 2
Shape: Light
Duration: 1 hour
SUMMARY: Father and son saw a light at high altitude perform zigzags, abrupt starts and stops, and 90 degree turns.
I am 18. I was going to bed at 0430 when I looked out the window at the sky. I saw what I thought to be a satellite or the space shuttle slowly and steadily track across the sky from the southeast towards the northwest. I watched this object for a minute and than it stopped. I thought it was strange, but I thought it might be a satellite that had moved itself into a geosynchronous orbit or possibly the space shuttle that had done the same. Than it started to move strangely. It started to do little zigzag patterns and move very slowly and rock back and forth. It also did a couple of 90-degree turns. I watched it for about 10 minutes, than I went and woke up my dad. He came and we watched it zigzag and turn and start and stop and do spirals. The spirals seemed strange, to me, because it would start out with a big, slow circle and than the circle would begin to tighten and it would speed up until it had come to rest where it started. I think I can best describe the spirals as looking like a planet falling into the sun. All this time the sum of its movements were bringing it towards the west (towards us). My window faces east. At one point I thought I saw a small light separate from it and head Southwest. It was so faint that I wouldn't have thought anything of it except that my dad asked me if I saw it. My dad went back to bed about 0515 and by 0530 it was out of sight over my roof. I thought about going outside to watch it some more, but I thought it was too cold and I had watched it for an hour anyway, though I hope I didn't miss any big exit.;)

Occurred: January 15, 1997; 1830hrs.

Location: Alturas, CA

Witness: 1
Shape: Disk
Duration: 3 minutes SUMMARY: pink/reddish glow with a gray round craft there was only one. It swayed back and forth and disappeared.
 It was Jan. 15, 1997 approx. 5:30pm Pacific Time with a cloudy sky. I was lying on my trampoline in my back yard when I saw a light, which at first I thought was the sunset, but it was the wrong direction for the sunset. Then I saw a craft in the middle of the glow. It just swayed back and forth. Then it was just gone.

Occurred: April 8, 1997; Approx. 1945hrs.

Location: Sonoma, CA

Shape: Chevron
Duration: 4 seconds
Extremely massive matte grey/black boomerang with no center section; jets scrambled; no lights, possible aura appears while I was taking a Jacuzzi.
I was in my Jacuzzi looking at the sky. Solar flares had been in the news starting April 7 and it was soon thereafter.
Exactly overhead I noticed an aberration moving across the sky. A solar flare I thought. But after watching it I noticed the well-defined shape and that it had a couple of distinct characteristics. It was shaped like a long boomerang, but wasn't quite as angled, and it spanned a substantial section of the sky. It looked matte black or dark gray. The entire inside edge was very sharp and the center section (perhaps 5-10% of the entire thing) seemed very hazy whereas the rest of the object was well defined.
It moved from south to north to the west of Sonoma as if following the I-5 Corridor ((sic. See below. PD)) It continued north moving very fast, it was about 4 seconds until I couldn't see it anymore as it began to merge with the just-after-dusk sky. It was very high up.
I ran into the house to get my sister and she came out and we both looked. Within seconds we saw two planes together coming from the north. When they got to about halfway between where I first and last saw the object, one of the planes split off sharply and headed west. The other plane continued south. The planes appeared to be at the same height as the object and like most planes that high up, they just appeared as spots of light. In contrast the object had spanned an entire section of the sky that is why I thought initially it was a visual aberration from solar flares. (That could have something to do with an aura, but I didn't specifically see one.)
I went inside and figured the size by triangulation. I think it came out to somewhere between a mile or 2 wide. Somehow the size, height and speed had combined to make it feel very heavy and deliberate, like it was doing surveillance. It made me feel as if I had seen it, but it had also seen me. As if I'd inadvertently and unintentionally had a conscious exchange, like when I catch someone's eye.

I didn't Jacuzzi at night for months afterward so I wouldn't be available for another encounter. Too chilling.

((NUFORC Note: Witness indicates that the date of the incident is approximate. PD))

Occurred: July 4, 1997; Approx. 2200hrs.
Location: Redding, CA
Witness: 1
Shape: Lights
Duration: 3-5 minutes
Summary: Hi...this is ((name deleted))...916-((number deleted)). I spoke with you regarding the lights that dropped an object over Redding on the Fourth of July. Here are some computer drawings ... not too good I am afraid, since I am not a graphics artist, however they should give you some idea of what we saw. I did include some maps of the area we were in and some maps with locations in "x". Thanks, ((name deleted))

Occurred: July 4, 1997; Approx. 2200hrs.
Location: Redding, CA (northeast; out over Millville, approximately)
Witness: 3
Shape: Light
Duration: approximately 5 minutes
Summary: Two amber lights moving at the same distance apart beginning at one end of Redding and ending in the north eastern direction above Redding. The lights dropped another WAY brighter light that was white to the ground. Then the most awesome part was that they just disappeared. My husband, self and six year old daughter witnessed this.
While driving East on Lake Blvd. in Redding, my husband pointed out a set of two lights that were not normal in formation nor in appearance for an airplane. They were, from where we stood, approximately a quarter of an inch big each, amber in color, moving slowly north by north east? When we pulled over to the side of the road after the first minute or so we realized they were not airplanes nor were they any other conventional type of air vehicle. These were two lights going next to each other, approximately two inches apart, in unison as if hooked at the middle. We could not differentiate if they were indeed hooked or not, but if hooked together by something this object would have to have been huge. While sitting there freaking out at these lights, seeing them not blinking, or anything else, moving slowly, they headed north. After a period of approximately 3-5 minutes, the further away of the two lights moved upward and dropped a giant white light from its bottom and this light was five times the size of each of the orange lights and dropped to the ground completely. It was a huge white light that was so bright that it hurt our eyes. The two other amber lights, still virtually in unison went out at that point and one popped sort of red for a second as they went out then disappeared completely. We are awestricken! It seems that even though we stood near our car atop a big city hill gaping and screaming it's a ufo that nobody else stopped, however, someone had to have seen these...they were too obvious.

Occurred: July 5, 1997; 0130hrs.

Location: Florence, OR

Witness: 2+
Shape: Triangle
Duration: Approx. 3 hours
A very large triangular craft with 3 lights on each point appeared. The blue green translucent type color filled areas between each light. Low hums were audible as well.
Until that July, none of us had ever seen a Flying Object that was Unidentifiable. We have been camping together for many years now, so with all the late night fireside chats the need to Identify has occurred. It was around 1:30 am when I first noticed a odd lingering greenish light moving SLOWLY over the treed hills. At first I thought it was smoke from fireworks, BUT when in full view the shape was too triangularly defined. I asked for an opinion and received confirmation that it was MORE! The last log was burning low enough and approaching closer a STARTLING sight it was. Then my husband had to shine a bright flashlight at it and within 5 minutes. It made a very clear change in direction OUR WAY. It reached the lake that we camp on and I began to PANIC, It was HUGE!!! (3\4 or more the size of a football field) Uncontrollably shaking with tears in my eyes I went for witnesses. When we got back to the lake it was nearly over us. It had three lights; one on each corner the front light was the brightest. The area between each point was a translucent greenish\blue color. It made a strange two-tone subsonic type low humming sound. Heading North West it moved in a rhythmic pattern with a soft up\down appearance at the front of it. I was SOOO scared, all in total AWE we had to sit. At about 3:00 am it was over the sand dunes, but thankfully past us. Although still in our sights everyone was very tired and wanted to retire for the night. I could not take my eyes off of it. I watched it from our motor home window. At around a 3:45/4:00am it disappeared with a sheet lightning or Giant camera FLASH. On our way home we stopped by a favorite place to dine in a nearby town Mapleton OR, we noticed that the paper mentioned it. When we returned from our vacation. We reported our "sighting" to the Unsolved Mysteries & Sightings Websites. This was in 1997 and I can't seem to get it off of my mind. If anyone else has seen or heard similar reports, PLEASE POST IT!!! Sometimes I feel so alone in the effect it had on me. In ways I wish that I could pass it off as a Military thing as I have other peculiar flying crafts, but I DONT BELIVE THAT! As far as our personal backgrounds... My husband I own two businesses, I manage both of them, my spouse is a ((deleted)) Training Tech. for the State of Oregon, others hold professional positions. Upon request, one of us can provide image docs.

Occurred: July 14, 1997; Approx. morning

Location: Medford, OR

Witness: Several
Shape: Sharp-turning object
Duration: ?
Summary: Sighting was reported in July 15, 1997 edition of The Mail Tribune, the Medford, OR daily newspaper. Object flew in sky and made sharp angled turns. Reported by several different groups including Medford police department and Medford airport, which was closed at the time of early A.M.

Occurred: July 22, 1997; Approx. 2215hrs.
Location: Hart Mountain, OR (near Lakeview)
Witness: 18
Shape: Triangle
Duration: 3-5 minutes
Summary: Isosceles triangle, reflecting light, moved at 45-degree angle from ground level to Milky Way over 3-5 minutes at about 22:15 on July 22, 1997.

One isosceles triangle, reflecting light, moved from behind Hart Mountain in western sky at approximately 45-degree angle steadily and rapidly. We watched until it disappeared into the Milky Way (approximately for 3 to 5 minutes). An anthropology class of 17 Linfield College adult degree program students and professor observed this event. Students are from all levels of career and live in Oregon, Washington and Hawaii. None have ever observed a sighting as this. Except for stars and campfire there was no other source of light nearby.

Occurred: September 9, 1997; Approx. 2130hrs.
Location: Roseburg, OR
Witness: 2
Shape: Light
Duration: 2 hours

Obscure Light bouncing around a valley. Light changes intensity, brightness, and color. Disappears, reappears different location.

On 9/9/97 at about 9:30 p.m., a friend and myself were watching the most severe thunderstorm to hit our area that summer. Shortly into our weather session, my friend says, "Hey, do you see that?" I looked where he was pointing. The object was zipping into this valley about 5 miles away to our window. We were located about 10 miles northwest of Roseburg OR., USA. The object would be real dim, then brighten intensely, brighter than Venus, but pulsing also, sometimes every color of the rainbow, it seemed. At one point, the light went down out of sight, then a pause, then it came shooting up out of the valley like a squiggly glowing red sperm, then disappear. Then it would reappear from somewhere else, moving a different direction. Keep in mind; a pretty violent thunderstorm is blasting lightning down, rather close to us at times. This object was below the cloud deck. At one point it appeared to come very close to us, I swear, I could make out some sort of lighted windowpanes, with a darker one in between. I assumed the object was close at that point, possibly within a few hundred yards. In any case it must have been rather small, perhaps less than 15 meters in size. Well, this went on for about 1.5 hours and we simply weren't watching the lightning anymore, we didn't even care that it was getting within 10 miles of us. Ok your thinking, ball lightning right. No. I would have settled for that, but then something remarkable happened. My friend was looking around for the object to reappear somewhere else, like it was doing, then I saw it, a perfect white cone of light beaming across the bottom of the thunderclouds. There was very little thunder, and no rain at our location, and we could hear no aircraft of any kind. This "spotlight" came from the object, which had reappeared again. I yelled to my friend, but it disappeared just then. It happened again, and we both saw it. This time, the object was near us and it beamed that perfect white light right above our heads about 100 feet and "scanned" the hillside right behind us. I have been an amateur astronomer and weather observer for well over 15 years, and I have never witnessed an event like this, in fact I have to say, it has been the most troubling event in my life. I also want to point out, at one time it appeared that the one object went into the valley, then came out with another one attached by, "energy" Something strange out there, and I think these strong storms have an adverse affect on them.

Occurred: October 20, 1997; Approx. 1930hrs.
Location: Klamath Falls, OR
Witness: 2
Shape: Formation of 5 lights
Duration: approx 8 minutes
Five lights appeared in the sky all the same distance apart, in a straight line.
My husband and I were traveling about 40 miles south of Klamath Falls, Oregon, on the evening around the 20th of October, 1997. The night was clear, and in the sky in front of us a light appeared, followed by four more lights, all the same distance apart, all in a row. They stayed stationary for about 3 min. then they all went out, at the same time. My husband and I were trying to figure out what we had seen, when the lights appeared again on the far left of where they had been before. They appeared in the same manner, they came on one after the other. Same distance apart and in a line. Off again, and almost instantly, they were to the far right of us, in the same line, in distance apart. The lights were a golden color, not like a regular color of light. While they were in the sky they were stationary. I would have told someone about this earlier, but it was just to strange. And I didn't think it was important, until I saw a show on TV last night about Aliens. I thought even though it happened over a year ago, I felt like telling someone. Thank you for your time. ((Name deleted)).

Occurred: October 22, 1997; Approx. 1500hrs.
Location: Milford, CA
Witness: 1
Shape: Disk
Duration: 5 seconds
Silver saucer shaped craft seen clearly over Sierra Nevada Mountains for a short time.
I am former USAF Metals Processing Specialist. I have worked on many types of Aircraft, military and civilian. So I know my aircraft by sight and sound. I have never seen anything like this before and I got a very good look at it.
It was shiny silver, like polished aluminum, completely round along the horizontal plane and yet about 1/3 that size in the vertical. To me it seemed about 50 feet in diameter. What really shocked me was its flight characteristics. I flew maneuvers' I have never seen any plane accomplish and it was completely silent. It I was not looking directly at it... I would have never noticed it.
It came over the mountaintops over the edge of the ridge, so it could be seen clearly from the valley where I was standing. It then abruptly turned a 180 and headed directly back over the ridge. But what was really amazing was during the 180 turn it also completely flipped over on it axis! So it was basically heading back over the ridge on its bottom...or upside down, if there is a top and bottom.
Anyway, this short sighting was very profound to me. Since I know planes and this was nothing I have ever seen and the propellants produced no noise, whatever it used. That too was shocking.
So there it is...I know what I saw and it was very clear, though very brief.

What really shocked me later; because I live in a small community we tend to know all our neighbors. Where I live they are mainly cattle ranchers. They run them on open range up there on the mountaintops. We had two cattle mutilations reported up there over the years... I really was blown away when I learned this...because it seems just like other reports. I found that to be very strange.

Let me impress upon you.... I am a very honest person... I strive to be all my life...this was no hoax. You can easily verify my story about the mutilations...the local law went up there and documented the cases. It was later in our local papers.
((NUFORC Note: Witness indicates that the date of the incident is approximate. PD))

Occurred: January 8, 1998; Approx. 2000hrs.
Location: Chico, CA
Witness: 1
Shape: Oval
Duration:
Reported by my girlfriend's son, seen over North Chico. He described a very large craft with several lights, not blinking.
Not much more to report--see summary. Ben, the son, is only 10 years old, but he's a gifted artist and drew us a sketch of what he had seen. I can send his sketch on request.

Occurred: March 9, 1998; Approx. 1940hrs.
Location: Butte County, CA
Witness: 2
Shape: Light
Duration: 20 seconds
A single star-like object moved in straight line, west to east, traversing overhead from horizon beginning to horizon ending within 18-20 seconds. Object stopped suddenly on 4-5 occasions for 1-2 seconds intervals and resumed course. Est. speed 25-30,000 mph. Similar witnessing by 4 Nat'l Guard members of 184th Air Assault while on patrol within 80 miles of this location approximately 8-9 weeks early. Saw 4 similar lights that departed in opposing directions after travelling for some distance in a high-speed straight-line formation. It took radical high-speed 90-degree turns.

I was setting up my telescope (huge 12.5 Meade light bucket) with 20 year old son-in-law. Night was especially clear and we were using my wide-angle lens for moon viewing. While I was focused on moon, SIL said, "Hey what's that!" I turned, followed his pointing and coming overhead was a fast moving object. It appeared slightly slower than a falling comet, but faster than any jet. It was a single dot of white light that went in a straight line, but stopped, started and repeated this several times while I wheeled my telescope around to catch a glimpse of it. My SIL kept it in constant view by his naked eye and I aligned the Dobson mount and pointed the scope skyward at the object. It was nearly at the end of the horizon when I believe I caught a glimpse of it. It was only a fraction of a second, but I believe it was longer than it was wide and bright metallic. No other detail could be determined. While explaining this to a friend the next day I was advised he was 1 of 4 on patrol at approximately 0400 hrs. in foothills about 90 miles away. All 4 guardsmen observed 4 star-like objects flying high overhead until they shot away in different directions. They were stunned. In both cases we believe the object/s was extraordinarily high and there was no noise. Sorry no other detail, just the pinpoint appearance of a distant stars moving fast! Very fast vehicle I ever saw. FYI: Beale AFB close by both sightings. Also I am former pilot. This was not a comet, strongly doubt it was a satellite, NOT a balloon or aircraft. Served aboard navy missile tracker. I know what rockets and satellites look like. Have no idea what this was. I am not at all interested in having my name linked to some UFO sighting... no thanks, but I will at least share this story for what its worth. It's a 100% true, unembellished fact. In closing I would not have believed the Guard story if I had not seen this first, even though the witness it's extremely credible. He seemed as astonished as I was. He was in Nevada County at the time, just above Yuba County! Where Beale AFB is located. Beale is a U-2 base and former SR-71 base.

Occurred: March 13, 1998; Approx. 1730hrs.
Location: Florence, OR
Witness: 1
Shape: Cigar
Duration: 2 minutes
I observed a cigar shaped craft sitting in the southwest sky about 40 degrees up. It didn't appear to be moving. I had the feeling it was a tourist craft. I went home to look at it closer with binoculars and seek other possible witnesses, but by the time I got home I forgot to look again. ((NUFORC Note: Date is approximate. PD))

Occurred: April 4, 1998; Approx. 2300hrs.
Location: Northern California, CA
Witness: 2
Shape: Light
Duration: 4 Minutes
The light made no sound then split into two lights.

I just found this site. If I'd known about it then I'd know better the date and maybe the location. My husband and I were driving on I-5 in Northern CA on our way to Southern Oregon for vacation. The sky was dark. It was to the hills to our left, in the fields to our right. A bright light (like from a police helicopter) flew very, very low over the highway just in front of us. It made no noise at all. No real shape we could see, but Just a huge ball of light. It crossed the highway and I watched it go into a field near a house, which had on a yard light high on a pole. It was too dark to see what was in the field around the house but when the light passed the house and went over the field it split into two lights, which seemed to dance over the field. I asked my husband to pull over so I could continue to watch but he said there was no place to pull over and to stop would not be safe (because of possible other cars coming up behind us). You'd think the splitting of the light would be the oddest part but the quiet was the strangest part to me. ((NUFORC Note: Witness indicates that the date of the sighting is approximate. PD))

April 22, 1998; various cities, WA www.nicap.org

UFOs sighted at sub base and nuclear storage facility.
Nuclear Connection Project
NCP
Established, July 1, 2003
UFO OBSERVED OVER SUBMARINE BASE AND NUCLEAR WEAPONS FACILITY
April 22, 1998
Synopsis: From between approximately 2120 hrs, and 2130 hrs. (Pacific Daylight Time), one or more objects, variously described as a blue-green sphere, an oval of blue-green light, or a disc with six lights on it, was reported from multiple points throughout western Washington State. The object was reported to change direction of flight on at least three occasions, to execute a rolling maneuver on at least one occasion, to come to an almost complete stop in two locations, to reverse its general course of travel from south to north, and to change its appearance, including its color, on at least two occasions. One reliable observer witnessed the object to pass slowly over the center of Bangor Submarine Base on Hood Canal, approximately 20-30 miles to the west of Seattle, passing directly over the underground bunkers where nuclear weapons are thought to be stored. In excess of a dozen seemingly independent and reliable reports regarding the alleged sightings were received by the National UFO Reporting Center between April 22 and May 11.
Summary of Witness Reports:
1. Snohomish County, WA: A 16-yr. old male and his mother, located approximately 15-20 miles north of Seattle, witnessed a blue-green object suddenly descend from the sky, hover, and then accelerate to the south-southwest and disappear in that direction at approximately 2120 hrs. (PDT).
2. Poulsbo, WA: Larry Swanson, a former federal employee and computer specialist, who resides approximately ¼ mile from the northeast corner of the Bangor Submarine Base, allegedly witnessed a brightly lighted, disc-shaped object glide silently over the center of that military facility at approximately 2122 hrs.. The object tilted somewhat, so the witness was able to see the ventral side of the object quite clearly, he reports. A computer-generated graphic, provided by the witness, is shown above.
3. Auburn, WA: A man out walking his dog witnessed a perfectly circular, blue-green ball of light descend from north to south in the western sky at 2120 hrs, at a very high velocity.
4. Puyallup, WA: A man driving to work witnessed a striking blue-green ball of light descending at a 30-35 degree angle from north to south in the western sky at a very high velocity.

5. Shelton, WA: A young father witnessed a blue-green ball of light streak across the sky to the west of his location.

6. Oakville, WA: A mother and her two sons, driving west on State Route 12, located to the west of Interstate 5, witnessed a "disc shaped" object, imbedded in a cloud of blue-green light, streak from north to south in the western sky. The object appeared to "follow the contour" of the horizon, dipping down suddenly, and then rising up again during its rapid flight to the south. The object was reported to generate a peculiar "tail," which streamed off the aft end of the object for a fleeting instant. All three witnesses believed they observed three lights on the top of the disc, and three lights on its ventral side, as well.

7. Steilacoom, WA: A woman witnessed a blue-green ball of light streak from north to south in the western sky at 2123 hrs.

8. Vancouver, WA: A husband and wife witnessed a blue-green ball of light move slowly from north to south in the western sky. Very peculiar appearance to the object, they reported, with a peculiar looking "mist" surrounding the object.

9. Portland, OR: A young musician witnessed a blue-green ball of light move from west to east in the northern sky from his vantage point in the city of Portland, OR.

10. Vancouver, WA: A male Ph.D. in computer science, driving north on Interstate 5 witnessed a blue-green ball of light, that seemed to him to descend vertically ahead of his automobile and hover momentarily above the highway. The witness added that he had experienced some very unusual emotional reactions to the event.

11. Yakima, WA: A law enforcement officer witnessed a blue-green ball of light moving at a very high velocity from south to north for approximately 2 seconds. During his sighting, the object suddenly changed its color from green to red. The sighting occurred at 2130 hrs, the witness thought.

Occurred: June 1, 1998; Approx. 2300hrs.
Location: Rancho Tehama, CA
Witness: 1
Shape: Disk
Duration: 10 minutes
1998 Rancho Tehama Ca.
When sitting out side about 11p, enjoying the starry night sky, I watched a bright saucer shaped bluish colored large UFO appear out of the atmosphere and descend over Yolla Bolla Mt. toward San Francisco Ca it took about 5 minutes for the craft to travel 200 miles and disappear from my sight over the mountain ridge. I was alone. ((NUFORC Note: Witness indicates that the date of the sighting is approximate. PD))

Occurred: June 1, 1998; Approx. 1100hrs.
Location: Kelseyville, CA
Witness: 9
Shape: Triangle
Duration: 10 minutes

It was low alt. very close, slow, had a spotlight, and looked like a black triangular stealth aircraft that my wife, myself and seven of my friends saw when we were out on a place called Panarma Drive. A spotlight coming from this flying vehicle catches our eyes. Be it was the famous "black triangles". At the time I had never heard of these ufo's So this thing flies over our head only like 75 feet up so low that it would've been illegal for an airplane. it is also going very slow i would estimate about 25/30mps too slow for airplane, yet it makes absolutely no sound! So it is slowly and silently hovering past us, and shining it's spotlight all over the place as if searching for something, and even shines it's spotlight on us then I found out later on by ufo television specials that this same craft same shape and silence, is spotted in other places. Then just about a week ago I decided to look up ufos on the internet, when I discover that this same ship with constant features; looks, silence and ability to move slow, has been seen many times in many parts of the world and there are even website sections about that. Then last but not least, I bring up topic UFOs with a friend of mine at work, and he said he had a UFO encounter, but he wanted to hear mine. So I told him, and he was astonished and said that was exactly like his UFO encounter back in 1979! Amazing some people claims it goes fast, changes shape, move back and forwards rapidly. It didn't do any of that for me, just quite and slow.
((NUFORC Note: Date in 1998 is approximate. PD))

Occurred: July 15, 1998; Approx. 2100hrs.
Location: Medford, OR
Witness: 1
Shape: Circle
Duration: 30 seconds
Huge, round, white light hovering over freeway overpass
I was traveling south on an I-5 freeway overpass through the downtown area of Medford. There was a huge circular white light above me to the left, not too high above the treetops, which lined the side of the freeway. It appeared to be the round, flat, glowing white bottom of an object, and I could not see the shape of it above the bottom surface. It was just sitting in the sky, unmoving, unblinking. I estimated it to be the size of a football field, large enough to hold a neighborhood of houses. There weren't many cars on the freeway, and they, like I, were continuing down the freeway, unable to stop. I don't know if anyone else saw it, and because of my rate of speed, it disappeared out of my view quickly. It certainly could have been there longer than 30 seconds, but I could not look back because of traveling on the freeway, and could not stop because there is no shoulder on the overpass.
((NUFORC Note: Date is approximate. PD))

Occurred: August 16, 1998; Approx. 2315hrs.
Location: Gridley by Chico, CA
Witness: 1
Shape: Flash
Duration: 3 to 5 seconds
Brief period of light flashed in my car window about 1/4 mile to right and probably no further than 2 to 3 miles in front.

First noticed greenish streak in car window that was dropping almost straight down rather than perpendicular to the area as a meteor might come in at. It had greenish blue sparks from it. Tried to find it for the next few days without any luck.

Occurred: September 13, 1998; Approx. 2120hrs.
Location: Weed, CA
Witness: 1
Shape: Fireball
Duration: 6 - 8 seconds
Very bright blue-green fireball moved downward then went out.
I was driving southward on I-5 about an hour or so north of Redding when I noticed a bright flash outside from my left. I though it was a CHP car coming up from behind me because the flash was about the color of a police car light. When I looked out my driver's side window, I saw a very bright blue-green light about 45 degrees above the eastern horizon. It moved straight downward for about six seconds then appeared to burn out. It didn't have a tail and it seemed to move slower than a shooting star or meteorite. It was about 1/4 to 1/3 the size of a full moon. It was distinctive because of it's blue-green, almost bright cyan, color.

Occurred: October 26, 1998; Approx. 1700hrs.
Location: Redding, CA
Witness: 1
Shape: Oval
Duration: 1-2 seconds
I was watching kit of pigeons fly, saw out the corner of eye, noticed it and it was gone within a couple of seconds
I had put my kit of Birmingham roller pigeons out at about 4:30 pm, they were getting chased around by a hawk and decided to go up fairly high, normal when predators chase them. I wanted to watch them as they were rolling around but I couldn't see the colors very were, up about 1500 feet I'd guess. I put down the bino's as it was hard to follow them, sore neck. They turned and went northern flying in a counter clockwise turn. I saw a shiny object out the corner of my left eye. I turned to look and it was a semi oval object just sitting there, not moving. It was like a dull chrome color, but kind of hazy like when you see a jet airline flying by. It might have looked hazy from the reflection of the object but not sure? It was only there for a little over a full second then a bright silvery vertical flash appeared (reminded me of something like on star trek, when they transport), followed by a flash similar to a "mirror" flash and it was just gone. I pulled up my bino's to see if I could see anything in the vicinity and there was absolutely nothing. The object like I said was kind of oval, but more like a circle with one flattened side to it, the flash appeared from this flattened side while it vanished. A guy I work with said his sister had seen a "UFO" the day before i did, this was after I had mentioned to him that I had seen one that he said his sister had mentioned to him a day earlier that she'd seen one. I haven't learned the details of her sighting yet. That's it.

Occurred: October 29, 1998; Approx. 0935hrs.

Location: Yuba City, CA

Witness: 1
Shape: Fireball
Duration: 1 second

Large blue/silver sphere with green halo travelled across sky approx 20 deg. arc leaving a white trail. Event lasted 1.0 sec.

While sitting in my duck blind on 10/29/98 I observed a large blue/silver sphere with a green halo streak horizontally across the sky followed by a wide exhaust trail, whitish in color. The event lasted 1.0 second and was gone. It moved as fast as a meteor. I was looking north and could see the white top of Mt. Shasta about 200 miles away. The object looked about the same size as the visible top of the mountain. It travelled from west to east and covered about a 20 deg. arc. The sky was clear with a few small clouds, no wind, bright sun, and ambient temp about 70 deg., and nothing interfered with full vision of the event. My eyesight is very good, corrected to 20/20.

Occurred: November 6, 1998; Approx. 1755hrs.

Location: Bend, OR

Witness: 1
Shape: Formation
Duration: 15-20 seconds

50 years old-Vietnam Era Air Force Veteran
5 objects were in straight line following one another. 3 objects in front were as bright as surrounding stars; 4th one was as bright as Vega; 5th one was as bright as Jupiter.
5 points of light observed approximately 20 degrees of arc in the northern skies.
I had just stepped out of my truck after arriving in Bend at my girlfriend's parent's house and the brightness of these lights caught my eye because I hadn't recalled ever seeing a constellation in that area of the sky before. There had been snow showers in the vicinity and the skies were in a broken condition with cirrus clouds also in the vicinity.
Objects were first observed almost stationary. Then, after a few seconds, I noticed they started moving in a straight line towards an approximate 330deg direction. The speed they were traveling was about twice as fast as a satellite would be observed moving but much slower than a meteor entering the atmosphere.
All objects remained equidistant from one another, approximately 1 finger-width between each. They were all white in appearance, the same as the surrounding stars. No sound, fire trail, strobe lights, running lights, flaring, or fragmentation was observed.
The front 3 objects were as bright as the surrounding stars. The 4th object in line was a little brighter, and the 5th object was approximately the same brilliance as Jupiter.
I had time to call my girlfriend over to my side of the truck to also observe this event. We watched the few remaining seconds, as they seemed to fade out of sight at approximately 40deg of arc due possibly to the high cloud obscuration.
I am 50 years old, a Vietnam Era Air Force veteran. I was a weather observer in the Air Force and have been an amateur astronomer for approximately 15 years. I have been staring at the heavens for as long as I can remember but I have 'never' seen anything like these lights before. My girlfriend is 47 years old and has a master's degree in psychology. She also has never seen anything like this before. (I might add that she says she 'doesn't' believe in flying saucers).

I reported this event to the Pine Mountain Observatory message center outside of Bend at approximately 19:15 11/6/98. They called me back Saturday, 11/7/98 and took the information. At that time there had been no other reports of the sighting in any other part of the state or area. The local newspaper, The Bend Bulletin, also had received no calls or reports on the news wire.

Occurred: November 9, 1998; Approx. 2335hrs.
Location: Roseville, CA
Witness: 1
Shape: Light
Duration: 5 minutes
Red light was seen in the eastern sky with a likeness of a laser pointer. It hovered, made a bobbing motion and then faded away.
I was standing in my back yard, and was facing east. The Light seemed to hover for a few moments, it made a bobbing motion, and it flared in and out couple of times. It faded in to a small light, about the size of a very distant aircraft navigational light, and appeared to leave at very high rate of speed to the northeast.

Occurred: November 12, 1998; 0200hrs.

Location: Bend, OR

Witness: 1
Shape: Disk
Duration: 5-10 minutes

Observed a disk-shaped object completing unbelievable maneuvers to the SE above the city of Bend, Oregon. Object hovered, did split second U-turns, appeared to be 'playing,' or 'dancing.' There was no sound emitted. The only light from the craft seemed to be reflected from the moon. Object when first sighted was moving towards the west.

On the night of November 12, in Bend, Oregon, I was outside of my home observing the moon and the planets currently visible. My roommate had asked that if they were at all visible to let him know (it has been overcast a lot here lately). It was clear. I went in and we both came out. I was pointing out a neighborhood light that was in violation of an Oregon code because it was not subdued (any amateur astronomer can relate,) when he pointed to an object in the sky and asked me what it was. I first saw the object heading due west and was about to say it must be an airliner when I saw it do an unbelievable U-turn. Astonished, we both watched this 'craft' do maneuvers that would be impossible for any known aircraft to perform. I am a veteran, and used to work while A-10's and a variety of other aircraft flew overhead. What we saw seemed to be so unhindered by gravity that I can only describe its behavior as playful, like it was 'dancing' in the air. It performed multiple U-turns, and would accelerate to incredible speeds and then come to a complete halt in what seemed milliseconds. It made absolutely no sound at all. I would estimate it's distance from us to be anywhere from one to three miles away, and it's size as possibly quite large. I did see a very distinct 'disk' shape. The thing eventually seemed to vanish, and we remained outside for about another hour watching for it before we gave up and went inside. I am an Archaeology student and my roommate is a Geologist. We are both well-grounded in science and the tradition that believes empirical data adds to the human understanding of our surroundings and ourselves. If this comes off sounding like a disclaimer let me just add that our 'data' goes against everything we have been thought to believe. I did not want, nor wish, to see such a thing. Nevertheless, I, we, saw the damned thing. For myself it calls too much into question. But there it is.

Occurred: November 17, 1998; Approx. 1830hrs.

Location: Chico, Ca

Witness: 1
Shape: Triangle
Duration: 6 minutes I was surprised to notice a large black triangular-shaped-wingless object cross the road directly in front of me flying about twenty feet from the ground. I was traveling northbound on Hwy 99. The road curved ahead of it and me was and at this moment when I noticed what I thought to be a falling star. I glanced at this object briefly and then turned my attention back on the road. Out of curiosity, I looked back to where I first noticed the object and did a double take when I noticed that it was still there. Not only was it still their, but it appeared to be even closer than before, and I was given the impression that it was traveling at an enormous velocity. As I observed this object with a keen interest, I noticed a long blue contrail behind it. This contrail appeared similar That of NASA's proposed Ion propulsion system, except for one thing; their now appeared to be what resembled orange fireballs shooting out the back of this object, and through the contrail. This was very unsettling because I thought I was witnessing an airborne tragedy, a plane on fire, plummeting to the ground. Indeed this object appeared to be loosing altitude with great abandon, and I was surprised once more to see the object bank hard and change directions approximately 180°'s changing from a Southeasterly direction to more of a Northwestward direction. The object's speed increased greatly after It had modified its waypoints. I passed an orchard and momentarily lost site of this object. I thought that was the last that I would see of this object, until it the road directly ahead of me flying about twenty feet from the ground. The triangular object was now flying very slowly, and its distinct triangular form was highlighted by small red lights along what appeared to be the sides of the object. These small red lights were aligned in rows along two opposite sides of the object. There were three small bright white lights, one on each corner of the object. In the center of the object was a large red glowing light about three feet in diameter, not typical of any aircraft I know of. It crossed the highway heading towards an open residential area with many houses. I crossed over a bridge and turned off of the highway onto a small side street. My investigative nature beckoned me to follow this object, which is exactly what happened. I was driving along the small side street, and the object was positioned completely parallel to me.

 I clocked it with my speedometer; it was traveling at approximately 25 MPH. I rolled down my window to observe the sound which I was sure an object of this proportion (about 60 feet long) would make flying at such a low altitude. I was stunned as I concluded that this object made absolutely no sound. It did not rumble or make any vibrations. I could not hear the wind whistling through its wings, because it did not have any. I lost site of the object as it again changed course and ducked below the tree line. With the information, which I was able to ascertain from this encounter, I can conclude that what I observed was not an airplane. It was certainly an object, It was flying, and I failed to identify it with any object which I am familiar with. It is with sound mind and being, that I certify everything stated above as truth, and the facts stated herein reflect the best of my knowledge and recollection of this encounter.

Occurred: December 15, 1998; Approx. 1710hrs.
Location: Oroville, CA
Witness: 1
Shape: Light
Duration: 20 minutes

A pair of lights joined by another light. They separate back into pair and single.
Three craft, which I first thought to be satellites, forming a close encounter over the Sacramento/Marysville area - 50 miles to the South. As I watched it was apparent that two of these craft were flying together as a pair up from the south, and a third craft was approaching the encounter area from the West. Clear skies, some very high clouds, good lighting, about a half hour after sunset. The craft pair was arranged so the larger of the two was always to the left of the smaller. All three craft went through continual complex fast changing lighting arrays. Their lights weren't synchronized, quite bright, brilliant colors, or a bright white light disc. Sometimes the lights would double or form arrays of four, all flashing differently, sometimes with different intensities. This didn't appear to be the light signature of the satellites, which are visible, here every clear evening, and their movements implied airplane, but on an out of atmosphere scale, which meant that the lights in their fantastic array were very brilliant. The smaller craft almost always was two tightly bound points of scintillating lights. Often times there were four. Only rarely did it become a single point of light. There did not seem to be an structure associated with the light arrays, yet they maintained an integral formation as if there were a structure. It appeared several times that it was presenting a changing profile to the viewer as it moved around, that perhaps a cause of the observed relative movements of the lights. As they converged the solo craft from the west executed a turn to the north and began flying with the two coming from the south. I then thought I was watching airplanes, but there never was a sound, and the scale of movement was wrong for an airplane. The solo craft cruises around the pair who turn to the east. There is movement between the pair, but it is subtle and their overall position remains exactly the same to this observer throughout their entire flight! The solo light is, what one could almost call, animated, especially as the scale of this encounter became apparent. The solo craft flew north above a cirrus changing vectors a few degrees several times, at one point appearing to reverse course as it circled back over an area - they were complex very large scale maneuvers in a patch of sky the size of my hand at arms length - never leaving a contrail or noise. As it did these high-speed maneuvers, it reached its highest point as it turned westward at about 10:30 as my arm would point to it. As it slowly went into the west glow I could see that it was still maneuvering but it was so far away that distance only indicated that this light or group of lights was not an airplane and was way off planet as compared to the usual couple of dozen satellite crossings. Eventually this brilliant cluster of glittering jewels was lost in the evening glow. As I caught up with the pair they had traveled across the sky in an enormous turn from the east to north over in the Reno area. That city is almost exactly East from this observer.

They traveled north 15-20 degrees then I saw that they were just slowly getting closer together in appearance which meant that they were now traveling to the Northeast and they maintained that till they faded into the night sky a hands width above the horizon. At their zenith in the East they seemed at about 10:00 in the sky as pointed to by my arm. There was never a sound. The scale of these executed movements would seem to require high altitudes. This was a very grand encounter, very majestic, and it would seem, when postulating actual physical craft, quite an extraordinary expression of speed and power. During this 20-minute dance the land and sky were very quiet. No jets and no local noise at all, it was quiet. Any sounds caused by these craft would have been audible, if they had been aircraft, as exampled by similarly placed jet flybys. This happened at 5:10 to 5:30 pm Pacific Standard Time on 15 December 1998.

Occurred: December 21, 1998
Location: Rancho Cordova, CA
Witness: 1
Shape: Unknown
Duration: 5 minutes
I heard like a whooshing noise that was high pitch and it hurt my ears I walked outside my back door and saw nothing but I took pictures...my friend was in side and did not hear anything or see anything...I didn't think anything of it for awhile until I got the film developed and it brought back that night in my head...I have taken a few pictures since that time and will be getting those developed soon...will let you know what I find them as well...it was strange that night also it was last week sometime and it was almost like a Morse code going on with blinks here and there some long some short...I am a psychic and see a lot of things with my third eye that others cant see...the last picture I took of the pictures I am sending you in of something big going over the house there was no noise anyone heard but I did.

Occurred: January 2, 1999; 2212hrs.
Location: Marysville/Beale AFB, CA
Witness: 1
Shape: Sphere
Duration: 30 to 45 seconds
Observed a spherical green fireball, which changes colors to red and white and very bright. it came from the NNE and then stopped over an area I think may have been between Marysville and Beale A.F.B. it could have been over the flight line itself for all I know because it was night and had nothing to gauge scale by. Then it headed west like a bat out of hell. Others must have seen the object but I never heard anything in the local paper. it didn't change in size really until it came west which is the direction I was from it but it was still at a distance so I can't tell you how big/small it was. To me it was just a light in the sky. The people in this area have often heard rumors that the "aurora project" was posted at the base but no confirmations other than that of retired personnel I have spoken with. I would have called my dad who was the maintenance supervisor on the sr-71 when he was active duty but it happened too fast. i told him about it today when I decided to report this.

Occurred: January 4, 1999; 2024hrs.

Location: Redding, CA

Witness: 3
Shape: Triangle
Duration: 10 Minutes

I went to get in my car. I looked up and SE while getting into the car. I saw an object with two red lights and one green light. These lights were not flashing. It was traveling NE parallel to Interstate 5. It passed overhead. While viewing it I could see no stars between the lights, only a large dark triangular shape. I drove my car NE to a friend's home and continued to view the object traveling slowly over Redding. It veered west toward Mt Baldy. At my fiends home it was still visible. She and her neighbors came out and viewed it with binoculars. Our Internet services also locked up at about the same time. I could not get on the Internet for over two hrs. It was 8:24PM PST when it went over me. It was larger then the full moon. I heard no sound from the object. As it went west I saw an amber light or glow also. At no time did any lights blink.
I'm a Genealogical Researcher and homemaker. I have raised five children and been married 39 years. I have some layman's knowledge of some of the constellations. I also do some research for Historical Projects. My friend is a Pharmacy assistant and her husband is retired. I don't know the background of her neighbors.

Occurred: January 8, 1999; 1925hrs.

Location: Paradise, CA

Witness: 1
Shape: Diamond
Duration: 25 minutes

1st sighted a bright white flashing light about 6 times the size of the surrounding stars. A mile further observed a 2nd separate, independent light. The 2nd light split and became two lights moving together, with no sound I could hear. The 1st light remained constant. Pulled over and stopped again, and the 2nd light turned and became three lights in a row with the middle one flashing. It turned more and I saw a fourth light creating a diamond shape. Then the object moved behind the hill I was parked by. I continued to drive up the mountain and observed the 2nd light occasionally through the trees. I stopped a third time on top of the mountain when I could see the 2nd lights again. It made a deep rumbling sound and was in a diamond shape with a darker blackness between the lights, which blocked out the stars behind it. It moved to the south.

At 19:15, while driving N on Hwy 70 at the Hwy 191 exit to Paradise, I observed a flashing white light, brighter and bigger (maybe 6 times bigger than the surrounding stars), in the sky above Paradise. I looked at the digital clock in the car. About one mile further, I saw a 2nd light NE of the 1st. At first I thought it might have been a reflection in the windows, so I pulled over, stopped the engine, and rolled down the window. It was a dark clear cold night. The 1st light was still constant in the same place in the northern sky above Paradise. The 2nd light split into two lights, fairly horizontal in the sky. There was no sound. I proceeded to drive NNE on Hwy 70, sighting both the 1st and 2nd lights the whole time, and turned W on Pentz Rd. Here, (4.8 miles from the Hwy 191 cutoff) I pulled over a second time, turned off the engine, viewed the 1st light in the same place in the sky, now left of my position, constantly flashing on and off. The 2nd light seemed to turn and the left light split into two, making three lights in a row with the middle one flashing. It continued to turn, while hovering in the same place, and I saw a fourth light creating a diamond shape, with the middle flashing light of the first three slightly ahead of the other two. It was now 19:25, I was parked facing WNW and was viewing the lights by ducking my head a little and looking out and up of the passenger side front window shield to the N. I held up my hand, and the four lights, if they were at the outside corners of an object, would make the object about as long as my thumb (maybe 1.5 inches). The object then moved behind the little hill I was parked beside. I continued W on Pentz Rd, nearly driving off the road to keep the object in site! Pentz Rd then turns and heads NE, up the mtn. I lost sight of both the 1st and 2nd light at times as the road climbs the mtn. On top of the mountain on Pentz Rd, (at 6.2 miles from the Hwy 191 cutoff on Hwy 70) I again sighted the 2nd object. I stopped the car (third time), turned off the engine, and rolled down the window. The car was facing N, and the object was in the sky to the W of me traveling in a S direction. Now I could hear the wind in the trees and another deeper rumbling sound that seemed to be coming from the object. It was about three inches wide in the sky and there was a darker blackness between the lights, which blocked out the stars behind it. I viewed it for perhaps three minutes before it went behind the trees. It was now 19:40 and I continued on my way home to Magalia. On Jan. 9th, my husband and I traveled the route again at the same time as the sightings on Jan. 8th, and took the mileage readings. Unfortunately, there was a pea-soup fog in the valley, so at the first two places I stopped we could not view the sky.

Occurred: January 8, 1999; Approx. 1930hrs.
Location: Chico, CA
Witness: 1
Shape: Unknown
Duration: 3 minutes
This is amplifying information to the Call in of an observation from Paradise, Calif. yesterday. The airport at Chico is used for practice approaches by aircraft from Beale AFB for the U-2 and SRC71. The Aurora replacement for the SRC71 is currently being stored at Beale for short periods of time, in training situations.

This is not meant to downgrade the observation of the Lady from Paradise, but I think it is important to note that Beale has a ready crew of ground personnel and equipment for Chico and the runway is capable of taking large aircraft. I have 26 years of experience in the distant early warning system, tower and radar approach control and carrier approach control and have observed UFO activity as early as 1957 over the North Atlantic and reported it to NORAD by single sideband direct communications. I have observed many "large military" aircraft on practice approach to Chico since 1979 and I am aware that it is a designated dispersal point for certain types during national emergency situations. There are old military silo launch sites in the Chico vicinity and some may still be active. Happy hunting from one who knows first hand that you are not on a wild goose chase.

Occurred: January 8, 1999; Approx. 1930hrs.
Location: Redding, CA
Witness: 1
Shape: Unknown sound
Duration: 20-30 seconds
I think I heard the loud ufo that the people on art bell are talking about.
I hope I'm not wasting anyone's time but I think I heard that ufo that the lady is talking about. I thought it was my dogs growling at each other, and yelled at them. They didn't stop growling, so I got up to yell again to stop them, and they were just sitting there looking at me. I could still hear the noise, and disregarded it as an airplane from the airport a few miles away or some hot rod or truck noise. Now I'm listening to art bell and there was a sighting in paradise, roughly 65 or so miles south of here. The sound I heard was in the same time frame that they are talking about. Hope this helps some? Can find me nights on #artbell on ((Personal data deleted.))

Occurred: January 10, 1999; Approx. 0300hrs.
Location: Paradise, CA
Witness: 2
Shape: Unknown
Duration: 2 min
Humming sound outside--unlike any other I've heard around here.
I woke up around 0245, couldn't sleep so I went into living room to read for a while. I heard a soft humming sound outside that had two long periods and one short period of duration. Unfortunately, I did not go out to look. Unfortunate because I understand a lady may have seen what I heard on the same night--tape played on Art Bell Sunday night. She was one ridge to the west in Paradise. We have had other instances of ufo here too. Had a bright square shaped light in the garden area a year ago--didn't go out to investigate it either--both my wife and I saw it.

Occurred: January 30, 1999; Approx. 2330hrs
Location: Red Bluff, CA
Witness: 1
Shape: Formation

Duration: 18
As they flew over in a triangular formation, a white thing struck the lead craft, and scattered them.

Occurred: February 2, 1999; 0643hrs.
Location: Redding, CA
Witness: 1
Shape: Other
Duration: 1 full second
I saw a very bright green ball fall from the sky, and then disappear over the horizon.
This morning at about 6:45AM, I was driving southeast on Buenaventura Blvd. I was noticing how beautiful the sunrise was, as it was just getting light over the horizon in front of me. Then I saw a very bright green ball shaped object fall from the sky, disappearing over the horizon. This seemed to occur just as the sun was coming up. The sky was just beginning to lighten up pink in color and I could also see the horizon of mountains, however the sky was still dark behind me. There were a few clouds above me, but none where I saw this object. This object was a very bright florescent green. I'm not sure how far away it was, but the mountains it fell behind are at least 60 or 70 miles away. It was about twice the size of the "lit" part of a streetlight. It didn't appear to have a tail. I am a homemaker with 2 children. I've done extensive traveling worldwide and also I've had the opportunity to fly various places since my childhood. I've worked as a secretary for my father's firm. His work is heavy construction.

Occurred: February 16, 1999; 0645hrs.
Location: Beale AFB, CA (2 mi. N. of)
Witness: 1
Shape: Circle
Duration: 20 seconds
VERY LARGE GREEN LUMINOUS GLOBE, seen hovering near Beale AFB.
While on a night shooting event at a Public Range, East of Beale AFB, I saw a Very Large Green globe hovering near South Main Entrance of Beale. Globe Hovered in my view about 20 sec. and seemed to generate a corona around itself, then moved up into some clouds, illuminating the clouds, then seemed to disappear, only to re-appear again, 10 seconds later, for only a few seconds, then it "phased out" once again.

Occurred: February 17, 1999; 2050hrs.
Location: Roseville, CA
Witness: 1
Shape: Unknown
Duration: 10 minutes
A bright red dot seemingly motionless in the sky, when viewed with 114mm Newtonian Telescope the bright red dot demonstrated constant velocity. The object was tracked for approximately 5 minutes, suddenly the object accelerated out of my telescopes ability to view it, not to the left or right limits of the telescope but directly away and only 1/2 to 1 degree to my right

A single bright red dot seemingly motionless in the sky, when viewed with 114mm Newtonian Telescope the bright red dot demonstrated constant velocity. The object was tracked for approximately 5 minutes, suddenly the object accelerated out of my telescopes ability to view it in les than .5 seconds, not to the left or right limits of the telescope but directly away and only 1/2 to 1 degree to my right.

Elk Abduction
Occurred: February 25, 1999; Approx. 1158hrs.
Location: Cascade Mountains, Washington State
Witness: Multiple
Object: Large Disc
Duration: 3-5 minutes
On this winter morning, fourteen forestry workers, employees of a large, unnamed company, were planting trees in the Cascade Mountains of Washington State about 20 miles west of Mt. St. Helens. Three of the men had been watching a nearby herd of elk in the valley below them all morning.
Suddenly, a heel-shaped object with two stripes on its back appeared over a nearby ridge and began drifting in a northeast direction. Initially, the three men thought it was something like a parachute, but it maintained a steady altitude, following the contours of the terrain below it.
As the object began to move toward the herd of elk, the three men called out to the other eleven members of the work crew. All fourteen men stood on the hillside and watched as the object floated down into the valley towards the elk.
The silent object was able to get quite near the elk before the animals noticed it. When they did notice it, most of the herd ran to the east, toward a densely wooded area. One elk, though, trotted off toward the north, down a logging road. It was to this lone elk that the object flew. The amazed workers watched as the object floated above the elk and then appeared to lift the elk off the ground with some sort of invisible force. The object then moved off, with the elk slowly rotating beneath it. It moved up the ridge, barely clearing the trees, and then down into the next valley, out of sight of the forestry workers. After a few minutes, the object then reappeared, apparently without the elk, and rose at high speed until it disappeared into the sky. The case was reported to NUFORC, and Peter Davenport of NUFORC and Robert Fairfax of MUFON Washington traveled to the site and interviewed the witnesses. For their report go to NUFORC. They also examined the body of a female elk that was found to the north of the site. It could not be determined if this was the same elk. Many of the witnesses had been with the company for years and they were generally deemed to be reliable. Sources: ufoevidence.org, MUFON, NUFORC.

National UFO Reporting Center Case Brief
March 11, 1999
Commercial Flight Crew Reports Being Burned by "Surge of Green Light"

Early Friday morning, March 12, 1999, Art Bell, host of the late night radio program, "Coast to Coast," reported on his program having just received a fax from an individual who identified himself as a crewmember of a commercial airliner. In his message, the unidentified individual asserted that both he and another member of the cockpit crew had experienced a burning sensation of the skin on their faces shortly after having witnessed a peculiar green light in proximity to their aircraft at approximately 2100 hrs, (CST) on Thursday night, March 11, 1999. At the time of the incident, the aircraft was in the northern mid-western United States.

On Saturday morning, March 13, The National UFO Reporting Center received a written report, apparently from the same individual, which details the events of the alleged incident. This summary is based on the contents of that written report.

Note: All personal and professional information regarding the individuals involved in the reported incident is deleted here. This practice is in keeping with the policy of anonymity that the National UFO Reporting Center adheres to, and also in response the crewmembers' request to remain anonymous and unidentifiable. For simplicity sake, the individual is referred to here as "he." SUMMARY: The individual described in his report the fact that both crewmembers of the airliner had been witnessing an "intense" display of the Northern Lights as they were descending for landing at a major metropolitan airport in the northern mid-western United States. The person adds that he had witnessed the Northern Lights on many occasions during his 10+ years of flying, so he was not paying a great deal of attention to the display. However, he comments that the display on the night of March 11 was of such intensity that it appeared to "stretch above and over the aircraft," and that the dramatic colors of the display—green, red, and blue—were reflected off the metallic surface of the aircraft. The crew member then describes how both he and the other member of the cockpit crew were witness to a short-lived "pulse of green light," which appeared to be "concentrated in a green 'ball'," and which approached their aircraft from the north, apparently at a very rapid velocity. Immediately following the event, all members of the cockpit crew began to experience and sensation on their faces, as if they had been "burnt by the sun." They reported this sensation to each other, at which point the crew immediately requested clearance to divert to a different altitude. After the aircraft passed through a thin layer of overcast, the sensation of burning that the crew members had experienced seemed to them to disappear immediately. However, on the next day, the author of the report noticed that the skin on his face was "red and sore." The individual reports that he has made arrangements to see a physician about the skin condition. Also, he requested assistance in investigating the case. The alleged incident occurred approximately 1-2 hours prior to a number of seemingly credible, and rather dramatic, UFO sighting reports were received from the states of Illinois and Ohio. (Please see related posting.)

Occurred: March 17, 1999; 2100hrs.

Location: Lake Almanor, CA

Witness: 1
Shape: Light 3 craft
Duration: 60 seconds?

I witnessed 3 crafts, in the sky, with strange lighting, and suspicious movements, tonight at about 9:00pm over Dyer Mountain, near Lake Almanor, CA. At 9:00 pm, out for an evening walk. I noticed what appeared to be an airplane flying southeast over Dyer Mountain. It was moving quite slowly, white and red lights flashing. I then witnessed it to completely change directions, suddenly, headed back towards where it came. Two more craft then appeared, towards the north, with red lights, not flashing. They moved suspiciously together, then apart, moving to and from, not in straight lines, neither vertically or horizontally. The first, seemingly larger craft that "joined" the two others, and they flew off slowly over the mountain. For an hour after this sighting, I noticed a remarkable number of planes flying over the area.1plane every 60 seconds. The sighting of the craft lasted about 60 seconds before they flew out of sight. Just an ordinary evening stroll turned quite curious. This was the first sighting of this type I have ever had.

Occurred: June 3, 1999; Approx. 2210hrs.

Location: Mount Shasta City, CA

Witness: 2
Shape: Light
Duration: 30min?

While walking dog along RR tracks my brother and I witnessed five fast moving craft in the eastern sky. Several of the craft headed for the Mtn.
Lights, which looked just, like stars (IE: not blinking) crossed over our visual horizon in the Eastern sky. We observed three commercial jetliners in the same field of vision while we were observing these lights. There is no possible way for them to have missed them. Concurrently we observed almost from the start of our paying attention several lights on the MTN itself that were in places were no vehicles could travel. From the distance we were observing there was in our opinion no way for the lights to be hikers (brightness, weather conditions in the area-high winds, cold temps) As stated three of the moving stars disappeared on the MTN. two of the moving lights crossed in parallel fashion while a commercial jetliner was headed north. Of the three commercial jets two went N>S, one went E>W OF the five fast moving lights three went ESE>E and disappeared on the MTN. The other two went north and south respectively. We were in near complete darkness (very little ambient light.)((Name deleted)), my brother was the one who called this sighting in, since I am visiting on vacation. I live in Seattle, I listen to Art Bell, I am very skeptical, but the facts remain: we saw exactly this. It is entirely true. I love my privacy, I am willing to talk to you in person or over the phone when I return this weekend but I am guarded about my privacy. You were the first person I thought of. Peace ((two names deleted)) (I am in agreement with this please call me at home. I have limited Internet access.((Number deleted)).

Occurred: June 23, 1999; Approx. 1135hrs.

Location: Grants Pass, OR

Witness: 1
Shape: Circle
Duration: 20 minutes
I saw a ball above a jet and it was still, then it went south and up till it was no longer in site I just moved to Selma or about a 40 min drive southwest of grants pass and I just saw the same type ball in the sky about two meters below the north star it is a silver and it can really move fast I went north then west at a high speed two state fire men where there to but they don't want to say anything about it, if oh it was 08/25/99 when I saw it again, sorry can't type to good, if you had a few people who it please e-mail me. Thank you much, Bob.

Occurred: July 12, 1999; Approx. 2210hrs.

Location: Chico, CA

Witness: 1
Shape: Light
Duration: 1.5 to 2 minutes
A bright, steady light traversed the sky over my area from SSW to ENE. It appeared to be at great altitude be cause the light seemed to be entirely reflected light. It was visible for approximately 2 minutes.
I am a disabled veteran who served in the United States Air Force for 8 years. I have been an avid aircraft enthusiast since a small boy. I am 42 years old. On the night of this sighting, I was out closing up my workshop for the night. It was very clear out with no moon or clouds. I was facing south while walking into the house. I always like to look at the stars and happened to notice a bright one straight up to the south. It was gold in color with the same brightness as Venus after sunset. It was then that I notice that it was moving to the east-northeast at a rapid speed. The light was constant and bright, no flicker. As it traversed across my field of vision it went behind the top floor of my home. I ran around to the front yard and continued to follow its path across the sky. By now it was to the east of me and it suddenly gave off a bright flash of reflected light as from a mirror. It then started to decrease in brightness as it disappeared over the Sierra Nevada Mountains. I note that the light remained steady as I viewed it from its aft end. No afterburners, or other lights were seen. I have seen the same kind of light twice before only traveling from due west to due east in the predawn hours.

Occurred: July 15, 1999; Approx. 0100hrs.

Location: Guerneville, CA

Witness: 1
Shape: Formations of triangle craft
Duration: 5 minutes
Woke up at 1am, went out onto deck outside bedroom and looked up at the sky through the trees. Saw 50 or more craft, flying west toward the coast, silently, with blue lights glowing from underside of craft.

Woke up suddenly and went out onto our deck through the sliding glass doors from our bedroom. Looked up through redwoods, and saw 50 or more craft flying west. The craft looked like three disks connected with a thin bar, creating a triangle shape. The underside of the craft emanated a beautiful blue glow, and no sound was heard from where I stood, about 2 miles from the hill they were all flying over. They were moving very slowly. For some reason, instead of continuing to watch them, I went back to bed before they had moved out of my vision. I attempted to wake my husband to show him, but I curiously could not! Nothing would wake him. It seems as if someone else would have seen this, as it was such a large amount of craft that I saw.

Occurred: August 4, 1999; 0400hrs.
Location: Grants Pass (Murphy area), OR
Witness: 1
Shape: Egg
Duration: 6 seconds
At 04:00 hours flying NE to SW, two white egg shape objects, was flashing off and on.
At 04:00 hours, I woke up to a flash of white light over our mountain top, about three seconds later I saw another flash, then I saw a egg shape white object fly over the mountain coming from the NE flying to the SW. After a second or two, another object with the same description followed behind and to the side of the first object. I saw the white light that seem to make up the objects flash on and off for about six seconds.
((NUFORC Note: One of several reports from same source in Grants Pass. PD))

Occurred: August 22, 1999; Approx. 2200hrs.
Location: Hwy. 199, CA
Witness: 3
Shape: Fireball
Duration: 30
A brilliant bluish/white ball of fire that looked at first like a comet streaking across the sky but than turned and came directly towards our car. It was round like a full moon. It became fuzzy/misty looking and streaked light in different directions as it disappeared behind some trees.
We were driving west on Hwy. 199 a very rural mountainous road. It was about 15 miles away from the first event my husband and I had witnessed. The girls and I kept our eyes peeled on the sky to see if we would see the first object again. We were looking toward the mountain range in the northeast off to our right. I suddenly saw a fuzzy white object that then began to streak brilliantly through the corridor heading east. It first looked like a comet or meteor but then it turned towards the direction of our car and quickly changed appearance to a huge bright blue and white fiery ball. It was coming right towards us. It moved very fast, and then it began to look misty at the top of the ball and seemed to streak light in different directions. I could see the light through the trees as it dissipated and then disappeared. Needless to say we were just totally blown away. The girls and I all had the sense that the ball was coming AT us. I felt like it was aware of us. We all described what we saw the same way. Our background is: My husband is a pastor. I am a homemaker. The girls are in high school and are A students. By the way it was a really clear and beautiful night, you could clearly see the stars. We are all just amazed at what we saw.

Occurred: September 1, 1999; Approx. 2130hrs.

Location: Trinity Lake, CA

Witness: 5
Shape: Fireball
Duration: Approx. 2 minutes
Witnessed a bright red ball of light floating across the sky, much slower than a meteor that broke up into several parts and left an orange trail, but never burned out.
On Wednesday, Sept 1, 1999 at approx. 21:30 I observed a very strange object floating across the sky. 4 friends of mine who were house boating with me at Lake Trinity, which is 30 miles north of Redding, CA., also observed it. My friends and I were sitting outside, on the lake when one of them noticed a very bright red ball of light moving very slowly across the sky. We quickly turned out all the lights to get a better look. The red ball of light appeared to be very low in the sky and was moving in a West to East direction. At first, it looked a bit bigger than the largest star in the sky. As it traveled, it got somewhat larger. Trailing behind it was a smaller bright white light that looked like a star at first, but it was travelling at the exact same speed and exactly the same length behind the red ball of light throughout our sighting. After about 20 seconds the red ball began to split and turn a brilliant orange color. It left a trail of orange pieces that continued to follow the main body of light. There were between 4 and 6 of these "tails" following the main body of light. We watched this for another minute to a minute and a half. It never burned out, but rather disappeared behind the mountains to the East of out boat. It was moving very low and too slow to be a meteor. When we saw this, we had just finished cleaning up after diner and were not drinking at the time. None of us have ever seen anything like it. We are all college graduates in out late 20's. We are a close group of friends who spend our vacation together on our annual house boating trip. I have a degree in Biological Science and I cannot explain what I saw that night.

Occurred: September 1, 1999; Approx. 2130hrs.

Location: Alturas, CA

Shape: Formation
Duration: 20 seconds
Bright lights in formation traveling south to north
Witnessed a single bright light entering earth atmosphere where the object then broke into approx 8-12 smaller bright white lights. The formation was traveling south to north. A fireball came down and hit the earth's surface. White objects continued in formation traveling across the horizon and then disappeared at one point. We then observed a green flashing light at the center of the formation much like a conventional aircraft beacon
.

Occurred: September 1, 1999; Approx. 2140hrs.

Location: Orland, CA

Witness: 1
Shape: Fireball

Duration: 2 to 3 minutes
Red or Bright Burning Flash of light and sparks
We have a Dairy in Northern California just east of Interstate 5. I was out side feeding the cows and to the east on the horizon level I noticed a Fireball that seemed to have sparks coming out the back or side, I thought at first it was a explosion of a electrical Transformer and when my husband came out of the barn I asked him if he say the explosion and he did not. I have never seen anything like this before.

Occurred: September 1, 1999; Approx. 2300hrs.
Location: Chico, CA
Witness: 1
Shape: Other
Duration: approx 30 seconds
a fast moving object appeared in the road while I was driving, blue and white lights, I stopped and watched, the object stopped and hovered, then took off in a flash.
It was a flat shaped object, lights on each end, alternating white and blue. I noticed it because of how fast it appeared to be moving. I stopped to see what it was. In a matter of seconds the object stopped and hovered in front of me, I was excited and scared, and went to turn my car around and get the heck out of there. As I started to move my car the object flashed and disappeared to the right. It moved at a speed that was just a flash, not even in my scope of estimation.

Occurred: September 1, 1999; Approx. 2100hrs.
Location: Medford, OR
Witness: 1
Shape: Changing
Duration: a few minutes
My sons caretaker was standing in the parking lot in Medford, Oregon last Wednesday and saw a bright orange/red ball coming over the Table Rocks that is north of the airport. The ball traveled south over Medford and broke up into 3 and then 6 separate balls of orange/light and continued south toward Ashland. He said they were as high as the planes usually fly and did not appear to be coming down but kept flying at the same altitude the whole time he could see them.

Occurred: September 1, 1999; Approx. 2125hrs.
Location: Coos Bay, OR
Witness: 1
Shape: Other
Duration: 15-30 seconds
Large fireball, faded to long cone shaped object moving from west to east about 15 to 20 degrees up from horizon.

I was walking south on the street in front of my house on September 1, 1999 at 9:25 and looked up to see a bright orange fireball in the sky. I followed it because it was such a bright object. While I watched, the fireball grew bigger, then faded to a long flat object with what looked like wings projecting from the top and bottom, like it was a very long tube like airplane on its side. As I was watching it move from west to east, I spotted what looked like a smaller large object move in the same direction, parallel with the large object, but lower in the sky, making it seem to be a great distance away. The first large object seemed to grow larger in size as it moved seemingly farther away. The second object continued to track the first. As I was standing in the middle of the street, there were trees in the area and I lost the object behind trees that were too numerous for me to track on. By the time I got to an area where I could see more sky, all objects were gone. This was no more than about 1/2 block and took place over about 30 seconds I believe. I was entranced, unable to believe what I saw, I almost pooh-poohed it but continued to watch and could not believe what I saw. I finally realized that it could not be a meteor as it was moving too slow for that. I am an elementary teacher in the coos bay school system and have been interested in these types of sightings for many years, but this is the first one I have personally seen. I was amazed and absolutely thrilled to have seen something out of the ordinary. I sent e-mail to KVAL-TV that night and never got a reply but did hear something on the 11:00 news that stated that it was a Russian space object. I forgot about it until I heard a broadcast on KWRO-am radio at approximately 10:15 on September 16th.

Occurred: September 1, 1999; Approx. 2125hrs.

Location: Florence, OR

Witness: 1
Shape: Fireball
Duration: 15 seconds
Orange, moving south, followed by small illuminated object below and slightly behind.
Continued burning until obscured by trees.
We live 3 miles from Pacific Ocean. Objects, were moving at consistent speed entire time. I can't really guess the altitude, but was able to see a second object both times I focused on it. Now sure the shape of second object but it was faintly illuminated, following closely below and slightly behind. I initially thought this was a meteor slowly burning up, but it continued across the sky.

Occurred: September 1, 1999; Approx. 2130hrs.

Location: Eugene, OR

Witness: 3+
Shape: Light
Duration: 5 seconds
White fuzzy circle, red streak, and fireball, accelerating at a very fast rate, gone...

I work at a warehouse in West Eugene, OR. At approximately 9:45, I overheard two employees talking about, "the strangest thing I have ever seen." They related the story to me. Looking due west, towards the Eugene airport, they saw a white light travelling in the sky, "at the pace of a usual airliner," traveling from northwest to southeast. While it was the same speed of usual aircraft, it just looked like a fuzzy light object. It continued to travel south, then developed a red streak behind it (did not mention if it was fire or anything) traveled for a few seconds, then burst into a large fireball. (I asked him if he was exaggerating, and he said no, "A real fireball.) Then the white object took off at an incredible rate, Southeast, then disappeared. We talked about it around 9:45 to 9:50. This is my first time reporting anything like UFO's or stuff, but my buddies saw it, were amazed, and were confirmed by another forklift loader on a separate end of the building.

Occurred: September 1, 1999; Approx. 2130hrs.

Location: Lake Trinity, CA (Approx. 30 miles north of Redding, CA)

Witness: 5
Shape: Fireball
Duration: Approx 2 minutes

Witnessed a bright red ball of light floating across the sky, much slower than a meteor, hat broke up into several parts and left an orange trail, but never burned out.

On Wednesday, Sept 1, 1999 at approx. 21:30 I observed a very strange object floating across the sky. It was also observed by 4 friends of mine who were house boating with me at Lake Trinity, which is approx. 30 miles north of Redding, CA. My friends and I were sitting outside, on the lake when one of them noticed a very bright red ball of light moving very slowly across the sky. We quickly turned out all the lights to get a better look. The red ball of light appeared to be very low in the sky. It was moving in a West to East direction. At first, it looked a bit bigger than the largest star in the sky. As it traveled, it got somewhat larger. Trailing behind it was a smaller bright white light that looked like a star at first, but it was travelling at the exact same speed and exactly the same length behind the red ball of light throughout our sighting. After about 20 seconds the red ball began to split and turn a brilliant orange color. It left a trail of orange pieces that continued to follow the main body of light. There were between 4 and 6 of these "tails" following the main body of light. We watched this for approx another minute to a minute and a half. It never burned out, but rather disappeared behind the mountains to the East of out boat. It was moving very low and too slow to be a meteor. When we saw this, we had just finished cleaning up after diner and were not drinking at the time. None of us have ever seen anything like it. We are all college graduates in out late 20's. We are a close group of friends who spend our vacation together on our annual house boating trip. I have a degree in Biological Science and I cannot explain what happened.

Occurred: September 8, 1999; 2057hrs.

Location: Redding Airport, CA

Witness: 2+
Shape: Light
Duration: 2:47

((NUFORC Note: The MiG-29 is not capable of achieving Mach 3.8. Moreover, we doubt whether the MiG-29 is capable of supersonic flight for 1.8 hrs and 1,400 miles without refueling. We are not aware of any MiG-29 in the U. S. that is privately owned.))
Chased light approximately 1400 miles in plane at mk. 3.8 for 1.8 hours
We took off from Redding at 20:34 doing touch and goes when a light appeared out of the west by northwest ski at an altitude of around 35,000. We noticed that it was flying straight and level with a heading of east by southeast we decided to take chase with object. The aircraft we were in was a MiG 29A capable of mk. 4.2 call me and I'll give an interview in full.

Occurred: September 9, 1999; Approx. 1235hrs.
Location: Mount Shasta, CA (Sand Flat)
Witness: 1
Shape: Fireball
Duration: 7 Seconds

After seeing disc 35 minutes earlier that night witness fell asleep to be reawakened. Large slow moving west to east green fireball 200 feet angle about struck trees, made a loud crack, and then vanished. I was awakened again to a very bright fireball slowly crossing Sand Flat west to east. It lit the entire Sand Flat area with a green glow. The fireball was about 200 feet angle, 300 feet north away from me, about 30 feet in diameter It appeared to be a perfect sphere and translucent, except for the core which was a solid yellowish orb about 3 feet in diameter. I could see a fine pattern like bicycle spokes radiating from the center 3 dimensionally. The fireball then vanished in the forest with a loud " crack" sound. The next morning I found a tree limb snapped off, and a large snag at the top of the adjacent tree, which was also snapped off. The batteries in my camera as well as my cell phone from this event were dead. I did manage to find a fresh battery and took photos (not available) of the injuries sustained to the trees that were obviously fresh.
((NUFORC Note: Date changed to 09/09/99, the presumed date of the sighting.))

Occurred: September 9, 1999; Approx. 2200hrs.
Location: Mount Shasta, CA (Sand Flat)
Witness: 1
Shape: Disk
Duration: 1 minute

Large disc floated over Sand Flat about 200 feet angle, north to south direction. Viewed 3 large rectangular lights (like windows) star field above it was blocked out. I was face up in my sleeping bag on the ground when I saw these windows with an aspect ratio of 7 to 1 horizontally curving around what appeared to be a large disc. It seemed at first that the angle of it was about 45o elevation due north and it had lights that blinked on and off in unison 3 times as it floated over me. I attempted unsuccessfully to awaken my son whom was with me but he would not budge. There was no noise at all to this event and it flew directly over my head and vanished. If the lights were on the outer rim of the craft, I would estimate the size of the craft to be 2-300 feet in diameter. I was stunned and the battery in my camera was dead after the passage of this craft. I had hoped for a good UFO sighting but gave up after a while and the CSETI group had moved to another location since all of us were awake the previous night stargazing.
((NUFORC Note: Date corrected to 09/09/99, the presumed date of the sighting.))

Occurred: September 9, 1999; Approx. 2200hrs.

Location: Mount Shasta, CA

Witness: 2
Shape: globe fireballs
Duration: 5 hours

Multiple anomalous lights, white flashes, orange fireball, wt/rd lights on side of Mt. Shasta moving and signaling. My son and I ran across Steven Greer, CSETI Group: We witnessed from sand flat, looking at Mt. Shasta 10pm to 2pm about 5 very bright white flashes like a flash bulb going off at high elevation. Multiple meteors leaving normal trails, white/red blinking lights both side below Shasta Peak for several hours. About 12am s strange anomaly of interference occurred to star field over about 5 degrees by 10 degrees for 30 seconds. It was like looking through a heat wave. (Cloaking distortion) About 1230am a slow moving orange fireball about 500 feet angle moving northwest. After it passed, a sound like a piston engine airplane was heard. There were no navigation lights at all. A camper reported a helicopter the previous day hovering by this location lasting for about 30 minutes on the face of Shasta in the daylight. There were also lights on the steep face of Shasta shifting from time to time and went out to very bright on occasion. We were also followed in the daytime by a White and Blue Bell Jet Ranger helicopter for an hour. I saw that night.

Occurred: September 24, 1999; Approx. 0700hrs.

Location: Chico, CA

Witness: 1
Shape: Other
Duration: 3-5 minutes

Small silver object, low flying in very strait line emitting what looked to be a thick bright billowy contrail the shape and length of a comet tai ended at a point and abruptly, no dispersion of smoke.

In car returning home the last 2 blocks, up in the sky what looked to be at the altitude of a small plane, but I could not make out the object only twice in its flight did I glimpse, the color silver. What was so odd was the tail of what seemed to be a contrail from the object immediately from the back end that was much thicker than the object itself and extended about 15 lengths of the object into a point, as though it were the tail of a comet. The contrail was thick and solid, but billowy looking and was very illuminated or bright, but it did appear like smoke that was coming from the object. It did not trail off in a long dispersed fashion, like it does from highflying aircraft that leave contrails across the sky. This stuff coming out of this thing was very contained in its appearance. The object traveled very slow almost 4 min in my vision while I parked the car in the drive went in the house, grabbed the phone came back outside and called my husband at work and proceeded to describe what I was seeing to him. It traveled as though I were seeing it from the side view but as it moved farther away it was not exactly perpendicular to me. The path was more like southwest to northeast. (If I have my bearings, correct) Clear sky and no other aircraft in the area I will send drawing by fax. ((Name deleted))

((NUFORC Note: The witness in this case is well known to NUFORC, and we consider her to be a very reliable source of information.))

Occurred: September 24, 1999; Approx. 2200hrs.

Location: Lower Lake, CA

Witness: 1
Shape: Triangle
Duration: 1 hour

At 10:00pm our dogs and the neighborhood dogs started to bark. I went outside to check why they were barking. I looked around and saw in the northeast a hovering triangle object. It's had bright lights coming from it. They would get bright and dimmer. During this time 8-12 Jets flew in the approximate area of the object. The object would move in circular and figure eight moves, while bobbing in the sky. It eventually moved farther away in the sky until it was just one light in the sky.

The object was close enough when I first saw it to see some detail. It had a triangle shape. Lights came from the three angles of its triangle shape. The light it omitted was white like a star, however at the angles of the object the light was a golden yellow color. The object kept moving away from me in a northeast direction in the sky. While hovering away it would move in circles and figure eight motions. While moving in these motions it would bob up and down quickly. A couple of jets were flying around it. Later many more jets flew over the area in the sky. They came from the east and west. The first few jets banked around the object. The rest flew straight across the sky. From the east and west. The last jet flew high above the object. The jets lasted about half a hour. Then they left. You could hear them and see their lights flashing on and off. When the object and the jets were in the sky together they seemed to dance all around the sky. We rarely hear or see more than one jet at night. To see that many in a half hour was bizarre. I also saw little dots moving fast around the sky of the object. During this hour I saw two shooting stars or what I believed to be shooting stars. One went across the whole eastern sky, while the other one went across the western sky. Eventually, after an hour the object moved far off into the northwest sky.

Occurred: October 11, 1999; Approx. 2110hrs.

Location: Ashland, OR

Witness: 1
Shape: Triangle
Duration: 30-45 seconds

I was looking into the sky towards the high east watching what appeared to be a passenger jet cross the sky when my peripheral vision caught sight of a luminous amber object coming from the southeast and headed in a northwest direction. At first I thought it was a big bird of some sort the way it was gliding in the sky. It was relatively low in the sky especially compared to the jet I had been observing. The object passed silently overhead and it was like the shape of a boomerang. It wasn't glowing or lit but yet it was oddly illuminated like it was reflecting the light from the valley floor. (I-5 is nearby and Ashland and Medford are well lit) I have seen many other craft and a strange light over the past 40 years but this was very unique to my experience with UFO's. In my judgment it was around 3,000 ft. in altitude and moving at a pace that allowed it to remain in my vision for nearly a minute as it passed over, veered right sharply and then took off rapidly due east. I can't get that eerie glow out of my mind. If I hadn't been looking it would have been nearly invisible.

Besides the 10/11/99 incident, on the morning of 10/ 14/99 at 4:15 AM a gigantic fireball was seen from my home in the southeastern sky. It was not a shooting star but a gigantic golden fireball streaking across the sky. We have been seeing many strange things in the sky here lately. Southern Oregon is being chem. trailed at least once a week for months now and it is so thick sometimes the blue sky is completely hazed over within minutes.

Occurred: October 15, 1999

Location: Weed/Yreka, CA

Witness: 1
Shape: Fireball
Duration: 3 minutes
Fall 1999 sighting of green and gold fireballs in N California
From your records I am sure you will be able to determine the exact date and time after reading my account. The following is from my sent email dated 111401."I have two purposes in writing. First, to describe my own sightings of some of these 'fireballs' in the north central California area in late fall early winter 1999/2000. I did not record the events other than mentally. At that time I was driving truck on a weekly run from Portland OR to southern California. The first of my fireball sightings occurred, as I was northbound on I5, about half way between the Weed airport and Yreka, in northern CA. It was between mid Oct and mid Nov. It could have been on a Tues or Thursday night, but was most likely Wed night. I do not recall the exact time, but did make a mental note of the time. I think it was between 2300 and 0100 Pacific Coast Time (GMT –8). It was a large green pulsing fireball that was emitting golden sparks. Relative size to me was most likely slightly over two inches in diameter. I used objects in near field vision within and on the truck, and middle field vision along the freeway as references. Horizontal vector distance was most likely much less than 20 miles from me, based on the distance to hills on the western edge of this relatively North/South valley.

The golden sparks were associated (possibly not emitted) with it on the sides and backside of its traverse. None of the golden sparks moved downward toward the earth so were most likely not influenced by earth's gravitation or else the sparks extinguished quickly enough so as to appear to not be influenced by earth's gravitational field. I have seen several greenish and other colored meteorite fire trails one of which lit up the sky and exploded over the San Antonio Texas area in 1967 when I was stationed there in the Army. This was definitely not a meteorite. I rolled down my window to listen and observe directly. It was soundless and left no after image of light in the sky or on my retina, nor any light streak, as do most meteor trails. In the 'tail' area, somewhat obscured by the golden sparks, were several (I think I remember counting either three or four) smaller objects. Each of these objects was approximately ¼ inch relative diameter with one larger than the others. When I first noticed the small objects they appeared to be slightly not in exact formation with the larger object. I observed them to maneuver independently into an exact single line path following in the path of the larger object. For! the res t of my observation, these objects appeared to follow and maintain an exact distance from each other and the larger object. In color, they appeared to vacillate slightly with the predominate frequency being a yellow-gold different from and paler than the golden sparks. The frequency shift ranged between reddish and bluish around the periphery of the entirety of each object with enough transparency to note the predominant central yellow-gold. Because of the pattern, flight path, and general and specific appearance, at that time I believed them to be manufactured objects (this includes manufactured plasma balls) under some kind of either external or internal control. The direction was roughly slightly east of south, but possibly south by southeast. The track seemed to be parallel to an average altitude above the earth. I observed them as they appeared from over the hills to the north of me until I could no longer see them as they passed out of my field of view as limited by my direction of travel and position in the truck. They could have been visible for as long as 30 seconds.

Time is relative and subjective and even more so in situations such as this. I think they were at quite low altitude because during part of the observation they appeared between the hills and me to my west. I was listening to the Art Bell radio talk show at the time and tried several times to call in on my cell phone, but never got through. Others were calling in to the Art Bell program with incident reports. The objects were seen from at least as far north as Portland OR. It was reported that jets were scrambled from a National Guard station at Beaverton OR airport, but never caught up to the objects. Observers reported that somewhere over central CA the fireballs made an orderly sharp-angled turn to the east toward Nevada. I vaguely remember hearing another day that they disappeared or 'exploded' somewhere over Nevada. On two or three Sunday evenings or nights sometime between then and early March 2000 I observed similar greenish fireballs coming from relatively west southwest of my position as I was southbound on I5 within 100 miles north of Sacramento and 50 miles south of Sacramento. These incidents each occurred between 2100 and 0200. On one occasion I saw three fireballs on the same relative directional track, each within one to five minutes of each other. I saw no golden sparks or other small objects during any of these observations. Each of them appeared to be relative size approximately half or less of the Oct/Nov object, and much further distant. My observation period was very short as they traveled quickly from the western horizon over my head toward NV. I remember hearing later of one report from a woman someplace in Castro Valley CA reporting a similar observation but do not know if it could have been the same as one of my sightings. I am certified in NLP and Clinical Hypnotherapy and was in practice for a few years. Part of NLP is about learning to be a good observer and make fine distinctions. I am also a snowboard instructor and used to climb a bit, both of which provide good perspective on outdoor distance and orientation.

Occurred: October 22, 1999; Approx. 0610hrs.
Location: Klamath Falls, OR
Witness: 1
Shape: Sphere
Duration: 3 seconds
Object was seen just as it descended over the horizon, behind local 7,000 foot mountains, to a heading of approximately 355 degrees, almost due North, at a high rate of speed, but slower than a meteor. It descended 15 degrees in approximately 3 seconds. Object was observed in the city limits, however, I do not believe there was enough light to diminish an ionization trail. It was exceptionally bright, and lacking a standard color format, it was the color of a BLUE Christmas tree light. I am an aviation historian, and amateur radio operator, and I specialize in "bouncing" signals off of Meteor trails. I know of no substance that burns in this color, so will discount a Leonids meteor, which was prevalent at this time. Object was quite large, and no signs of diminishment in size, such as a meteor would display.

Occurred: November 18, 1999; Approx. 0320hrs.
Location: Redding, CA
Witness: 2
Shape: Other
Duration: 15-30 seconds

Large, boomerang shaped craft.... no lights, completely black...no noise. Flew over us...north to south direction (approx.)..Very fast

While lying on my roof watching the meteor shower, my friend and I noticed a large, boomerang shaped object flying at a high rate of speed directly over Redding, CA. This object was large; completely black...no lights of any kind showed, and made no noise. The craft never strayed from its single course, and disappeared in the horizon when the city lights were too bright for us to see it anymore. It came from the north...flying south...maybe southwest. The whole experience lasted no longer than 30 seconds from the initial sighting to losing track of the object.

Occurred: December 20, 1999; Approx. 0400hrs.
Location: Brown's Valley/Marysville (near Beale Air Force Base), CA
Shape: Disk
Duration: 30 minutes+-
Object appears to remain stationary 20-30 minutes, has many greenish-blue flashing lights some rapidly moving white lights. No noise. Some small white lights entering and leaving larger craft. Has appeared each night for at least a week. Occasionally turns and disappears. Possibly object is observing BAFB. Several unit seem to combine as one.
I am a WWII aerial observer and navigator. Flew in B-24's in the western Pacific.

Occurred: December 28, 1999; 0200hrs.
Location: Shasta Lake, CA
Witness: 2
Shape: Light
Duration: 5 seconds
A friend and I were driving from Tacoma to Los Angeles for vacation. We were about halfway over the Lake Shasta Bridge, and off to the east I saw a green glowing light hovering in the sky. It hovered there for maybe 3 seconds, and just shot across the sky and was gone. The rate of speed it traveled at seemed faster than a shooting star to me.

Y2K 2000: SECTION 10
2000-2012

There was a great deal more sightings in central and coastal California and Oregon than are listed in this book. We used the sightings to validate a simple point. The patterns of UFO activity here indicate what is happening everywhere, and not just on earth but all about the universe; life finding and becoming acquainted with itself. Life does this routinely, for food, familiarizing itself with the occupants that it shares territory with, and ultimately the orchestra that is available, as ones own allies.

The Regional sightings we listed, for our area, follow a pattern from south to north. Some year's activity was sporadic, attributed to weather changes and the human climate that distracts our focus from our sky watching, as politics and economics influence our attentions. If we are hungry or jobless, or challenging an epidemic, our focus out of necessity becomes narrow to find a specific solution.

Keep in mind that it is estimated that only 1 in 10 sightings are reported as it is. Again be mindful at the abundant sightings from the 1980's, 90's and on of BLACK TRIANGLE Craft. Although we feel that there are some genuine Extraterrestrial Black Craft, the appearances of these Black Craft consistently follow the creation of the SR-71 Stealth Bombers and Lockheed Secret Dirigibles all beginning to occur in the 1980s. Now we have the DRONE aircraft in the skies in the new millennium, and can speculate since the drones are taking a commercial role, that there are newer secret drones being used for military use. In fact, it is suspect that many new drone craft are black, wedge, solar powered (as the original is) and off the radar for public scrutiny. Another point to note is that most sightings were reported 3-5-10-20 years later. We removed the notations in many cases to save room, but on average most witnesses wait long yearly durations if they are not immediate.

Occurred: January 2, 2000; Approx. 2000hrs.
Location: Oroville, CA
Witness: 2+
Shape: Light
Duration: 10+minutes

These lights (egg shape) put on a dazzling show they would go in a long oval circle probably a football field in length the egg shape lights were going around in pairs of two they would go around about three to four times then they would all join in as one in the center of what use to be the circle I'm here to tell you it was a brilliant show in the sky these lights kept spinning and twirling we (my family) pulled over twice on highway 70 to watch these lights do there thing I even was going to stop to ask a CHP officer who was stopped on the side of the highway if he was seeing the same thing we were but my wife was to overwhelmed to talk to a police officer at the time this was something that my family and I will never forget I've never seen anything like this on any TV program it was definitely a rush watching the lights in the sky over Oroville California that evening one more thing these lights were in perfect order not one was out of sync it was like a rehearsed choreographed dance that was fit to be on Broadway
See summary description
((NUFORC Note: Sounds like a 4-light, articulating advertising light. No report received from the CHP officer. PD))

Occurred: January 17, 2000; Approx. 2140hrs.

Location: Orland, CA

Witness: 2
Shape: Triangle
Duration: 10 minutes
Small triangular object flying low to ground flashing green then gold light.
It was January 17, 2000 around 9:30-9:45 p.m. I was riding my bicycle next to my friend Dave who was walking as we headed to a friend's home at the rear portion of the mobile home park we lived in at the time.
We were perhaps 100 feet away from our friend's home when we both noticed a large, bright green light blinking on something that was moving through the air away from us. It couldn't have been more than 50 feet above the ground as it moved slowly from our left (south) diagonally away from us toward our right (northwest). It was approximately two football fields away from us when we first saw it. All we could see was the extremely large, bright green light blinking on and off for 1 to 1 1/2 seconds at a time.
I realized that I wasn't able to hear any sound coming from whatever it was, so I asked Dave, "Can you hear it? Can you hear it?" To which he replied that he couldn't hear anything from it either.
At that time, I ditched my bike and we both began running toward the object which had by then stopped flashing the green light and began flashing a gold light 10 to 15 feet forward from where the green one had been. The gold light was as large and as bright as the green one had been, and it was definitely more captivating.
Just as the two of us had reached a full-speed run, the object abruptly changed direction and began flying directly towards us. There was no curving turn like an airplane would have to make, no sweeping change like a glider, the object just stopped going away from us and began heading immediately toward us. I think even a helicopter wouldn't have been able to change direction of flight as quick and immediate as the object did.
Then we were scared! We turned and ran back to our friend's house. We pounded on the door for him to let us in, and he opened the door just as the object flew directly over our heads.
At that time, there were no longer any lights flashing, but we could hear a faint humming noise. As the object flew over, all we could see was its black silhouette against the night sky. We both agreed that it was triangular in shape and small like an ultra-light.
We do not believe it was an ultra-light. We honestly don't know what it was, but it was the very first time any of us had seen what we believed to be a UFO.

Occurred: May 24, 2000

Location: Mount Shasta, CA

Witness: 2
Shape: Teardrop
Duration: UFO-Strange object caught on web-camera
UFO / strange object caught on web-camera

Peter, My wife checks a web-camera that is focused on Mt. Shasta, California -everyday. It takes still shots and then puts them together in the form of an .Mpeg files (size 520K). We used to live up there and we love the mountain and surrounding scenery. It brings back lots of memories. ANYWAYS, when ((name deleted)) checked the .mpeg movie for yesterday (5-24-00), there was a strange object to the right of Mt. Shasta in the film -near it's beginning. Each frame is taken 10 minutes apart from each other for a daily movie of 102 frames. It looks to me like an upside-down tear shaped object that is spinning. You see it in frames that have no clouds and in frames with clouds. AND it is moving in a northeasterly direction while the clouds are clearly moving in a Southeasterly direction. Go to the below link: http://www.((URL deleted))l Cursor down until you see "Daily Mpeg Movies" and select "Yesterday" (If you read this on 5-25-00) OTHERWISE, choose 5-24-00 I have saved this .mpeg file and will email it to you as well.((name deleted))
((NUFORC Note: We have not been able to open the file yet, but that may be due to our computer system. If we are able to open it and inspect the image, we will attempt to comment here.))

Occurred: July 30, 2000; Approx. 2338hrs.

Location: Ashland, OR

Witness: 2
Shape: Disk
Duration: 30 seconds

Large object just hovered over the vehicle for about 30 seconds.
Myself and a female friend ((deleted)) looked out of the car window to observe a large saucer just hovering about 100 feet off the right side of the car. It then took off after what seemed like an eternity. I don't know how long it had been since my watch stopped along with the car. I estimate it was only 30 seconds although it seemed much longer.

Occurred: August 26, 2000; Approx. 2200hrs.

Location: Trout Creek Recreation, OR

Witness: 4
Shape: Light
Duration: 2 minutes
A bright large round light that moved in horizontal, vertical and circle, and left in rapid speed.

On Aug 26, 2000 my daughter, son, his girlfriend and myself were sitting outside our camp at the Trout Creek Recreation Area campground, which is North of Madras, Or about 20 miles on the Deschutes River. We had spent a afternoon of fishing and were enjoying seeing the stars as we were away from the light pollution of the cities, We were concerned about the fire in the Ochoco National Forest and were aware of the airplanes that were flying and we also were seeing shooting stars. But to the North of us about half way up from the horizon we could see this very large light which was stationery when we first noticed it , however there where airplane lights flying towards it several maybe as many as three or four. All of a sudden the large light very quickly moved up and down in a vertical line, several times, very rapidly, then it moved side ways in a horizontal movement, then it went in a circle, a very tight small circle and then just took off at a amazing rate of speed to the northwest and disappeared there was no way the planes could even begin to follow. Observer's, a Medical Technologist for a hospital laboratory, a Dental Assistant, 2 are avid fishermen and use to observing the night sky.

Occurred: September 15, 2000; Approx. 1400hrs.

Location: Eugene, OR

Witness: 1
Shape: Other
Duration: 20 seconds
Late Fall, 2000: Eugene OR Bowl / Cone Two-Part UFO Sighted in Broad Daylight
I was outside smoking a cigarette on the back patio in my apartment complex. It was a cool autumn day, with clear blue skies and no clouds, and lots of sunshine. There was a bit of wind, as I recall. I looked up and saw a white puff of cloud appear, like a firework...but there was no sound.
Then, a white hemispherical shape came out of the smoke, with a white/metallic cone underneath it, pointing down, and seemingly separate.
I thought to myself, "This looks like a model rocket nose-cone with a parachute, I remember making those." However, I saw no indication that a model rocket had been launched (they leave smoke trails.) I slowly realized that this object was actually much larger and further away.
My estimates are that the bowl shape was 100ft across, at a height of 750ft to 1000ft above ground. I puzzled over why it was hovering in the air.
Then, to my surprise, the cone began to rotate around underneath the bowl shape, clockwise if you were looking up at it from the ground. It acted as if it were connected to the rim and remained pointing down. There was a hazy dark cloudy spot between the cone and the bowl shape, like a field of some kind. The cone spun around at a rate of 3 to 4 times per second.
Then, the whole unit turned on its side at right angles to the ground, and proceeded due south at an estimated acceleration of 1000km/s (100 Gs!!!). The cone continued to rotate around the rim as if attached to the bottom of the bowl the whole time it did this. It was on the horizon and out of sight in less than 4 seconds. Then, mysteriously, there was a second puff of smoke where the first had been, as if something closed behind it.
This happened in broad daylight!
((NUFORC Note: Witness indicates that the date of the event is approximate. He provides an illustration of what the object first looked like. PD))

Occurred: September 18, 2000; Approx. 2110hrs.

Location Redding, CA

Witness: 1
Shape: Flash
Duration: 30 seconds
Saw 3 flashes disappeared over horizon 30 seconds and very high rate of speed!!
Dark clear night, one bright flash, focused in that area 10 seconds later, second flash occurred, when second flash dissipated saw one faint orange pin point, focused on pin point another 10 seconds past and pin point omitted third flash this sighting occurred from beginning to end in 30 seconds, then disappeared over the horizon. It is difficult to judge altitude, but was not in outer space. Note: I am pilot with extensive aviation background/knowledge.
((NUFORC Note: We spoke with this witness, and found him to be quite credible. He has extensive experience in the aviation industry, and emphasized that he had never seen anything similar to what he was witness to on this night. The object apparently appeared to be moving extremely fast. PD))

Occurred: November 3, 2000; Approx. 2235hrs.

Location: Fortuna, CA

Witness: 1
Shape: Oval
Duration: 30 minutes
Bright orange oval object, located over Hwy. 101, Fortuna, California
Object located over Hwy. 101, near Main Street Exit, northbound traffic lanes, object was oval in shape, and bright orange with no other lights noted, no sound or noise heard. Object was stationary in the sky, estimated altitude 750 ft. Observed object for approximately 30 minutes, then I had to leave the area (was on my way to work in Eureka). This object has been seen before--first sighting January 17th, 1998, and has been observed by myself on 2 other occasions, with the last sighting February, 2000. On this occasion, the object was not seen to emit "sparkly" objects from the bottom of the 'craft'; however, this phenomenon has been observed on 2 other occasions.
((NUFORC Note: Because of recent UFO reports from California, we requested additional information. Our request, and the subsequent response, is below:
Dear ((name deleted)),
Thank you very much for the report! I am interested to know who the other witnesses were. Would they like to submit reports, as well? Did they leave the scene at the same time you did? Thank you very much for sharing the information with our Center!
Cordially,
Peter Davenport
Peter B. Davenport, Director: National UFO Reporting Center
PO Box 45623
University Station
Seattle, WA 98145
director@ufocenter.com, http://www.UFOcenter.com
Hotline: 206-722-3000 (From 8AM to Midnight Pacific preferred)

Occurred: February 21, 2001; Approx. 2045hrs.

Location: Rogue River, OR

Witness: 1
Shape: Disk
Duration: 15 to 20 minutes

I saw a disk or saucer and it was glowing. It made the sky brighter than what the sky usually is. It also looked like it had a beam coming down from it. Like in a movie where the spaceship was abducting something, everything was quiet but when the disk or saucer went away all the frogs started to make noise again and every thing else that made noise in the mountains. The disk or saucer moved really fast. For an example the disk was on the left than like a blink of the eye and the disk or saucer was on the right. The disk or saucer was on the top of the mountain but it was hovering like a helicopter but there was no noise at all. The disk was so bright it lighted up the trees around it so you could see them. The light was a bright green. There was also a type of haze around the disk. Like a force field around it like in a space movie. When the disk went behind the mountain the disk looked like it got lower like it was going to land. It also got brighter as the disk went behind the mountain. That was the last time I saw that one disk. I saw a second disk but this one was smaller like a dime and didn't have a glow. This disk was to the right of the mountain where I saw the first disk. This disk that is as small as a dime was really fast to. It would move way to the left then back where it started and then down but back to where it started. The disk kept on doing that.
((NUFORC Note: We have spoken numerous times with the witnesses to the events they describe, and we find them to be exceptionally credible and serious minded in their report. The entire family was witness to the event, which apparently was quite dramatic. PD))

Occurred: February 22, 2001

Location: Rogue River, OR

Shape: Diamond
Duration: 10 min

Bright diamond shape light northwest sky right above a ridge line in the back of our house, we watched the light pull apart from the sides and pull back together we saw this happen a few times then it would pull apart from the bottom all a sudden a bright red light came from the north, it came right next to the bright light it stayed there for 5 to 10 seconds then it zigzagged then it arced to the south fast real fast and it was gone. The bright light stayed there and would go up and down and side to side all sudden it was gone.
((NUFORC Note: 2nd sighting in two days in this area. Same witnesses. I spoke with the witness after the second sighting, and his report was quite convincing. He is an outdoorsman, lives in a rural area, and he appears to be a good witness. We suspect the report is quite accurate. PD))

Occurred: February 22, 2001; Approx. 2015hrs.
Location: Rogue River, OR
Witness: 1
Shape: Diamond
Duration: 10 minutes

Bright diamond shape light at first it didn't move. Then it started getting lighter and brighter and smaller and bigger. Then it started moving down a hill. Then it stopped. It then got lighter, brighter, smaller and bigger, And went back up the hill. It stop at the top of the hill then it went back down the other side. It kept on doing it over and over.

Occurred: February 22, 2001; Approx. 2017hrs.
Location: Rogue River, OR
Witness: 1
Shape: Diamond
Duration: 10 minutes

The shape of the craft was a diamond. The craft had circular lights that were bright. It had a glare coming from the craft that was grayish white. The craft had a beam coming from the bottom of it and it looked like there was something in the beam. The thing in the beam was a ball shape. It was as big as a soccer ball. It was stationary. The craft split in have a couple times. If you spread your index finger and your middle finger apart that's how far apart the craft split apart. The craft stayed sprayed apart for 2 to 3 seconds then would go back to one craft. Then did this for a couple minutes. Then out of nowhere a red disk shape came zooming across the sky, stopped over the craft for a second and zoomed off into the sky.

Occurred: March 1, 2001; Approx. 0200hrs.
Location: Chiloquin, OR
Witness: 2+
Shape: Triangle
Duration: 2-4 minutes

Three UFO's seen flying over the Klamath sky.
While we were driving along Modoc pt rd back to my parents house we saw an extremely bright light hovering about 2-300 feet in the air in the middle of an empty field. When the car got close to the light, just close enough to see the triangular shaped outline, it took off at high speed heading south towards Klamath Falls. We drove the rest of the way home. When we got out of the car, we looked back in the direction that they went, and we saw some craft heading north at extremely high speed, this craft looked to be the shape of a lopsided triangle and had about five to seven varicolored flashing lights along the outline. Soon afterward, an identical craft appeared heading the same direction. This one however was flying in a high-speed zigzag pattern. When the craft were overhead, all of the neighborhood dogs began barking. (This may or may not be a result of the craft flying overhead) We watched the craft until they disappeared over the north horizon. I have grown up around Kingsley air force base, and have seen all manner of jets and helicopters and these behaved like none of those that I had ever seen.

Occurred: March 13, 2001; Approx. 2200hrs.
Location: Roseburg, OR
Witness: 1
Shape: Unknown; Large bright lights
Duration: 10 minutes
Large object was travelling North with 2 Large Yellowish Bright lights and other lights on object, with NO SOUND. Alt. <2000 feet, vel. <100mph
My second major sighting in this county since Sept. 1997 Very large object passing directly overhead and slightly to my east. 2 LARGE and BRIGHT yellowish lights (steady) visible as well as a couple other dimmer blinking (unsure but seem to recall red or green perhaps). Object was well within 2000 feet altitude and made absolutely no sound travelling due north at slow speed, <100 mph est. perhaps half that. Total spread of lights at least as large as apparent full moon, appeared to be one craft. I was trying to convince myself it had to be a plane, but after a minute or so I realized ANY local terrestrial traffic would be plainly audible at this range! Also I suspect there may have been other light on the east side of craft not visible to me. Had to make this report, as it appears others north of me had possibly seen same object. NOTE: very uncertain of exact time of event but certainly was after 9pm local and before midnight. We have a HOTSPOT of sorts here in SW Oregon. KEEP LOOKING UP FOLKS THE BIZZARE IS THERE!!!
((NUFORC Note: Please see reports from Puget Sound, and other areas, for this time and date. PD))

Occurred: March 14, 2001; Approx. 1830hrs.
Location: Redding, CA
Witness: 4
Shape: Light
Duration: 30 seconds
Unknown object cruising over town suddenly changes direction and accelerates up and out of sight, witnessed by 4 people.
While driving to town to watch a movie with the family, my son pointed out a very bright light in the sky, moving just below the tops of the mountain range that surrounds our city. All four of us were captivated by the extreme "whiteness" of the light and the husband pulled to the side of the road for a better look.
The light moved at a good pace horizontally for about 20 seconds longer then suddenly veered upward and shot out of sight in an instant. I actually saw "tracers" following its sudden departure upward. There is no way it could have been any sort of known airplane or helicopter. It certainly was not a bug or a spotlight reflection. It was totally silent and was close enough I am sure we could have seen the vehicle if it had been daytime. No meteor could have possibly fallen and ricocheted back into space at this sharp of an angle.
The color never changed or distorted, it was there and then it simply was gone.
((NUFORC Note: Witness indicates that the date of the sighting is approximate. PD))

Occurred: April 3, 2001; Approx. 1540hrs.

Location: Corning, CA

Witness: 1+
Shape: Circle
Duration: 5-7 minutes

Small circular object seen over our house, traveling north, then it turned back toward us, then started moving slowly to the east. It was round and silver-white in color with a bit of dark on the bottom. A military type jet flew over traveling in the same direction, then a minute later another jet going in the same direction flew by. About 4 minutes later one of the jets flew back in the southerly direction and disappeared in a large cloud. We feel that the jet was looking for the UFO.

Addendum:

The original statement sent to you earlier today stated that we had seen one "craft" at approximately 15:40 hours in the sky above our house, which traveled north at a slow rate of speed. There was no sound, and a slight rotation of the object. It disappeared into the clouds. About 20 minutes later another, or perhaps the same craft, appeared in the sky to the east of our home. It too was the same shape, but this one performed a acrobatic maneuver, before it too disappeared into the clouds. We DO have videotape of both events. If you can send a snail-mail address, we can send you a copy of the video. Also, exactly 6 months ago today on October 3, 2000, at approximately the same time of day as the previous event, we videotaped another craft over our home. The same description of the object was seen.

Occurred: April 3, 2001; Approx. 1540hrs.

Location: Corning, CA

Witness: 1+
Shape: Circle
Duration: 40 minutes

The original statement sent to you earlier today stated that we had seen one "craft" at approximately 15:40 hours in the sky above our house, which traveled north at a slow rate of speed. There was no sound, and a slight rotation of the object. It disappeared into the clouds. About 20 minutes later another, or perhaps the same craft, appeared in the sky to the east of our home. It too was the same shape, but this one performed a acrobatic maneuver, before it too disappeared into the clouds. We DO have videotape of both events. If you can send a snail-mail address, we can send you a copy of the video. Also, exactly 6 months ago today on October 3, 2000, at approximately the same time of day as the previous event, we videotaped another craft over our home. The same description of the object was seen.

Occurred: July 11, 2001; Approx. 2300hrs.

Location: Grants Pass, OR

Witness: 1
Shape: Egg
Duration: Seconds

Large egg shape UFO, solid (no blinking lights) white light with blue tones in it. The UFO was traveling north to south at a very high speed. No sound heard from the UFO. Animals (dogs and owls?) in the neighborhood were making noises when object was going over.
((NUFORC Note: One of several reports from same source in Grants Pass. PD))

Occurred: August 14, 2001; Approx. 2148hrs.
Location: Grants Pass (Murphy Area), OR
Witness: 2
Shape: Light
Duration: Several minutes
Large bright light traveling low at a slow rate of speed with no sound
My husband and I were going to bed when we saw out our bedroom window a large bright white light (not flashing) traveling low in a straight line north to south, at a slow rate of speed. I ran outside with a spot light and flashed it about six times, and it seem to burn out slowly before my eyes. No sounds could be heard from it. We have been having a lot of sightings in this area since moving here 8 years ago.
((NUFORC Note: One of several reports from same source in Grants Pass. PD))

Occurred: September 6, 2001; Approx. 1030hrs.
Location: Florence, OR
Witness: 1
Shape: Circle
Duration: 30 seconds
Bright red fire ball in front yard
While taking photos in my yard a bright red burning ball of fire appeared in my camera lens. I lowered the camera and no more than 10 feet from me was a ball approximately two feet in diameter; with an outer glow that extended it to about 3 feet. The shape was almost a perfect circle. I did not feel any heat irradiating from the ball, yet it was burning bright red and emitting heat waves. It remained about 20 seconds, and then started to rise vertically, very slowly to about 60 feet, where it remained motionless for a short period. It then made a 90-degree turn and then accelerated out of site. I have a color photo of this event. I called the police to report it and they had another report of a fireball a block away.
(FOLLOW-UP INVESTIGATION REPORT: ERIC BYLER, Oregon UFO Review, <Oregon UFO Review@home.com>.
Hi Peter,
I just got back from an investigation at the coast where I got the picture you see here along with the negatives. I am still working on the report and have to make one more follow up call with him to clarify something, but the long and short of it is that the photographer is a ((deleted)) year old man of (partial) American Indian heritage by the name of ((deleted)). He lives alone with a rather large ((dog)) in a very nice home on the coast of Oregon in a small town named Florence.
((He)) spends a lot of time on his yard as it looks quite good. On the day of the picture, he was out with his camera...a ((high quality SLR camera))mm 1:3.5 - 4.5...using 35mm 400 film... a very nice camera that he chooses to keep on automatic for the simplicity of pointing and shooting. He had already made a number of pictures when he approached the plants you see in this picture.

According to his testimony, as he was looking through the viewfinder and was about to shoot, this image appeared in his viewfinder. He quickly hit the shutter button and dropped the camera to see what this was. He estimated the object as being a couple feet wide and glowing like fire with an outer lighter glow that emanated like a heat wave. He pointed out however that as close as he was, only a few feet away, he could feel no heat. He thought this as odd.

After a few seconds, the object began to slowly rise until it reached tree level which is quite high as you can see in the picture, the whole time just watching in awe thinking, "'what in the hell is this?" Then he realized, hey, take another picture. He brought the camera up to began lining it up with the object when it did a 90 degree turn and shot away from the house traveling west towards the ocean. He immediately took the film down to have them developed to see if he indeed had got the object on film. I have checked out the negatives and the package with the number of pictures that were developed and it does appear that he stopped taking pictures in his yard by the time he got to the picture in question. My thought is if this were just a reflection of some sort on the lens then it probably would have only shown up when the pictures were developed and he would have continued taking pictures. The fact that he did indeed stop taking pictures at number ten (the roll goes backward from 24 to 1) and then developed the pictures tell me that he may have indeed seen something of a nature that points away from simple lens reflection and does add up with his story.

I spent some time in the house with ((witness)) and I found him very easy to talk with. ((Witness) is a big man but soft spoken. His house is very nice with lots of reference to his heritage seen in the artwork on his walls. He has three ((medals)) permanently displayed on a side table as he was in WW2. A wall in another room had a number of framed certificates of achievements and education showing his many accomplishments over the years.

In his main den where we talked there was a large volume of books on shelves. I saw no books pertaining to UFO material as many were related to American Indian heritage

((Witness)) does not have a computer for Internet access but does have basic television.

He has no idea what he saw. He did contact the police who came over and took a report. The following day he says he talked to the assistant fire chief while in town and he (the chief) mentioned hearing of a similar report on his scanner that day but remembers it as being on 24th street, rather than 23rd where ((witness)) lives. I am following up on this. I did contact the state police who reports that they have had similar reports like this over the past few months in other parts of the coast but did not elaborate. This is being followed up on as well.

Just in case the photo doesn't come through, go here...

http://www.oregonuforeview.com/florence.html ... to view it. Sincerely, Eric Byler - Oregon UFO Research.

((NUFORC Note: Our gratitude to Eric Byler, Oregon UFO Review, for the excellent investigation and follow-up review. PD))

Occurred: November 18, 2001; Approx. 0300hrs.

Location: Chico, CA

Witness: 1
Shape: Triangle
Duration: 1 minute
Huge slow flying triangular object blocking out sky during meteor shower on California's I-5 outside Chico.

I stopped to watch meteor shower at a rest stop on I-5 just outside of Chico California. When I went between two big trucks to block out some bothersome lights I saw out of the corner of my eye what looked at first like several meteors traveling together. When I looked in that direction I did not see any meteors. As I started to look in another direction I noticed something moving across the sky blocking out the stars. That was how I determined the shape. It was triangular, very large, moving slowly in an easterly direction. There were no lights and no noise. It was hard to judge the size of the object but I estimate that if it was flying low (as in 2 or 3 thousand feet it would have been the size of at least a football field. It disappeared behind the truck trailer and I was unable to find it again when I moved around to the other side of the truck.

Occurred: December 15, 2001; Approx. 1810hrs.
Location: Grants Pass, OR
Witness: 1
Shape: Triangle
Duration: 5 minutes
Very large triangular shaped UFO sighting, North-West, outlined in blinking red lights
I was driving to Eugene from Medford and happened to look up to the sky and through the clouds I saw a very large blinking red light. I thought it was just an airplane as I couldn't see it clearly at first, but I kept looking at it as I was surprised in the large amount of area it was lighting up. It was moving from North to West, now very rapidly, still blinking, then I saw its huge enormous size as it was blinking and realized that this wasn't an airplane, and had to be a ufo, no doubt in my mind. Triangular shape, very large, moving down West closer down into the mountains, outlined in red blinking lights. I saw it out of the window as my mother was driving down the freeway.

Occurred: January 9, 2002; Approx. 1930hrs.
Location: Springfield, OR
Witness: 2
Shape: Teardrop
Duration: 45 seconds
A mother and son from Springfield Oregon reported seeing a UFO from their car. The boy was the first to notice the object. "What's that mom?" by which she replied, "I'm not sure son but I've never seen anything like it." As they drove directly underneath the object, which hovered at around 600 yards overhead, they could clearly make out a craft of some sort which was described as being like a huge upside down black rain drop with windows containing bright white lights inside. The entire sighting lasted approximately 45 seconds. Both a report was filled out at the Review as well as a phone call made to the Research Hotline regarding this sighting. An investigator is being sent out to follow up on this sighting with a visit so as a more accurate location where this object hovered can be determined.
Eric Byler - Oregon UFO Research.
((NUFORC Note: Our gratitude to Eric Byler, Oregon UFO Research, for sharing the report. Their website is: oregonuforeview@home.com

Occurred: February 14, 2002; Approx. 1915hrs.

Location: Merlin, OR

Witness: 1
Shape: Sphere
Duration: 20 minutes
Multi colored light in the sky ....
Went out side to the car coming back from the car looked in the sky and something caught my eye it looked like a star, but was changing colors red blue purple green orange got my wife to come and look at it and she said it was something other than a star. I went and got my camcorder and filmed about 20 min. of it on the TV it looked like the earth spinning out of control and it got big and then small like a light bulb way off the get big again and you could see the colors change when it was big. You could also see little lights come from off the screen into and towards the object that we was looking at then leave again. Don't know what it was but sure was neat to watch. …((NUFORC Note: First of two reports from the same source of the same incident. We post both versions here. PD))

Occurred: February 14, 2002; Approx. 1915hrs.

Location: Merlin, OR

Witness: 2+
Shape: Circle
Duration: 20 minutes
Strange light that changed all colors in the sky
Went out side to look in the sky to see how clear it was and something caught my eye it looked like a long light but changed from red to orange green blue purple yellow. I just thought it was my eyes so I went in to get my wife she saw the same thing that I saw. So went in to the house and got my camcorder and filmed about 20 min. of it. While watching it on the TV you could see it get large and small to a solid white light and get big again. It kind of looked like the planet earth spinning out of control at one point. You could also see little lights coming in from the west and exiting the picture .It was in the south part of the sky at about 45 degrees. I have never seen any thing like it before. Neither have any of the 7 other witnesses
((NUFORC Note: First of two reports from the same source of the same incident. We posted both versions here. PD))

Occurred: March 6, 2002; Approx. 2350hrs.

Location: Burney, CA

Witness: 1
Shape: Egg
Duration: 6-7 minutes
Captured on video a very bright, fast moving and color changing object.
In clear night sky, one object was seen. The object moved fast, up, down, and sideways. The object changed colors while moving. Colors seen were a very bright, white, blue, green and pink. The object appeared to have two tops, which isolated separately. The object became smaller as it flew away and disappeared.

This sighting was captured on video for about two minutes. The camera was purchased in Germany, it is PAL / secam system which is not compatible with American systems. So, viewing can only be done on the camera screen. The camera is digital 8mm high E.

Occurred: April 2, 2002; Approx. 2230hrs.

Location: Lookout, CA

Witness: 1+
Shape: Circle
Duration: 30 minutes
It was above the cemetery, you could see the circular shape of it and it was spinning. There were white and blue lights on it they would change colors. Every once in a while, there would be a light flash in the middle of these circles. Very quickly it went away and went out of sight behind the mountains. A few of the town's people called and asked me if I had seen it and we all saw the same thing! It wasn't just my eyes playing tricks on me.

Occurred: May 10, 2002; Approx. 0300hrs.

Location: Bend, OR

Witness: 2
Shape: Light
Duration: 1 minute
A friend and I were camping when he said, "Look at that." I looked up. I didn't see it at first. Then it came out from behind a tree. It was a huge ball of light. It was a steady light it did not blink. We had a fire going, and flashlights. It came right over our heads. There was no sound at all. Just as it was over our heads I blinked my flashlight at it. It rose in elevation very fast. It then went on for about 10 miles. Then it shot for the stars. The light got smaller & smaller until it was gone. Just before it was gone it was about the size of a star. It was a clear night. WE were in a clearing by river. The whole sighting lasted about 45 sec. to 1 min.
((NUFORC Note: Witness indicates that the date of the alleged event is approximate, even though it apparently had occurred only three days earlier. PD))

Occurred: June 18, 2002; Approx. 2257hrs.
INTERNATIONAL SPACE STATION (ISS) Sighting

Location: Paradise, CA

Witness: 1
Shape: Light
Duration: 3 minutes
I saw a pulsating light, marrow white in color traveling north to south around 2300 hours

((name deleted)). Well, I was outside taking a break looking about the stars. Next thing I know, there was this pulsating light moving in a straighter north to south direction. It was a bit closer parallel to the west. I know it wasn't a satellite because they don't have lights. I also really don't believe it was a standard aircraft because they have lights other than white. They have red and/or blue lights. Furthermore, even IF they did have just a white light, it wouldn't pulsate nor would it be semi-dull, bone marrow white as such what I witnessed.
I know it seems crazy that I saw two sightings 23 hours apart, but I know what I saw. ((NUFORC Note: We do not know what the witness observed on June 18, but the ISS was visible from California at approximately 21:13 hrs. (Pacific) on June 17. Please see following URL for information about ISS. PD)) ISSS mistaken identity

Occurred: September 10, 2002; Approx. 1630hrs.

Location: (Near) Mount Shasta, CA

Witness: 1
Shape: Light
Duration: 5 minutes
Mount Shasta light phenomena observed in brightly lit meadow in broad daylight for 5 minutes
I saw a single stationary light change from red to orange to yellow to green and finally blue. This occurred in a brightly lit meadow near the summit of Mt. Shasta at about 4:30 in the afternoon. The light was intense, pure, and very small (about dime sized). It appeared in front of the trunk of a Spruce tree not more than 100 feet away, slightly above head level (about 6 feet up). The light would slowly transition from one solid color to the next, and then it would remain an intensely bright pure color for about a minute before slowly transitioning to the next color. I approached the tree and examined the trunk, limbs and foliage. I could find nothing that could have emanated that sort of light. I toyed with the idea of pitch oozing slowly out and while solidifying changed the reflected color, but the pitch I observed was cloudy white and in huge dried clumps. The pitch was to the right of the location of the light I pinpointed while using binoculars (I observed the light with the naked eye and binoculars). I had a camera and didn't use it; I couldn't see the light in the viewfinder. The light could only be viewed in one direction, and as I approached the tree, the light would disappear. I have no idea of where the source of the light was in relation to the distance between me and the tree except that it was in front of the trunk about 6 feet above ground level as viewed from ground level.
((FOLLOW-UP REPORT BY SAME WITNESS))
With regard to the lights on Shasta, unfortunately I have no theories as to what the lights were. Here is the whole boring story:
I was asleep in the meadow when I opened my eyes to find a red light shining into them. I closed my eyes, thinking "Oh its just a reflection from tensile on the tree", thinking better of that, I reopened them to find the light changing from red to orange. I sat up and thought that was strange. When I sat up the light disappeared, so I lay back down and discovered that the light was directional. I had to be careful how I moved to keep the light in view. The 'now orange' light was intense and pure and I couldn't write it off as a reflection. I stared at it for a while, enjoying the beauty of it; then it started transitioning to yellow. Well that really piqued my curiosity. As the light grew in intensity and purity to a very bright lemon

Yellow, I took out my binoculars and isolated the position on the tree where it appeared to be coming from. I stood up to investigate but discovered that I was ill. I had a hard time going to the tree as waves of nausea were sweeping over me. As I approached the tree, I lost sight of the light as I was no longer in it's line of site, for the moment I was sure that an examination of the tree would yield the answer. At the tree, I couldn't find anything that could cause it except for maybe pitch, but the pitch was in the wrong location and none of it was fresh. I went back to my original position on the ground and found the light transitioning from yellow to green. I admired the green light for a while and tried to take a picture of it, since I couldn't find it in the viewfinder I gave up. I tried again to find the source, to no avail. When I returned again to my original position, the light had transitioned to blue. Again I tried to approach the tree this time keeping the light in sight, but as soon as I'd get close, I'd loose the light. When I returned to the original position to regain it, the light was gone. Although I said the whole thing lasted 5 minutes, it could have been longer. I had time to sit and stare at it, take out binoculars and stare at it, run back and forth to the tree, and finally tried to get a picture of it. The camera I had was just a digital camera-not my camcorder. My camcorder would have been able to zoom in on the light no problem as it has a fantastic zoom. The camera I had with me doesn't have a very good zoom. I regret that I didn't take pictures anyway, as it may have shown up in the picture. At this point, my guess would be pretty far out at to what it could be. I have the impression that some unseen person or thing was standing there shining lights in my eyes. ((END FOLLOW-UP))

Kentucky Train Collision with Disk UFO, 2002

Paintsville, Kentucky -- At exactly 2:47 a.m. on January 14, 2002, while working a coal train enroute from Russell, Kentucky to Shelbiana, Kentucky, our trailing unit and first two cars were severely damaged as we struck an unknown floating or hovering object. I know it was 2:47 because my watch froze, and to this day shows that time.

Along with my watch the entire electrical systems on both locomotives went haywire.
Approaching a bend near milepost 42 in an area referred to as the Wild Kingdom, for the many different types of animals spotted there, my conductor and I saw lights coming from around the way.

This ordinarily means another train is coming and will pass on the other track. The outlay of the area is this, the river, #1 track, #2 tracks and a straight up mountainside, carved out for the laying of these tracks. I killed our lights as not to blind the oncoming crew.

As we rounded the corner our onboard computer began to flash in and out, speed recorder went nuts, and both locomotives died. Alarm bells began to ring and that's when we saw the objects. Apparently scanning the river for something. At least three objects had several "search" lights trained there, the first object hovered about 10 to 12 feet above the track.

It was metallic silver in color with multiple colored lights near the bottom and in the middle. There were no windows or openings of any kind that we could see. It was 18 to 20 feet in length and probably ten feet high.

With both engines dead as we rounded the corner we made little noise and the first object did not respond in time, I estimate that we hit the object at 30 mph with 16,000 trailing tons behind us. It clipped the top of our lead unit then skipped back slicing a chunk out of our trailing unit and first two coal cars. The other objects vanished.

Our emergency brakes had initiated due to the loss of power and we stopped approximately a mile and a half to two miles after impact. Our power restored after we were stopped and we notified our dispatcher, located in Jacksonville, Florida of what had happened.

We were told to inspect the cars to see if they'd hold the rail and try to limp into milepost cmg 60 which used to be the Paintsville yard which is no longer in full operation. We checked everything out and the cab of the rear locomotive was demolished and smoking, the second two cars looked as if they had been hit with a giant hammer, but looked like they'd hold the rail.

We pulled into Paintsville yard at approximately 5:15 am. The huge overhead lights lining the yard were noticeably dark and the only lights came from what we assumed were railroad officials vehicles parked near the end of the track. We pulled to a stop and began unloading our grips off the wounded train. We could hear what sounded like an army of workers immediately tending to our train.

Vehicle doors slamming, guys running by in weird outfits and lights glaring from all directions, the one thing missing was railroad officials.

A guy named Ferguson shook my hand and asked me to follow him into the old yard office. We did, once inside they, and by they I mean I have no idea who these people were, began to ask us hundreds of questions, they then told us for our own protection we'd be medically tested before we could leave.

I asked repeatedly to talk to my road foreman or trainmaster and not only were these requests denied but they confiscated my conductor's cellular phone.

Hours later we were led outside the old yard office and the strange things continued to happen, the 2 locomotives and two cars were removed from the rest of the train we had brought in and my only guess was parked 4 tracks over under a huge tent like structure buzzing with activity. We were led off the property and told, due to national security, our silence on this matter would be appreciated.

We were then put in a railroad vehicle and taken to Martin, Kentucky were we went through questioning again with railroad officials and were then drug tested. After all of this we were sent on to Shelbiana, where we took rest for 8 hours and worked another train back to Russell. Working back we passed by Paintsville, no sign of the engines, cars, tent, people, nothing.

Thanks to Peter Davenport

Source and references: NUFORC-http://www.nuforc.org/

Image © http://terraserver-usa.com/image.aspx?T=1&S=11&Z=17&X=855&Y=10465&W=1

Peter Davenport

Occurred: November 30, 2002; Approx. 1745hrs.

Location: Yuba City, CA

Witness: 1
Shape: Triangle
Duration: 2 minutes
Three large lights in triangular formation spotted near Yuba City, CA on 11/30/02.

It was dark and overcast, at about 5:45pm on Saturday, November 30, 2002. I was driving north on a busy stretch of Highway 99, 5-10 miles south of Yuba City, CA. Roughly 1-2 miles ahead of me, I saw a very large white/yellow light hovering at about 1,000-2,000 feet in the air, approximately 100 yards off the highway. As I approached (about 1/2 mile away), I could see three large, distinct lights in a row, and one small red/orange blinking light on the left (west). I assumed the blinking light was a landing light on a wing, so I briefly dismissed it as a plane or jet. But the object continued to hover (as far as I could tell, it never moved), and as I began to pass underneath it, I realized that the three large lights were arranged in an equilateral triangle formation. I couldn't make out an outline of the object, because the three white/yellow lights were too bright. Unfortunately, there was too much traffic to find a place to safely pull over and observe the object. But there were 3-4 cars in front of me and 5-6 behind me, so I have to assume that someone else saw what I saw. I know that Beale AFB is roughly nearby, so who knows what it was. I suppose it could have been a Harrier jet, but as near as I could tell, it wasn't making any noise. I've seen hundreds of military and civilian aircraft in my life, and I've honestly never seen anything hover like that before. I am a senior at Chico State University in northern California, and to my knowledge, I am not insane.

Occurred: December 6, 2002; Approx. 2030hrs.
Location: Chiloquin, OR
Witness: 3
Shape: Sphere
Duration: 45 minutes
6 rotating, flashing spheres of colored lights, in night sky above Chiloquin Ridge, Oregon
Looking to the southeast of Chiloquin Ridge we saw 6 spheres of flashing light just below the tree line. These balls of light seemed to be rotating and flashing red, blue green, and gold lights. For 10 or so minutes the lights bounced and moved erratically in the sky, after which it appeared they were flashing to each other for several minutes, and then proceeded to form a v formation and flew in sort of circle for another few minutes before separating and began bouncing and moving erratically again, finally moving farther out in the sky. Also, while the lights were in the tree line there was also a bright light below them, backlighting the trees, making the forest floor, and Manzanita bushes glow. We watched this for about 45 minutes before these lights stopped flashing and moved very quickly toward the south and disappeared, the bright light in the forest below went out, leaving us in pitch-black darkness. My husband and my self are in our late 40's college educated, myself now a full time house wife, my husband holds a managerial position at a local company, our son 13 is in Jr high school. The three of us observed these lights after going out to close up the barn and put the dog up for the night.

Occurred: December 24, 2002; Approx. 2100hrs.
Location: Susanville, CA
Witness: 3+
Shape: Light
Duration: 3 minutes
North Eastern Calif. Fiery object emits fire like objects that float down to ground.

On December 24, 2002 I pulled up to one of my friends house, him and another friend said they were going up to the construction site of a new hospital where my friend worked as the night security guard. (Only person at the site at night) The security guard had called terrified, babbling, and literally crying that a UFO was hovering over his office (the guard is 6'2 and weighs around 200 lbs.) I immediately drove to his work located about a mile and a half from where I was. On the way up I saw an orange light in the sky but thought that it probably was just a airplane and didn't notice when it must have disappeared.

When we got up to the hospital the security guard was still hysterical and was trying to tell us what it looked like and what it was doing. We kind of joked with him for about ten minutes when we all went into his office that was located in a modular next to the hospital. I called my father who is the chief of one of the fire departments in town and asked him if he would call dispatch and see if anyone had reported anything. One of my other friends called his girlfriend on the office phone. The guard sat down in his chair and said to me I was looking out this window when I first saw it, as the security guard pointed out the window to show me where, he said, "Oh my god there it is again".

When I looked out the window I saw an orange light that was bright orange. It was somewhere around six hundred feet off the ground and about a mile away. I could not pick out a shape of the object only that it was emitting a light that looked like fire and that it was coming closer. I immediately called my dad who lives up on a hill about the same distance away and told him that I didn't know what this thing was and for him to try to see it from his advantage. Then the UFO started emitting little tracer fireballs that looked like they were floating down (swaying back and forth). It must have dropped around 8 of these. I then became frantic when my friends started screaming in terror as this object is slowly coming towards us. My parents immediately hung up on me when they heard my friends screaming to come to where we were. At this time I felt that it was time to leave so we ran to the cars. As we were running a car came over the horizon and the UFO disappeared like turning off a light bulb in a dark room.

Friends of ours and my parents and grandfather arrived shortly and we stayed there for a while but it never came back. The security guard said that when saw the UFO the first time a car stopped alongside the highway to watch. But never heard any other reports. Would love to know if anyone has seen anything similar in the North eastern California, South eastern Oregon, or North west Nevada area

References:
1. Volume II, The UFO Evidence, A Thirty-Year Report - Richard Hall, 2000
2. CUFOS report, Mark Rodeghier,
3. Ted Phillips, Center for Physical Trace Research.
4. Joan Woodward, Animal Reaction Specialist.

2003- 2005

Once again, the previous year spouts off with countless sightings from, Sacramento to the Oregon Coast, from Reno to Susanville, Santa Rosa, Napa and of course, Mount Shasta. We have, for brevities sake, listed the significant and plausible sightings. We removed satellite sightings, plane and meteor sightings as much as possible. NUFORC is inundated with sightings of all flavors. Many new objects that are known are mistakenly reported.

Occurred: January 21, 2003; Approx. 1149hrs.
Location: Bend, OR
Witness: 2
Shape: Disk
Duration: 15 minutes

A single, stationary fairly large saucer/disc with rotating lights that changed colors from red to green, blue and white that hovered f

Object was seen at 2:00 a.m. 01-28-03 in the southwest skies over Bend, Oregon, USA. It appeared to be located approximately in the direction of Mount Bachelor. We watched it for 15 minutes until it dropped below the rooftop line of our neighbor's house.

There was one single disc/saucer type object observed that was stationary for the fifteen minutes we watched it. Skies were clear. It was stationary although it was dropping some in altitude to eventually drop below our neighbor's rooftop that was our viewing area and certainly moving away from us as the vision got smaller in that fifteen-minute period. It was as if it was going sideways as it did get smaller and did drop in altitude a little, but appeared stationary at the same time.

It was fairly large as seen at a distance, (hard to determine how far away it was) had a rotating outer circle of lights all the way around the saucer and the lights would change colors from green, to all red, to all blue and to all white. Most of time, I observed more all red, or all green and this light rotated the whole time. I was thinking to myself, just like they make it seem in real life or in the movies.

We had time to get our glasses and also binoculars to take a better look. With the binoculars it was strange to only to be able to see a white undefined mass. Kind of like a thin white mass/cloud-like in the darkness. I thought that was strange. Obviously you could see it with the naked eye, but there was some sort of field that made it unobservable while trying to use another type of viewing apparatus. Thought about taking a picture of it, but didn't. Glad there were two of us that viewed it, so there is no question that we saw what we saw.

After it dropped below the rooftop line of our neighbor's house, we didn't pursue it any more. As far as backgrounds of the observers - I am 49 years old and a professional by occupation. The other observer is a 20-year old college student who was home from school for a few days, and had been studying late. He originally spotted the thing and brought it to my attention. Interesting, yet kind of scary.

I never believed in those things before, but I certainly have to change my stance on them now. Either that, or our own government is experimenting with some type of contraption in the wee hours of the morning when they knew most people would be asleep.

((NUFORC Note: Witness elects to remain totally anonymous. The object seems inconsistent with a celestial body, since a star or planet in the southwest sky would appear to "rise" as the Earth rotates to the east. However, the star, Sirius, which is very vivid during the winter months, would be in the southern sky at 0200 hrs. We do not know what the witnesses observed. PD))

Occurred: January 25, 2003; Approx. 1800hrs.
Location: Corning/Redding/Richfield, CA
Witness: 1
Shape: Light
Duration: About 3 hours
Three bright lights seen in many California towns

I saw the three lights from many different towns in the California valley. The witnesses included siblings, parents, grandparents, and me. We saw them at first around 6:00 P.M. and they were three circular lights that flew in circles. The two dimmer ones followed the brighter one and they were going pretty fast. They went in the same path until around 9:30 when the two brightest ones disappeared. The dimmest one kept going for about ten more minutes before disappearing. We know they weren't planes or anything like that because we went to different towns to make sure and they were always there and we were too far away for them to be searchlights.

Occurred: February 16, 2003; Approx. 2015hrs.
Location: Chico, CA
Witness: 2
Shape: Fireball
Duration: 6 seconds

Driving home in the evening headed east, I looked north out the van window and saw what appeared to be a rocket entering the atmosphere. I called to my wife to look and we looked at this shooting streak of light for a good 6 seconds. I proceeded to head to the Mountain terrain where it would be lost out of sight behind the mountains and clouds. I in a frantic state turn the radio to a news station in expectation of the beginning of an attack from N. Korea or something of this nature. To my surprise got nothing. When I got home called the police to report it, I was 1 of 2 in 20 mins to report. I call the local news TV station and again was only 1 of 2. I believe it could be part of the space shuttle Columbia.

((NUFORC Note: We spoke with this witness, and she sounded quite lucid and sober-minded. He contacted Chico, CA, 9-1-1 facilities immediately after his sighting. PD))

((FOLLOW-UP CORRESPONDENCE FROM WITNESS))

Dear Mr. Davenport,

Thanks so much for the on going updates. I spent some time since our submission visiting the Web site to familiarize myself with it. You have a very interesting and tremendously challenging job! I have a whole new awareness and appreciation of the job sorting out UFO's etc. It definitely requires a certain personality, intelligence and integrity.

Our experience has definitely changed our life. I always thought it was other people who saw those types of things (i.e. comets and things flying in the sky). It felt really good to know that there are people that we could call after having that kind of experience. So you might pass it on to the powers that be that the arrangement that you have with the local authorities in our experience was a highly effective one from our perspective.

Best Wishes to you.

Occurred: February 16, 2003; Approx. 2022hrs.
Location: Chico, CA
Witness: 2
Shape: Fireball
Duration: 3-5 seconds

A fiery ball of light descending from left to right and high to low in front of us changed colors from white, to yellow to green.

My wife and I were traveling in our car in a north by northeast direction on Palmetto Ave. in central Chico at 8:22 PM PST (per the clock in the car). In a very surprising instant we both observed what appeared to be an astrological event in the eastern sky. A large, bright object that seemed to me to be a fireball was descending rapidly from left to right starting at about "12 o'clock" and disappearing at about "5 o'clock" in our field of vision. The object basically followed very slightly curved path approximately 30-40 degrees from a perpendicular line drawn from the horizon directly in front of us.

The size of the object appeared 1/2-2/3 the size of the full moon overhead. It changed color while in view, from bright white to fiery yellow to a brilliant iridescent blue/green. My first thought was that it was a large shooting star or meteor, but then I was shocked to see it change its size (it become larger) and color. About half way through its descent it begin projecting a tapering trail as its color changed to bright yellow, almost as if there had been an explosion and acceleration in speed. The trail remained yellow as the object then took on a beautiful greenish-blue color just before it disappeared from view. It was not clear whether the object disappeared over the horizon or simply self-extinguished.

I estimate the entire elapsed time of viewing was less than five seconds.

((NUFORC Note: Other reports from multiple points in the western U. S.. We spoke with this witness, and he sounded to us to be exceptionally sober-minded, sincere, and credible. PD))

Occurred: February 16, 2003; Approx. 2027hrs.
Location: Chico, CA
Witness: Numerous
Shape: Fireball
Duration: seconds
Telephoned Reports:

NUFORC received telephone calls from the Chico, CA, 9-1-1 facility, as well as from the California Highway Patrol (Susanville), reporting that those facilities had received several reports each from citizens, reporting the dramatic fireball seen traveling through the night sky, apparently generally from north to south. The county fire facility near Susanville, CA, also reported to Chico 9-1-1 that their office also had received reports.
((NUFORC Note: We express our gratitude to Chico 911 and the California Highway Patrol for having taken the time and trouble to share the reports and information with our Center. PD))

Occurred: February 16, 2003; Approx. 2040hrs.
Location: Chico, CA
Witness: 1
Shape: Sphere
Duration: 5 seconds
A bright glowing blue sphere shaped object with a fiery tail falling to the earth.
One object seen falling to the earth Bright Blue in color, sphere shaped with a fiery tail.

Occurred: June 9, 2003; Approx. 0350hrs.
Location: Burney, CA
Witness: 2
Shape: Circle-Orbs
Duration: 10 minutes
Formation of 3 flying orbs flew over the skies of Burney California.
The objects were travelling toward the direction of my apartment in a formation of 3 with 2 side by side at the front and the other trailing below and to the left of the others. At first we just thought they were stars, but the stars around them were not moving. They were extremely high maybe even outside of earth's atmosphere and they were flying in perfect synchronicity with each other until they reached almost over our heads, then they all turned left (keeping perfect formation) and disappeared out of sight. All was quiet at that time of the morning and we heard no sounds from the objects.

Author's Note: There were several space shuttle launchings and landings witnessed by people who either knew it was ours, or reported it as an UFO. They have been deleted. There are also mistaken UFO reports of naval satellites that fly in formation at high altitude. The International Space Station also orbits around the globe and has been mistaken for a UFO unless the observer knows it is there.

Occurred: July 4, 2003; Approx. 0946hrs.
Location: Placerville, CA
Witness: 2
Shape: Formation

Duration: 2-3 minutes
Triangle formation of lights, Looked like moving stars about Placerville.
We looked up and say three lights in a triangle formation. They did not seem like they were all one craft. They moved slowly across the sky and looked jittery. What I mean by that, was when compared to a plane or something else, the planes moved smoothly...not jerky. Anyway, we followed them for a couple of minutes, they were straight above us, the lead one sort of disappeared, the next one, then the last one. They were the same size as the stars. It just looked like 3 stars moving together, same light and everything. Anybody know what it was? ((NUFORC Note: We suspect the witness observed the three satellites, which fly in formation, and which are thought to be part of the U. S. Navy "NOSS" surveillance system. PD))

Occurred: August 5, 2003; Approx. 0430hrs.

Location: Ashland, OR

Witness: 1
Shape: Oval
Duration: 5 seconds
Camping at a high mountain lake & happened to see 3 oval electric blue objects traveling directly west to east at a fast rate, the leading one, then at a distance, one directly behind the lead with a slightly smaller one at about a 5 o'clock angle, yet very close to the second, when they were directly over head, the 2 in the rear went directly opposite directions north and south at a VERY high rate of speed, the lead continued at the same speed, west-east until out of sight ((NUFORC Note: Date is approximate. PD))

Occurred: August 7, 2003; Approx. 2200hrs.

Location: Newberry Crater National Monument, OR

Witness: 2
Shape: Light
Duration: 1 minute
Bright light pretending to be a star in big dipper
I was reading over some of the sightings and was stunned when I read the report from 4/10/08 (surprise). So I decided I should report an almost identical sighting that occurred while I was camping in Oregon.
I was visiting a friend there that told me to stand in front of him and he pointed at the big dipper specifically the star in the upper left corner that begins the handle portion of the dipper. We were camping inside of a dormant volcano so we were up high enough to have a panoramic view of the big dipper was lying just above the horizon with the handle portion on the left in the sky.
He was talking to me for a minute or so and suddenly it was like a very bright star was superimposed over the star in the upper left corner of the cup portion of the big dipper. This light began to drift slowly through the big dipper and it crossed over the star in the bottom left of the cup portion.
((NUFORC Note: Source of report indicates that date of the event is approximate. We believe that one would have to rule out the possibility of a satellite, possibly an Iridium satellite, before concluding that the object was a genuine UFO. Just a guess. PD))

Occurred: August 24, 2003; Approx. 2200hrs.
Location: Mount Shasta, CA
Witness: 1
Shape: Light
Duration: 2-X about 10 seconds
Two occurrences of lights appearing to play leapfrog on eastern slope of Mt. Shasta
Three or four lights appeared to play leapfrog. At any given moment there were three stationary lights in the sky, however, as one light went out another light would instantly reappear just opposite the two remaining lights. I would estimate that the lights were about 20 to 30 yards apart from one another. The sequence occurred about five times for a period of about 10 seconds. It was a moonless night with the only nearby town far in the distance. The lights appeared to be less than two miles away and fairly near to the hilly horizon. On this night the lights were brighter than Mars and about the same color of yellow. We waited approximately 30 minutes and another event of the same phenomenon occurred in a distant location. This second occurrence was higher in the sky and at a different angle to the first. The leapfrog event continued about 5 times for a period of about 10 seconds. We were camping on the eastern slope of Mt. Shasta at an elevation of about 8000 feet.
((NUFORC Note: Witness is a trained medical professional. We suspect he probably is a very good witness. PD))

Occurred: September 1, 2003; Approx. 2025hrs.
Location: Grants Pass, OR
Witness: 2
Shape: Other
Duration: 3 minutes
We saw a lobed orb over Grants Pass, Oregon, on 9/1/03.
We were watching a jet going north to south over Grants Pass, OR. We noticed two flashes in the area of the jet. A few moments later, Debi, trying so look at the jet through binoculars, accidentally spotted an object following it. It looked like a sphere made up of oval lobes. For a few minutes it followed the jet, then it changed to an easterly course. She followed it into the east, when she lost sight of it.
The next day, ((name deleted)) contacted the nearby Medford International Airport, to see if they had noticed anything following the jet. They took her name and number, saying they would get back to her, which they never did. Later we called and left a phone message for MUFON, but we never heard anything back from them.
A few months prior to witnessing this orb, our 12-year-old granddaughter, who was living with us at the time, sold us she had a dream that she was taken out through the ceiling of her room. She thought it was just a dream, but the nest day we saw a group of perfectly round bruises on the area of her back near her coccyx. She did not remember any play or other activity, which could have caused such uniform, round bruises. We took her to her doctor, but he was unable to identify any possible cause. He and his colleagues had never seen anything like them.

A few weeks later, my wife had a similar dream. I think she needs to share it with someone. She was taken through the ceiling into a bright, oval room. She was on a metallic table, and a device emerged from the ceiling. Three bright beams of light came from it, focusing on her stomach. Then, she awoke back in our home, on her bed. After she shared her dream with me, I noticed a large deep bruise on the right side of her naval, with three hole-marks in a triangular shape. This made us wonder if it was a dream, or did it really happen? I am a psychologist and I do testing for the State of Oregon, and my wife is an intelligent, mature woman. We are not given to wild imagination or conspiracy theories.

We have reason to believe that aliens have been abducting other members of her family, including her mother, a sister and at least one brother (who won't speak about it). If anybody would like to know more details, we would like to share them.

Occurred: October 18, 2003; Approx. 2015hrs.

Location: Clearlake, CA

Witness: 1
Shape: Unknown
Duration: 30 seconds
Strong bass sound that cycled up and down from the sky. No crafts in sight.
It started with my bedroom window rattling and a deep hum that seemed to cycle up and down. I went outside to observe where the sound was coming from and my attention was drawn to the sky. It was a clear sky above me with no signs of any craft of any kind. The sound continued for about thirty seconds and stopped like someone turned off a switch. This is the first time I've heard anything like this before. Please review this and tell me if anybody else reported this incident.
Thanks for you time. ((Name deleted))
((NUFORC Note: One of three seemingly credible reports from same source. PD))

Occurred: February 3, 2004; Approx. 0305hrs.

Location: Medford, OR

Witness: 1
Shape: Unknown
Duration: 3 minutes
Fast moving object disappears into night sky
I am a Commercial pilot and was on the ground at the Medford (MFR) Oregon airport at 3:00 am waiting for transportation. I looked into the clear sky to the north over towards the Rogue Valley VOR (navigation facility) and saw an aircraft, estimated at 10,000 feet, maneuvering at very different angles of turns, circles, and other zigzag patterns. The aircraft had a red/orange tint with no flashing lights or strobe lights. This seemed very odd. I thought for sure it was an F-16 from the Klamath Falls base doing some fooling around. I looked for another plane in trail (most travel in pairs), but there was nothing else.
All of the sudden, the aircraft started climbing straight up and disappeared into the night sky. I watched it as long as my eyes could see it. It never returned as I stood on the ramp looking for about an additional ten minutes.
There was no noise. I have never seen an aircraft move as quickly as this one.

I departed the airport and contacted Seattle Center for my clearance. I asked them if they had any traffic on radar over the last 15-20 minutes in the vicinity of MFR. They said no. I asked him if there had been any military traffic. Again the answer was no.
I think I may have seen my first UFO.

Occurred: February 11, 2004; Approx. 2000hrs.
Location: Yreka, CA
Witness: 1
Shape: Other
Duration: 15 seconds
Observed UFO traveling south to north, high altitude, v-shape, 7 maybe 8 lights in a v shaped pattern.

Occurred: February 20, 2004; Approx. 2330hrs.
Location: Roseburg, OR
Witness: 2
Shape: Cylinder
Duration: 5 minutes
Orange glowing cylindrical spinning object - Roseburg Oregon
 After we stopped the car, we saw that it was moving slowly in a straight line. Husband and I were driving southbound on hwy 99 to Roseburg (approx. Winchester) I noticed a glowing orange light about 600 feet above the ground. I asked him to pull over and see w southbound. As we got out of the car, it was above us. It had stopped moving. We noticed it was cylinder shaped. It was rotating like a spinning globe, with a tilt. After 2 min of watching this, it started moving in the same direction as before. Just then a cop car pulled up to see what we were doing. As my husband was talking to the police officers, I continued to watch it slowly move and flicker, almost like it was burning out. Then from the bottom of it, a glowing orange small object fell. As my husband was trying to explain and show them where it was, it had vanished. They then joked with us, thought we were lying. Then drove off. ((NUFORC Note: Witness elects to remain totally anonymous. PD))

Occurred: June 20, 2004; Approx. 2300hrs.
Location: Ashland, OR
Witness: 1
Shape: Light
Duration: 1 hour
Strange distant lights were moving in zigzag patterns from south to north, one after another approx. 1 minute apart.

Very clear sky with no moonlight. At approximately 11: 00 P.M. (Pacific Time) I looked up into the southern sky and saw very small light objects moving one after another from south to north across the night sky in a slightly "zigzag" pattern as if from one imaginary point to another in a northerly direction at uneven speeds. Unlike any kind of aircraft I've ever seen. They appeared to be barely visible and at an unusually high altitude for any aircraft that I am familiar with. They darted like "insects across a glass window" one after another about 1 minute apart, and along different paths which, in my mind, ruled out satellites. I watched for almost an hour and came back out around 12:30 am and there was no sign of them. I have no idea how long this had been going on but would like to know if anyone else witnessed this event. Thank you. I have never filed a report like this before now.

((NUFORC Note: We spoke with this witness via telephone, and he sounded quite credible to us. However, when we explored his sightings, it seemed to us that he might have been witnessing satellites passing overhead. Given that the witness observed the objects as they flew to the east, we suspect that at least some of them may have been satellites. However, please see other reports from Washington State for the same time and date. PD))

Occurred: July 21, 2004; Approx. 0000hrs.

Location: Lake Shasta Resort, CA

Witness: 2

Shape: Fireball

Duration: 7 seconds

I was out for a walk at midnight, saw an object burning and disappear behind the mountains. Friend and I were out for a walk. We began talking, and looked up. We saw a huge object, engulfed in flames, with a trail behind it. It disappeared behind the mountains. If you compare it with the full moon, it was about the size of it, or bigger.

Occurred: August 18, 2004; Approx. 0000hrs.

Location: Weed, CA

Witness: 2

Shape: Changing

Duration: 6 hours

We saw big and bright object that changed shape and form in sky, happened 4 nights in a row, and moved erratically.

First we saw what appeared to be a big bright star in the east but realized it was more. It was moving side to side and up and down and around in the same area. Then we noticed there was a red light also to the left of the big one. We got the camcorder out and digital camera and zoomed in on the big bright white one but it really was round with green around it and appeared to be spinning inside with different colors, (red,black,white and orange).At times the object when zoomed in appeared to look like a planet but then changed shape. We also have lots of pictures and video. So far from the pictures we have taken there are 9 different forms each night. The object is a different shape in each picture taken. The pictures taken each night are all similar in shape and color. We have seen them 4 nights now in a row in the eastern sky.

((NUFORC Note: Probably Venus. PD))

Occurred: August 20, 2004; Approx. 2200hrs.
Location: Grants Pass, OR
Witness: 1+
Shape: Circle
Duration: 30 seconds
Three bright red cylinders changing to very bright white lights before disappearing at high speeds
Initially, there was a very loud low rumble that soon showed twin red balls moving very fast for several seconds before disappearing very suddenly in a northwesterly direction. After about 5-10 minutes the sound returned and there was a red cylinder shaped object in the north. Suddenly, the cylinder became a very bright light like a searchlight had been turned on. Another bright light in the south and another in the west answered this beacon. These lights stayed on for 10-20 seconds in a stationary position and then suddenly all three became red balls moving very quickly and disappearing to the west. After another 10-15 minutes the same red cylinder was observed very high in the sky moving from west to east and the white light appeared three more times but at a very high elevation.
((NUFORC Note: We received a telephoned report from this individual, and he seemed quite credible to us. We have asked him to invite the other witnesses to submit reports, as well. PD))

OBVIOUSLY CELESTIAL BODIES, UFOs to the UNINITIATED
Occurred: August 22, 2004; Approx. 1152hrs.
Location: Susanville, CA
Witness: 3
Shape: Changing
Duration: All night
Minister and police spot ufos in Susanville, CA
For the past seven days, my wife and I have been observing every night 3 to 6 craft in the sky. They change colors (red, green, blue,) and when we observe them through the binoculars we notice they change from saucer to oval shaped. We can see like portals on the top outside of the craft through the binoculars. With the naked eye they sway from side to side. A green light or red light blinking it changes. Just a few nights ago my wife and I were in Skyline Park in the desert in Susanville and we were observing them. A police car pulled up and the officer walked up to us. She asked us what we were doing and we told her and pointed out a ufo. We handed her the binoculars and she looked at it. She looks back at us with a stunned look on her face and then pulled her maglite and flashed the craft. The craft flashed back at us and it moved, the officer then looked at us and said: "Did you guys see that it moved"! And then said again to us: "You guys are freaking me out"! And she took off! We noticed that there would be two of these objects by the moon and the rest in a circular position in the night sky around the Susanville area. Another thing I would like to add is that the objects will black out when we try to see them in our telescope, when we look away they turn their lights back on. They don't seem to mind us looking at them through the binoculars. Yesterday my wife and I flashed one of the closest ones with a spotlight and it moved its way toward us, so my wife runs in the house to avoid it. We have currently had a small storm and that night with the police officer there was a storm in the distance about a mile, it seemed like they were observing the storm. These ufos are still out here right now. Sincerely Reverend ((name deleted))

((NUFORC Note: We suspect the witnesses probably were observing celestial bodies, probably "twinkling" stars. PD))

Occurred: October 15, 2004; Approx. 0145hrs.

Location: Grants Pass-Murphy area, OR

Witness: 2
Shape: Unknown
Duration: 2.5 hours
Fighters give chase to UFO within President Bush restricted air space.
October 15, 2004 at 1:45 am Friday morning, I was going to bed and noticed a orange white UFO flying behind my home, coming from the North flying South East. I mentioned to my houseguest what I was seeing. We both filmed a huge UFO that looked to be chased by fighter jets. It appeared to me the fighter jets were having a hard time catching up to the UFO. This continued to happen several more times until I fell asleep at 3:00 am, waiting for it's return once more.
President Bush was in another town, during this time so the Air Force was at full alert status of any flying objects in the area.

Occurred: December 9, 2004 Approx. 0004hrs.

Location: Chico, CA

Witness: 1
Shape: Triangle:
Duration: 5 seconds
Observed flying triangle with five orange lights as it flew silently and slowly in an eastwardly direction over city.
Shortly after midnight I witnessed one large triangle-shaped craft/object fly silently over Chico State University almost directly overhead from where I was standing. The object was visible as it flew through the low, blowing, broken overcast. Five orange lights were visible running around the circumference of the object, and none were particularly bright. I would estimate its altitude was between 1,500-3000 feet, and its speed between 50-80 mph. (quite slow) The direction of travel was east.

Occurred: January 11, 2005; Approx. 2200hrs.

Location: Arbuckle/College City, CA

Witness: 2
Shape: Light
Duration: at least 10 minutes
Red flashing lights were hovering over a body of water near us.

We live in a small farming community; there is a three-mile strait away between our town and the next. We were driving out to a friends house when over a canal we observed between 6 and 8 red lights in an almost strait, yet the rim of something round looking, line - hovering. The lights are not there normally and were flashing one at a time. It was visible almost the whole drive and until we entered our friends home. No sounds came from the object. It looked like it was about a mile away over a waterway that is a run off of the Sacramento River. There was a car ahead of us that may have also witnessed the lights. We didn't have a camera with us or we would have taken pictures to send you.

We are friends that drive this road almost daily. We are both students at a near by college. my friend is a former military member. This is not the first time we have seen things before and have had a local sheriff's officer verified or at the least acknowledged past sightings.

Occurred: January 12, 2005; Approx. 0145hrs.
Location: College City, CA
Witness: 3
Shape: Disk
Duration: About 45 minutes

Two sightings occurred about 5.5 hours total between the two. 3 people in the country skies saw oscillating lights in the sky.

Driving home from family/friend's place, headed north on Grimes-Arbuckle road - the object was off in the east (just as my husband and buddy described) red lights oscillating at random, blinked in the sky. The arrangement of the lights indicated that there might have been a disk or saucer like shape to the actual object. After driving down the road about .5 to .75 miles, we turned and headed back toward Arbuckle. It was still visible as we hit the town and cruised about a mile through it, north bound, as well as about an additional 2 miles north of town as we headed to turn east onto Hahn Rd. All the while, it continued to flash a great distance out, as it stayed pretty much just a little north of our vision - indicating a great distance between us and the object or that it was moving with us, more than likely it is the former explanation. After about 2 miles down Hahn, we headed south on Grimes-Arbuckle Rd. Again; at this point we all acknowledged seeing some new unexplainable bluish-green lights along the 2047 drain (a run-off of the Sacramento River). This was the area at which my hubby and friend saw the red flashing lights earlier (see other report). This area is about 1 mile from where we were, at that point. We then headed back to Arbuckle after having scaled about 6 miles of road around a big square of fields and orchards. Both my husband and friend have seen objects/lights in our area and elsewhere. I have unfortunately seen many objects/lights and the like through out my life, within the central and northern areas of California. There is also a somewhat extensive encounter record for my family. All three of us are college educated and were of sound state of mind at the times of the sightings. No one was intoxicated nor on any type of drugs.

Occurred: January 12, 2005; Approx. 2215hrs.
Location: Chico, CA
Witness: 2
Shape: Light
Duration: 10 minutes

Large bright star-like light, mostly orange, travelled from N or NE towards South, basically in a line, but with an erratic motion: pausing, moving up and down randomly, etc. There was no aircraft noise, no flashing lights, and no red/green/white lights that would normally be seen on aircraft. Light faded out as it moved towards the south, seemingly moving farther away. Observers: two 30-year-old women.
((NUFORC Note: Witness elects to remain totally anonymous. PD))

Occurred: March 7, 2005; Approx. 2000hrs.
Location: Alturas, CA
Witness: 3
Shape: Triangle
Duration: 20 minutes
Newspaper Article to March 7th Sighting
This report is in response to an email that my dad sent to you and you sent him an emailing asking him and me to fill out this report.
My editor, and I work for a local newspaper was so interested in my UFO encounter (on March 7th, 2005) that he wrote an article about it in our March 10th issue. He wrote the article off a paper I wrote describing what had happened. It describes better than I can, the events that happened that night. I have used *** on last names and street addresses. My editor later wrote a follow-up editorial on April 7th, 2005. Here is the article as follows. If you would like me to send you an actual paper, that is available also.
What was in that March 7th sky? So you think your Monday was a bit taxing do you? Elaine ****, Webmaster and graphics artist at the Modoc County Record and her family were jolted by some very strange skies over Alturas.
She said it all started about 8PM when she saw something odd from her living room window, "I saw a large hovering aircraft with five lights, shaped like a boomerang," she said. "I stared at it for at least five seconds before jumping up and looking out another window. I didn't see it from there, so i shut off the lights, but didn't see it again." Elaine said she ran outside and the object was gone, but several jets were flying in different areas of the sky.
"I called my dad and told him I thought I just saw a UFO," said Elaine. "He told me to come over to his house, so I bundled up my son and raced over to their house on East *** Street. When I arrived my dad and sister were outside looking at the 12 or so jets flying back and forth over the night sky." Then things got weird. "A big bright ball of light appeared from the north," said Elaine. "I pointed it out to my dad and we watched it come towards us. As it did, it morphed into a triangle. As it got closer to us, we could see the belly and notice it was indeed a triangle with little red lights creating the shape.
"On the right side was one green light. It flew over our heads flying in circles and turning at sharp angles. It's flight pattern defied logic. The lights were steady, not blinking once." The craft flew south towards Likely and then it turned back, appearing to head straight for the family.
"Once it was over our heads, it flew back and forth, like a bird flying in all directions. At that time, we noticed a second one to the north. It was white. Once the two triangles met, they started a dance in the sky, gliding here and there," Elaine recalls. "Then, all of a sudden, out of nowhere, one of the triangles suddenly burst into a bright ball of light. I thought whatever it was, it was going to crash land, but it broke away from the ball of light. The lights hovered over **** Street near the Alturas Cemetery, just sitting there until they slowly disappeared." Elaine said the triangle then shot off towards the north, followed closely by the second one, hovered a minute or so, then disappeared.

"There was no sound at all when the burst of lights appeared and when the triangle appeared," said Elaine. "It was pretty freaky, but we all saw it." There were no reports to the police or sheriff's office and the military has not returned calls. If anyone saw what Elaine saw on Monday night, contact the Record at and let us know.

Occurred: March 7, 2005; 2000hrs.
Location: Alturas, CA
Witness: 3
Shape: Triangle
Duration: 30 minutes
This is clearly happening here and to quote Mr. Holloway.... "There are UFO's in Modoc County."
I will respectfully refer you to the Editor of the County's Newspaper, Rick Holloway, in Alturas, CA. The phone number is (530) 233-2632. I believe this Well Known; straightforward and extremely respected man is probably your best contact person. I will also have my Daughter respond to your request to complete a "form" as well.
After my daughter called me on the 7th of March, 05 @ 8:00 PM and informed me that she had just seen a ufo up close I told her to grab my grandson and come to my home, she was very upset. I went directly outside and looked across the sky towards her home. I soon became aware of red triangular crafts flying in the sky everywhere.
I had one fly / float over my head and it was very distinct and clear. Very fast and able to stop in flight, expel a white 'flash' of something, looked like plasma.
At that point I grabbed my two daughters and ran in the house. My wife was also present, but stayed inside the home with my two-year-old grandson.

Occurred: March 8, 2005; Approx. 2100hrs.
Location: Yuba City, CA
Witness: 2
Shape: Triangle
Duration: 5 minutes
3 Triangular objects, bright white lights with 2 round red "tail lights", slow speed, low altitude
3 bright white lights forming the shape of an obtuse triangle, with approximately .25 mile between each object.
The top object and the one on the right slowly began heading south.
The object on the right passed us first- we were on the balcony and it was south west of us, maybe 3 city blocks laterally away and .5 mile high, traveling at approximately 30mph. At first we heard a faint noise-almost like a prop. Plane- then that went away as the object neared us. There were white lights that would at times combine with a blue and a red. The object was white (or silver?), triangular in shape, but the bottom of the craft bowed down like the hull of a boat. The back of the craft had two round red lights (like tail lights on a corvette!) the size was approximately 18 ft. long and 14ft. wide. The two crafts passed over identically (about 3 minutes apart from each other) kept the same speed and then when almost out of sight, gained altitude and continued going southeast.

The 3rd craft (which would have made the left corner of the triangle) then proceeded forward and went over EXACTLY as the others did. THIS appears to be the same sightings of the stealth

Occurred: May 28, 2005; Approx. 0124hrs.
Location: Willits, CA
Witness: 1
Shape: Circle
Duration: About 2 minutes
Black spot, killed digital camera
I was walking through my yard when I heard a high-pitched sound, sort of like the sound you hear when you first turn on a TV,
I looked up (that's where the sound came from) and I saw what I thought was a jet plane, we see planes a lot in this area, they fly so high that all one can see is a sliver dot.
But this on was pitch black and it was moving extremely fast from north to south.
I ran to my house and got my camera (a Sony digital). I got back to find that the object was almost out of sight and now it was moving in the opposite direction, it had changed course.
I got it in the viewfinder of the camera and took about 25 pictures in rapid succession. I would have I was about to take more when the camera registered that the flash memory card was full, this card was a 512mb and could hold almost 180 pics.
I watch the object until it was out of sight and then I went into my house and looked at the pictures. All of them showed a bright white light with a trail of smaller lights behind it.
All other pictures I have taken sense then with that camera have had a blue spot in the center
((NUFORC Note: Witness indicates that date of the incident is approximate. PD))

Occurred: July 3, 2005; Approx. 1400hrs.
Location: Springfield, OR
Witness: 2
Shape: Disk
Duration: 1 hour, 30 minutes
Saucer shaped reflective object moving and hovering with chasing Government Jets and a Helicopter/Photos taken
We were working in the yard this afternoon. At 2:00 p.m. we heard and spotted a huge Coast Guard helicopter flying from the west into the east, as if it came from the Oregon coastline. Right in front and above, at approx. 20,000 feet of the helicopter was a bright reflective saucer shaped object moving very, very slowly. It was back dropped by a clear blue sky without a cloud behind it. It seemed to be reflecting the sun in a type of pulsating, rhythmic display of extremely bright light. It may have just been pulsating with intense light. It was moving very slowly and then stopped moving. At approximately 2:12 the UFO did a vertical ascent and disappeared. The helicopter did a sharp turn right/south when the object stopped moving and did a 180-degree reverse path heading straight back west. The helicopter turned south, running along what seemed to be the west side of the Cascade Mountains. We live on the east side of Springfield as you start up the pass to cross the Cascades.

Within two minutes we saw a large white jet coming from the south (maybe Klamath Air Force Base) heading north. It had radar unit on top of it. It was headed north into the exact spot from where the UFO hovered and disappeared. That jet was first noted at 2:35 p.m.. Another radar jet appeared, following the same south to north path that the first jet traveled. At 3:14 a third jet appeared following the same path from south to north into the same space where the UFO disappeared. At 3:17 and 3:19 the same type of jets appeared also, following the same path. Using our digital camera we took several pictures of the UFO. We now have them downloaded from our camera and burned to CD's for safekeeping. Note: Today is the day that the USA NASA probe will hit a comet at 10:52 p.m. West Coast time.

Occurred: July 12, 2005
Location: BEALE AFB, CA
Duration: 3-5 minutes
Witnesses: 2
Number of Object(s): Cigar shaped Single craft, 75-100ft. long, 30 feet wide, dark with rows or windows or lights
Distance to Object(s): 150-200 yards
Full Description & Details
My date and I had just returned to my place after an evening out (no drinking or drugs). As I was stepping up on the curb to go into the house I noticed a dark object moving slowly back and forth at what appeared to be tree top level a few blocks behind the house. At the time I was active duty Air Force stationed at Beale AFB and felt I had a pretty good idea of the type of aircraft that operated in that area. What really caught my attention was the close proximity of the object with no sound. I asked my date if she saw it and she said it was probably a helicopter. I was sure it wasn't so I ran into the house and got my binoculars. When I got back it was still hovering in the same manner. As I could see it better now, there were what appeared to be windows down the side, like you would see on a commercial jet. My mind was racing trying to explain what I was seeing, as it turned there seemed to be a white light like landing lights or a spot like actually illuminating the tree tops which kind of gave me some idea of how close it actually was. When it turned away I thought I saw tail exhaust (2 like the SR-71) and then as I was watching it turned toward the sky and started to ascend. This is when I became convinced that it was really from another world, it instantly zoomed up to the size of a star and was blinking blue and red high in the night sky. If I hadn't been watching it though the binoculars, I would never been able to follow it and after it went up it probably wouldn't be noticed by anyone as anything but a distant star, planet, or satellite. Many times that night I went out to see if it was still up there and it was. After that night I started watching and noticed that there were several of such lights up there and some would move in unnatural patterns. Another thing of interest that I have noticed, even though I haven't seen then that close since, when I start watching them they seem to know I'm watching and move farther away. I don't bring this up very often because most of the time I get the response like I must be delusional. I have been on the runway during an SR-71 launch and have seen it turn skyward with exhaust flames rolling off the runway as it went full and went almost straight up. This object was so much faster that I can't begin to understand the physics that would keep it from ripping to pieces. I know the published speed of the SR-71 is mach 3+ but I can tell you that when I was in a KC-135Q on a refuel mission, we flew on a straight path for over 4 hours at more than 450 knots when they told us the SR-71 that we were to refuel had just taken off. By the time the boom operator and myself walked to the back of the plane and got into position in the boom pod. The SR-71 was pulling up behind us. That plane had covered in about 5-10 minutes what had taken us over 4 hours.

Personal Background: Crew chief on, B-52 bombers from 1973-1976, KC-135Q tankers 1976-1978, instructor on the tankers until discharge in 1980. Currently working as a building equipment mechanic for the U.S. Postal Service.
Other Comments: I had another sighting after this one that I may go into at another time.

Occurred: July 31, 2005; Approx. 1610hrs.

Location: Eugene, OR

Witness: 2
Shape: Other
Duration: 2 -3 minutes

On Sunday, July 31, 2005 at approximately 16:10-16:15 a transit supervisor and a bus operator were observing two ospreys flying over the transit facility in Glenwood, just east of Eugene Oregon near the Willamette River.

The birds flew off, and then the operator noticed what looked like another bird but at a very high altitude. The supervisor spotted it as well. It was a craft of some kind. It was translucent white in color surrounded by a misty teardrop shaped aura. It looked similar to a plane with swept back wings but neither of us felt is was a plane. It was going north to south at a very high altitude. It was so high it was barely visible; it seemed to blend into the blue sky.

A passenger jet was also traveling in the same direction at a lower altitude; the normal cruising altitude for commercial jets in our area.

The white craft overtook the jet like it was standing still. It streaked to the south and then switched direction and shot straight up and disappeared from sight in a matter of seconds.

We continued to watch the sky as the jetliner disappeared from our view. A minute or so later, we then spotted another craft flying along the same trajectory as the first UFO. It streaked across the sky and then shot straight up and disappeared in exactly the same manner as the first one did. The sky was clear at the time, except for some contrails overhead. The craft were very hard to detect: they were very well camouflaged. If we hadn't been looking up already we never would have seen them.

Both the supervisor and the operator had seen a UFO previously; one in 1989 and the other in the 70's. We didn't report the sighting to any civil authority.

Occurred: August 2, 2005; Approx. 0000hrs.
Location: Kelseyville, CA
Witness: 6
Shape: Other
Duration: 1 hour, 30 minutes
My BF and I were sky watching last night and at around 12am I saw a bright red blinking light to the right of me. I went to my stairs on the right side of my deck to get a better look. It was right above the tree line in a northwest position, I'm guessing it was about 20,000 feet up. At first I thought it may be a planet and stood there watching it for about 10 minutes and it kept blinking red and also other bright colors like white, orange, green. It also was moving, kind of like a swaying, it kept going down and backs up again. My bf was standing there looking at it with me, He also couldn't figure out what it might be. I then decided to go inside and get my dad to come out and look at this object and see if he could figure it out. He came out and looked at it and also thought at first it was a planet, but then he noticed it was really bright and flashing all those colors and told me to go to his truck and get his field binoculars. I gave them to him and he took a look and almost automatically after viewing his words where "wow" it's got lights on it. I took a look and it had three white lights in the middle and red blinking ones on the outside and the rest looked green, orange whitish, it looked sparkly made a drawing of what it looked like through the binoculars. We made my mom and two brothers come out side and look at it too. It then went behind the trees so we decided to go down the street a bit and get a better look at it. When we got there it must have traveled at some sped because it was way much further away to the north west than it was when we were observing it at the house. But still you could see through the binoculars the red flashing going from bottom to top of the object and the sparkle. Couldn't make the shape out anymore though. Then my dad said your not going to believe this and said there was another one, it was in the northeast sitting there. Looked just the same as the first one through the binoculars and it was closer. We stood there watching them both and the first one slowly moved to the northwest till it was a little dot and the same happened with the second.

Occurred: November 13, 2005; Approx. 0000hrs.
Location: Cottonwood, CA
Witness: 1
Shape: Other
Duration: 2-3 minutes
A Bright light in sky appears before our very eyes.
I was driving south on Highway 273 in Anderson, heading for the I-5 on ramp when I saw a bright light put in the South West sky. My ten-year-old daughter was all wrapped up in a blanket lying down on the seat of the van. I said, "Look at the bright light. What the heck is it? Is it a meteor?" I was fascinated with the brightness of it. The night was so dark, the clouds were thick and fairly low. There were no stars to be seen. I could see a shadow of the moon form time to time and then it would be dark. As we came onto I-5 south, there are a few hills and I was driving rather fast trying to get around them so I could see the light again.

It was much brighter than any light I can remember seeing. I hurried around one hill and there was little space in between where I could still see the light before we came on to the other hill that had blocked our vision. I hurried around that hill and here we could see the light real clear. I kept saying… What could it be, my daughter sat up in her sat watching it with me. I said is it an airplane? I thought and said it can't be… its too bright. The light was not round, it was more of an oval shape from what I could see. I was swerving all over the freeway. It was almost as if this light that first looked like a meteor out of the sky, was coming to us as we were driving. Next I looked over and it was just hovering over in one spot. Kind of like a helicopter. My daughter watched it and got a much better look at it. I asked her is it a helicopter? Before I could even finish my sentence, I said… There is no way it could be a helicopter because the light was brighter than any I've seen and it was way too big. All of a sudden it started dropping straight down, from where it was hovering over, not moving to straight down. I said… is that thing dropping down or is it just my eyes cause I'm driving? I told her I was going to pull over, keep going, keep on going, don't stop! It scared the holy crap out of her. All the sudden I looked over and it was as if this aircraft thing had turned. We were now passing each other. I n o longer saw the bright light but when I looked over again… what I saw was huge! Ata glance it kind of had the shape of an old military helicopter (the grasshopper looking kind) but much bigger. Bigger than anything I have ever seen that could fly. It had rows of lights, which appeared to look kike it had different levels to it. The easiest to even try to describe it would be my thought of standing on a city street in the dark night and looking at the five to eight story building with lights on in some of the rooms on each level. I told my daughter I have to see this better. I remember looking behind me on the freeway and I asked my daughter…can't anyone else see this? People have got to be seeing this! There were other cars further behind us, I could see their headlights. Being I almost went off the side of the road on the freeway, I told her I am getting off at Gas Point, the street were getting off at. Then I had to pull over. I told her were would be further away and we would be ok. As I exited off on Gas Point, I instantly started looking for it.

There were a few trees in the way and as I got passed them, I looked all over for the lights and they were nowhere to be seen. Poof... just gone, not a trace, I have shared my experience with a few people. Most of make a joke out of it. My biggest question is did anyone else see what we saw? I called around to see if there had been any reports of a sighting. Wow! It was like I had the plague. I went to the airport and funny as it sounds, the lady at the des said the weirdest thing happened to her... IN the middle of the day, she was out riding her horse and a beam of light came over her and followed her for a short way and then it was gone. But, she had no info. And could not tell me where to go to find out if anyone else had seen this. I called a radio station, a News station and left a message at the planetarium. Called the tower at the airport...nobody. I say nobody wanted to hear anything about it. Each passed the buck and never even asked so much as a description. I was very... I don't even know how to explain just how I felt. That's when I got on the Internet, and started reading one persons description that was almost word for word what was said, and seen by my daughter and me! After reading through quite a bit of information, I realized a few things I would like to list... where we saw this ting drop down from a stand still, is directly across from an auction yard. (now is the time all the ranchers are moving their cattle from the higher altitudes to the areas with more feed, and the auction yard is very active at this time.) There are quite a few lines running through the area and quite a few of these electrical looking towers all through the area it was seen in. Right over the hills I spoke of, being almost directly Northeast of our sighting ins Redding Airport. Other than that our surroundings are all mountains. We sit in a big valley. Cottonwood is a rather small place. There's a lot of land spread out. I just wonder what the heck brought that big ole thing here. Keep in mind I was driving. My youngster was watching the whole time. She got a much more detailed picture of this thing. She said... If it weren't for the lights on it, I would not have been able to see the shape of it. Quite an experience I won't forget what seemed like twenty minutes was actually two minutes.

((NUFORC Note: We believe that this is a serious report, but we don't understand how just one person can be witness to this alleged event, and yet there are no other reports. There is one other similar report from Davis, CA, for this date, but we cannot be certain that PD))

Occurred: December 2, 2005; Approx. 1930hrs.

Location: Burney, CA

Witness: 1+
Shape: Triangle
Duration: 45 minutes
Thousands of strange objects streaking across the night sky
We observed objects coming from the north and moving southward and sometimes changing to a path of southwest. They appeared to be in groups of 20-30. They started coming out of the north in a horizontal line to somewhat of a v shape and looked like groups of moving stars. We saw several passenger jet move across the sky while we were watching the objects and judging from the speed of a jet these objects were going 8-10 times faster than the jets. I got out my binoculars and looked at them as they came over and they had a bird like or plane like shape. In the binoculars they looked slightly illuminated or white. These groups came one after another for at least 45 minutes that we watched. I have lived here for 50 years and I have never seen anything like this in the sky. I don't believe these were birds on migration because there were too many of them and there was no sound. Also as I was watching one group approach one of the objects fell from the formation and had the appearance of a meteorite except it fell much slower. If anyone can tell me what this was please let me know. E-Mail ((e-address deleted)) Thanks

((NUFORC Note: We believe that this is a serious report, but we don't understand how just one person can be witness to this alleged event, and yet there are no other reports. There is one other similar report from Davis, CA, for this date, but we cannot be certain that PD))

Occurred: December 12, 2005; Approx. 1930hrs.

Location: McCloud, CA

Witness: 2
Shape: Unknown
Duration: 1 hour
Huge flash of light and then 4 UFOs enter into our atmosphere
At approximately 7:20 pm my friend and I got into the hot tub to enjoy the cool evening before going in to eat dinner.
10 min later as we were both looking at the western sky I saw what appeared to be a huge flash of light in the sky - literally reminded me of a worm hole effect..
When the light subsided (approximately 10 seconds or so) there were (4) UFOs that came in one right after the other. Moving extremely fast they came in and were in a group - then one moved to the right and took up a position not far from a star. The other three moved to the left and spaced them evenly apart and also became motionless at that time.
They went from being very bright to almost fading out completely and then they would all start to come into brightness again. A type of synchronized strobe effect was taking place with these (4) unknowns.
We watched for approximately 1 hour. 3 disappeared altogether from our sight as they went behind the tall pines in front of us. The 4th stayed in place not moving - just changing in brightness from time to time. The interesting thing about this was that the 4th unknown moved with the night skyline. If we had not seen these 4 unknowns come into our skyline from the flash of light, we would have thought that they were just stars after they took up positions. Exciting to say the least... From our position we estimated that they were probably staged over the Pacific Ocean as the crow flies from us.

2006 to Present

Occurred: January 24, 2006; 2315hrs.

Location: Oroville, CA

Witness: 1
Shape: Teardrop
Duration: 45 seconds Object seen in Kelly Ridge

It was right about 11:15 pm when I went out onto my front steps to have a smoke. Moments after I sat down on my stoop I noticed a faint shimmer of light at about a forty-five degree angle from the horizon immediately in front of my position. After focusing directly on the object i realized that it was not a star or satellite. I sat staring for probably thirty seconds when i actually realized what i was seeing. I stood up and as I did a second craft came into my field of vision and literally chased the first object away. It seemed like where the first object was located was less than a mile from my house. It also seemed like something fell off of both the first and second objects, possibly landing in the golf course in my neighborhood. I then stood for a few minutes, finished my smoke, and then went inside.

Occurred: March 2, 2006; Approx. 0900hrs.

Location: Medford, OR

Witness: 1
Shape: Disk
Duration: 2 minutes +\-
Object seen over Medford, Oregon

Witnessed a dark colored, round object at roughly 50,000 feet altitude, traveling south at a speed of approximately 500-600 Knots. Object was headed south when first observed but, performed a high speed, high G turn and altered heading to south-southwest and continued flying in that direction until obscured by a buildings roofline. Around 2 minutes later a commercial airliner passed through same area as the UFO (route is well traveled from Portland to San Francisco. I have over five years experience in the navy as a radar operator and 20 years experience as a merchant marine captain and can state that this object did not perform like a jet aircraft would. UFO's turn would have been impossible for any aircraft currently in US inventory. It was a rounded, 90-degree turn at a speed of 500 knots plus that took perhaps 4 seconds total. It left no vapor trail, even though the aircraft that went through same area moments later did leave one. Object appeared much smaller than any aircraft normally flying at that height, roughly 1/10th the size of the airliner that was nearby.

Occurred: March 18, 2006; Approx. 1900hrs.
Location: Junction City, OR
Witness: 2
Shape: Other
Duration: 45 minutes
Perfect silver ball floating
my daughter was getting ready for bed when she saw a plane outside her bedroom window. I was sitting a the computer, when she came running the living room and says daddy there a plane outside my window. So i got up to see what she was taking about and to the left of her window I see this silver ball floating above the ground if I had to guess I would say it was about 50 ft maybe a little higher. I got my binoculars and took a better look and it was perfectly round and I had my girlfriend was standing right next to me and I had her look and she said the same thing, that it was a ball floating in the air. I call the airport that's just down the road and asked if someone in the tower could see it and told them where to look. And when he did see it he said he did not know what it was either So I called the state police and told them what we'd seen and that they would send someone to check it out we watched for another 15 to 20 mins and then blink it was gone. I believe it was a device for scanning. I believe it was from space and a u.f.o. I was in the service growing up and I've seen a lot of weird things and this was one of them. ((NUFORC Note: Witness indicates that date of incident is approximate. One of three reports from same source. PD))

Occurred: April 3, 2006; Approx. 2144hrs.
Location: North Bend, OR
Witness: 3
Shape: Circle
Duration: 34 minutes
 Multiple crafts traveling at high speed and emitting flare-like blasts.
My husband had stepped outside of our home at 9:44 pm on the evening of April 4, 2006. It was a clear evening, no clouds or breeze and a crescent moon. I was inside and heard what appeared to be a very loud engine noise. My husband came in the house and told me to come outside.

A short distance off, in the southern sky, was a very large disc shaped figure with red, randomly blinking, and lights. It was enormous. We watched as the object continued to fade in the distance.

A few moments later, it came streaking, at a very high speed, back in our direction. This time the object was very high up. It was very loud although it was traveling faster than sound. The object passed overhead about 9 seconds prior to our hearing it.

There were huge intermittent bursts of flare-like (yellow-orange) releases in the sky both coming from the craft and sometimes in the sky where there didn't appear to be a craft. When the sky lit up from these blasts, there was an aurora around the light. Random "sparks" from the blasts reached the vicinity of our house. It was similar to seeing the sparks from a professional fireworks display showering down, although it came from a very high altitude.

At one point in the eastern sky we witnessed two crafts, again, traveling at a very high speed. Our teenage son had joined us outside and observed these events also.

The clear sky quickly became cloudy with a light cover although the "clouds" weren't moving and there were no winds present at ground level.

We'll keep the video camera charged in the future.

((ADDENDUM FROM WITNESS))

I filed a report just moments ago. I'm sorry, but I filled in the military time incorrectly. It took place at 9:44 p.m.

((NUFORC Note: We have altered the time above to 21:44 hrs. PD))

Occurred: April 5, 2006; Approx. 0831hrs.

Location: Shingletown, CA

Witness: 1
Shape: Disk
Duration: 1 minute

I saw a massive blue object floating in the sky for about 60 seconds then spilt in 2 then went back to original shape then sped up and vanished. After it vanished, there was a bright light off in the hills. That's all I saw

((NUFORC Note: Witness elects to remain totally anonymous; provides no contact information. PD))

Occurred: April 9, 2006; Approx. 1635hrs.

Location: Orland, CA

Witness: 1+
Shape: Sphere
Duration: 1 minute

ORBS was watching over a launch of their own.

On 04/09/06 at approximately 16:35 hours 3 friends and I seen a double rainbow so I snapped seven pictures in less than a minute. I used the digital camera view screen to look at the rainbow and snap a few photos and then zoomed in and took a few more.

I later on downloaded the pictures at first we thought that rain drops were on the lens when the pictures were taken but as we looked at the next set of pictures we were lost for words and explanations. The pictures revealed orbs and a object that resembled something launching from ground level and heading skyward and the orbs seem to watch over this thing. The funny thing of all this is we never seen this with the naked eye it was captured with my digital camera.

I have shown the pictures to some friends and no one knows what it is, a friend of mine that was a sheriff for many years as a evidence man and investigator seen the photos and said you have something there I do not know what it is but you have something there.
I have saved the memory chip and place it in a safe place. I watched UFO HUNTERS and was told by friends to contact you and let you examine the evidence.
I want to say that we never seen any thing till we looked at the photos.
SEEING IS BELIEVING these pictures are worth a thousand words and are better seen than told.

Occurred: May 7, 2006; 2134hrs.
Location: In orbit/Space Shuttle
Witness: 1
Shape: Circle
Duration: 10 seconds
Lights pass under STS 121 -The circular "hoop", has 6 nodes which are glowing and strips of light seem to link each node or light.
While watching the Space Shuttle STS 121's progress on the NASA website.
At 21:34pm on the 05/07/2006 I captured these images. (Taken with a digital stills camera) The circular "hoop", has 6 nodes which are glowing and strips of light seem to link each node or light. It passes under the Space Shuttle! Any ideas!!!! I have already sent the images to Director National UFO Reporting Center Seattle, WA I was advised me to post them here for your opinions! Best wishes
((NUFORC Note: Witness elects to remain totally anonymous; provides no contact information. PD))

Occurred: May 9, 2006; Approx. 0520hrs.
Location: Yuba City, CA
Witness: 1
Shape: Light
Duration: About 5 minutes
Bright lighted aircraft forms a cloud and disappears!
Is seem at first to be a an air craft with a spotlight, but, only this light seemed to be much brighter than normal, within a minute or two the object disappeared in to the sky leaving the light that was seen fading into a cloud then a fog, the weird thing was that there was no clouds or fog any where in sight. There was also no trace of any kind of explosion or anything falling from the sky.

Occurred: May 10, 2006; Approx. 0100hrs.
Location: Somes Bar, CA
Witness: 1
Shape: Light
Duration: 8 seconds
Witnessed airplane sized light triple in size, shrink back and speed away at incredibly high rate of speed

While camping on the Offield Saddle near Somes Bar, CA., on May, 10 2006 at approximately 1:00 A.M. during clear, almost full-moon night, witnessed airplane-sized craft triple in size, quickly shrink back and then shoot north in straight-line and disappear at very high rate of speed.

Occurred: August 13, 2006; Approx. 2300hrs.
Location: Redding, CA
Witness: 1
Shape: Egg
Duration: 2 minutes
Egg shaped orb slow moving north towards Mount Shasta moved across the sky for a minute then made a right turn and disappeared at an extreme rate of speed.

Occurred: August 20, 2006; Approx. 1330hrs.
Location: Redding, CA
Witness: 1
Shape: Circle
Duration: 5 minutes
Saw about 20 bright objects kind of like a swarm of bees, only they moved in jerks and zigzags Was a beautiful clear day, and decided to relax in the back yard. Suddenly about 70 degrees up a red got shot out of the blue. It kind of moved in an erratic motion. Then it started to fade away as it went back up.
Suddenly another object appeared, and another. Finally there were about 20 of them. I tried to count them but it was difficult because of their pattern. They make right hand turns or even reverse themselves without slowing down. Sometimes they would suddenly be ahead of where they were going. Wasn't sure they just appeared there or moved so fast I couldn't see it - kind of like the snap of a rubber band.
They kept coming closer and after a few minutes started to move away. One kind of hung back, and as the others faded away, it shot to the northwest at an incredible speed. This may sound strange, but it left a hole in the atmosphere. You could make it out for a few minutes.
Best I can say for color, it's kind of like looking at a 100-watt light bulb on a bright sunny day.

Occurred: August 20, 2006; Approx. 0200hrs.
Location: Bend, OR
Witness: 4+
Shape: Oval
Duration: 45 minutes

I came home from work and a neighbor called me over to view what they had been watching. I said "it's a bright star" My Neighbor said watch for a few seconds and that's when it started moving around the sky in a side/up movement, paused then moved straight down and then strait up. When the craft stopped moving we noticed small bright drops falling from the craft every two seconds if I remember it was about four of them. At this time we woke up two other households. They witnessed the same movement we saw earlier with the up and down side to side. Before the craft vanished it turned a bright orange color and then vanished.

I saw the same thing a few months later in Sunriver Oregon on New Years Eve working the late shift. About five minutes after 12:00am in the NW part of the sky, the same location the first sighting was seen I noticed a star that should not have been where this was. I radio'd other people working "Want to see a UFO"! Most on the evening shift came running out and saw what seemed to be a UFO. A small group of the bar crowd witnessed the same thing. A bright object moving sporadically then dropping small bright lights out every two seconds, turned orange then vanished. This lasted about 10m minutes after we first sighted the craft.

The final UFO sighting was in the summer of '07. I had to respond to an alarm at one of the golf courses at approx. 1:00 AM. I showed up the same time as the local police, we checked over the clubhouse to find a false alarm. As we were wrapping thing up I saw a very fast moving bright light moving from West to East and asked the officers "What do you think that is"? She said a satalite, Jet, and by this time it had moved from the Cascades to Easten Oregon and the final response! "Don't know".

I'd like to know if anyone else has seen these objects during these time frames.

Warmest Regards Central Oregon Sightings

Occurred: August 24, 2006; Approx. 2045hrs.
Location: Medford, OR
Witness: 1
Shape: Flash
Duration: 30 seconds
One very large falling object seen North by Northwest of Medford Oregon. Falling object was silvery blue with a white aura and very short trail. No smoke or long trail common with an object like a meteorite. It fell quickly from high in sky down below the visible horizon. Approximately 15 minutes later 2 military helicopters approached the airspace near where it was seen while 2 F-16s were observed in a holding pattern outside that area.
 There was lot's of unusual air-traffic in the area for 5-6 hours after this event.

Occurred: August 31, 2006; Approx. 2130hrs.
Location: Mount Shasta, CA
Witness: 1
Shape: Triangle
Duration: 5 seconds
Low flying, dark, large car sized triangle with rectangle dim lights

It was just about dark when I was on my back porch hanging out clothes. I was looking up as if viewing the 3-story apartment house's roof to the NW of me. Right over my head flew a dark triangle just over my back porch and the roof of the 3-story apartment building. It was a short view due to the size of the building NW (next lot) of me and not quite fully dark yet. A more detailed map with drawing will be submitted as an update when I can. It came into my view as it passed over my head look toward and above the 3-story apartment building NV of me. It was a dark, triangle craft, about the size of an old Cadillac car. There were dim, rectangle shaped lights from the point to the back, but not across the back. Stranger yet was that there was one more light on the left size than the right side, like one light was out. It was about a 4/5 second sighting. I won't forget it. I screamed for my roommate to come see, and then passed him on my way back through the house to get outside to try and see past the bigger building. A Street light, made further viewing impossible, past the larger building. It was moving along 15/20 mph basically paralleling I-5 from North Mt Shasta, towards, if it kept along that flight line, to the NW of Black Butte, to the town of Weed, Ca. I would think that cars on I-5 might have been able to see it. I didn't think until later that I should have called Weed PD to look for it coming. There was no noise, no engine seen, and it was too slow and quiet to be any sort of military or regular small plane I've ever seen.
((NUFORC Note: Witness indicates that the date of the incident is approximate. PD))

Occurred: September 23, 2006; Approx. 0005hrs.
Location: Mount Shasta, CA
Witness: 2
Shape: Light
Duration: Split second
A light over Mt Shasta went within itself.
The light lit up the entire face of the mountain. So I can't say it was a ship; it's beyond any concept of a ship. As I sat on Mt. Shasta at the highest parking lot around midnight, I was with another man who was stargazing when I arrived. We both looked up at the peak of the mountain, and suddenly, within a split second, a circle of light came on and off, what appeared to be perhaps 1,000ft above the peak, in that split second, it totally lit up the entire face of the mountain, no shadows, and went off just as quickly. What was strange, I remember that when it went off, "It went within itself", swallowed itself. I know it sounds strange, but I've not told anyone but my wife. All I know about the man up there that night was that he was from Reno, NV. I noted his name somewhere. I could find it if necessary.

Occurred: September 24, 2006; Approx. 0400hrs.
Location: Yreka, CA
Witness: 2
Shape: Light
Duration: 30 minutes
The UFO hovered, not moving, changing color and was 10+ times brighter than any star.

At 4:10 A.M. this morning, for the first time in my life, I saw an unidentifiable object in the sky. My boyfriend woke me up to witness what he saw as he walked the dogs early in the morning. It was multi-color, extremely bright hovering (not moving) towards the south 0f Yreka ... towards Mt Shasta. The colors were changing. After a while, it had lightning type rays below it (like jellyfish tentacles) then showed a bright ring shape with haze. The colors were bright red, blue, white - changing often. It was too far away to determine any shape.

Occurred: October 8, 2006; Approx. 1830hrs.
Location: Eugene, OR
Witness: 1
Shape: Formation
Duration: 2-3 minutes
Formation of lights 4 white 1 red with amazing sign-wave pattern over entire craft.
I viewed a formation of lights that came in from the north east and traveled south, four white lights and one red, they hovered in the sky briefly then flew directly over head, there was no sound and no contrail, these craft had no visible navigation or anti-collision lights.
I was amazed as they passed over head the craft were brightly lit, and the light seemed to move around them in a sign-wave pattern like you might see on a oscilloscope or frequency generator, and they were all at the same frequency.
I am a pilot and licensed mechanic with a strong background in engineering and design, I have been viewing aircraft for thirty years I have never seen anything quite like this I had my camera and was able to get a couple of photos, these craft were faster than normal aircraft I only had time to point and shoot,
I think it is very notable that this sighting had taken place on a day with heavy (chem.-trail) spraying? Thanks.

Occurred: November 8, 2006; Approx. 1300hrs.
Location: Red Bluff, CA
Witness: 1
Shape: Triangle
Duration: 10 minutes
As I was going to work I saw a slow moving triangular shaped object in the sky.
It made no sound. Others must have seen it. This was daytime over a city of around 12,000.
((NUFORC Note: Witness elects to remain totally anonymous; provides no contact information. PD))

Occurred: December 3, 2006; Approx. 2215hrs.
Location: Mount Shasta, CA
Witness: 1
Shape: Oval
Duration: 2-4 hours
Mt. Shasta, CA pulsating, rotating lights at approx 40-50degree angle, east-southeast of Mt. Shasta, lasting at least 2-4 hours

In the eastern sky, more east than south east of Mt. Shasta, moving back and forth VERY slowly, unsure of that distance of movement, with beautiful lights outlining the entire structure which was many, many feet high-the lights stayed on, but pulsating in a pattern of red-pink-white-yellow-blue-green around the bottom of the craft and along the structural outline. It reminded me of the deep-sea jellyfish that glow-the outline and tentacles shine with their own light. I also observed it splitting into two separate objects, like watching a cell divide. The skies were totally clear, with a bright moon-one night short of a full moon. The craft had an oval shaped bottom, with a round, tall, beehive shaped body to the structure. The object was in the sky for a very long time, in fact was still there when I went to bed an hour later.

((NUFORC Note: We spoke with this witness, and she sounded quite sincere to us. However, we believe that she may have been looking at the star, "Sirius." PD))

Occurred: December 22, 2006; Approx. 2230hrs.

Location: Susanville, CA

Witness: 2
Shape: Light
Duration: Approx. 2 hours
Now I believe

On December 22nd, 2006 at Approximately 10:30 Pm while working as a "Tower" Officer at a Prison, a call came to me from another tower. The Officer asked if I was watching or noticed the "lights" beyond the Prison water towers facing north.

At first glimpse I noticed two fluorescent orange orbs of light equally distanced from each other, which appeared to be hovering. I grabbed a pair of Binoculars and hoisted my hands on window bars to steady the shaking of my hands.

Now a clear shot I watched as both orbs seemed to be in conjunction with the other, perfect match of each other's size... As one orb would appear to get brighter the other would dim then two smaller lights shot from the left orb.

After about ten minutes or so watching these objects I witnessed a smaller row of orange orb-like lights seem to have turned on, above the two bigger orbs, this was all one object I'm sure. These smaller lights seemed to "twinkle" as they spun in a sort of carousel motion. The closest thing to the looks of the object is a gigantic freightliner with running lights coming at you head-on.

This lasted for an approximate 2 hours. There is nothing but BLM land and forest there. Later that morning at 0600 Hours I was relieved of shift and drove up a single dirt road to where I thought this object was, my car lights began to putter dim and I turned around and took off!

I am in the Air Force and have seen many strange lights but this is the closest thing I have ever seen.

((NUFORC Note: Witness indicates that the date of the sighting is approximate. PD))

Occurred: December 26, 2006; Approx. 1952hrs.

Location: Red Bluff, CA

Witness: 3
Shape: Light
Duration: 3 to 4 minutes
Regular appearance of straight traveling lights

Hello, These lights have been coming around here for years on end, but for the last few years the lights have been coming over much lower then usual. In this past week of December 26th, 2006, at 6:52 PM and December 31, 2006 at 6:52 PM same time again, these Golden ball lights with no sound with no other lights came over My area at about 5 miles out at about 5,000 feet from a west to east direction over the North part of the City of Red Bluff, California. These lights came over while aircraft traffic was in the area at the same time. I know these aircraft seen these lights and the light on December 31, 2006, because one aircraft on the December 31, 2006 at 6:52 PM made a turn towards the light, I am sure it was to see the Golden ball of light better. The aircraft was a signal prop engine aircraft, civilian because I could hear that it was a, possible Cessna Type aircraft. The aircraft was at the same altitude and a half a mile if not close from the object. The golden ball light was bigger then the aircraft's light because if the aircraft had gone into the light or passed behind it you would not see the aircrafts lights at all. Two witnesses and I watched it as it went over the City of Red Bluff, California the light just went out in front of our eyes.

2007

TRANSCRIBED TEXT OF THE NPR INTERVIEW, JANUARY 01, 2007

Melissa Block: In the Chicago Tribune today, this tantalizing headline, "A Bird, A Plane, a UFO?," the report describes a number of workers at Chicago's O'Hare International Airport, who all say they saw more or less the same unexplainable thing on November 7th. A large disc shaped object hovering in the sky over an airport gate then shooting off into the clouds. Jon Hilkevitch wrote this story. He covers transportation for the Tribune, and joins us now. Thanks for being with us.

Jon Hilkevitch: You're welcome.

MB: And, Jon, you talked with 6 workers at O'Hare, what more did they tell you about what they saw on November 7th.

JH: Well it's very interesting because they all tell a very similar story from different vantage points at the airport. And they saw this gray, metallic object, just below the cloud layer, and this object was stationary for some minutes, anywhere from 5 to 20 minutes, according to different accounts.

MB: And how big was it, did they say?

JH: Anywhere from 6 feet in diameter, to 24 feet in diameter.

MB: That's a pretty big difference.

JH: That's a big difference, but, uh, these were some of the calculations made by some of the UFO watching groups, based on the witness accounts that were given.

MB: Uh huh. And then they describe it very suddenly taking off?

JH: That's one of the weird parts too. Uh, a very thick cloud layer that this craft was hovering below it, but then when it streaked away, it burst through the clouds, and unlike an airplane or a helicopter that would just slice through the clouds, this thing created this big donut hole, uh, vacuum, indicating tremendous energy force as it burst through the clouds and leaving this hole of clear sky that was there for several minutes, until the drifting winds pushed the clouds back together.

MB: Now, when you were talking to these workers at O'Hare, were they saying, ya know, your going to think I'm crazy but this is what I saw?

JH: Yeah, one of them, in fact, when he went to his manager in an email, the manager sent him an email back and said, "Hey, looks like somebody got a hold of your email account and is pulling a hoax." Um, but, this is anything but. That's what impressed me about this. All aviation professionals, very credible sources and they are very serious. They are not saying what they saw was a, you know, a spaceship from another planet, but it was unidentified, it was in restricted airspace, and they were concerned from a safety standpoint. That if this was something man-made they needed to get it out of there because they were having busy flight operations in the early evening hours.

MB: You know, in the course of doing your reporting on this story, you were working along with, in some ways, the UFO Reporting Center. Did that strike you as odd, in some way?

JH: Yeah, I mean, they kept wanting me to say this was a visit from some other world and further proof that, uh, we, on this planet, are visited regularly by other beings and, uh, you know, to me, I am not qualified, I wasn't a witness. What was interesting to me was that United Airlines, after receiving numerous reports, including from high-level management officials, decided to deny that they got any such reports! I don't know if they were, felt, embarrassment, or maybe people would think United Airline employees are kooks, or something, by reporting this. But it's very odd. The FAA, at first, too, said they didn't know what I was talking about when I went to them, they had no reports of this. It wasn't until I put a Freedom of Information Request in and they got back to me saying well, were going through the communication tapes from O'Hare tower and by golly, there is a lot of chatter here about this UFO. So I am waiting for the full communication tapes on the FOIA requests, as well as radar data and we'll see if there is some follow-up.

MB: Okay. We'll wait to see what happens. Jon Hilkevitch. Thank you very much.

JH: You're welcome.

DECEMBER 15, 2006 UPDATE: NUFORC has an unconfirmed report from a journalist investigating the O'Hare Airport incident of Nov. 7 (see full details lower on page) that the FAA and the airline involved have declared that the incident did not occur. We strongly suspect a cover-up may be starting to form.

DECEMBER 07, 2006 UPDATE: NUFORC has received a second report about this incident from a senior aircraft mechanic, who was taxiing a Boeing 777 at the time of the sighting, and who witnessed the object. His report is here.

ORIGINAL REPORT: The National UFO Reporting Center has received the following information from a single source (see below), who, for the time being, wishes to remain anonymous, and who prefers not to reveal for what entity he works. We have received documentation about the alleged sighting, which satisfies us as to the veracity of the report, and as to the credentials of the party reporting the incident.

We have delayed release of this case, principally because an investigation was begun almost immediately after our receipt of the initial report, and because we were hoping to obtain addition documentation about the sighting, before it could be concealed, or destroyed.

At approximately 16:30 p.m. (Central) on Tuesday, November 07, 2006, Federal authorities at O'Hare Airport received a report that approximately a dozen witnesses were observing a small, round disc-shaped object, metallic in appearance, which hovered over Gate C17 at that airport.

The object was first spotted by an employee, working on the ramp, who was engaged in "pushing back" Flight 446, departing Chicago for Charlotte, NC.

The employee reported to his supervisors that the object appeared to be almost directly above his location at Gate C17, it appeared to be perfectly round, and that its size was approximately equal to a U. S. quarter, held at arm's length. The object had a metallic appearance, according to the first witness, and it appeared to him to be spinning.

The first witness apprised the flight crew of Flight 446 of the existence of the object above their aircraft, and we believe both the pilot and copilot were witness to the bizarre object, as well. The witness also contacted his supervisors, who also witnessed the object, which was visible for approximately 2 minutes.

At the end of that time, the object was seen to suddenly accelerate straight up at a very rapid pace, and it "shot" through the solid overcast, which was at 1,900 feet at the time. The witness added that the object appeared to leave a "hole" in the clouds, where it had streaked upwards through the overcast.

Both the Federal Aviation Administration and Transportation Security Administration were apprised of the event at the time it was occurring, and FAA personnel in one of the towers at O'Hare may have witnessed the object, probably with binoculars. The FAA apparently reported that the object was not visible on radar, although that fact has not been confirmed at the time of this writing.

We hope to be able to release more information about the incident at some time in the near future. In the meantime, we would like to invite anyone who may have been personal witness to the event to submit a report of their sighting, using our Online Report Form. We would be most grateful if you would indicate in your report where you were located, at the time of the sighting, and what the object looked like, from your vantage point.

Occurred: January 1, 2007; Approx. 0000hrs.
Location: Windsor, CA
Witness: 2
Shape: Other

Duration: 10 minutes
An array of lights, formed in two parallel rows, moving across the sky into a cluster.
Shortly after midnight, my husband and I saw an array of approximately 12-14 small lights. There were six to seven in a row forming two parallel rows. They were moving across the sky, and then clustered together.
Once clustered, they all moved off into the sky where they couldn't be seen anymore.
We have never seen anything like it before!

Occurred: January 1, 2007; Approx. 0001hrs.
Location: Windsor, CA
Witness: 2
Shape: Circle
Duration: 5 minutes
Several yellow/red objects hovering in the sky, then they disappear.
My husband and I just saw the strangest thing in the New Year sky. We went outside of our Windsor, California home to 'ring in the new year' and saw dots in the sky. They looked like they could have been stars, but there were too many of them and the Milky Way does not look like that. Each of them was yellowish color and they were all identical. There was approximately 30-50 of them.
I told my husband to go in and get the video camera, but it could not record the dots. They just seemed to be hovering but then they would move into various formations.
Given there were some fireworks in the western skies, we thought about that, but there was nothing in the vicinity of the dots. Then we thought of the possibility of planes, but there were too many for that to be the case. We are astonished. What are they???
They slowly started to move from south east in various directions again, so we went from our back yard to the front yard and they were almost all gone accept for a few that were going into formation, four of them in a straight line. Then, there was nothing. It was like they just disappeared. No way to explain it.
So strange! Any thought as to what it was, or any other sightings??

Occurred: January 1, 2007; Approx. 0001hrs.
Location: Windsor, CA
Witness: 1+
Shape: Light
Duration: 10 minutes
Series of 17 unidentifiable lights cross the skies of Northern California
At midnight, Jan 1st of 2007, we stepped out on the deck for New Years and observed a bright Satellite moving southeast from the Northwest. Magnitude was as great as Mars at it's closest. Then 2 more appeared behind the roof into view, so we stepped out to see if there were more moving across the sky, 17 lights in total, all within 70 degrees of each other moving southeast from the Northwest. All were the same brightness and color, a bright Orange to reddish.
As the first 3-4 moved through the 12 o'clock (meridian) they started to change course. A few slowed then turned about 90 degrees and out of the path, then a few faded and made a left 270. 2 actually stopped, as noted against a clear sky of stars and a bright moon. All in all, about 7 or 8 "danced" overhead, the rest continued on course.

At this point I recalled we had just bought a new video camera so I ran in, grabbed it and shot the last few minutes of the "satellites" continuing across the sky. At this point, pretty much all those that had not faded were continuing SE bound.

Since moving into the country from the city 5 years ago, watching satellites and meteors has become one of my families evening pastimes. We are quite adept at this pastime now and can identify the difference between Satellites, Space Station, Etc. and Aircraft. And I have spent the better part of my 36 years, watching aircraft, As well, I am employed as an ((deleted--aviation specialist.)). So something like this event really stood out.

There were no contrails visible as they would have been with the moon as a backlight. There were no strobe lights or red and green wings lights as associated with A/C. There was no sound associable with these "satellites". The time for the "Non Dancing" lights to traverse the sky was about 9 minutes, the same as a satellite or the Space Shuttle.

A quiet check of the net did show that the prevailing winds aloft were moving in the same direction.

"Lighted Weather Balloons? ROV's?" I truly hope someone else who might have noticed these lights will also report something.

Occurred: January 1, 2007; Approx. 0010hrs.
Location: Windsor, CA
Witness: 3+
Shape: Formation
Duration: 13 minutes
Swarm of Amber spheres

Myself and a few friends were visiting another friend of ours in Windsor, CA. for new years eve, I believe its north of San Francisco and 30 minutes from Bodega Bay.

A little after midnight on New Years Eve, our whole party went outside to ring in the New Year. When we looked up we noticed what looked like a whole swath of stars that were amber in color were travelling in a column, we watched them for about 13 to 14 minutes. I happened to have a pair of binoculars with me and managed to hand the binoculars to everyone in our party. From what I saw, when I had my binoculars were these objects were spherical in shape and were slightly pulsing in 1-second intervals. Some were swirling around one another. At the end of the 13 - 14 minute window I saw the objects gather in a swarm on the horizon and split off, one part of the swarm going right the others going left with only a couple going straight up and disappearing. A few minutes after the sighting I noticed a couple jets in the vicinity.

Our party had one amateur astronomer and one fireman. Our astronomer states they were not stars or airplanes. Please contact me if anyone else has reported anything of this sort in the area. Every person in the party is still stunned from this experience as am I; we need to know that we are not crazy.

Occurred: January 25, 2007; Approx. 2100hrs.
Location: Medford, OR
Witness: 1
Shape: Cone
Duration: 5 minutes

A cone shaped object with a long strip of lights that changed color, seen over Medford.

I was driving down a rural road in Medford, Oregon when I noticed 2 lights in the sky ahead of me. It is a very clear night.

I pulled over to see what they were. One was obviously a plane with a blinking light on it. The other one however was what seemed to be a sort of cone shaped thing with one white light on the bottom and a strip of lights on the top part of it that would change colors from red, blue, green, and yellow.

The plane flew near it and then the object changed direction and flew over me (west to east) until it was out of sight.

The plane then flew in the opposite direction and then away from me (north).

The strange object seemed to be much bigger than the plane.

Screen grab from the UFO video

Screen grab from the UFO video

Screen grab from the UFO video

Object observed and videotaped between Eagle Point and White City, Oregon-3-20-2007 about 11 AM.

With my eyes it was first seen as a small, shiny, white dot. I got my camera out and tried to zoom in and couldn't find it. Then I saw the shining again and zoomed in and started recording and it was dark again and then shiny again and then dark again before it was blocked by a tree and I couldn't find it again, about 16 seconds of video.

It was just above the mountains and when I was recording it, it seemed to be moving south. When I watched it on the TV and on computer it appeared to stay in the same place above the same cloud feature. We were on Modoc Rd. headed towards Medford, Oregon and it was east of us towards Eagle Point or White City.

I thought maybe it would turn out to be a plane when I watched it on the TV screen but I still couldn't tell what it was exactly. The truck was bouncing me around and it was hard to keep it on the object and we couldn't pull over because we were working. My husband was with me but he was driving and couldn't see it.

There were no airplanes or helicopters in the area. It was a cloudy day, but no rain. There were no electromagnetic effects, missing time, or smells associated with the sighting.

http://www.theblackvault.com/wiki/index.php/Oregon_(3-20-2007)

Occurred: March 20, 2007; Approx. 1100hrs.
Location: White City /Eagle Point, OR
Witness: 1
Shape: Cigar
Duration: About 10 minutes
Unknown over the hills of Eagle Point and White City
My husband & I were headed to Medford. It was very cloudy but not raining and I'm not sure if it was windy or not. We were on hwy 62 getting ready to turn onto hwy 234 when I saw a shiny dot in the clouds just above the mountains. I tried to zoom in on it then and was not able to find it in the viewfinder. My husband couldn't see it. I couldn't find it or see it from hwy 234 but I kept scanning the sky. When we were on Modoc rd. I saw it again. It was going from shiny to dark to shiny to dark and I was having a hard time finding it in the viewfinder. My husband couldn't find a good spot to pull over with the load we were hauling and he still couldn't see it. When I was able to find it and start recording it I was leaned over from the passenger side of the truck towards my husband looking between his face and the wind wing, the window was up. It was more of an oval or a cigar shape when I was zoomed in on it and I thought that when we got home and viewed it we would be able to see wings and be able to identify it as a plane. I thought it was moving south while I was watching it but when we viewed it on the TV it seemed to be staying in the same place. I stopped recording when some trees came in view and I couldn't find it again. I don't think it disappeared, I think it just went dark and I couldn't see it again.

Occurred: March 20, 2007; Approx. 2111hrs.
Location: Chico, CA
Witness: 1
Shape: Sphere
Duration: 5 seconds
Extraordinarily fast moving spherical object with pulsing light...
On the evening of Tuesday, March 20, 2007 at 9:11 P.M. in Chico, California, I was walking with my dog NE on Satinwood Way, just having turned N. from Eaton Road. I looked into the sky and there was a spherical object moving at an extraordinarily fast speed (faster than anything I have seen with my naked eye) on a diagonal down towards earth. I would say it was about 100 feet off of the ground.

There was a light pulsing from the object that was somewhat like a strobe light although the lights did not move like a strobe light. The light was pulsing and was very bright, almost incandescent. I only saw the object for approximately 4-5 seconds and then it disappeared as if the light was suddenly turned off. The object was seemingly gone and I did not see it again although I waited for a few minutes. My dig did not notice and there was no one else around on this dark cul-de-sac.

Occurred: April 30, 2007; Approx. 1104hrs.
Location: Chico, CA
Witness: 2
Shape: Circle
Duration: 10 seconds
Three circular objects moving fast and without sound in a formation pattern
My wife and I were leaving an event at the Chico fair grounds and as we were walking to our car I noticed 3 small circular lights in the sky, they were heading north than made a sweeping turn to the west. They appeared to be larger than a star, somewhat flat and circular, and their lights were reddish in the center and whiter around the edges. The lights did not blink or pulse and was more dual than bright. As they approached from the south, they were moving quite fast then slows down, the two objects in front seemed to be joking in the sky then moved very close to one and other as the third in the rear followed. They moved very swiftly and made a big turn to the west then ascended upward until they disappeared. There was no noise and they moved like nothing I have ever seen.

Occurred: May 28, 2007; Approx. 1838hrs.
Location: Cottonwood, CA
Witness: 1
Shape: Changing
Duration: 15 minutes
Not a sound was made from ship but it stopped over us and we could just about throw a rock at it!
It started out with one light in one spot not moving, and then it made a sudden movement left up then back to original spot. Not a noise at all for being as close as a helicopter but no noise. Then it started moving towards the house very slowly but coming and it was close enough or closer than any helicopter would ever get, still no noise. then it stopped on a dime and didn't move for one minute, then it started west slowly as another figure started toward our house as it slowly came up on our house we stood waiting in aw, then it stopped right above our house for about 5 min as we seen the other still going west, then the one above our house slowly went towards the other one. From the time they were there till they left two jets like streams in the air. Not one noise was made but they were close as a kite would be / never in our lives have we seen anything like it and probably never will again.

Occurred: June 15, 2007; Approx. 1200hrs.

Location: Susanville, CA

Witness: 2
Shape: Circle
Duration: 39 seconds
Small while orb flying rapidly over farmland, we were driving and both saw it, the orb was not a kite, or small plane or radio controlled aircraft toy, this moved rapidly in straight line then quickly shot off into the sky...we were stunned, this was no human machine.
((NUFORC Note: Witness indicates that date of incident is approximate. PD))
((NUFORC Note: Witness elects to remain totally anonymous; provides no contact information. PD))

http://www.greatdreams.com/red-bluff-ca-may-2007.htm

WHY IS RED BLUFF SUCH A HOT SPOT FOR UFO / CATTLE MUTILATION ACTIVITY?

Sent: Tuesday, May 15, 2007 1:45 PM
Subject: Crop circles reported in Red Bluff, Tehama County, California
Importance: High
AUTHOR'S NOTE: The entire sighting is compliments of the sources within, including MUFON.

Mark Fussell from the Crop Circle Connector forwarded me an email from a reporter from the *Record Searchlight* regarding a reported crop circle located in Red Bluff, Tehama County, California. A competing newspaper the *Red Bluff Daily News* wrote the original report. Here it is, quoted in its' entirety:

> "Three crop circles could be seen in the field across the street from Helser Chevrolet on Adobe Road in Red Bluff Friday. Red Bluff police said they were aware of the circles but had not received any calls on the matter and were not investigating."

The report included two photos:

Daily News **photo by Rebecca Wolf**

The field is located alongside the Sacramento River

Steve Moreno (a founding member of the ICCRA) who is Director of Psi Applications, and Ruben Uriarte (ICCRA) who is also a MUFON CA state director. They are on their way to do an on-site investigation; so more details will be pending. Right now, the only details we have is that the crop circle was first reported on Friday, May 11, 2007 and that there are three circles in the field. Lots more work to do…

Red Bluff, California is notable as being the location of where Ishi, "the last wild Indian in North America," came out of the wilderness (just to the west of Red Bluff) back in 1911; Ishi was the last remaining member of the Yahi tribe. There were Indian mounds that were excavated in Red Bluff (the Tehama-Red Bluff Mounds) back in 1907, and there are more than 250 recorded ancient settlement sites along the Sacramento River in Tehama County. Red Bluff is located about 40 miles south of Mt. Shasta and 40 miles west of the Lassen Volcanic National Park.

More details to come…stay tuned.

All the best,
Jeffrey Wilson, Director
ICCRA – Independent Crop Circle Researchers' Association [International]
jeff.wilson@asmnet.com **or** jeff.wilson@adelphia.net
http://www.cropcirclenews.com/
734-891-2689 (cell)

I did some research on the internet for Red Bluff to see if any other crop circles have been made in the area and found this:

01/13/2006 -- 31st Cattle Mutilation on Red Bluff, California Ranch on Linda Moulton Howe's website.
http://www.earthfiles.com/news/news.cfm?ID=1197&category=Environment

"We've had a young calf dropped through trees
because he had his legs intertwined with tree branches." - Rancher Jean Barton

In Red Bluff, California, the Jean and Bill Barton ranch has suffered 31 cattle mutilations over the past decade. A few miles north, a rancher in Anderson, California, had cattle mutilations in 1999 and 2004.

January 13, 2006 Red Bluff, California - For the past half century, unusual deaths that law enforcement call "animal mutilations" have been reported around the world. Both domestic animals and wild game have been affected, especially horses and cattle. Ranchers and law enforcement have long been puzzled because animals are found with the same pattern of hide and tissue removed usually without blood from the head, sexual organs, and vaginal/rectal area. There are no signs of struggle or tracks around the dead animals, not even the animal's own tracks. That peculiar fact provoked law enforcement long ago to wonder if the mutilators came in and out of pastures from the sky?

Many ranchers have also speculated that their mutilated animals have been cut with lasers because of the bloodless nature of the excisions. In fact, pathology exams over the years have confirmed in some mutilations that cuts *were* made with heat energy of some kind. But another puzzle is the lack of carbon residue at the heated cut lines. Earth life is carbon-based and normal laser surgery always leaves some black carbon on surgical excisions that can be seen under a microscope.

<u>January 17, 1997, Barton Ranch - Pure Hemoglobin Particles On Mutilated Bull</u>

Biophysicist W. C. Levengood in Grass Lake, Michigan, thinks a very complex set of energies are involved. He began studying soil and grass samples from mutilation sites. A similar pattern of respiration changes in plant cell mitochondria were repeatedly confirmed around animal mutilation sites, including grass collected near the rectum of a mutilated bull discovered in northern California on the Jean and Bill Barton Ranch in Red Bluff, on January 17, 1997, south of Redding.

Dead and mutilated 2,000-pound bull discovered January 17, 1997,
by owners Jean and Bill Barton on their winter ranch near Red Bluff, California.
BLT Research, Inc. field investigator, Jean Bilodeaux, collected tissue, grass and soil samples,
including unusual hardened dark particles from bull's testicles and chest near excision.
Photograph on January 18 © 1997 by Jean Barton.

The Bartons were checking their cattle from horseback at their winter ranch when they found their 2,000-pound Black Angus bull dead and mutilated in a remote field strewn with volcanic rocks which can only be reached on foot or horseback.

Field investigator Jean Bilodeaux discovered an unusual substance on the bull's chest and testicles that was black and hard. Biophysicist Levengood confirmed the substance was pure bovine hemoglobin that could only be produced by complex laboratory centrifuge processes. He told me:

"To do this, you need to break down the cell membranes to the erythrocytes and leukocytes to remove the hemoglobin molecules. That requires a laboratory procedure with very precise biochemical steps. It's totally incomprehensible how the hemoglobin could be removed in the middle of the night out in the middle of a pasture and be separated from all the other cellular components and some of it sprayed on the dead animal."

Dehydrated fragment from black particles found on chest hide of mutilated bull discovered in Red Bluff, California,

January 17, 1997. This photomicrograph (450x) shows uninterrupted "mud crack"-type patterns and no indication of cellular structures such as erythrocytes and leukocytes. Analytical chemistry analysis determined this is pure bovine hemoglobin. Photomicrograph © 1997 by W. C. Levengood.

August 7, 2004 - Anderson, California Ranch
North of the Barton Ranch in Anderson, California, on August 7, 2004, a dead and mutilated female calf only two or three days old was found dead with neat, bloodless excisions that resembled an earlier mutilation. A veterinarian took photographs of that older cow and told the rancher there was no explanation for how the animal's jaw flesh, eyeballs, udder, rectal and vaginal tissue had been so precisely and bloodlessly removed. Then in August 2004, the newborn heifer suffered the same fate.

Newborn female calf found dead and mutilated mid-day on August 7, 2004, on the Harry and Carole Hawes ranch in Anderson, California.
Photograph © 2004 by Carole Hawes.

January 4, 2005 - Barton Ranch
Five months later on January 4, 2005, back south on the Red Bluff ranch; Jean and Bill Barton found another of their bulls dead and mutilated at the bottom of a steep hill.

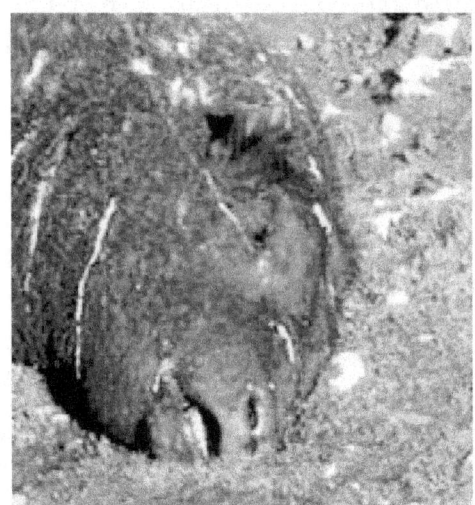 

**Bull found dead with bloodless excisions January 4, 2005, on the Barton Ranch.
Photographs © 2005 by Jean Barton.**

January 5, 2006 - Barton Ranch, 31st Cattle Mutilation

One year after that on January 5, 2006, the Bartons found yet another bull at the bottom of the same steep hill, surrounded by rain water and mud, but no signs of a death struggle or any other tracks.

Another bull found dead with bloodless excisions January 5, 2006, on the Ranch, in same remote and rocky field. Photograph © 2006 by Jean Barton.

Bull was discovered January 5, 2006, with its right jaw stripped of all flesh, the tongue was removed, the right eye was gone, the right ear was cut straight across at the ear tag, and the rectum was cored out. Photograph © 2006 by Jean Barton.

Occurred: June 18, 2007; 1200hrs.
Location: Shafter VOR (in flight), CA
Witness: 2
Shape: Cylinder
Duration: 10-15 seconds. While flying at 8500' Object appeared at my 11 o'clock position and moved at a high rate of speed to my 5 o'clock position beyond my vi
I was flying 310 degrees just SE of the Shafter VOR, coordinates 35:29.07 119:05.84.
I observed a bright magenta colored cylinder spinning vertically moving at a high rate of speed in the opposite direction. It appeared to pass me from 11:00 o'clock to 5:00 o'clock at my altitude at approximately 1/4 mile when at my 03:00 position.
I realize the actual distance would affect my perception of size. If the distance estimate were correct, It would be about 6-12' in diameter - thinking back, at that distance, it would have to be larger than that. All of this is guess work because I doubt I saw it for more than 10-15 seconds. The object was not hazy. It was very bright and clear. It appeared to me to not be very wide on its circumference, to have a slightly angled up surface about 1/3rd of the way toward the center and then flat to the center.
At the time I exclaimed about it to my wife who was riding in the back seat of our C177 TRG. I didn't think she had time to see it but when I asked her the next day she said she did see it. When she writes up what she saw, I will send it to you.

((NUFORC Note: We spoke with this witness via telephone, and we found him to be exceedingly sober-minded about the sighting. He stated that he had never seen a UFO before, and that he hesitated to talk about the incident.
We express our gratitude to the FAA for forwarding the case to our Center, and for providing the pilot with our telephone Hotline number. PD

Occurred: August 11, 2007; Approx. 0045hrs.
Location: Grenada, CA
Witness: 7
Shape: Fireball
Duration: 3-4 seconds
Huge, neon green fireball with orange halo travelling very fast perpendicular to earth in northern CA
Two members of the band, myself a newspaper publisher and our keyboardist, a school teacher, were driving home heading south on I-5, going 70 mph, at about 12:45am, when we had a huge, bright neon green fireball appear, going incredibly fast from behind us and disappear behind the mountains ahead of us all in about 3-4 seconds.
It had a reddish-orange halo around it and appeared to be at least as large as a house if not much larger.
When we arrived home 20 minutes later in Mount Shasta, two more band members who were 10 minutes ahead of us on the highway were waiting and excitedly asked, "Did we see it?" and described the same thing. The fifth band member called, he was in a third vehicle and was on another road, old highway 99 to Gazelle, and had also seen it.
((NUFORC Note: Witness provides illustration of her sighting through the windshield of her car. PD))

Occurred: August 16, 2007; Approx. 2308hrs.
Location: Redding, CA
Witness: 1
Shape: Light
Duration: 5 seconds
Single light making 90 degree turns, starting and stopping instantly without acceleration.
I was staying at the Holiday Inn Express in Redding, California. At approx. 11:08 PM, I was in the outdoor pool floating on my back looking up at the sky. Near the constellation of Cygnus and Lyra, I saw a faint light (much like a faint star) moving south. There was no way for me to tell it's altitude nor it's speed, however, its actual movement was puzzling. At first it was travelling south and passed by just west of Vega; it's speed was constant. Suddenly it stopped instantly for a short moment (maybe for 1/2 of a second), then instantly returned to its original speed in an easterly direction. When I saw it do this, the first thought that came to my mind was, "am I really seeing this? Is this my imagination?" Yet still, I was able to continue following the object through the sky. And for a second time, it stopped instantly without any negative acceleration for about 1/2 second. The object then resumed it's speed instantly without any acceleration almost due south before disappearing out of sight still high up degree-wise in the sky. I spent another 10 mins looking up but did not see anything more after my first sighting.

SPACE SHUTTLE SIGHTING CONFUSED
Occurred: August 19, 2007; Approx. 2110hrs.

Location: Chico, CA

Witness: 2
Shape: Light
Duration: 2 minutes

We saw two bright lights traveling across the night sky at a high rate of speed in some type of formation.

At approximately 9:10 PM on Sunday August 20, 2007, my son Matthew and I were sitting outside on my dock talking and admiring the night sky.

Looking to the North West Matthew asked, "What or those two lights up there? Planes?" I looked to where he was pointing. It was to the two bottom stars of the big dipper. There were 2 bright lights traveling across the sky from west-northwest to the east-southeast. They were not flashing. They did not have any color like an aircraft light. These were single lights. The distance apart was approximately 3 inches at an arms length. They did not change direction nor did the space between them change from what I could see. They were traveling much faster then any aircraft that I have ever seen and the lights were much brighter. The total time they were in my view until out of view was no more then 90 seconds.

As they passed overhead I did not here any engine noise as from an aircraft. I told my son that it is probably the ISS or possibly the shuttle but today I looked up their orbit at the time and they were nowhere near the west coast of California last night.

We could not identify them as any kind of craft, just two bright lights traveling across the night sky at a high rate of speed in some type of formation, and probably the ISS and Space Shuttle, which had undocked. PD))

((NUFORC Note: August 19 fell on a Sunday.

Occurred: September 11, 2007; Approx. 2140hrs.

Location: Medford, OR

Witness: 3
Shape: Triangle
Duration: 5 to 6 seconds

A triangular set of three dim or orange lights were seen moving very fast and silently across the night sky over Medford, Oregon.

On the night of Sept. 11, 2007 my son, his girlfriend and myself were looking in the sky. At approximately 9:40 pm a triangular shape of three lights appeared directly overhead going from north to south traveling very fast and without a sound. It seemed to come into view not quite over head like a satellite does when it come out of the shadow of the Earth from the sun, but then continued consistent until it went behind the neighbor's tree line. It lasted about a five to six second count. The lights were not bright. They were dim or orange in color.

Occurred: September 15, 2007; Approx. 1300hrs.

Location: Roseburg, OR

Witness: 1
Shape: Sphere
Duration: 20 seconds
Silvery sphere observed flying in a non-linear path; I'm convinced it was a UFO.
I was startled to hear a report of a UFO sighting that happened in September in Albany Oregon. The sighting that I had was during midday on a very bright sunny day. The silvery (mercury like) single colored object was traveling from the SW to the NE in a non-linear path. From almost overhead I followed it to a place fairly well before the horizon. It was a single object that took about 20 seconds to disappear into the sky. My first inclination was that it was a balloon, but following it fairly high in the sky I was positive that it was not.
I would guess the elevation to be 10,000 feet or below. It did not leave a con-trail and did not resemble anything that I had seen before in my 46 years of life. The path was not an ordinary North - South commercial jet flight path.
When I saw it, I was very awe struck; so much so that I remarked to my wife just moments after it disappeared that I had just witnessed my first UFO.
I'm not one to be fanatical about such things but am now convinced that I did see a UFO.
((NUFORC Note: Witness indicates that the date of the incident is approximate. PD))

Occurred: September 17, 2007; Approx. 2000hrs.
Location: Oroville, CA
Witness: 1
Shape: Triangle
Duration: 30 seconds
Silently flying south at very high rate of speed a triangular shape with three very bright lights in each corner zoomed overhead.

Occurred: September 17, 2007; Approx. 0545hrs.
Location: Placerville, CA
Witness: 1
Shape: Triangle
Duration: 2 minutes
Triangle with flashing lights in Placerville
It was about 5:45am and I was driving west on Hwy.50 from Pollock Pines to Placerville. As I was leaving to town of Camino I saw a flashing light in the sky. At first I thought it was a plane because there is an airport close by. As I got closer I could see it was triangle shape with flashing white and red lights that would turn green. As I drove into Placerville it went behind a hill. Then I saw it on the south side of the Hwy while I was in the middle of Placerville. Both times I saw it, it did not move.
I am 51 years old and retired. I drive someone to work in Placerville most mornings.

Occurred: September 20, 2007; Approx. 2000hrs.
Location: Coos Bay, OR

Witness: 1
Shape: Cigar
Duration: 10 -13 minutes

The cigar shaped object, possibly hundreds of feet long, flew slowly along the Pacific coastline. October 12, 2007 around the middle of September I sighted an unusual object in the western sky. The reason for the untimely reporting is simply because I had no idea that there was even a place to report these things to. It wasn't until I encountered a second sighting of a completely different object (which I will put in a separate report) sometime during the first week of October that I started researching for a place to report. Our local airport manager gave me phone number and web site.

I live about 700 or 800 yards from the Oregon coastline and the western sky is over the coastline and the Pacific Ocean in general. Approximately 8 PM, I was standing on my front porch looking at the recently darkened sky when I noticed movement to the northwest, which appeared, initially, aircraft like a white and a red light in the sky, moving, at first glance, typical aircraft. As I looked closer I thought to myself, "That's sort of odd."

What I saw didn't really look like an airplane. I thought possibly a blimp but I could hear none of the familiar noise that travels with the average blimp. As the object moved slowly toward a more westerly position from my porch, my view became clearer and the object started looking nothing like an airplane or a blimp. Looking due west from my front porch, I first sighted the object at approximately a 45-degree angle to my right in the northwest sky and seemed to be running parallel to the coastline. Our local airport would be at approximately 35 degrees assuming due west was 90 degrees and zero was to the right. Often time, jets leaving the airport take a similar flight path. As the object reached its due west position, I could tell it was many times larger than any of the largest jets I've seen here or even the blimps that fly over the area fairly often. It also now appeared to be further out than I originally thought. The object continued in a southerly direction until it reached approximately 120 degrees (my left) where it started to angle slightly more eastward and also began to angle more upward into the sky. Until this point, it had been traveling more or less horizontally along the coast. It continued this upward path until it flew out of sight.

When I first sighted the object, all I could see was 2 lights, a very bright white light at the leading end and a less bright red light at the trailing end. The white light did not appear to be particularly directional and seemed to create a soft glow or haze near the leading edge. The red light was probably 30-40% dimmer but crisper with less haze factor. As the object got to a more westerly position its shape became more apparent. It looked to be tube like with the white light fairly centered on the leading end and the red light seeming to sit on the upper side of the trailing end. Although we do get some thin to thick fog along the coastline, this night was clear and so fog was not the simple answer for why the upper and lower edges of the tube did not appear particularly crisp. The crispness of the tube edges seemed to fade in and out slightly and if, indeed it was further out than it appeared, then thin, not quite visible fog could have been the reason for this. The size of the object was kind of hard for me to tell but was quite large, maybe hundreds of feet long. I would say that it was probably 8 to 12 times as long as it was thick. The side clearly appeared to have the roundedness of a tube with the light that seemed to glow very dimly along the side from the leading end, almost highlighted like. I thought possibly (?) reflected light from the town. At no time did I ever hear any noise or sound come from it.

When the object reached its roughly due west position is when I called my wife out to see it. She came out and caught the last 3 or 4 minutes as the object angled up and away from the coastline into the upper part of its path where it flew out of sight toward high, thin clouds. It did not appear to reach the high clouds and seemed to just disappear 20 or 30 miles out to the south and east. The entire time of my sighting was probably 10 to 13 minutes. It flew quite slowly for the first 7 or 8 minutes and seemed to speed up some (to not quite as slow but not fast by any means) after its flight path changed upward and inland.
((NUFORC Note: Witness indicates that the date of the incident is approximate. PD))

Occurred: September 20, 2007; Approx. 2000hrs.

Location: Coos Bay, OR

At 8:00 p.m. PDT a cigar-shaped object, possibly hundreds of feet long, was viewed for 12 minutes from Coos Bay, Oregon as it flew slowly along the Pacific Ocean coastline. At the same time (9:00 p.m. MDT) three black triangular craft manoeuvred across the sky from the horizon and then hovered over the witnesses' location for two minutes in Kingston, New Mexico. (Source: Peter Davenport, National UFO Reporting Center, Seattle, September 2007 webpage, reports uploaded November 28, 2007).

Occurred: September 27, 2007; 0434hrs.

Location: Kennedy Space Center, FL

Witness: 1
Shape: Unknown
Duration: 1-2 seconds

UFO Fly-By At The 'Dawn' Probe Launch Object flies past in upper-left of screen in the launch video at different times in different versions but always just after the words: "We've exceeded the speed of sound".

Launch videos available on YouTube:
http://www.youtube.com/results?search_query=dawn+spacecraft((NUFORC Note: Witness elects to remain totally anonymous; provides no contact information. PD))

Occurred: September 28, 2007; Approx. 0620hrs.

Location: Williams, CA

Witness: 1
Shape: Chevron
Duration: 15 minutes
Bright motionless chevron shaped light and two orange orbs.

I was driving to a job site during the early morning hours of September 28, 2007. I was traveling west on California Highway 20 when my attention was drawn to bright light approximately 45 degrees above the horizon. The bright light was positioned to the south of my location. My first thought was the light was the sunlight reflecting off a highflying jet as the sunrise had not yet occurred. However, it soon became apparent the light was not moving. I then pulled off the road and noted two other cars parked with the occupants of these cars looking toward the light. The light appeared bright and had the shape of a chevron or "V" shape. Directly below and to the left of the light (southeast of my position) were two small orange orbs. The height of the light and the orbs could not be determined but it was above the broken cloud cover. I observe the light to remain motionless for
Approximately 5 minutes. It then rotated slightly and exposed a diffused red light on what appeared to be the rear area of the craft. I continued to observe the light for an additional 5 minutes before leaving. The two other cars and their occupants had left prior to my departure. I continued to observe the light as I drove towards my job site only loosing site of it when I arrived at the job site. The surrounding mountains prevented viewing the light from my job site location.

MUFON CMS Case #14395
CONFIDENTIAL:
Occurred: September 29, 2007; Approx. 1145hrs. (Pacific Daylight Time)

Location: Miller Mountain, CA-Klamath National Forest

Witness: 2
Shape: Disc
Duration: 5-10 Seconds
Event Witness testimony
"My husband and I were deer hunting on Miller Mountain. It had snowed a bit the night before up on the top, which made us happy because we could now spot fresh deer tracks in the new snow. All of the top dirt roads are dead ends and about 1-2 miles long. The snow clouds had cleared and we had a sunny sky with puffy clouds. We decided to take one of the "back side" western facing roads that take you along the steepest part of the mountain, pretty much straight up and down and heavily forested. We were happy that morning to notice that there hadn't been any other vehicles up on the top back roads yet that day, because there were no tire tracks in the fresh snow. We proceeded to take our time along this road with my husband driving. We came over a rise about half way to the end and my husband said, "What's that?" I was looking out my side window for deer tracks and turned my head forward to look out the windshield ahead. I noticed a "very shiny and bright" metal object maybe 8'-10' wide, in the road ahead approx. 50 to 75 yards away. Looking at it, the first words out of my mouth were, "Oh, there's a pickup ahead of us." It looked like a new, extremely shiny chrome pickup roll bar with two lights on top…but I could not see a "pickup", only what I thought was a roll bar. Then my husband said, "No, there are no pickup tire tracks in the snow ahead of us and we were on a dead end road." As I leaned down to get the binoculars and come back up to look it was gone. I said, "Well, let's go to the other end just to see". We drove to the end and indeed, there were no fresh tire tracks in the snow at the end, either. We talked about what we had seen, and came to the conclusion that no pickup could have gone over the edge of the cliff or gone up the side of the mountain…it had to have been a UFO. I asked him what he saw as I was getting the binoculars off the pickup floor, he said, "it just kind a floated up a bit off the road and disappeared." We did not see it again, nor did we stop and get out of the pickup where it had cloaked itself..

Investigator Notes:
December 15, 16, 17, 2008.
Playing phone tag. December 15-17, 2008. Witness ----------- apparently owns & operates a ------ --------shop in Weed, California.
Emails exchanged. Mainly regarding sending motivational emails to witness for landing location information. Delays in obtaining information were due to Christmas Holidays. January 5, 2009 Witness on January 5, 2009 finally emailed position of UFO landing (or near landing) location (position exact) (see jpeg attachment of satellite image-'Google maps')-Latitude 41 degrees 37'55.36N, Longitude 122 degrees 14'10.82"W, Elevation 2097 Meters (6880 feet).
Miller Mountain is roughly 12 (14) miles due north of summit of Mt. Shasta, and 15 (20) miles NW of Weed, California.
Notes: Witness –used Google Maps. I checked and verified witnesses position-but Investigator used 'Google Earth'. My position of same exact spot as seen visually on map display reads: Lat. 41degrees38'3.44"N, Long. 122 degrees14'11.86"W Suggests a possible slight or negligible variance between 'Google Maps" & 'Google Earth' .
No GPS coordinates available by my Google Earth programming.
Never have spoken (yet) with other witness-Mr. ------- -------- (-------'s husband). ---------- still abit reluctant to discuss sighting at present time.
Very interesting landing or 'near landing' CE-1 case. Witnesses did not 'see' from their truck close by minutes after sighting any apparent physical traces on ground (snow). No tracks or landing gear imprints seen. Witnesses keen experienced hunters familiar with this hunting area. So would they have been keen in seeing and animal (or perhaps entity?) tracks or footprints (in snow)? Both witnesses remained in their car, for their safety. Perhaps curiosity, but also some fear & trepidation of what else might be nearby remaining unseen perhaps kept them inside their seeming safe vehicle.

Occurred: November 5, 2007; Approx. 1818hrs.

Location: Lower Lake, CA

Witness: 3
Shape: Rectangle
Duration: A few seconds
Large lighted rectangular object that appeared, descended and disappeared in an instant
Just a little while ago, I was driving home headed southbound on Hwy 53 with my two daughters. The sky was dark and we just entered into the area of Lower Lake then I looked up because of a bright object that caught my eyes. It also caught the eyes of my two daughters.
We looked at the bright object, as it seemed to descend. My first impression was that a plane was landing. Then I thought, "Wait a minute, there are no landing strips anywhere near the position of that object!"
It was rectangular in shape by the outline of the bright white lights lining the shape of the object. They looked similar to very huge headlights like on a vehicle. It just appeared out of nowhere and seemed to descend before it made a sharp turn to the right disappearing lights and all.

SETI's large-scale telescope scans the skies

In northeast California, an array of 42 radio antennas is helping advance the search for extraterrestrials. With additional funding, the array will expand to 350 antennas.

by **Daniel Terdiman** December 12, 2008 4:00 AM PST

In a tiny California town town within sight of Mount Shasta and Mount Lassen is the Hat Creek Radio Observatory, home to the Allen Telescope Array--the only large-scale telescope fully at the disposal of the SETI project.
(Credit: Daniel Terdiman/CNET News)

HAT CREEK, CALIF.--From the perspective of an extraterrestrial, I wonder if there would be much difference between a human and a deer.

You might think that's an odd question, but on Wednesday, as I stood in an open plain here, at around 5,000 feet, with Mount Shasta visible far off to the north, a stunning blue sky, I watched a deer poking around at the base of what on its own would be an odd piece of astronomy equipment.

In fact, though, the 20-foot-diameter antenna the deer was investigating was just one of **42 identical units that make up the** Allen Telescope Array, **currently the world's first large-scale telescope meant for the full-time use of the** Search for Extra-Terrestrial Intelligence (SETI) project.

The ATA, as it's called, opened in late 2007 with these first 42 antennas. Designed to work in pairs, the antennas are intended to work together to mimic the stellar investigatory capacity of far larger single dishes. And the ATA is hardly finished. In fact, it is planned to eventually be made up of 350 of these antennas.

And while the famous Arecibo uber-antenna in Puerto Rico, with its 73,000 square meter size, has seven times the collecting area of the full ATA, the telescope here--the array in its entirety is a telescope--will be able to look at 2,500 times as much sky as Aricebo.

For my visit, resident astronomer Rick Forster took me around, explaining the history of the facility, as well as how it is used today.

Originally, the Hat Creek Radio Observatory--the official name of this facility--was a joint effort by UC Berkeley, the University of Illinois at Urbana, and the University of Maryland, called BIMA. It had ten 20-foot-diameter antennas that operated in concert to create a millimeter-wave radio interferometer.

A deer investigates the ground around one of the 42 antennas in the Allen Telescope Array.(Credit: Daniel Terdiman/CNET News)

But eventually, that project moved on, and now, Hat Creek is the home to the ATA, and for the same reason that BIMA was here: it is one of the few places in North America that provides astronomers hoping to scan the skies with little-to-no terrestrial radio interference.

That's because the facility is bound by the Cascades on one side and a fault scarp to the east.

Of course, for the folks who live here, that means no cell phone service, and they're pretty much out of luck for listening to the radio or watching broadcast TV.

But since what these scientists want is to do serious astronomy, it's fair to say that's a trade they're willing to make.

To be sure, however, their hope for radio silence is dashed by the ever-present broadcast satellites that scream overhead. And those mean that there are a series of frequencies that simply aren't available for scanning.

The Allen Telescope Array has been funded so far mainly by Microsoft co-founder Paul Allen, and it's hoped he will be the benefactor for the future, as well.

Forster said that the original 42 antennas cost around $25 million to put up, and that while no additional funding has yet been acquired, they are in negotiations for the money to expand the array to about 128 antennas.

You may be familiar with other large-scale radio observatories. Perhaps the most famous is the Very Large Array, in New Mexico, which comprises 27 giant dishes.

Like ATA, SETI makes use of the Very Large Array. But the chief difference is that the SETI folks only get to use VLA once in a while. At ATA, however, they are always on. And that means, the SETI folks think, that their goal of tracking down E.T. is now getting a serious jump start.

McDonnell Douglas studied UFOs in the 1960s

2008

Occurred: January 1, 2008 Approx. 2120hrs.

Location: Yreka, CA

Shape: Cylinder
Duration: viewed over Oberlin Rd. in Yreka as I was returning from picking my wife up from work. I thought that it was a low flying aircraft but then it disappeared then re-appeared about a mile away. Then it was out of sight. All this took place in a matter of seconds.
((NUFORC Note: Witness indicates that the date of the event is approximate. PD))

January 08, 2008 Possible UFO Sighting Reported at Yreka County, California

By BRAD SMITH-Daily News Staff Writer
Tuesday, January 22, 2008-Stephenville, Texas residents aren't the only ones reporting unidentified flying objects above their community. A few weeks ago, Yreka resident Mary Hay reportedly saw a strange aircraft fly above her while she was at the Valero gas station on South Main Street.
According to Hay, she was purchasing some kerosene and saw something moving across the sky. It happened on Jan. 8, sometime around 8 p.m., she said.

Describing the object as 'slightly wedge-shaped,' Hay said it had three rows of lights underneath it, the middle row longer than the others. The object flew from east to west, 'then disappeared very quickly,' she said.

This is the first time that something like this has happened to Hay.

Since her sighting, she has told some of her family, friends and co-workers. While some have scoffed at the notion of UFOs and aliens, others have told her that they have seen the same thing and that she shouldn't feel "weird" talking about it.

'I didn't even know about the (Stephenville) Texas UFOs until a few days ago,' she said. According to the Associated Press, several people have reported seeing UFOs since Jan. 8. The sightings have made national news, even causing CNN's Larry King to do a show about them. Ruben Uriate, director for Northern California's Mutual UFO Network, said that Hay's sighting is 'one of many.'

'There's been a spike in reports,' he said. He speculates that the Stephenville sightings have raised awareness and that 'people are looking up at the skies, paying more attention.'

Uriate said that Hay shouldn't feel uneasy talking about her claim. Many pilots, law enforcement and military personnel have stepped forward, he stated, reporting their experiences.

For more information on reporting a possible UFO, check the Mutual UFO Network (www.mufon.com) and the National UFO Reporting Center (www.nuforc.org) Web sites.

Source and references:
http://www.siskiyoudaily.com/articles/2008/01/22/news/5912208news2.txt

Occurred: February 17, 2008; Approx. 1800hrs.

Location: Grants Pass, OR

Witness: 1
Shape: Light
Duration: 5 seconds

A bright light was falling from the sky over Grants Pass, Oregon.

At about six PM West Coast time, I looked in a Northern direction, the sun hadn't completely set yet, and I did not see any stars in the sky in that area. A bright light, similar to a shooting star, yet brighter, and seemingly closer, shot straight down into the horizon.

((NUFORC Note: We spoke with this witness via telephone on two occasions, and he seemed to us to be exceptionally capable as an observer. PD))) it was not Chicken Little) author BW

Occurred: March 1, 2008; Approx. 1400hrs.

Location: Chico, CA

Witness: 2
Shape: Light
Duration: 1 -2 minutes

Witnessed unexplained multiple lights/orbs traveling east to west in clear daylight. First impression was that they might be balloons. However, a smaller group of two broke off from the others and traveled in the opposite direction at a rapid speed.

My neighbor was outdoors at the time, so I called out to him to confirm the sighting. He could not distinguish what the objects where, so he went to his garage and brought out binoculars. We both observed the objects through the binoculars. The objects appeared to be indistinguishable (reflected?) lights or orbs.

Unable to clearly make out forms we came to differing conclusions. He believed that what we observed must rationally be birds flying at high attitude. I could not come to a definitive conclusion, fore I have often in my 50 plus years observed birds in flight and never did they appear like this. I only know that I witnessed glowing lights or orbs in an asymmetrical formation. To me, they are unexplained.
((NUFORC Note: Witness elects to remain totally anonymous; provides no contact information. PD))

Occurred: March 6, 2008; Approx. 0728hrs.

Location: Eugene, OR

Witness: 1
Shape: Disk
Duration: 2 minutes
A black, round, flat, disc was seen flying twisting spiral pattern over the Eugene Oregon morning sky.
I went outside to smoke my morning cigarette this morning (Thursday 3/6/2008, @ 7:28 AM exactly) when I noticed something out-of-the-ordinary flying in the sky to the east of me. At first I thought it was a bird, but when no wings were moving I focused in on it with my eyes and figured that it had to be a loose 'weather' balloon, or any type of balloon for that matter, ascending into the sky with the help of the wind. Then I realized that there was no way it could be a balloon either. As it continued to travel toward me (to the west) and further North, I was staring at it as its path advanced very quickly throughout the sky. I could see it moving like no other 'balloon in the wind' would ever move. It was completely black and its shape was changing as it traveled through the air to the north before disappearing in the light layer of clouds that was layering the sky over the valley this morning. It was clearly a thin (relative to my perspective), perfectly round disk of some sort, and while it flew, it spiraled and twisted like a flat Frisbee, a CD, or a flat rock would spiral if you were to throw one into the air. It was twisting on its axis while flying through the air (could not tell if it was spinning or not) very gently and softly but moving very fast though the sky. The sighting lasted for about 2 minutes…from the moment I spotted it directly East of where I was standing, until the time it disappeared due North from my position. I could see that it had a perfectly round shape when the angle was right, and I could also see its flatness when it shifted just enough on its side in the middle of its flight. Its flight pattern was just like that of a bent Frisbee when thrown upside down…you know, it spiraled, sort of rotating on its lengthwise axis, as it flew, kind of twisting as its path advanced. If you've ever thrown a CD into the air, it was flying almost exactly like that, but proportionately, probably not as fast as a CD flung in the air. It had very slow, smooth almost floating dynamics that made it seems like it was hovering almost…but it seemed like it was moving at a constant rate throughout the sky. I could see its round profile and its flat profile as it spun/spiraled, depending on its position in flight relative to my point-of-view. Its color was black, but it may have been that it was being silhouetted or shadowed or the like. It was a flat, round disk, twisting and spiraling through the sky about as fast as it would take a small airplane to fly across that much airspace. If one side was the "top" of the disk, and one side "the bottom," I must have seen the top and the bottom about 6 to 8 times each. That is about how many rotations it made on its axis. Its size was not determinable, and the altitude at which it was flying was also not determinable, so I cannot make any postulation as to any of those details.

I have never in my life believed in any sort of inexplicable flying objects of any sort. I always thought that people that believed they saw UFO either had overactive imaginations or were just ignorant in terms of existing aircraft, or a combination of many other factors. I do not wear glasses and I have perfect 20/20 eyesight in each eye. This was my first time ever seeing what in my mind, I guess, HAS to be a 'UFO', meaning there is no way I can explain what I saw based on my own first-hand experiences. I have never believed in "UFO's" simply because I have never witnessed, first-hand, anything that I would qualify as such...there always seems to be an explanation (and there probably is one in my case, with my sighting this morning too! I just don't have one right now!). I always said that unless I see one myself (or a Sasquatch for that matter) I couldn't logically claim to believe in them. This is my first ever 'UFO' sighting. I even caught about 30 seconds of what I thought would be good footage using my cellular phone's video application...to no avail. Immediately upon recording the sky, I could tell that there was nothing-visible showing up on my phone's screen. Nothing showed up during the recording process and there is nothing at all on the footage I have saved on my phone...just the hazy morning sky...no black spot, no nothing. The object must have been too far away to register on my phone (a 2 Mega-pixel camera-phone). That is all I have to report. Conveniently, I am making this report here because I happened to see your website on TV; a History channel special on "the grays" which ran last night, and that is how I knew that this is the place to come if I ever needed to make a report. Well, wouldn't you know it...the day after that is the day I have my first UFO sighting.

Occurred: April 21, 2008; Approx. 0107hrs.
Location: Redding, CA
Witness: 1+
Shape: Egg
3 egg like objects in the Redding sky!!!
Last night in the early morning, I was outside with my friends, walking our dog. All of a sudden the dog started going crazy. It got off the leash and took off running up and over this big hill that was in front of us. When we got to the top of the hill we looked and there we saw them!!!! They are what looked like 3 big illuminated eggs in the sky. We got so scared that we ran back home but when we went to wake up our parents and tell them what happened they wouldn't believe us. I had to report it if no one else would believe me and my friends.

Occurred: April 21, 2008; Approx. 1500hrs.
Location: Redding, CA
Witness: 1+
Shape: Egg
Duration: 4 minutes
3 egg like objects in the sky.
Last night, in the early morning we were out walking around my home with friends. we were just walking our dogs and all of a sudden, we heard a loud Swirling noise and our dogs went crazy. They ran far and were jumping at the sky. We looked up and there were 3 helicopter-like objects and had really strange designs on them. They started coming toward us and we ran back to the house. Then this afternoon, we went to see if there was any evidence of it. We went back and there was this strange spherical object with the same designs as the helicopters had.

Occurred: May 5, 2008; Approx. 0500hrs.
Location: Medford, OR
Witness: 1
Shape: Light
Duration: 3 minutes
Bright white-blue light in the sky, the light moved, no sound, no other light. Object changed colour, speed, and course quickly.

I was leaving my front porch for a walk this morning, it was dark, and I noticed a bright light behind me. This light was not usual for the time of morning.

At first I thought it was the moon because of the purity of the light, but the moon is not in the west at 5 in the morning. I turned and saw what looked like an extremely bright star; however, this star was moving from the southwest at a steady pace no faster than a jet-plane.

I watched the object move overhead and checked for running lights, flashing lights, and sounds. There were no colour lights besides the white-blue light, and no sound at any time.

When the object passed over me it light seemed to rotate from front to back and spin. As it spun it became brighter and cleaner to me. Following the moving light, the object-changed course. It also changed from the white colour to an orange-pink colour. I realized what the object was once it passed overhead, having seen one many years ago. The shape, however, was different. The tear-shape was not exaggerated. There was only one (I searched for others). As far as my background, my father and I have had previous contact with a craft like this. My father (who died last winter) has notes on his experiences on the Ore. Calif. border in the seventies, and my own occurred near the same region. The same kind of light moved very fast from Mt. Shasta.

I have also seen one being chased by Military aircraft while on a hunting trip. Do people who see UFOs often have more than one experience?

My father claimed he had a stand off with the light and took it very serious that is why I'm reporting this to you now.

Occurred: June 8, 2008; Approx. 2136hrs.
Location: McKinleyville, CA
Witness:
Shape: Other
Duration: 3 minutes
Bright light making quick, fast turns across the sky
A very bright light, larger than any star, moving in straight lines diagonally across the sky. Appeared out of nowhere, vanished into the horizon. no sound, definitely wasn't a plane.
((NUFORC Note: Witness elects to remain totally anonymous; provides no contact information. PD))

Occurred: June 8, 2008; Approx. 1815hrs.

Location: Oregon (in-flight; 15 min. Mt. Hood), OR

Witness: 1
Shape: Circle
Duration: 15 seconds
3 white circular objects flying closely together
Sighting happened June 6th, 2008 between 6:00-6:30pm. I was flying North to Portland, Oregon in commercial aircraft. It was perfectly clear outside, sunshine, and not a cloud in the sky. As I was looking down outside at the beautiful mountains and green forest below I saw what appeared to be 3 circular white objects flying very close together in a perfect formation below the aircraft traveling very fast in the opposite direction. They must have been 10,000 to 20,000 feet below the aircraft so they looked very small. Each object was positioned a little ahead of the other in a perfect formation flying at the same speed.
((NUFORC Note: We spoke with this witness, and found her to be an exceptionally capable witness. PD))

Occurred: June 20, 2008; Approx. 0200hrs.

Location: Cloverdale, CA

Witness: 2
Shape: Sphere
Duration: approx. 45 seconds
Bright red sphere moved radically in a clear starry night sky and we got the feeling that we were being watched after it disappeared.
My boyfriend and I were standing on a levy down by the river looking at the starry night sky. I looking out over the mountain ridge and noticed a reddish sphere in the sky moving left and right along the ridge so I said "what is that?" because there was no noise coming from it and it also moved unlike anything else I had seen before. I pointed to it and I had to ask my boyfriend again what it was and he said he didn't know. When he said he didn't know what it was either the red sphere jet back and forth in a couple of different directions in a radical fashion and then it blinked out and disappeared. We both felt as if we were being watched and got spooked out and went home.
((NUFORC Note: Source of report indicates that the date of the incident is approximate. PD))

Occurred: July 13, 2008; Approx. 2035hrs.

Location: Grants Pass, OR

Witness: 1
Shape: Cigar
Duration: 3 minutes
Strange object seen in Grants Pass, Oregon
At about 20:30 hours on July 13, 2008, I was standing out side and was talking and looked up saw what I thought to an aircraft, as I looked on this did not look like and aircraft, it was to low and where I live, aircraft fly higher up.

As I watched the object move from north to south, I watched it and there seemed to be an aura or what looked to be a haze around it. The object was moving kind of slow, as I walked down the driveway to watch this object, was when I noticed more the haze or aura around it. It seemed to all of a sudden disappear as I was watching it. When I walked back up the driveway, I looked up again and saw the same object moving in the same manner as I have described. I again walked down the driveway and watched the object and this time it had disappeared.
My friends are trying to tell me that perhaps I seeing a meteor or a comet these objects would have left a trail of some sort. I know what I saw and in no way it was a meteor or a comet.
((NUFORC Note: Witness elects to remain totally anonymous; provides no contact information. PD))

Occurred: July 31, 2008; Approx. 2345hrs.
Location: Red Bluff, CA
Witness: 1
Shape: Circle
Duration: 1 minute
Light In The Sky
On July 31, 2008 at 11:45 pm Pacific time, I was sitting on my back porch and noticed a large ball of light orange in color something about half the size of a full moon in the sky. It appeared to be floating around a small area and they suddenly faded away. There was not any blinking lights or noise from the object while it was in the sky or after it faded away. We live in Northern CA. in the Redding, Red Bluff area out in the country.
((NUFORC Note: Witness elects to remain totally anonymous; provides no contact information. PD))

Occurred: September 9, 2008; Approx. 1835hrs.
Location: Medford, OR
Witness: 1
Shape: Cigar
Duration: 4 minutes
Cigar shaped object viewed through binoculars over Medford
At about 6:35 pm on Tuesday September 9, 2008, I was on the back balcony of my apartment when I noticed an object flying up high in the sky, traveling from south to north. My first thought was that the object was a airliner, however the speed of the object seemed slower than usual and there was no visible vapor trail, even though there were a few light clouds in the upper atmosphere.
I ran inside and grabbed a pair of binoculars and continued to observe the object. The binoculars are 10x50 and through them the object appeared to a little over 1/4 of an inch in size. The object was reflecting light from the setting sun, which highlighted certain details. It was cigar shaped and displayed no wings, tail or visible lights. The object appeared to be metallic and was bright white in color along most of its surface with a hint of red towards the rear. There seemed to be three large circular impressions that ran along the length of object. There was no discernible propulsion that could be seen.
I watched it though the binoculars for about two and half minutes before it flew out of my field of vision.

Occurred: October 21, 2008; Approx. 2000hrs.
Location: Arbuckle, CA
Witness: 1
Shape: Light
Duration: 5 seconds
Light like Venus disappears in 5 seconds
Back from the store, I look up because I saw a beautiful light, I was thinking is Venus but Venus was more to the west; and then I look back to the light it was fling vertical and disappears in two seconds.
((NUFORC Note: Witness elects to remain totally anonymous; provides no contact information. PD))

Occurred: November 1, 2008; Approx. 2345hrs.
Location: Redding, CA
Witness: 4
Shape: Sphere
Duration: 2 minutes
An Unidentified Flying Object with bright orange light was seen, with brother and two other friends on November 2008 over Redding.
On the night of November 2008, forgot what the exact day. I was star gazing with my reflector telescope south east of where I live approximately about 11:45 pm. My brother had came out to star gaze with me and just as he had looked Northeast approximately about 40k-100k ft. (don't know how high, but no higher than a regular airplane). I had told my brother that is not a moon, because a moon does not appear where it should. This unidentified aircraft looked like a full moon large circular with one bright color (orange), the light was huge, not glimmering, shining, neither blinking, it was bright light. That night there was four of us, me my brother and two other friends. The unidentified flying object sits in the air without a sound and interestingly the object moved/hover. And I said oh Sh*t, it's moving. It had moved just a bit at the moment, and then we were stunned. This object had no sound, moving very rapidly but slowly towards us and then turned direct! ion towards the nearby airport. We had thought it was a airplane, all of the sudden the craft flew over us towards northwest of our home and stopped for a few seconds. The object started moving again, now going south. As the object was moving away further, the very last thing it did was took off just like we see in the movies Star Trek, it just zoomed and disappeared.
((NUFORC Note: Witness indicates that date of sighting is approximate. PD))

Occurred: November 2, 2008; Approx. 2045hrs.
Location: Marysville, CA
Shape: Light
Duration: all night
13 different flashing lights surrounding my house for the past month

I was sitting outside of my house at around 10:45 pm and noticed a flashing light not much bigger than a star but noticeably brighter, as I looked at it I noticed that it was flashing from red to green to white and repeated. it was then that I noticed there were several others, 12 that I could count, surrounding my house just a little dimmer than the first one I saw. I've seen them every night since then. It's kind of scary, but at the same time its really fascinating
((NUFORC Note: Stars or planets?? PD))

Occurred: November 20, 2008; Approx. 1930hrs.
Location: Antelope, CA
Witness: 2
Shape: Rectangle
Duration: 2 minutes
Rectangle with chasing lights on the bottom
My husband and I were pulling out of a Wal-Mart parking lot and we seen what looked to my husband as a banner like the ones pulled my airplanes but then we noticed it had lights on it that were like the chasing lights, there were about 4 rows of yellows lights on it, then it just disappeared.

2009

Occurred: February 11, 2009; Approx. 2300hrs.
Location: Rancho Cordova, CA
Witness: 1+
Shape: Light
Duration: 13-16 minutes
Little army of star like lights pass over Sacramento County! It gave us quite a scare.
We were looking in the sky like we seldom do and we noticed several lights some whitish blue and some light yellow. At first we saw three moving across the sky not in any formation or nothing. Naturally we keep our eyes in the sky and we noticed even more coming up from behind the others. As odd as it sounds we have seen these objects before but, only one to three at a time, it was quite a scare to see them moving and scattered across the sky above us. Star like lights some brighter than others but all traveling in the same direction, like a little army of them.
((NUFORC Note: One of three reports from same source. PD))

June 7, 2009 - Roper Poll
A nationwide survey by the Roper Organization has uncovered the following: when asked what they thought UFOs were, 25% thought they were alien spaceships, 12% thought they were secret government programs, 9% said hallucinations, 19% said UFOs are normal events that are misinterpreted by witnesses, and 7% said travelers from other dimensions.

Occurred: June 30, 2009 (Entered as: 2009)
Location: Oroville, CA
Witness: 1
Shape: Cylinder
Duration:
UFO caught on Google Maps.
I was looking at property to buy on Google maps. In the picture that Google took there is a very curios ufo in the sky. There is a exhaust plume and a craft, the curios thing is that you can see an exhaust area where the ufo seemed to appear out of nowhere and then accelerate. Anyone interested in seeing this should go to Google Maps, type in Lumpkin Road, Oroville, Ca.
Use the satellite view and go to the first bridge over the lake, then go to street level and point north and up. It is VERY visible and very intriguing.
((NUFORC NOTE: Witness indicates that date of the sighting is approximate. PD))

Occurred: July 5, 2009; Approx. 0100hrs.
Location: Redding, CA
Witness: 4
Shape: Disk
Duration: 03:00
Rotating discs with colored lights seen ongoing for 3 weeks in July 2009 over N California

Beginning late in the first week of July, a number of brightly lit and multicolored discs were witnessed in the vicinity of Redding, California. The discs were situated approximately 25-40 miles from the area of witness and clustered at the north and west coordinates in small groups, sometimes stacked on one another. Near Mount Shasta and in the general vicinity of highway 5 is the most prominent location for these discs in the last 3 weeks. They have also been seen due west over the rugged mountains of Trinity County at approximately the same distance away from the witnesses. A rough estimation of the distance away is 25-40 miles and elevation above the ground is estimated at 10,000 to 25,000 feet. All sightings have occurred right around 01:00 hours and most have persisted until 03:00 hours or later with the witnesses all having seen their fill by that time. It is the duration of constant hovering combined with the entirely unusual colored lights that! Makes a very convincing argument the objects are not helicopters or airplanes. All four witnesses were satisfied with this conclusion and have agreed there is no "normal" explanation for what has been seen.

In particular about the craft, they look to be rotating discs oriented horizontally or sometimes tilted at an angle of around 30 degrees. The discs emanate four flashing colors of bright light identified as green, orange, red and blue. The lights flash in a rotation around the perimeter of the discs in an obviously attention getting and unnatural display. The discs have been best described as "Simon" electronic games in the sky rotating. The "Simon" is a game that is still available and consists of four colored lights arranged around a circle that are hit in succession to form a pattern. Surely most people can recall one of these from a television ad at least and visualize the same four colors flashing around the edge of a disc hovering in the night sky. There are sometimes but not always accompanying craft but normally one flashes more brightly than the others or the others seem to hover near or above a main disc. The discs do not make fast movements but have been seen to move drastically in the sky in a period of only 15 to 20 minutes. Witnesses would see and identify the disc, then go indoors for a short while and come back out to see the disc had moved a hand's span in the sky in only a short time. There has been reliably one disc in the northern sky near Mount Shasta and also similar discs in the western sky over Trinity County, a bit south of Whiskeytown Reservoir. One of the first sightings involved the disc in the northern sky moving in very small rapid circles side to side but maintaining the same altitude. Almost like the tip of a pencil drawing small curly-Qs over and over in the same place. This was much more pronounced than the usual "shimmer" of a star in the sky that results from the atmosphere refracting the light. The discs were definitely making rapid circling movements while rotating their flashing lights constantly.

These sightings are complete textbook UFO instances, lasting so long that one has plenty of time to view the object and make a very detailed observation. These objects are definitely not standard aircraft and resemble nothing seen before by any of the four witnesses.

((NUFORC Note: Source indicates that date of event is approximate. PD))

(NUFORC Note: We believe that stars would have to be ruled out, before we could conclude that the witnesses had observed genuine UFO's. PD))

Occurred: July 6, 2009; Approx. 2333hrs.
Location: Redding, CA
Witness: 1
Shape: Changing
Duration: 3 minutes
Large orange shape changing object emits smaller orbs, winks out reappearing instantly in other locations

Last night, here in Redding, California, Monday 7-6-09, I took my two dogs out one last time before I headed off to work an overnight shift. It was about 11:33 PM.

I looked up overhead and spotted a brilliant orange object, somewhat circular in shape. It was about half the diameter of the holes punched in three-ring notebook paper held 30" from the eye. I estimated its altitude at 25K feet or higher. From my position near the intersection of the Shasta View Drive and Tarmac Road, at 40*34'49N X 122*19'26W, the object appeared to be at 85 degrees above the horizon to the SSW. Judging by its altitude, it was probably directly over the southwestern margins of the city limits of Redding, California.

At first, its coloration was steady and unchanged; the object was stationary. But as I watched, it began to change shape, with several smaller orange objects coming from it, rapidly separating from the main object to a short distance then momentarily flashing and winking out, never to be seen again. This happened about four different times.

The main object would alternately appear as a circular shape then dim slightly from a bright orange color to a bright red and back again, and as it did so, it took on the appearance of multiple spheres clustered together; very much like the oft-cited theodolite camera footage shot near Area 51 some years ago.

But then the large object quickly winked out itself, going from bright orange to red, then to a brownish hue and vanished completely. It was completely out of sight for about a second and would suddenly flash and reappear in another location a short distance away, where it again stood motionless. This on-off-on and sudden relocation event happened about six times. Held at arm's length, these "jumps" would appear about three inches from point to point. At an altitude of 25,000 feet altitude or more, this represents a jump of about a half-mile or more, (transitioned instantly).

I debated whether to run and grab my high-powered binoculars, but was afraid that the object would disappear before I could return to view the phenomenon. The object moved in several directions in this disappear-reappear sequence and finally, as I was on my cell phone yelling for my girlfriend out in Palo Cedro to go out and look, it began moving in a south easterly direction and vanished just as she was just stepping outside. The sighting lasted about three minutes.

I saw the International Space Station (ISS) pass overhead moving in a northeasterly direction about an hour earlier, so I know this was neither a sighting of nor misidentification of that.

After checking with the administration at the Redding Municipal Airport, I was told they have a recently installed ARTS-3C radar system that is "slaved" to the TRACON System at the Oakland Bay Airport, but they have no provisions for recording the returns locally. I checked the Internet for live radar feeds, but I guess there aren't any available to "mere civilians". Maybe it might be possible to find out if there is some ATC radar tapes that could be examined for a signature on this.

Occurred: August 24, 2009; Approx. 1430hrs.

Location: Medford, OR (outskirts)

Witness: 3
Shape: Light
Duration: 7 minutes
Strange flashing whitish light that moved like a plane but then stopped and stood still for five minutes.

My brother, mother and I were driving along North Phoenix Road North towards Medford. We were in between Medford and Phoenix, perhaps the outskirts of Medford, around Campbell road, when I noticed a light flashing in the sky in front of us and a little to the right (East). It was flashing in a strange way, so I pulled to the side of the road. We got out and watched a UFO move steadily in our direction but somewhat to the East over the mountains. At first I thought it might be a plane, but the entire craft was flashing, as I said, in a strange way. It was about 2:30 pm, so it was broad daylight, and the light needed to be fairly bright in order to flash that way. The way that it was flashing gave me the impression that various parts of the craft were rotating. The whitish light was continuously emanating from it, but in different intensities and from different parts of the craft. So, the craft continued along and passed us, continuing South and a little East,! When it suddenly just stopped. I then knew definitely that it wasn't a plane. It sat there stationary, and still flashing, for about five minutes. We watched it for those five minutes until we had to go (we were late for something), so I have no idea how long it stayed in that position.

Occurred: September 8, 2009; Approx. 2000hrs.
Location: Paradise, CA
Witness: 1
Shape: Light
Duration: 3-4 Minutes
Two bright lights following each other seen in the sky when no stars were present
Two bright lights were seen flying over the West Feather River canyon branch in Paradise, CA. One light was dimmer than the other and they seemed to be following be following each other. There were no stars in the sky, and they could not have been domestic aircraft, because they were both solid and steady lights.
((NUFORC Note: We believe that the Space Station and Space Shuttle were flying in formation, at this time. PD))

Occurred: September 18, 2009; Approx. 2115hrs.
Location: Redding, CA
Witness: 1
Shape: Disk
Duration: 8 Seconds
Soft orange glowing diamond shape with circular center
I saw one object flying north to south at a high rate of speed, traveling in a straight line. It was a soft orange glow, with a spherical center and conical "tails or wings" on each side. The sphere was dark in the middle. Seemed to have sort of a vibration appearance. It was completely silent. First UFO I've ever seen. I was pretty much a skeptic until tonight.

Occurred: September 20, 2009; Approx. 2130hrs.
Location: Springfield/Jasper, OR
Witness: 1
Shape: Triangle

Duration: 1-2 minutes

Big, hovering triangular object with multiple lights lit under it gliding at low altitude on a clear, calm night

I was on my way to work on jasper road, right after 58th street. I had just gone past t traffic lights on 58th street and was now slowly accelerating on jasper road...after the lights; the road is just a little elevated so you have a wide and open view of the sky. As I was driving, something came to my eye. I noticed a triangular shaped object, with six amber lights or so and a middle yellowish-blue blinking one in the middle. What made me notice it is that it was gliding/hovering at pretty low altitude, steady speed and it caught my attention because there's no airports close around and it didn't make sense to me why there would be a low flying aircraft right there. I pulled my car to the side and since I had a wide clear view, I decided to look a little closer at whatever was slowly hovering ahead of me...and by no means was it a helicopter because there were no blades whirring and no noise coming from it. I had just slept for 7 hours before getting up and ready for work! so I was fully rested, awake, sober and straight minded at that point. I watched this thing slowly glide while emanating no noise from it for a good minute and a half or so. If I could approximate, this thing was probably 1/2 a mile away but from that distance and by what my eyes were telling me, it looked pretty huge. This just happened 2hours ago so forgive me for the way I'm writing all of this down but I'm still awed, dumbstruck and scared from what I saw, I will never EVER forget that moment as long as I'm living and breathing on this earth/

Occurred: November 1, 2009; Approx. 0148hrs.

Location: Redding, CA

Witness: 2
Shape: Changing
Duration: 30-45 minutes
Changed shapes with bright multi-colored lights all over/around it.

I first saw the object as I walked out on my porch to get some wood for my fire. It looked like a plane at first; I thought it was just coming in to the airport, but then I noticed it wasn't moving and also it was flashing green, blue, red & white lights off and on, but in no particular pattern. I ran back inside and yelled for my friend Mary to come out and take a look and I grabbed a small pair of binoculars on the way back outside.

I looked up and it was no longer a "strip" of flashing lights, but was now an oval shape with the same colored lights on and around it. They seemed to be twinkling instead of flashing now. I handed the binoculars to Mary and asked what she seen.(I didn't want her to be influenced by what I described, and also wanted to see if she could see anything at all.) She described the exact things I was seeing. It continued to hover there and I called several friends to get some one else to acknowledge this "object", but no one answered.

We then drove down to our local Shell gas station/convenience store to get the cashier to verify the sighting of this object. She hadn't used binoculars before and could not really get a good look at it; she did state that her son had seen a UFO a couple days earlier, though.

I called the local Police and the cashier called the Highway Patrol to see if anyone else had reported anything and they both said "no". It was not in the sky when we got home and we did not see it leave, fly or disappear; We called this number provided by the Redding Police Dept. and the recording gave us this web site address.

The entire incident lasted approx. 40 minutes I am including sketches of what we saw, with descriptions and will email it ASAP.

Author's Note: The following report in my opinion is one of our NEW "CLASSIFIED" helicopters that now have been used in movies like "Avatar" and other newer Sci-Fi movies. It is not a UFO but I would not put it past the military to put out "disinformation" to keep the focus off our newest technology. Included for your consideration.

Occurred: November 8, 2009; Approx. 2017hrs.

Location: Chester, CA

Witness: 2
Shape: Other
Duration: 4 minutes
Helicopter like body, jet plane wings, and miniature jet
I am an 18 yr old male, high school graduate, college student. I have always been somewhat skeptical about UFOs, but until the night of November 8th 2009 (I believe) I saw something completely incredible!
I was walking out of my girlfriend's house to kiss her good night at around 8:12 pm. We looked across the lake up towards the mountains and noticed what seemed to be two sets of headlights (meaning four headlights total) coming through the sky (I thought it was coming down the mountain but that was about 6 miles away) my girlfriend too thought it was a car way off in the distance, so she walked inside. I stood there for a few moments noticing the lights getting closer and closer. I noticed a man jog by as I was staring at these lights, which were now only 100 yards away from me, and getting closer. A sound of a VERY HIGH PITCHED jet engine came closer and closer (it was so high pitched it was almost a whistle) by this time I was looking straight up in the sky at this thing directly overhead me. It had a helicopter shaped body but with wings that were slanted downwards at a slight angle, and at the end of these wings were large "fans" which appeared to be angled down causing the fans to act as propellers. But under the main body was a jet (meaning I saw a small flame shooting out of the bottom which was maybe 2ft long) the object in question seemed to be about 30 ft long with a twenty-foot wingspan and was only flying 50 FT OVER THE TREELINE. It proceeded to keep its pace which seemed to be about 100-150 mph I then ran into my girlfriends house screaming and yelling when I told them what I saw and heard that were amazed because they couldn't hear this thing fly over. In all my experience living around planes, I have NEVER seen anything like this…the guy who was jogging said he thought it was a low flying B12 bomber or something…but it wasn't that big. I don't know if this is alien craft but it looked man made…but either way it was! Unidentified. There were no blinking lights just one light on the bottom and the four "headlights". This is not a joke or scam. You may Email me at ((e-mail address deleted PD))

National UFO Reporting Center Sighting Report
Occurred: November 27, 2009; Approx. 0530hrs.

Location: Weaverville, CA

Witness: 1
Shape: Light
Duration: 1 hour
Bright pulsating stationary light, changing positions quickly.

I saw a bright light in the sky on November 27, 2009 between 5:30 a.m. until 6:40 a.m. I was woken up at 5:30 a.m. by my dog doing one of those low growl barks dogs do. I noticed it was lighter than usual outside, but dismissed it as me not being a morning person. My dog would not stop making that low sound so at 5:40 I got up to put her and my other dog, outside. By then, the brightness from outside was gone. My patio door faces North and it was clear outside except a very few horizon clouds.

When I got to the patio door I noticed a really bright 'star or Venus' in the sky. It seemed strange because I have never seen 'Venus' so high in the sky, nor a 'star' so bright. The light was stationary and very bright, twinkling two different pulses. I looked at it for about 8 seconds. When opening the patio door, I looked down for about 5-6 seconds or so to keep the cat from getting out; as I closed the door, I looked back at 'Venus', however, it had moved visually 5 feet over to the West, and 2 feet to the North! I know it was in a different position because when I first saw it I compared the position in the sky to the plant on my patio railing. I looked at it for another 10 seconds or so to determine that it was definitely stationary in the sky.

I closed the curtain, because my dogs will stare in the patio door instead of doing their business, and went over to a window on the same wall and two feet over from the patio door (taking about 2 seconds to do that). I could not see the light anymore. There are no large objects on my patio, which would block out the area where I saw the light; in fact I have a very clear view of the sky from the entire North side of my house. I got right up to the window searching for the light all over the sky and horizon. Did not see anything but a few regular stars.

About 3 minutes later I let my dogs in and saw the light further down on the horizon, just above the clouds and more toward the West visually about 1 foot above the horizon, above the town water tank. I had looked in that area minutes before and did not see it there. It appeared further away because it was about 3/4 the size as when I first saw it. I went outside stood there for approx. 5 minutes watching it. Getting cold, I went inside and watched it for 45 minutes. I watched the light stay in exactly one position, still very bright and still twinkling 2 different pulses. There were tiny flashing lights that would pop up all around the sky near the light at different times, about 8-10 of them. At one time a cloud passed in front of it, but it was still there once the cloud moved. The clouds slowly went away, and now it was completely clear. After 45 minutes I walked over to grab a chair to sit in to watch it, when I returned seconds later, the light was gone. I right away started to e-mail my Mother, when I noticed my computer was 2 hours off. I had not been that way the night before because I always check the time while I am e-mailing someone. What was strange about it was that the light would be stationary, and then be in a totally different area in the sky within seconds. Unfortunately I never saw it actually move, I just happen to look away each time it relocated.

2010

2010 brings more disclosure of the newer black craft designed by some of our more illustrious Military Contractors, like Lockheed and McDonnell Douglas. They have built black fast moving dirigibles that appear or copy some of the ETs black triangle craft. Confusion continues to misinform the public. Some of the newest technologies, we suspect are reverse engineered technologies the scientists have worked with since the first UFO was seized by the community that does that stuff. This first 2010 entry illustrates such a technology, but it does not imply the triangles UFOs are all terrestrial. Worth mentioning is that the area between Sacramento, Grass Valley, Reno, Vacaville and Davis are constantly having verified sightings from 1950 to 2012. With Several Air Force Bases that UFOs constantly observe and visit, many official UFO reports are made but are classified. Since those areas are so far away from our focus of Northern California, those sightings are available at: MUFON, NICAP, NUFORC, CUFON, and other websites.

Occurred: January 1, 2010; Approx. 0000hrs.
Location: Grants Pass, OR
Witness: 3
Shape: Triangle
Duration: A few minutes
Triangle shaped craft with three red lights at points hovering over Grants Pass, Oregon.
Late on January first, two friends and me were driving in Grants Pass, Oregon. I noticed out the passenger window two red lights hovering in the sky. It was hovering in the air, much too low for any aircraft that I would normally see. It was dark, and cloudy, but the moon still kept in light enough to see. I watched it for a few more seconds, confused as to what I was really seeing. I pointed in the sky, and asked my friends if they saw that was well. As we kept driving, we could see it from underneath and to the side. The UFO was a wide triangular shape, with two red lights that sort of blinked at each end on the back. At the front tip was another light, and underneath there seemed to be another light closer to the front. The craft was dark, and thick. It was close enough that I could see the wings were the thick side angles. The craft was sleek, like a jet, but wider then anything I have seen. We all grabbed our cameras and pulled the car over. As we did though, the craft (still hovering in the same spot) begin to rise. We got out of the car and the craft was higher in the sky. Then, it turned to the right while staying in the same spot, about 90 degrees, and begin flying away. It moved away from us fast, and was too far to take a picture. Within minutes we had seen this craft hovering above the city, then turn to face the right in the same position, and leave. No military type craft can hover in one spot and then turn on a dime like that. Another strange thing was, there was no noise that came from the craft, it was silent.

Occurred: January 2, 2010; Approx. 1800hrs.
Location: Sacramento, CA
Witness: 2
Shape: Triangle
Duration: 4 minutes
Triangle shaped object spotted over Sacramento River.
On the evening of January 2, 2010, my friend and I were driving East over the Capital Mall bridge from West Sacramento across the Sacramento River into downtown Sacramento. It was a beautifully clear night. I noticed lights over the I-5 freeway that seemed to be too low in the sky to be a plane and they didn't appear to be a helicopter. As we approached the object appeared to be hovering. I pointed it out to my friend who was driving and said, "What the hell is that!" At that time we were approaching a stop light on the corner of 3rd and Capital Mall and it was clear the object was moving but very slow.
At that point with all the lights from the freeway and building etc, all we could make out were the lights on the object that I think were red and white.

As we turned left heading south on 3rd Street, the object passed overhead slightly to the rear of the truck and we both clearly saw the triangle shape. It headed west across the Sacramento River toward the Port of Sacramento, which is located in West Sacramento.

We both saw the triangle shape clearly as it passed overhead and we could see a white light in the center of the triangle. Once it went over the river it tilted to the right and headed in a Northerly direction. There were no news reports of any kind the next day. It boggles the mind that something that big and that incredibly close went unnoticed.

BACKGROUND INFO: My friend and I are both 54-year-old females. She retired from State Service and lives in Idaho but is currently working as a retired annuitant. Prior to retiring, she was an investigator. She's a little more closed lipped about it then I am but when I told other friends she substantiated what we saw. I am an Equal Employment Opportunity (EEO) Officer; my job entails investigations as well.

Occurred: February 15, 2010; Approx. 2300hrs.

Location: Sutherlin and Roseburg, OR

Witness: 2
Shape: Light
Duration: around ten seconds each time
The same UFO spotted in small Oregon town

On this particular night a close family member and I were driving out to the lesser traveled roads of our town. He simply just loved driving, its what he liked to do, so I knew we were going to really be going out into the country. Around nine or ten we were still just talking, when he kind of dozed off. I had not noticed what he was looking at until he started yelling that he saw a ufo. Naturally I didn't believe him. Too far fetched for me. He told me that he saw a strange green light stay still in the sky fore a few brief moments, then zigzag back in for rapidly a few times, then take off again not to be seen.

Later in the night I had to go to the town over to get to my girlfriends house, around the time I got on the freeway I noticed a green light. I watched in amazement for a bit eating my own words. I immediately called him and told him I witnessed the exact same thing. I am open to the possibility of it being a military, gas, etc. I have never seen any aircraft move like that, nor a meteor that stays put in the sky for seconds at a time, zips around, and then vanishes.

Depiction drawn by witness.
LITS Case # 06-4172010-01
MUFON Case # 22911 Occurred: April 17, 2010

Location: Red Bluff, CA

Witness: 2
Shape: Boomerang

I received an email after a witness commented on the Bellville, TX Boomerang report that was posted here in September 2009. It seems that recently on April 17, 2010, he and a friend saw the same type craft over Red Bluff, California. Could this be one of ours and if so, what... or is it something else?

E David wrote in comments:

My friend and I saw this exact same object two days ago, April 17, 2010, in the night sky over Red Bluff, CA! You describe the object perfectly, that is exactly what we saw! I went online to research "v-shaped" or "boomerang" shaped UFOs and saw the image from this site that led me to this article. The object moved in a completely straight line, and your description of the light shining off of it as a spotlight would off of clouds is spot on, that's what I thought it reminded me of too, but there weren't any clouds in the sky, it was clear; no spotlights circulating the sky either. It was as if the cloudy atmospheric substance was flowing from the front, over the wings and the rest of the object, which was a dark boomerang shape against the night sky, devoid of lights. Blew our minds! I'm glad we're not the only ones who have seen something like this, everyone thought we were nuts, but we know what we saw, and it was this same UFO.

Later in an email, E David wrote to me and sent a drawing (shown above) of what he saw:

I recently replied to an article on your site from September 15, 2009, where a v-shaped, or boomerang type,

UFO was reported seen over Belville, Texas. The description given in the article is quite close, almost exactly, to what my friend and I saw over Red Bluff, CA, on April 17, 2010. It was flying away from us, in a straight and steady path, silently, with this weird cloud-like substance flowing over the craft from the front to the back. At first glance I believed it to be a strange, angular cloud, moving, but the movement of the object was impossible for it to be merely a cloud, and as I watched it further I could see that an object was underneath the cloud-like substance. I drew a picture last night of what we saw, and have attached it to this email.

SEEN OVER RED BLUFF

Cloud-like Boomerange UFO Seen In Bellville, Texas

I wish to extend my thanks to E David, for sending in this report. Since this posting, E David has submitted an excellent report to MUFON. You can read the complete report here:

MUFON Case # 22911

http://lightsinthetexassky.blogspot.com/2010/04/cloud-disguised-boomerang-shaped-ufo.html

Occurred: May 2, 2010; Approx. 1600hrs.

Location: Bend, OR

Witness: 2
Shape: Disk
Duration: 2 minutes
Silver disk over Oregon
I was with my friend on her front porch, with a clear view of the sky. It was a semi-cloudy day, and we noticed a silver disk moving across the sky, very quickly. It was making a very unusual sound, which I can't describe very well. We ran inside to get her mom, but by the time we made it back outside, it had disappeared.
We were very unsure of what it was, but were definitely sure it was not a plane or jet.
At the time we saw it, we had her radio playing, but when it appeared, the radio stopped playing, even though it was plugged in.
((NUFORC Note: Witness indicates that the date of the sighting is approximate. PD))

Occurred: May 13, 2010; Approx. 1600hrs.

Location: Eugene, OR

Witness: 1
Shape: Circle
Duration: 8-10 minutes
Circular object that looked like star in blue sky moved slowly south, stopped, changed directions and went north.
Viewed from backyard while watching chem. trails in blue sky. Object looks like star when sunshine reflects from it, and like a ring when not. Object moved very slowly on same path as chem. trail only seemed much higher. It moved unlike any airplane going south at first, and then stopping completely for several minutes, then changed directions and began moving north on the same track it seemed to come from. It showed no trails or exhaust of its own. Object then got further away until we could no longer see it. We did capture digital images and video.

UFO Sighting Report! DUNSMUIR

Occurred: May 26, 2010; Approx. 2200hrs.

Location: Dunsmuir, CA

Witness: 2
Object: 4 bright objects
Duration: several minutes

On May 28 2010 around 10PM, my girlfriend and me saw 4 star like objects in the sky that were odd colour and were moving all around the sky. The objects would move up and down, side to side, two of them moved in sequence till one blinked out and the other headed closer to the mountain, and made snake like patterns in the sky. They were moving at a high rate of speed and would also disappear as if a light made been turned off then they would appear again. At one point we saw one of the UFO's appear just over the mountain top, it begin to fly closer towards us and it looked as if it was aware of us looking at it and it approached us. As we felt uncertain of how closer it was going to get to us we started to get frightened and started walking to near by homes. The object went over our heads and blinked out! We also saw one of the other objects appear over the mountain ridge, we could see the object in between 2 houses, it looked like a big star and it seemed out of place. I told my girlfriend to remember the two houses so we would be able to tell if the object moved. We moved further down the street and my girlfriend saw a star that looked like the other one directly across from the other one and as we stood looking at it, it slowly moved up higher in the sky well we watched it move up we suddenly remembered the other object between the 2 houses and went back when we got there it wasn't there anymore. And earlier that day about 10:20 about 10 huge all black UN-marked coppers flew over the town, they appeared military and flew low over the town. Dunsmuir is a small town that has never had that much air traffic and even when there is an air craft in the sky its usually only one and if its flying low it 99% of the time looking for some kind of drug bust in the mountains. We were unable to take a picture or video.

Original post:
Ufo sighting in Dunsmuir, CA 5/28/10

http://www.ufoworldnews.com/ufo-sighting-in-dunsmuir-ca-52810/

Occurred: June 10, 2010; Approx. 2130hrs.

Location: Red Bluff, CA (Dribble Creek Subdivision)

Witness: 7+
Shape: Changing
Duration: 2 Hours

NEAR HIGHWAY 36- Red Bluff, California (Dribble Creek Subdivision) I saw a bright round light sitting in one spot for a least one hour in the sky about 3 or 4 miles away at about 2 miles up in the sky and I thought it w as a star but it wasn't. I tried to focus and get a better look but the light was so bright and powerful it hurt my eyes to look at. I couldn't focus on this object as hard I tried.

I was outside of my house and I called 6 other family members to observe it and they all agreed it wasn't a star but a unknown object. I stayed outside to observe the object to see if it would move or change but it didn't move at all and I tried to take a picture of it but it was too far away to capture a good picture of it. I have a picture of the light but it is small. As I stood out observing the object other people came out of their houses and I could hear people talking about the object in the sky. I also called a family member driving towards the house! And they could also see it. The object slowly started to move down and it seemed to get closer. As it seemed to get closer maybe within 2 miles it looked like it was pulsating light coming from it. I started to worry and I went into my house. When I went into my house I heard airplanes and helicopters flying by my house which is uncharacteristic of this time of night or even flying in my are. An hour later I went out to see what it was and it was gone. I would say this is the most eerie thing I have observed in the night sky in my life.
((NUFORC Note: Witness elects to remain totally anonymous; provides no contact information.))

Occurred: June 24, 2010; Approx. 2140hrs.
Location: Redding, CA
Witness: 2
Shape: Light
Duration: 5 minutes
Low flying ball of bright light - gliding across the sky
After returning from an evening walk, my husband and I stopped on our driveway to observe the stars and moon.
We were facing south and right at "12:00" position was a ball of light that was moving towards us. The object was bright and much, much lower than the stars - we thought it was a plane or helicopter. We observed the object trying to identify the flashing red/blue lights that accompany aircraft. This lighted object had none.
We watched as it "glided" through the sky in a northern direction. It would drop then continue, drop then continue - it never stopped but it took a long time to cross the sky. We watched for a good five minutes until we had to walk out into our street to see it (now behind our home) suddenly, it was gone. It never darted away; it didn't drop like before - it was simply gone.
It looked to be following the route of our freeway.
Less than one minute after it vanished, a jet-type plane flew overhead in the direction of the lighted object.
((NUFORC Note: We will check with the witness, to confirm the time. The stars are not visible at 17:40 hrs. PD)) (((Venus? PD))

Occurred: June 27, 2010; Approx. 0000hrs.
Location: Hat Creek, CA
Witness: 2
Shape: Diamond
Duration: 10-15 seconds
Two sightings of the same object within 22 hours of each other in Northern California

These are two events occurring within 21 hours of each other. At approximately 12:40am Sunday the 27th my friend and I went outside to talk and we were seated facing west. I was looking up to the southwest a bit when a bright flash of light with a diamond shape object in the middle just appeared. It came in really fast through what I call a portal, with a bright flash of light. AS it entered it stopped in the sky almost as fast as it came in and floated slowly from my left to right, a south to north movement as I could tell.

After the initial bright flash of light as it broke into our atmosphere and as it began it's glide/descent, the light quickly faded but not light instant on instant off. It just dimmed but as the celestial craft dimmed it also began to cloak or transform into a small light behind the foliage of the tree. I could see the small light just either go away from me or disappear altogether. My friend witnessed some of the event but not all; from the dimming glide to the small light. This event took about 10 seconds or so.

I was in total awe but the initial reaction I felt was that stir in your stomach and that tinge of fear that comes from seeing something so celestial you just know it is not of this world. I thought. "Oh my God" and I exclaimed 'what is that!?!" which was then when my friend pulled the focus. I could see the object/craft so clearly as it entered our airspace, and I say that because I have seen many things concerning outer/inner world phenomenon like fireballs screaming across the sky, meteors striking our atmosphere so hard I could see the outer layer of earth's atmosphere and even sightings of UFO's floating in the sky. This brightness entered our earth and came to a halt as if it was the easiest thing an aircraft could do.

The second sighting came at 9:35pm the very same day, Sunday 27th and lasted more or less about a minute. My friends and me jokingly said, "let's sit outside and lets see them again." We looked in the same position and my friend says," here they come" but in a joking manner because he saw a bright star coming at us. He said it again, and then again and I realized the pitch in his voice changed and this object was moving straight towards us.

It was moving, no gliding across the sky from west (southwest) to east (northeast). It was in perfect view and I could see the diamond shape craft/celestial with a light that was so bright and . . . pure. I mean, I can't really describe it because I have NEVER seen such beautiful white bright light like this celestial emanated. It was smooth as glass and it was so bright and white there was this hue of blue in the center of the diamond shape... incredible beauty of light! It was so . . . pure It was too early for a star to be this bright. Even Venus was dim because there was light in the sky still. This was as pure brightness and a color of light I had never seen. As it moved across the sky it was unworldly in its movement. By that I mean unlike any aircraft we know today. Not like a plane or even a highflying jet aircraft. They ALL move with a worldly rhythm but this moved so effortlessly across the sky at a pace that is not of this planet. It glided without movement or turbulence of any kind and the brightness of the light NEVER wavered. It stayed beautiful and bright throughout the course of its flight until it went out of our vision. It moved from west to east and it was SILENT and PERFECT! NO SOUND came from this celestial at all. I ran from one yard to another yard to follow the flight path and I tried videotaping it but I could not capture it. I'll look again but it was dark on camera.

I know something big is happening. It just falls into place with what I know is occurring with the planets. My psychic ability tells me that we are in a transformation for the entire human race but now I realize this is on a much larger scale than even I had foreseen.

I am still in awe and I know this is not just another sighting, I just know that. Seeing something move across the sky is one thing; watching a diamond craft/celestial enter our airspace, and come to a halt like it did with such ease is another. Maybe it didn't know, or care, that I saw it enter but it wanted us/me to see it later that evening. It was clear and not fuzzy, it was close and not far away although I cannot tell the size from its flight. It looked HUGE when it entered the portal early morning but much smaller when we saw it later. I/we saw it clearly and it was, for lack of better words, beautifully perfect.

Occurred: July 4, 2010; Approx. 2300hrs.
Location: Orland, CA
Witness: 2
Shape: Flash
Duration: 1 hour
Satellites flash and then disappear and ones that change direction.
My husband and I live in a rural town 100 miles north of Sacramento. We like to stargaze and took our recliners out into the pasture where we could get a good view of the entire night sky. We had our radio. This was at 10pm sometime between 10 and midnight we saw three strange flashes. There were three incidences: two we both saw and one my husband alone saw.
The first one my husband alone saw. He said what he saw was an extremely bright flash of light, right between the two bottom stars of the big dipper. By time he said, "oh wow!" and pointed, it had disappeared. It wasn't streaking like a meteor. He said it just got bright and disappeared.
A while later, maybe 20 minutes to a half hour later I was watching a satellite cross the sky. I pointed it out to my husband, "I see one!" It was directly in front of us. We both were watching it. As it crossed, it suddenly got extremely bright. So bright the light from it almost hurt my eyes the way high beams from an oncoming car would. It lasted no more than two seconds and them dimmed. After it dimmed you could no longer see the satellite at all. I thought maybe it had burned up in space. There wasn't an object to track across the sky any longer. I thought maybe because of the brightness I just couldn't see it anymore, but it was really gone! My husband said that was what he had seen earlier.
We saw a third one which wasn't nearly as bright as the first one I had witnessed. The third one happened the same way. Visually tracking a satellite across the sky, this one was further to our North (our right hand side, our chairs were facing West).
It brightened, and then disappeared completely from the sky.
After hearing reports on the radio last night, July 6th, about others seeing something unusual, I was compelled to write. We sat out briefly last night and something unusual again.
At around 10:20 we saw the same thing again. While tracking a satellite across the sky we saw it brighten significantly and then disappear. This was not as bright as the first one I saw on July fourth.
Maybe 15 minutes later we both saw what looked like two satellites traveling North one behind the other quite close (I have NEVER seen that before). We watched as the one behind then gradually changed direction heading east as the first one continued north.
I am 52 years old. I don't drink and I have been watching satellites cross the night sky since my early twenties when a friend pointed them out to me while camping. I know the difference between a high flying plane and a satellite. I have watched hundreds of satellites cross the sky and I have NEVER EVER seen this kind of thing before. I have watched them disappear behind the Earth's shadow but nothing as strange as this.

Occurred: July 8, 2010; Approx. 1930hrs.
Location: Redding, CA
Witness: 1
Shape: Circle
Duration: 13 seconds
After reviewing a video of clouds, I saw four UFOs.

This is no joke. I was walking my dog at the park when I saw this unusual sunbeam above the cumulus cloud.
So, I took several photos and a video of it. When I checked the video the next day, I couldn't believe what I saw.
Not one, but four UFOs can be seen within the First 13 Seconds of the footage.
They are very small and translucent, flying close to the clouds.
You'll need to repeatedly play then rewind the first 13 seconds of the footage to clearly see them. The first three moves linearly while the fourth moves haphazardly.
UFO #2 and #4 seems to originate behind the clouds.
Study the still photo showing the sequence and trajectory of each ufo first here:
<a href="http://farm5.static.flickr.com/4136/4785400959_df24566447_b.jpg" rel="nofollow">farm5.static.flickr.com/4136/4785400959_df24566447_b.jpg</a> Then review this video again. I am not really sure what I've captured in this video. I'll leave that up to you experts in this matter to decide. This report is filed with MUFON.
<a href="http://www.mufon.com" rel="nofollow">www.mufon.com</a> To Download the Original 75 MB Raw Footage Video I uploaded to MUFON You can find it here by following the instructions: 1) Go to this site: www.mufon.com/mufonreports.htm 2) The heading will read: "MUFON Case Management System - SEARCH" 3) Choose this date, "July 8 2010", In the spaces provided for: Date of Event (within date range): 4) Choose, "California" from the choices provided for: Event State (USA): 5) Choose, "Shasta", from the choices provided for: Event County: 6) Click on the "Submit" button 7) In the heading of, "MUFON Case Management System - SEARCH RESULT", under "Attachments", located at top right, look for my downloadable video titled CIMG2468.AVI.

Occurred: August 10, 2010; Approx. 0152hrs.
Location: Medford, OR
Witness: 1+
Shape: Triangle
Duration: 15 minutes
Looking North we saw a triangle shaped object in the sky, rapidly changing colors from blue, red, green and white, almost looked like it was spinning, at first thought it was some aircraft, but I have never seen a plane that big and bright before. Then it started moving to the left (west) just a little, stop, move some more to the left. It wasn't a type of movement that any aircraft would make. It kept moving left in weird jerky movement, until it had moved behind a tree and we could no longer see. Later farther west we saw something similar, but it was smaller, and this time it was moving right, but slower.

Occurred: September 3, 2010; Approx. 2200hrs.
Location: Red Bluff, CA
Witness: 1
Shape: Circle
Duration: 5 minutes

I was driving down highway 36 towards Fortuna out of the Red Bluff city limits and we saw some three bright lights to the right of my car as we were driving. As I kept driving initially I discounted it as maybe stray car light behind us from a distance. I checked the mirror to look for a car behind me and saw nothing. I looked over again and I saw something with three bright lights following my car and I got scared because what ever it was followed us up to the subdivision I live. What ever it was kept up its speed with my car and seemed to float with no sound coming from it. I was getting really worried because we thought it was coming after us. As soon as it approached where the entire house are it took off out of sight. This is the second time I have seen weird activity in this area.
((NUFORC Note:
Witness elects to remain totally anonymous; provides no contact information. PD))

Occurred: September 7, 2010; Approx. 2300hrs.
Location: Fort Jones, CA
Witness: 2
Shape: Triangle
Duration: 8 seconds
Large Silent Metallic Black Triangular Object
UFO Report 07/09/2010 23:00 at the time and date noted, I was sitting outside on our deck, the night was totally clear and stars were vivid. My wife came out to see where I had gotten. We both looked up at the edge of our roof and exactly at that moment, a HUGE black triangular shape object appeared and moved off in a W/N/W direction. There were no lights and it made no sound. We could see the shape of the craft against the backdrop of the stars and it had a metallic/black color. From the time it appeared till the time we lost it was approximately 8 seconds. It was HUGE.
My wife and I looked at each other and asked if we had seen what we had seen, amazing experience.
Scott Valley CA

Occurred: September 10, 2010; 2200hrs.
Location: Bend, OR
Witness: 3
Shape: Triangle
Duration: 5 minutes
Bend Oregon has sighting of UFO on Friday September 10th at 10 pm.
Object entered field of vision from north at rapid rate of speed. Object stopped without slowing down and abruptly changed directions, accelerating instantaneously and traveling in the opposite direction, instantaneously stopping and accelerating again at speeds beyond the laws of physics, then stopping again without slowing down and performing a number of gravity defying maneuvers. Then it propelled out of site, again without noise or the appearance of acceleration or deceleration. The object displayed bright white lights in the form of a triangle with non-blinking red lights on what appeared to be short wings.

Three of us observed the object and concluded based upon our combined experience that what we saw was not something manmade, obviously a UFO. Since we have been examining the night sky for a repeat performance, but we agree it was probably a once in a lifetime event.

Occurred: September 10, 2010; Approx. 2300hrs.

Location: Bend, OR

Witness: 3
Shape: Triangle
Duration: 5 minutes
Large Black Triangular UFO
At 23:00 hours, my wife joined me outside to watch the night sky, as we both looked up, we saw a huge triangular shaped UFO pass over the roof of our house, I would guess at around 20,000 ft. even at that height it was the size of two football fields. It was going WNW towards the Pacific Ocean. There was no sound, no lights, no vapor trail, and no sonic boom, even though it was gone over the horizon in 6 seconds, the large size of it made it appear to be in slow motion. It was flat black, but a bit luminescent. I am a retired Hwy Patrol Officer and a credible witness. My wife and I couldn't believe what we had just witnessed.
((NUFORC Note: Witness indicates that the date of the sighting is approximate. PD))

Occurred: October 13, 2010; Approx. 2245hrs.

Location: Etna, CA

Witness: 2
Shape: Flash
Duration: 25 minutes
UFO Rotating Red Green Yellow Lights
Sky Clear.
While watching TV, my wife noted what appeared to be a very bright star, we know our sky well on the ranch and hadn't seen this before. We went outside and took our binoculars, what we saw was rotating lights, Ruby Red, Green and Yellow. It wasn't distortion as they were sequential. We watched this object for at least 20 minutes. At that time, the lights became brighter, more intense and then it suddenly disappeared, it didn't fly off, it was just gone. I had seen this exact same object in 1997 on the Alaska Hwy, when a missile shipment was going to Alaska, The drivers in the convoy had also mentioned seeing it when they returned.

Occurred: September 13, 2010; Approx. 2300hrs.

Location: Fort Jones, CA

Witness: 2
Shape: Triangle
Duration: 6 seconds
Large Black Triangular UFO

At 23:00 hours, my wife joined me outside to watch the night sky, as we both looked up, we saw a huge triangular shaped UFO pass over the roof of our house, I would guess at around 20,000 ft. even at that height it was the size of two football fields. It was going WNW towards the Pacific Ocean. There was no sound, no lights, no vapor trail, and no sonic boom, even though it was gone over the horizon in 6 seconds, the large size of it made it appear to be in slow motion. It was flat black, but a bit luminescent. I am a retired Hwy. Patrol Officer and a credible witness. My wife and I couldn't believe what we had just witnessed.
((NUFORC Note: Witness indicates that the date of the sighting is approximate. PD))

Occurred: October 20, 2010; Approx. 0010hrs.
Location: Roseburg, OR
Witness: 1+
Shape: Rectangle
Duration: 3 minutes
5 round white lights hovering over I-5.
Driving south on I-5 in Roseburg at 1205 at night I saw 5 very large round white lights in a long row floating in air about 300 feet up. They were even with top of hill and extended out of river. I could not understand what they were since it was silently hovering there. Others were pulled over on side of highway looking at it. As I got closer it started to float north above I-5. As it passed over me I could not see the stars or clouds between the lights so I believe it was a long solid object with the 5 lights. The lights did not illuminate the ground or send out a beam like a car light.
It continued to head north above the highway.
I am in law enforcement and my husband is an ex air force pilot. We have observed planes of every description as a hobby and I know that this was nothing either of us has seen before.
((NUFORC Note: Witness indicates that the date of the sighting is approximate. PD))
((NUFORC Note: Report from law enforcement officer. PD))

Occurred: October 25, 2010; Approx. 0900hrs.
Location: Yuba City, CA
Witness: 1
Shape: Cigar
Duration: 2 minutes
Cigar/Cylinder shaped object seen over Yuba City, California
Approximately at 9am I stepped outside to sit on my back porch when I noticed an object that was cigar shaped approaching from the North and heading south.
I walked out into the yard to get a better look at it. I could immediately tell it wasn't an airplane. It lacked the usual appearance of the commercial airline planes I see periodically; tail, wings, exhaust stream, markings of any kind, windows. It was a gleaming white yet translucent. Each end was "rounded" and the center was a different shade, which made it look like it had a "band" around it. There was absolutely no sound coming from this craft. The weather was calm; no wind, blue skies, and a bright morning sun in the east.
Since I had been uploading vacation photos from my camera into my laptop that was on the kitchen table, I had my camera in an accessible location. I ran into the house and grabbed the camera that took less than 10 seconds.

Upon returning to the back yard, it was evident that this object was moving faster than any commercial airliner. I decided to put my camera on "consecutive" shooting but the craft was moving so fast and had changed it's course from a straight North to South path to an ascending southerly path and was gaining altitude with such speed that I was losing visual in a matter of seconds. It went from a size where it would have taken a nickel to cover the entire craft to a pinpoint.

I was only able to catch one pretty clear photo.

Within just a few minutes of seeing this craft, I heard multiple helicopters in the area but never got a visual on them.

My background is 20 years as an Insurance Agent and Mutual Fund Investment Adviser. I am currently retired.

Occurred: October 31, 2010; Approx. 2215hrs.

Location: Dunsmuir, CA

Witness: 2
Shape: Circle
Duration: 30 seconds

Big, bright light that left tracers behind it, which abruptly stopped in the sky until it later blinked out of sight.

My fiancée and I were driving south on I-5, around Dunsmuir, California. At around 10 PM we both witnessed a bright, circular light on the southeast horizon. We were listening to the radio and both noticed the light in the sky.

After 10 seconds it disappeared and my fiancée interrupted the silence asking, "Did you just see that?" I quickly replied, "YES!! I DID!" Then about another 10 seconds later to the east of us we noticed the light flickering then it dipped down in a swirling motion, leaving traces of light in it's path and it disappeared again.

My fiancée told me that when she first saw it she thought it was a giant shooting star, because it had a trace of light following it. She was excited because she had never seen a shooting star, but then the light abruptly stopped in the sky and stayed there until it blinked out of sight (and then later appeared to the east like I described).

We know it was not a plane, it was much bigger than a star, an! d it was far too high in the sky to be a street light.

Whatever it was, we cannot explain. She never fully believed in UFO's until our experience.

Occurred: December 28, 2010; Approx. 2038hrs.

Location: Colusa, CA

Witness: 1
Shape: Light
Duration: 1.5 hrs

The object appeared to the west of the constellation Orion and was amber-orange in color.
**Attn: I just submitted this same report only the date was wrong! I apologize for the inconvenience Location: Colusa, California Time of Sighting: Noticed an oddly colored star in the very center of the sky from my view, slightly to the west of the constellation Orion; 8:38pm PST. At 10pm, object was still in the sky.

I was outside looking at the stars, and as Orion came into my view I thought what I witnessed was the International Space Station flying by. As I watched the object, it moved north, south, and then to my amazement did a full circle in the sky. It returned to its position to the west of Orion. The object was a slightly off color, amber-orange.

I went inside my house and got out a pair of binoculars to see if I could get a better understanding of what I just saw. Once I looked at the object through the binoculars it appeared to be in our atmosphere, too close to is a star. And once more, startled me as it moved in the same pattern while I was looking at it through the binoculars. I was under the impression it was trying to disguise itself as a star. I felt the urgent need to report this to some one. I used Google and found your website.

I am a student at AIU. I am getting my master's degree in IT/Internet Security in 3 years. I have lived in the country most of my life. I have only seen one other irregular star while living out here. And that one blew my mind so much I was scared to go outside at night for weeks. That time, an object similar in color, appeared to be flying in front and around the moon. It looked like it was heading straight for me, and then when it got a short distance away from the moon, the object split into two objects and continued on their separate ways. That was several years ago now. I did not know that I could report such incidences.

((NUFORC Note: If the object was to the right, i.e. west, of Orion, it could not have been the star, Sirius, which is located to the left of Orion, as one faces that constellation. PD))

Occurred: December 29, 2010; Approx. 0145hrs.

Location: Redding, CA

Witness: 1
Shape: Other
Duration: 20 seconds
Dark, V-shaped figure with 7 lights seen above Redding, CA through thin clouds, making drastic movements.

On December 29, 2010, I witnessed a dark V-shaped object flying above thin cloud cover in Redding, there were 7 lights in total, 3 on each side of the V and one at the very point. It was flying south, made an instant change to the west for a few seconds, then shifted back southward. There we geese flying nearby and I've never heard a more panicked group of them in my life, let alone at almost 2 a.m.

2011

Occurred: January 6, 2011; Approx. 1830hrs.

Location: Eugene, OR

Witness: 2
Shape: Triangle
Duration: 3-4 minutes
Huge triangle with 3 lights

What I thought was a star I realized was actually moving, so I pointed it out to my daughter and told her it must be an airplane super high up. She says, "but daddy there are 3 and are in a big triangle." sure enough that's what I saw also. The lights were so far up they looked like distant stars. They moved across the sky in this pattern. It could maybe fit in the big dipper. It was headed NW and was not visible after just a few minutes and held this pattern and the distance between the objects which did not change. When looked at as a whole it looked like a single huge object moving. Would have been hard to see without someone pointing it out.

Occurred: January 14, 2011; Approx. 2200hrs.
Location: Davis, CA
Witness: 2
Shape: Triangle
Duration: 2 minutes

 A Black craft in triangle shape was in a big field between residential neighborhood and country, Approx 15-20 yards wide, about 150-200 ft up, and about 200 yards away from our car as we were passing. White lights were around edges of craft, with one light blinking, traveling about 15-20 MPH. I rolled down the window and could not hear sound from craft over passing wind. We did not stop because I was too embarrassed to suggest it.

Occurred: January 2, 2011; Approx. 2230hrs.
Location: Crescent City, CA
Witness: 2+
Shape: Sphere
Duration: 5 minutes
a burning orange sphere of light moving up towards the stars
On Jan 2 at 10:30 pm my family observed a sphere shaped object that appeared to be on fire. It moved away on a 45-degree angel and changed shape. It appeared to have flames around the bottom and bright light in the middle with florescent green. It zigzagged and moved up into the sky until it just disappeared. We took video of the object. We were watching TV in our living room, which has a large sliding glass door. We saw the strange sphere and run out on our porch where we watched it and videoed it.

Occurred: March 30, 2011; Approx. 2100hrs.
Location: Chico, CA
Witness: 1
Shape: Triangle
Duration: 15 seconds
In a clear night sky, stood out very clearly, triangle v shape lights.
From the west, it turned to a northeast direction at high elevation but very clear in the sky.
Appeared to be a very large slow moving and almost playful in its motion.

Occurred: April 8, 2011; Approx. 0300hrs.

Location: Corning, CA

Witness: 1
Shape: Other
Duration: 4 seconds
Unidentified object reported in northern California.
I am a maintenance worker at a local casino I work from 12am to 8:30 am. I was finishing putting away some equipment when the sighting occurred. In a brief window above the casino the object was moving from south to north. The object consisted of three circles in a triangle formation-rotating counter clockwise. It only took around 4 seconds and my boss was to far behind me to get him out there to see it because it was over so quickly.
((NUFORC Note: Witness indicates that the date of the sighting is approximate. PD))

Occurred: April 23, 2011; Approx. 2035hrs.

Location: Orland, CA

Witness: 1
Shape: Light
Duration: 1 minute
Yellow-white light hovers before speeding off into the night sky
I look at the sky every evening, and in this way, I'm more familiar with what should be in the sky and where. I went outside after supper and noticed a yellowish-white light at 37 degrees NE and just above the tree-line in the night sky. I stared at it for a long moment because I knew that the light had not been in the sky on previous nights. Then I realized that the light was moving ever so slowly towards the NE. I only took my eyes off of it long enough to look back behind me and call one of my sons to come out. When I turned my head back, I saw the light speed away towards the east faster than anything I've ever seen before.

Occurred: May 11, 2011; Approx. 0335hrs.

Location: Red Bluff, CA

Witness: 1
Shape: Sphere
Duration: 10 - 15 Seconds
I am working graveyard shift this morning, and I clocked out for lunch Approx 03:15 am. I did my daily walk around work, and I found a nice quite & dark place to lie on the pavement so that I may gaze at the stars. Well it was pretty hazy/cloudy up in the sky, with clouds rolling in from the West. These were the type of Clouds that ripple, the ones that have been reported as HAARP Generated. Although it was somewhat cloudy I could still see stars. After awhile of gazing, all of a sudden two dim orange objects drifted over my head, going from North to South. They looked to be in the clouds or above them because they were pretty blurry. They were probably about 300 feet apart. (It's hard to tell being on the ground but that would be my guess) They were both travelling at the very same speed, which was faster than your normal airplane or jet. They had no blinking lights and no noise. One of the Orbs seemed to be rocking back and forth coming closer to the other and then back away. They did not take off super fast

and disappear like most UFO sightings; they just kept on drifting by at a steady speed until I could no longer see them.

Occurred: June 16, 2011; Approx.
Location: Redding, CA
Witness: 1
Shape: Light
Duration: 20 minutes

A Bright light came from the west and travelled east. This light moved in a straight line almost zigzag in the air and then went on a steady pace never changing colour or direction from west to east almost like it was heading somewhere. It just kept on moving till I could not see it anymore.((NUFORC Note: Source of report elects to remain totally anonymous; provides no contact information. Source indicates no time of the incident, but our suspicion is that the sighting may have been caused by an over-flight of the International Space Station. PD))((NUFORC Note: Strangely, witness indicates that the date of the sighting is approximate. PD))

Occurred: June 26, 2011; 2100hrs.
Location: Willits, CA
Witness: 1
Shape: Oval
Duration: 2 minutes

The red, orange, gold slow-moving crafts appeared and vanished June 26, 2011 Sunday night. About 10:00 p.m., outside of Willits in Northern California, I sky watch every night and had just gone out, facing west, on this clear dark night, when a beautiful bright, sparkling object appeared towards the south. It silently moved at a height lower than our few commercial flyovers, quite slowly and steadily without variation in altitude, headed north-northwest. I of course assumed it was a plane—and from my angle it did seem to be a flat elliptical shape, but I knew the colours were wrong: red at the rear, gold at the front, with the lights moving and interplaying into brilliant orange and a coppery gold at the center. No separation between the lights and colours. They weren't flashing but instead moving and pulsing, really lovely.

After only a few seconds of watching this, the lights all suddenly and uniformly dimmed and got hazy and changed character and it looked like a tiny version of a nebula cloud like I've seen in Hubble photos—but then I was distracted by a sudden narrow streak of dull white light that shot down, from some distance southward behind the "craft," straight down to the ground (which I thought well, this could be a meteor, just without luminosity which isn't uncommon). Then the original "craft" immediately got bright again, back to red and orange and gold, and travelled only a short distance more when it just disappeared. I waited and searched the sky to see if it reappeared but it didn't. Then about 30 seconds later, another one appeared from almost the same point of origin, again out of nowhere, again silently and fairly slowly travelled the same flight path, but this one was all gold light and the light was more steady and the "craft" seemed a bit smaller. When it got to the point where the first one had dimmed, it vanished. All of this took between 1 – 2 minutes. Nothing more occurred. I had not myself heard of any UFO sightings with orange as a key colour but I've never Googled and see there are some, including multiple sightings in Washington in 2009.

Occurred: August 3, 2011; Approx. 2100hrs.
Location: Mount Shasta, CA
Witness: 2

Shape: Unknown
Duration: 1 hour or so
The lights on Black Butte in Mt. Shasta, CA were in an area not accessible to humans or animals in symmetrical lines. My boyfriend and I were driving back to our house in Mt. Shasta, CA 96067 from Weed, CA at about 9:00PM on 08/03/11 when we got to Truck Village Drive (our exit south bound on I5) and noticed bright lights on the mountain Black Butte (a pilot cone that is almost straight up and down). We stopped the car and got out to look and realized that the lights were not coming from anything we could explain, and it was nowhere near the trail that goes up the mountain. It was about 300 yards up from the base of the mountain in a rock valley (lava rock) that is not accessible to humans, or animals. The lights were in a symmetrical line and very bright, and white, and not moving or flickering. We raced home (about a 5min drive from the Truck Village Drive exit) and grabbed our binoculars. We returned to the same spot and the lights were still there. When viewing through our binoculars we could see that there was a perfectly horizontal line of about 4 lights and directly in the middle going diagonal (the line falling from left to right) were three more lights very clear. It kind of looked like a cross of some sort. We stayed there for a long time looking at it through the binoculars and it did not move. We got home at about 9:45PM and called the local police and sheriff's department to see if maybe they heard anything about those lights, if there was a lost hiker by chance, etc. but they didn't know anything about it and acted like we were wasting their time. My boyfriend and I have lived here for 12 years, have hiked that mountain before many times and have never seen anything like it before. We can't explain how it got there or what it was, except that it most likely is a UFO. The lights were exactly the same size, brightness and just too symmetrical to be done by anyone. ((NUFORC Note: We spoke with the other witness, and he sounded quite credible to us. He has hiked the area in question, on many occasions, and he is certain that there could not have been any kind of vehicle in the area where the lights were, given that it is a very steep, danger-covered area. PD))

Occurred: August 8, 2011; Approx. 2300hrs.
Location: Medford, OR
Witness: 2
Shape: Light
Duration: twenty five seconds
My friend and I were laying in a grassy hill in Medford, Oregon. We were on vacation, just counting bats and stargazing. The stars were almost all visible, and the warm summer breeze swirled below them. As a third bat flittered close by, we were both completely calm and relaxed. Then we saw it. I say "it" because I am still not positive what it was that we saw. It was just a single light, much like the satellites that you some times see, or a plane, or helicopter. But it was no satellite, was no plane or helicopter. It was for sure a UFO (unidentified flying object), and possibly and extraterrestrial flying spacecraft. All it was a single light, and if looked just like a star. But it was very faint, and then grew brighter and brighter, until it was as bright as the clearest stars. Then it started travelling around the sky, just a bit faster then a plane and much

faster than a satellite. It looped around the sky for roughly twenty-five seconds, and then eventually flew out of our sight. It was no shooting star, for it moved around the sky much longer than one would. It kept the same speed through out the whole time we watched it. I have provided as many details as I can think of. ((NUFORC Note: Witness elects to remain totally anonymous; provides no contact information. PD))

Occurred: August 9, 2011; Approx. 1000hrs.
Location: Marysville, CA
Witness: 1-2
Shape: Changing
Duration: over two weeks
For the last two weeks we have witnessed two UFO's, on the outside of Beal Air Force Base about twenty five miles west of the north gate. This is what I assume but the UFO may have been further. We have observed them through the telescope, and they have radical changing of colours (Blue, yellow, white and red) and constantly zigzag. They appear to be following the constellations. They also appear to be growing. We have observed this many times in the past and in other locations. We would like to know if they are on radar, because we know that they are in our atmosphere. Please contact us.

STRANGE OBJECT PHOTOGRAPHED OVER SISTERS, OREGON, SEPTEMBER 16, 2011 We received a sighting report, and several interesting photographs, from a very experienced airline pilot, who inadvertently captured a very strange, high-speed, object with his camera, as he was flying his Grumman "Albatross" from Seattle to Los Angeles. We post below two of the photos he forwarded to NUFORC, and request that anyone who can identify the object please contact our Center. In our opinion, the object appears to be reminiscent of an SR-71, but we are not aware that they are painted in international orange.
http://www.mufon.com/Nuforc_frame.html

Occurred: September 16, 2011; Approx. 1116hrs.

Location: Sisters, OR (in-flight)

Witness: 1
Shape: Triangle
Duration: 5 seconds
Orange Dart-like craft passes our aircraft at close proximity in flight
I am an airline transport pilot with over 14,000 flight hours in unique aircraft such as the HU16b I am flying in these pictures I have been flying for 40 yrs I am also a certified A&P mechanic and an IA or inspector with over 25 yrs maintenance exp on planes like these. The Albatross we have is used as a test bed for Satellite broadband Internet; we are one of the few companies in the world that has airborne satellite systems for high-speed Internet. The plane is full of sophisticated equipment for this purpose as it is the original test bed for this.

I have better than 20/20 vision and I usually can spot any single helium balloon of 15" diameter from a great distance and I am trained to do so. I see them all the time. That being said, the object was literally a blur and I didn't see it as it was moving in the same direction as the plane. The order of the pictures attached here is the series of the 3 that I took with my iphone as I was just trying to get a shot of "Sisters" and Mount Jefferson in the Bend area.

As for the pictures, the images cannot be a balloon cluster as this thing passed us!! Look at the lake on the first picture as the object comes from right to left, go to the next shot, it's in frame, look at our relative movement passing the lake then the last shot as the thing is hidden by the wing and our location relative to the lake again. We were cruising at 11,500 ft at approx 200 mph. the balloons would have to be fast!! That thing was a streak; it went by as nothing but a blur. Only the luck of the camera could tell me that it was some sort of object of quite some size, as our wing is nearly 50 long and that float out there (40 ft away) is 10 ft long. I estimate that it is about the same size as our plane or longer that would make it approximately 70-80 ft long. I would also estimate it to be within 1000 ft of us laterally. The object made no noise that I could discern, nor did it leave any sort of contrails or exhaust paths. It appears to be banking around us for a split second then gone. Our plane is traveling southeast at approximately 200 mph at 11,500 ft We were in the vicinity of Bend or Sisters Oregon.

Location: Redding, California

On the evening of September 30, 2011, while enjoying his hot tub, a man in Redding, California observes a crescent-shaped object flying fast and erratically across the sky. He claims the sighting was quite disturbing; that it "was not normal in any way and in fact seemed more cinematic and unreal." This report contains minimal corrections. —SW

Occurred: Saturday, October 1, 2011; 2030hrs.MUFON Case # 32238

Date: 2011-09-30
Time: 20:30
Status: Not yet assigned
City: Redding, California
Shape: Boomerang, Unknown
Duration: 00:00:12
Distance: Unknown
Summary: Visually this was very disturbing and very real
Report: I was sitting in my hot tub watching the stars, jets and satellites passing overhead; we're in northern California in Redding right under the jet corridor from southern California and the Northwest U.S.
Many nights have been spent observing the skies. Last night 9-30-11 I caught an image of something flying south at a very unusual high rate of speed...enough to actually make me stand up.
The object was crescent shaped with lights moving through what appeared to be liquid without a specific geometric pattern. It came from the northern horizon at a very high rate of speed approx. 30 degrees off the horizon heading south.
What got me to stand up in the hot tub was that it began to turn east quickly in a 45 degree arc covering nearly one third the horizon line before it quickly snapped to a southern flight at a higher rate of speed in a slight "S" pattern heading due south where I observed it until the end of my observable horizon line.
The object was very visible, flew in what I felt was a frantic pattern and emotionally caused the hair on my body to stand up simply because what I saw was not normal in any way and in fact seemed more cinematic and unreal.
I assumed seeing something like this would be more interesting instead of eerie and somewhat threatening. This object possessed an immeasurable power that I can't relate to which stirs the soul more than ever.
I called my wife and brother and couldn't find the words that would convince them that what I saw was very very unnatural and disturbing. As for size it seemed to be somewhere between the altitude of passenger jets and myself. if this were the case then the object would be quite large. If it were at the altitude of passenger jets I would then assume its size to be massive.
What bothered me were the motion, silence and hesitation of flight. It moved with purpose and this bothers me. http://lightsinthetexassky.blogspot.com/2011/10/crescent-shaped-object-observed-over.html

Occurred: November 24, 2011; Approx. 1825hrs.
Location: Roseburg, OR
Witness: 2
Shape: Light
Duration: 5 minutes
Amber / Orange lights evenly spaced in a line moved north quickly and disappeared in the clouds
My mother had just returned home and she burst into the house yelling, "Come outside and see the lights!" She sounded nervous and excited. She ran outside and pointed out 5 lights that were moving quickly north. My mom said that when she was driving, there were at least twice as many lights, arranged in two groups. One group was arrayed in a wedge shape, and the other was in a line. She had tried calling us on her cell phone, but she couldn't find it in the car. She saw a man that had pulled over at a nearby intersection who had gotten out of his car to watch the lights. The remaining lights were all in a north-south line, more or less evenly spaced. The lights were a strange amber / orange color, unlike any we had ever seen on an aircraft. The lights did not blink, as they remained a constant glow. One by one the lights flew north rapidly and disappeared into cloudbanks. We couldn't hear any noise from the lights.
It was dark with a high cloud cover and intermittent showers.

Occurred: December 29, 2011; Approx. 2200hrs.
Location: Anderson, CA
Witness: 4
Shape: Fireball
Duration:
On December 29, 2011 my daughter, Son in law, granddaughter, and I were returning home at about 2200. While we were unloading the car my Son in law said look at that and he was pointing at the sky. It was overcast and below the clouds were three round orange glowing objects in a triangle formation. I had my Sony Cyber shot camera in my pocket. I took it out switched it to video and pressed the start button. After about a minute, one of the orbs moved into the clouds. It was at this time I noticed the camera wasn't recording so I pushed the start button again and recorded for about a minute and when I stopped the recording the orbs went into the clouds. We took another load into the house and when we came back out they were back. I took 5 photos and 1 minute of video and they left again. Having a small camera the videos and one of the shots are a little shaky. The orbs were about a quarter mile away and looked about ½ inch across. Orange in colour lighter in colour in the center. They were in a triangle formation and moving in a slow clockwise rotation then one of them moved into the clouds at which time they stopped rotating. When they returned there were only 2 of them from the time we first saw them until they went into the clouds the last time was about 20 minutes.

Occurred: December 31, 2011; Approx. 2155hrs.
Location: Roseburg, OR

Witness: 1
Shape: Light
Duration: 3 minutes
Three lights were seen, forming a triangle. At approx 9:55 pm pacific time, I saw 3 very bright orange lights forming the shape of a triangle. They were stationary for around a minute, then two of them took off. The last one moved for a few seconds, stopped, then slowly left in the same direction as the others. The lights appeared to be round. I could see stars in between them.

Occurred: December 31, 2011; Approx. 1930hrs.
Location: Medford, OR
Witness: 1
Shape: Unknown
Duration: 5 minutes
Orange ball of light, moved north, then South, then West. Lasted about 5 minutes and maybe less. After it moved about, it just disappeared. I've never seen anything like it. Definitely was not an airplane or a helicopter or a satellite or even a meteor or an asteroid; I know because its movements were too jerky and random to be anything like that. After the jerky movements, it just vanished. There was no explosion or anything like that. Just vanished. The sighting was not extremely close to earth. It was relatively high in the sky, however was not so small to be a star or planet but it was not close enough for me to distinctly make out its features and there was a sort of haze around it.

Occurred: December 31, 2011; Approx. 2359hrs.
Location: Anderson, CA
Witness: 7
Shape: Light
Duration: 5-7 minutes
Moving lights that changed colours and sped away at a high rate of speed. 1-1-12 Report of UFO in Northern California skies, witnessed by 7 people Dec. 31st, 2011 at about midnight. On Dec. 31st 2011 at my residence, at approximately midnight, 7 witnesses, including myself noticed 8 reddish/orange lights in the southern sky that appeared to be large stars that moved slowly in a westerly direction. The light from them seemed to twinkle and they would sometimes fade in and out. They hovered still for a few minutes, then they started moving west, slowly twinkling/fading in and out, then they disappeared. All 8 seemed to be at a similar distance from the horizon, but not in a parallel line (approximately where the sun would be at 3-4 p.m. at that time of year, but in the south). The weather was calm with no clouds visible. There was no commercial airline traffic visible. In retrospect, I realized that I live in the flight path for commercial air traffic from the south into the Redding Airport. It seemed odd that there were no flights that night. After they disappeared from the southern sky, 3 more of them appeared in the North Eastern sky, slowly rising from the horizon. They behaved the same as the other's, except that after a few minutes of hovering and flying around in the same area, they started moving south along the horizon, slowly moving higher in the night sky. When they got about where the sun would be at 11a.m. at that time of year, they changed colour to a white/opaque (like a ghost) and shot away in a southerly direction very fast. 2 of them seemed to follow each other in a southwest direction and 1 went south by itself.

2012

Occurred: January 1, 2012; Approx. 1208hrs.
Location: Yuba City, CA
Witness: 3
Shape: Light reddish orange lights on New Year's morning
Duration: 10 minutes
My brother and I were watching the fireworks just after midnight on New Year's and my brother said look at that light it was a reddish, orange light, Then I could see more light's coming from behind a tree in a side ways vee formation, I ran back into my house to get my mom to see the lights by the time we ran out side there was about 15 to 20 lights they were spreading out low in the sky, and then they stopped. My brother ran to git his camera but his pictures did not look to good the lights looked like stars. The lights came in from the south and were headed in a northern direction until they stopped right in front of us and over the fireworks we were watching that were down the street from us, it was like the lights were watching the fireworks to. The lights stayed there for about 10 to 15 minutes and then they faded out, and then 20 minutes later we all saw another single reddish orange light north of us it stayed sill for about 4 minutes then it faded into the clouds. It was just off the 99 & Lincoln area.

Occurred: January 1, 2012; Approx. 1208hrs.
Location: Anderson, CA
Witness: 3
Shape: Disk
Duration: 45 minutes
At 12: 30 AM on new years morning in Anderson, California, 7 witnesses saw 11 pulsing lights and when the lights would go out they became discs. We noticed several red pulsing lights to the south above the tree line. The lights were moving slowly to the right, some were higher from the tree line but both high and low each individual light would become stationary for a moment and then continue. The lights disappeared from the right as if going behind a curtain. Immediately to the west, over a 20- 30-minute span, lights began rising one, two, or three at a time. Two might start close together but usually separated with one ascending at a faster rate, and drift to the right or left as if following the gravitational pull of the earth, some meandering around on their own accord until the pulsing lights winked out. It wasn't until the last three that we noticed that once the pulsing lights winked out, nearly transparent discs continued though the night sky, and just like the lights the discs would pause stationary one moment, then continue in a seemingly impossible trajectory. I focused in bewilderment on one of these discs after ascending as a light continued as a disc travelling from west to south until it too disappeared.. I witnessed this along with 6 other people on new years morning just after midnight. None present were under the influence of alcohol or drugs and all agreed that, what was witnessed was beyond any explanations we could think up.

Occurred: January 26, 2012; Approx. 2120hrs.
Location: Redding, CA
Witness: 1
Shape: Oval
Duration: 20-30 seconds

White light, rotating blue lights on it in the night sky Redding Ca. On the evening of January 26th, 2012 at approximately 7:20pm while driving home from work, I observed a bright white light in the night sky to my left side as I drove up a local street. I was heading up Highway 273 north up what is called North Market Street or to locals Miracle Mile, in Redding, California. As I got approximately 2/3 of the way up the hill, I caught sight of a large white object to my left, or the western side of the road. It appeared to be of an oval shape, well above the tree lines and had rotating blue lights around it, resembling a carnival ride. It was heading in the same direction I was, north, and as it got above the buildings at the top of the hill, it seemed to make an immediate sharp right turn. I was almost at the top of the street when this occurred, and as it made the right turn it seemed at accelerate, then suddenly it was gone. I have served in many countries while in the military, and ridden in many types of military helicopters and airplanes, but nothing I have ever seen resembled this. The way it disappeared as it accelerated was what caught my eye, as the night was clear and I had an un-obstructed view of the dark where I had seen the "whatever" it was. I have been inside burning aircraft to remove dead bodies, have seen many things in my life which were amazing and extremely curious, but I have to say that this was without a doubt one of the scariest things I have every witnessed. I felt more fear on the rest of my drive home than I can ever recall feeling, and even tonight while driving home the same route at the same time I felt a chill as I drove up Miracle Mile. I cannot for sure state how big the object was, but if I had to make a guess, I would say it was maybe 20-30 feet, basing this on roughly how high it was in the sky and in relation to the traffic lights at the top of Miracle Mile. Thank you for taking the time to read this. Sincerely, ((name deleted)) Redding, Ca.

Occurred: February 3, 2012; Approx. 0530hrs.
Location: Weed, CA
Witness: 1
Shape: Light
Duration: 1-2 minutes

A very bright disc with gold/white light, flying saucer was flying low over Lake Shastina area. An extremely large ball of white/gold light-- brighter and bigger than any star, yet star-like in nature, was flying very low and in total silence from the horizon behind Lake Shastina and travelling towards Mt. Shasta. The night sky was dark and clear as the moon had set already. There was no noise anywhere at five-thirty in the morning; dead quiet. It made it's way within a minute to the Shastina cone of Mt. Shasta, faded to a dim spot, and seemed to drift down into the mountain; jaw dropping. I don't see things like this... ever. To be honest I see satellites but this was just too bright and golden colour-- and too big. There was no way a plane could have passed over without being heard by me-- it was too low. Today, one day later, a big black helicopter flew over-- never seen one of those here, either, and it was loud.

Occurred: March 10, 2012; Approx. 1845hrs.

Location: Redding, CA

Witness: 1
Shape: Fireball
Duration: 5 seconds

An orange fire-ball was seen flying through the sky, pretty low, and then I seen it disappear behind the trees.

((NUFORC Note: Witness elects to provide no contact information. PD))

Occurred: April 8, 2012; Approx. 2035hrs.

Location: Marysville, CA

Witness: 5
Shape: Light
Duration: 45 seconds

4 hovering lights at very low altitude South of Marysville

My husband and I, along with our 3 children, were travelling South on Hwy 70. It was approximately 8:35 P.M. and we were about 2 miles south of Marysville. Suddenly we saw 4 lights in the sky right in view of our windshield. They appeared to be between 150 and 200 feet above ground. They were in a straight-line formation laterally, 4 across. There were absolutely no blinking lights and there were no discernable shapes that we could see surrounding the lights. As we moved south, the objects were moving North/East and they were passing us on our left side. My husband pulled over and I got out of the truck to better observe them. He remained in the truck and rolled down his window. As I got out and stood beside our truck I observed the 4 objects hovering on the other side of the highway. There was absolutely no noise coming from them. Even out of the truck I could not hear anything coming from the objects. They remained at the same altitude of about 150 to 200 feet above ground, although now they were a little further away (North East of where we were). At that point they broke formation, so that they were no longer in a straight lateral line. Some were higher and others were lower. Then 1 by 1 they blinked out of existence. They were gone. It all happened rather quickly, but 45 seconds was longenough to make those observations. It was very fascinating and at no point felt threatened or scared. In no way am I an expert in aircraft, but I have seen plenty of aircraft in my lifetime. My father was in the Air force. I have been on military base and have been to numerous air shows. I am well aware of the appearance and behaviours of planes and helicopters. I also crew for a hot air balloon pilot. I often observe the skies. My husband was in the Navy and lived on an aircraft carrier. We can tell you that whatever we saw was not something that we could identify.

Occurred: June 15, 2012; Approx. 2130hrs.

Location: Kelseyville, CA

Witness: 4
Shape: Fireball
Duration: 10 minutes
A Fireball changes color and size while sitting still and silent. We were coming home from the local Kelseyville high graduation when I noticed a large burning red and orange light right besides the red rotating light on the Mt. Konotci Mountain. My wife and mother and sister in law all noticed the object also. It sat still then came slowly at us; it dropped to about 1,500 ft because we could see if was about half way below the top of the mountain moving west. The object changed shape from round to oblong and back to round and changing colors from red to orange. Whatever this thing was stopped directly over my house and hovered. We all got out of the car and looked at it. It was silent and very still. After a few minutes it shrunk down in size and moved northeast and a unknown speed. It was just too small to judge its speed. We saw this same object over the same exact area about 5 years ago this time. It did exactly the same thing and then moved away from us.

Occurred: July 4, 2012; Approx. 2221hrs.

Location: Redding, CA

Witness: 1
Shape: Fireball
Duration: 3
Ball of light across sky in Redding, CA. Saw this after Redding, CA Firework Finale...went across the sky from North to South about 10:21 pm.

Occurred: July 4, 2012; Approx. 2300hrs.

Location: White City, OR

Witness: 1
Shape: Sphere
Duration: 5 minutes
Viewed 3 bright orange objects moving across the sky in a straight line. Saw a bright round orange light in the sky driving home (similar to large glowing embers but in formation). Got home and got out and saw two large orange objects moving south in a line, one passed the other at high rate of speed and then a third one came a 1 min or so behind. They all seemed to dim and brighten as they disappeared from sight.

Occurred: July 11, 2012; Appprox. 0430hrs.

Location: Redding, CA

Witness: 1
Shape: Light
Duration: minutes
White fireball streaking downward At 4:30 yesterday morning there was a glowing white, very

large and bright light on a downward path in the sky. It suddenly dropped backwards and down and then continued its path. I did not hear any noise at all. I live in the Shasta Lake area, minutes from Redding.((NUFORC Note: We spoke via telephone with this witness, and she sounded quite sober-minded to us. We suspect that she is a highly reliable witness. PD))

Occurred: July 13, 2012; Approx. 2125hrs.

Location: Roseburg, OR

Witness: 2
Shape: Light
Duration: 55 seconds
Light came up from horizon, stopped, turned and faded before reaching horizon again. Friday, July 13, 2012, at about 9:25 p.m., wife drew attention to orange ball coming up from the horizon SSE. It proceeded up for 15-20 degrees; stalled at this position; got dimmer; turned left; headed east approximately 2 degrees; dimmed more and headed back toward horizon faded to nothing well before hitting the horizon. Wife brought to my attention later, the distressed horse in the field below during the whole scene. Horse was silent before and after this event. This is not my horse it belongs to our neighbours.

Occurred: August 13, 2012; Approx. 2130hrs.

Location: Rogue River, OR

Witness: 2+
Shape: Light
Duration: 15 seconds
My friends and I were rafting on the Lower Rogue River in Southern Oregon on August 13th. That evening in camp, Wildcat camp GPS (N42 39.537 W123 38.301), we were sitting and talking between 2100 and 2200. Initially there was a sound that seemed like it was a jet plane flying in the distance. We then saw a white flash of light and looking up into the night sky we observed two orange spheres that were moving slowly. As we were facing them the light on the left moved towards the one on the right and they merged together into one light. The single remaining light just disappeared. The entire event lasted for approximately 15 seconds.

Occurred: August 15, 2012; Approx. 1400hrs.

Location: Yreka, CA

Witness: 1
Shape: Cigar
Duration: 1 minute
Light colored jet craft bulky, long cigar shape with no wings visible travelling at high speed at one to two miles high. One presumed drone aircraft jet proceeded southeast at altitude of approx. 2 miles. I was only witness from location in mountains of east central Yreka. Angle of sighting from vantage point made craft to appear at same angle of Mt. Shasta. Craft appeared to be large body almost bulky fat but long cigar like cargo ship but with no wings visible; Clear sky, no clouds, no vapour trails but possibly some haze that completely obscured very thin wings if

any. Other visual and audible clue such as estimated altitude made me expect to see wings but there were absolutely no wings or tails visible No other aircraft visible but sound of jet was faint but distinct; Travelling faster than the sound because of lag time behind visible craft; Could have been ventriloquist effect since area is corridor for jets high up, travelling at high rate of speed and appeared to be a light weight glider or dirigible. Profile was high and long but not clear how deep fuselage was hence reminded me of pictures of Predator drones but more regular cigar-shaped. When I spotted it, it seemed to recognize I was watching and moved away from me veering slightly to south each as if bumped by wind only momentarily.

Occurred: August 20, 2012; Approx. 2200hrs.

Location: Grants Pass, OR

Witness: 1
Shape: Circle
Duration: 1 minute or less

Stars appear, move and fade away. I have been seeing stars that stay put and then slowly fade to a small pinpoint, and then leave the area. I have also seen stars that are bright take off across the sky, flash, move some more, then fade to a small, almost undetectable light across the sky, then disappear. This is happening at or around 9:30 to 11:00 pm. I saw a star at the bottom of the big dipper handle, stayed for 1 minute...then fade to a smaller light and take off. This has been happening almost every night for 2 or 3 weeks. Moved rapidly out of sight; slightly white-aluminium color possible metallic. Hard to tell relative size because of unknown altitude, but impression was that it was large-bodied but very high up. Impression was that of a jet-powered Remotely Piloted Vehicle (drone) or piloted glider or lightweight craft with very long, thin wings that were not visible because of altitude. General impression was confusing somewhat because of conflicting phenomena. But my mind was clear and I was well rested in broad daylight without stressful circumstances.

((NUFORC Note: Witness indicates that the date is approximate. PD))

CONCLUSION

In concluding this book, there are several issues we hope to tie up. UFOs do exist beyond any point of rationalizing them away.

It should be obvious to the readers that there is enough eyewitness evidence to ascertain that there are UFOs in and about Mount Shasta. The overwhelming evidence from a vast cross section of people witnessed over the decades substantiates this premise. In just this region alone, a small slice of Americana, if joined to the sightings globally, our world is a massive hotspot. These sightings directly support the premise that we are not alone, and we have not been.

With the abundant ancient ruins inexplicably built beyond current sophistication, on every continent, under water, and supposedly on the Moon and Mars, there seems little doubt. The vortices, anomalies and massive libraries of data, as to these beings relation to us and we them, further supports our galactic connection. We should be excited!

Einstein and other physicists using Quantum Mechanics have proven the physical universe does not exist, making room for it to be considered of a spiritual nature, perhaps indicating we already exist in a spiritual realm. Einstein himself acknowledged a universal consciousness he referred to as God, not in the biblical or material sense. We could view UFOs no differently to any other unfamiliar or unidentified life form in nature we have not had the privilege of becoming familiar with; with healthy curiosity. Like the hummingbird, UFOs are fleeting, cautious, and suspicious. Would you not be around humans who out of fear, kill first and ask questions the victim can't answer later?

The last issue is of deceit, concealment and demonization of UFOs, their occupants and technology. This rationalization is more cunning than the rest of the conundrum, and poses the worst threat of them all. It is hard to swallow that propaganda and psychological subterfuge has been used on the masses, as we truly want to believe we still live in a free society where the open door policy is real. In most cases it is, and in most cases most of us are wonderful, kindhearted, warm, good, loving individuals who's intent is genuinely authentic.

We have bought to your attention the threat of those who are committing crimes against humanity, and the right use of taxes for everyone's benefit, instead of funneling it in to subsidies and contracts that directly feed the military industrial cartel which is bloated and filled with crooks, munitions makers, who the bottom line is war. The wealthiest companies do not pay taxes, they buy the politicians and manipulate how our tax dollars are spent, even though they pay none. They want more "deregulation" which means no oversight and no accountability. They help build the contraptions of a new millennium we don't get the benefit of and they are freebasing our economy. As long as they are mainlining these government contracts, they control any innovation, stifle any invention, alien or terrestrial, which if it threatens their profits and control, they will cease to let it prosper. It demonstrates how addiction works and how the lack of connection to one's own inner purpose destroys. Here are just a few examples of withheld products.

Powder Meds, a company the author tracked since 2007, invented and developed a microinjection system using a gold dust suspension, could and did successfully, in trials, carry insulin indefinitely, with no need of refrigeration ever, anywhere in the world, and in one ounce, carry enough insulin, to treat 30 thousand people. In 2010 Pfizer bought them out and buried the invention. Pfizer makes refrigerated perishable insulin making billions a year for them.

A Canadian fireman developed a one-inch thick fireproof brick from materials costing less than a dollar that he for demonstration, made into a helmet that withstood 3,500 degrees with a oxy-acetylene torch on his head, in front of many witness on live television where the temperature remained at 100 degrees between his skull and the inside of the helmet. That was on "Invent this" in 2008, and not a word about the guy has surfaced. He would have put 100 billion dollar industries out of business, for a penny on the hundred dollars and save countless lives, including the space shuttle tiles.

There are more the author has tracked: Kergy, who made the anaerobic still, that makes methanol from cellulose in hours instead of days, thousands of gallons instead of 10 gallons. I talked with the inventor. Devices now fully developed to allow mute people to talk, and deaf people to hear, the company is in Texas, and is also used for classified surveillance.

We have shown the ultimate hypocrisy of those leading the space programs; the "titans" as they are called, operate covertly. There are laws put in place to protect them under the guise of national security to black out their private "conflict of interest" intent in procuring and securing for themselves the secrets of our origins in alien technology, onboard earth and off world on the Moon and Mars. So, in their own self-serving belief of being the "chosen" few are more entitled than the rest of us, to our universal inheritances, to meet our predecessors, perhaps. This is utterly unacceptable by any measure of honor, and by its very nature causes mental instability. Since they assume they are more worthy, being in the boys club of the" brothers from an alien mother" with ruler, compass, jewel or dagger (sarcasm) that this entitles them to break all moral, national, global laws?

This secret space program wrapped up and enmeshed within the innocuous, purpose driven, manned space program only makes it harder to take. It should be blaringly obvious when 460 high ranking, prominent, respectable individuals, along with countless others in the name of freedom, would sacrifice their reputations to uphold the integrity of our highest ideals, and disclose the amazing truth of ETs and such, it must be authentic. This is something not to scoff at or take lightly. Readers, this is not conjecture, or a conspiracy, it is factual. Eisenhower passed the law, you can read the Brookings Act, read the Condon Reports and decide for yourself, if you want to let someone steal what belongs to us, from under our noses.

This is part of what has been disclosed in the book. Now there are things that no one can touch, steal, blemish or hide; universal knowledge and your own true self that is within you. No matter how you cut it, we can eliminate science, religion, myth and we will still come back to the fundamental question of our relationship with all that is. The beautiful thing is that the answer is in the same place the question resides: in yourself. The fact remains you can still think, still peer into the skies and see the universe presenting itself to you. We have come so far and the benefits are at this point, needed for our survival as a race. We have reached critical mass in population and you know the rest. We need to move into the future with efficient power, clean resources and a sane plan. There is too much at stake.

Life does exist in our universe in abundance. The likelihood of a similar race peering in our development is very possible, judging from our own behavior and curiosity about our environment, about science, of having questions answered and higher viewpoints embraced about our own kingdom. We are driven to explore to find and know ourselves.

Regardless of one's belief in God, or not, if one identifies as a scientist, theologian, spiritualist or whatever, there are cohesive, unified underpinnings between us all, not only ideologically but biologically, anatomically, atomically and all else. We are related by our origins, by the reality we are all made of the same stuff, cut from the same cloth and aligned and complete even if we are conscious of it or not. Science has helped confirm this through our studies of nature and the cosmos. We must then also have the same questions and interests. We are all " the stuff the stars are made of" as the late Dr. Carl Sagan said. In fact, we use identical principles to communicate as every other organism, chemical, compound and element in the universe. The definition of God needs to be re-examined and perhaps expanded. Perhaps the idea of mistaking a more evolved race as 'Gods" who we may in fact be related to, makes more sense now you have read this book. Less sophisticated tribes still mistake westerners because of their technology as gods evidenced by their attitudes, trust, statues and art, Icons, Deities and so on. In reality are we Gods to them, or equals? The Off-Worlders maybe are our older galactic siblings, instead of being superior to us. We are equal, but just not as well informed. We may be able to become what they demonstrate to us.

Our very drive is, to discover and examine the unifying principle behind all existence, you can be sure it's at the core of every explorer who has ever lived, no matter what part of the galaxy they are from. The same essence acts on their atoms. All of us are under the same principle: consciousness. Maybe we are all gods as well, what a concept that maybe there was never a lesser being. Perhaps these Off-Worlders have played out their role in helping propagate a universal plan in assisting in our growth; like a gardener.

Science can only examine behaviors and from that, we have learned important principles on how things work and how to use those principles to duplicate a desired result consistently. The purpose for why things exist have not been answered or proven empirically, like; why do atoms prefer to attract themselves with other atoms to create iron or gold?

One must admit that the current act being played out in our global theater contain some of the worst qualities of each of us. In essence if we are sharp enough, we will recognize and correct them in our own psyche and behave more universally and compassionately towards our fellow person. In doing so, we can share the wealth. As good tenants on this planet, we should be occupying ourselves with leaving the rental (our planet) in better condition that we found it so we get our deposit back.

If we are being groomed by what ever it is in this universe, to be civilization builders, then we got a lot of homework to do, and do well. This profession is so evident if you look around within and without. Our bodies and the organs comprise a collaboration and organization if you will, agreed upon from the atomic level, from the elements we blend, to the cells that are created to the stem cells that manifest the innate blueprints over and over again of the various organs, or instruments in this orchestra we call a body. It in itself is a civilization working in perfect synchronicity, that you never even think about it, it does it so well.

In our environment, the civilization/organization building is going on, evolving being learned and innovated upon. The ant colonies, the bee hives the plant world, the various animal worlds, which we need to have more respect for unless we lose them all out of greedy stupid carelessness, have a purpose we rely upon more than we are aware of. If we destroy them, we will destroy ourselves as well. On and on, order, intelligence, harmony goes on, and the examples show us that fascist civilizations fall, ones that supply every member equally, thrives, as nature does not scrimp when creating a flower in the wild, she manifests millions on a hillside. The spider effortlessly weaves geometric patterns all day long, and it seems too obvious there is order all about us that is not a slave relationship. Perhaps we should take a hint instead of being so miserly with each other, and realize that all we need to do is become more conscious of this and learnt o work with it, instead of fearing it.

Through all the assorted sophisticated information, facts, history and events illustrated in this book, by devoted research and scientific communities; it's obvious that the presence of Off-Worlders is in our midst. We must to the best of our ability, in the most authentic part of our heart, take a sobered look to its meaning. We must strive to find the thread that lets us feel akin to our seeming alienating universe, be it spiritual, scientific, or intuitive.

If these extraterrestrial beings were hostile, they would have over the course of our history, been able to seize this planet with little force, taken what they wanted, disseminate what development of civilization we have and be gone. It does not seem a logical choice for them to wait until we were more aggressively armed and knowledgeable that would cause greater collateral damage to this planet or them if these were the prizes. An advanced race of beings must by necessity have operational guidelines in dealing with visitations of other developing civilizations and it would be a confident notion that a peaceful collaboration would be more beneficial and preferred no matter what side of the situation you occupied. If anything, if we were observing a new civilization, how would we go about it? Would we seize their assets, or acquaint ourselves with them and establish a dialog and trust and then perhaps start trading as we have done responsibly at times in our own history, learn from them and teach as well. It seems most the mistakes made have been by us in our historical treatments of other cultures on this planet, disrespecting their culture with religious eliticism, then rationalizing that barbaric behavior to entitlement of their possessions, children, women, and their lives, since they were "savages". As much as we abhor tyranny, fascism, slavery, it seems that any other civilized race would also dislike the travesties of such inequalities and try to do something about it diplomatically, as we had tried in the Middle East.

We in fact, are responsible for enslaving other cultures, torturing the innocent, disrespecting and exploiting without forethought our own-shared resources. By default we should already know from those mistakes the loosing proposition of that behavior. If the Off-worlders ever appear collectively or just weave quietly among us, the impact of their visits, the symbolism of their presence and what they represent as our higher good, has affected us by making us think about, reflect on our presentation to the world, our behavior, and how we can improve our own lives to not be so selfish, destructive and live according to the highest ideals. Our world has demonstrated through people like Plato, Michelangelo, Mozart, Beethoven, The Beatles, Einstein, and Tesla and countless others, what we are capable of. When we all realize that we too carry the seed, the best democracy is put in action if we can allow ourselves to apply it. To realize that the simple action of procreation indicates our ability to channel creative energy, we can consciously inject that essence into ideas, mathematical equations, recipes, inspiration, musical scores, scientific breakthroughs, and the like. We can let beauty and goodwill spring forth from our hearts and minds through our creations to others in the name of love for all life.

To be in "THE ZONE" in the art we love, we may stumble across the way some of our more mature Off-Worlders think and operate, coming from a place of inspiration. We may just perhaps, in some way, be more evolved, more in touch with our spiritual self than anyone can imagine. It was once said our species is the jewel of the galaxy in one respect: unconditional love and its application. Remember that, it is our inheritance. Rather than this principle being an offshoot of alien knowledge, perhaps we are growing into a universal knowledge; we can share with more advanced cultures. If not, we may be able to teach them, not through religion, but through the applied science of love and acceptance. As in the movie 2001, A Space Odyssey, the monolith was used to accelerate being's thoughts. Perhaps this is how our Off-World brethren are grooming us to accept their presence. Perhaps once we feel our own confidence and competence behind the wheel in steering our lives, and begin to be true to our own heart, then less fear will be driving us to the brink of extinction, and we will accept the Off-Worlders as a matter of fact, like a neighboring Allied country rather than a mystery, and go from there.

Bibliography and Reference Material

Mount Shasta Companion Folklore/Native American/Traditional Prose Narrative History
http://www.siskiyous.edu/shasta/his/index.htmhttp://www.siskiyous.edu/shasta/fol/nat/index.htm
www.mtshastaspirit.com

Brill, E., J. The Nag Hammadi Library. San Francisco. Harper &Row 1988

RowBrunvand, J. H. (1998). The Study of American Folklore. New York: W. W. Norton.

Lind, M. (1996). The Next American Nation: The New Nationalism and the Fourth American Revolution. New York: The Free Press.

Melton, J. G. (1986) The Encyclopedic Handbook of Cults in America. New York: Garland Pub.

Thompson, S. (1955) Motif-Index of Folk Literature. Bloomington: Indiana University Press

Curtin, Jeremiah. 1898. Creation Myths of Primitive America in Relation to the Religious History and Mental Development of Mankind. Boston: Little, Brown.

de Angulo, Jaime and L. S. Freeland. 1931. Two Achumawi Tales. Journal of American Folk-Lore. Vol. 44:125-136.

Demetracopoulou, Dorothy and Cora Du Bois. 1932. A Study of Wintu Mythology. -Journal of American Folklore. Vol. 45 (178):375-500.

Dixon. Ronald. 1905. The Mythology of the Shasta-Achomawi. American Anthropologist, Vol. 10:2o8-220.

1910. Shasta Myths. Journal of American Folklore. Vol. 23 (87):8-37 and Vol. 23 (89):364-70
1905 Oliver.Fredrick, A. A Dweller on 2 Planets
1959-1969 Jung, C. Flying Saucers, New York, Signet
Borden, R.C., and T.K. Vickers. Preliminary Study of Unidentified Targets Observed on Air Traffic Control Radars. (Civilian Aviation Administration, 1952.)
Durant, F.C. Report of Meetings of Scientific Advisory Panel on UFOs [Robertson Panel Report]. (Central Intelligence Agency, 1953.)
Goodrich, Norma, l., Ancient Myths. (New York, Signet, 1960)
King, Godfre, R., The Magic Presence, (Chicago, St.Germain Press, 1974)

Kharditi, Olga,. Entering the Circle. (San Francisco, Harper, 1998

Hall, Michael David, and Wendy Ann Connors. Captain Edward J. Ruppelt: Summer of the Saucers 1952 (Albuquerque, NM: Rose Press International, 2000.)

Hall, Richard. Radar-Visual UFO Cases in 1952: The UFO Sightings That Shook the Government. (Fund for UFO Research, 1994.)
Keyhoe, Donald E. Flying Saucers From Outer Space. (New York: Henry Holt & Co., 1953.)
McDonald, James E. UFOs: Greatest Scientific Problem of Our Times? Address to annual meeting of the American Society of Newspaper Editors, Washington, D.C., April 22, 1967.
McDonald, James E. Symposium on Unidentified Flying Objects, House Science & Astronautics Committee, U.S. Congress, July 29, 1968. Comments on radar-UFOs, pp. 18-86.
Menkello, Frederick V. Quantitative Aspects of Mirages. (Air Force Foreign Technology Division, April 1969.)
National Investigations Committee on Aerial Phenomena, U.S. Air Force Projects Grudge and Blue Book Reports 1-12 (1951-1953); (Washington, D.C.: NICAP, 1968).

Pagels, Elaine, The Gnostic Gospels. (New York, Random House, 1979)

Project Blue Book Unknowns. (Original BB Unknowns are listed as (BBU + the number), cases added by Brad Sparks in his "Comprehensive Catalog of Project Blue Book Unknowns" are simply marked (BBU).

Randle, Kevin. Invasion Washington: UFOs Over the Capitol. (New York: HarperTorch, 2001.)

Ruppelt, Edward J. The Report on Unidentified Flying Objects. (New York: Doubleday, 1956.)
Also available online at www.nicap.org
Swords, Michael D. UFOs,
    1.   Volume II, The UFO Evidence, A Thirty-Year Report - Richard Hall, 2000
    2. CUFOS report, Rodeghier,
    2.   KEN PFEIFER WORLD UFO PHOTOS

Sitchin, Zecharia The 12th Planet New York: Avon, 1978,
Sitchin, Zecharia End of Days, New York, Harper, 2007

Archives
NICAP.org Chronologies, Selected Sightings used with Permission
MUFON.com Selected Chronological Sightings used with Permission
NFORC.com Selected Chronological Sightings used with Permission
Project Blue Book Archives used with permission, NICAP/MUFON
The blackvault.com and ufoevidence.com within NICAP chronologies
Cufon.com/ufoevidence.com/, linked, with permission

www.ingramcontent.com/pod-product-compliance
Lightning Source LLC
Chambersburg PA
CBHW080720230426

43665CB00020B/2561